Integrated Study of Marine Mammals

Integrated Study of Marine Mammals

Edited by Suzy Bullock

SYRAWOOD PUBLISHING HOUSE
New York

Published by Syrawood Publishing House,
750 Third Avenue, 9th Floor,
New York, NY 10017, USA
www.syrawoodpublishinghouse.com

Integrated Study of Marine Mammals
Edited by Suzy Bullock

International Standard Book Number: 978-1-68286-443-2 (Hardback)

Cataloging-in-publication Data

Integrated study of marine mammals / edited by Suzy Bullock.
p. cm.
Includes bibliographical references and index.
ISBN 978-1-68286-443-2
1. Marine mammals. 2. Marine ecology. 3. Marine habitats. I. Bullock, Suzy.
QL713.2 .I58 2017
599.5--dc23

Printed in China.

TABLE OF CONTENTS

Preface VII

Chapter 1 **A Critical Time Window for Organismal Interactions in a Pelagic Ecosystem** 1
Kelly J. Benoit-Bird, Margaret A. McManus

Chapter 2 **Mortality of Inshore Marine Mammals in Eastern Australia is Predicted by Freshwater Discharge and Air Temperature** 15
Justin J. Meager, Colin Limpus

Chapter 3 **Recovery Trends in Marine Mammal Populations** 25
Anna M. Magera, Joanna E. Mills Flemming, Kristin Kaschner, Line B. Christensen, Heike K. Lotze

Chapter 4 **The Sun, Moon, Wind, and Biological Imperative– Shaping Contrasting Wintertime Migration and Foraging Strategies of Adult Male and Female Northern Fur Seals (*Callorhinus ursinus*)** 37
Jeremy T. Sterling, Alan M. Springer, Sara J. Iverson, Shawn P. Johnson, Noel A. Pelland, Devin S. Johnson, Mary-Anne Lea, Nicholas A. Bond

Chapter 5 **Wintering Habitat Model for the North Atlantic Right Whale (*Eubalaena glacialis*) in the Southeastern United States** 58
Timothy A. Gowan, Joel G. Ortega-Ortiz

Chapter 6 **Probability of Detecting Marine Predator-Prey and Species Interactions using Novel Hybrid Acoustic Transmitter-Receiver Tags** 72
Laurie L. Baker, Ian D. Jonsen, Joanna E. Mills Flemming, Damian C. Lidgard, William D. Bowen, Sara J. Iverson, Dale M. Webber

Chapter 7 **Habitat Fragmentation and Species Extirpation in Freshwater Ecosystems; Causes of Range Decline of the Indus River Dolphin (*Platanista gangetica minor*)** 83
Gill T. Braulik, Masood Arshad, Uzma Noureen, Simon P. Northridge

Chapter 8 **Two Intense Decades of 19th Century Whaling Precipitated Rapid Decline of Right Whales around New Zealand and East Australia** 92
Emma L. Carroll, Jennifer A. Jackson, David Paton, Tim D. Smith

Chapter 9 **Migratory Movements of Pygmy Blue Whales (*Balaenoptera musculus brevicauda*) between Australia and Indonesia as Revealed by Satellite Telemetry** 105
Michael C. Double, Virginia Andrews-Goff, K. Curt S. Jenner, Micheline-Nicole Jenner, Sarah M. Laverick, Trevor A. Branch, Nicholas J. Gales

Chapter 10 **Remote Acoustic Monitoring of North Atlantic Right Whales (*Eubalaena glacialis*) Reveals Seasonal and Diel Variations in Acoustic Behavior** 116
Leanna P. Matthews, Jessica A. McCordic, Susan E. Parks

Chapter 11 **Humpback Whale Populations Share a Core Skin Bacterial Community: Towards a Health Index for Marine Mammals?** 123
Amy Apprill, Jooke Robbins, A. Murat Eren, Adam A. Pack, Julie Reveillaud, David Mattila, Michael Moore, Misty Niemeyer, Kathleen M. T. Moore, Tracy J. Mincer

Chapter 12 **Hybrid Speciation in a Marine Mammal: The Clymene Dolphin (*Stenella clymene*)** 140
Ana R. Amaral, Gretchen Lovewell, Maria M. Coelho, George Amato, Howard C. Rosenbaum

Chapter 13 **Satellites, the All-Seeing Eyes in the Sky: Counting Elephant Seals from Space** 148
Clive R. McMahon, Hamish Howe, John van den Hoff, Rachael Alderman, Henk Brolsma, Mark A. Hindell

Chapter 14 **First Long-Term Behavioral Records from Cuvier's Beaked Whales (*Ziphius cavirostris*) Reveal Record-Breaking Dives** 153
Gregory S. Schorr, Erin A. Falcone, David J. Moretti, Russel D. Andrews

Chapter 15 **Fine-Scale Variability in Harbor Seal Foraging Behavior** 163
Kenady Wilson, Monique Lance, Steven Jeffries, Alejandro Acevedo-Gutiérrez

Chapter 16 **Assessing Performance of Bayesian State-Space Models Fit to Argos Satellite Telemetry Locations Processed with Kalman Filtering** 176
Mónica A. Silva, Ian Jonsen, Deborah J. F. Russell, Rui Prieto, Dave Thompson, Mark F. Baumgartner

Chapter 17 **200 kHz Commercial Sonar Systems Generate Lower Frequency Side Lobes Audible to Some Marine Mammals** 189
Z. Daniel Deng, Brandon L. Southall, Thomas J. Carlson, Jinshan Xu, Jayson J. Martinez, Mark A. Weiland, John M. Ingraham

Chapter 18 **$\delta^{15}N$ Value does not Re lect Fasting in Mysticetes** 195
Alex Aguilar, Joan Giménez, Encarna Gómez–Campos, Luís Cardona, Asunción Borrell

Chapter 19 **Effects of Age, Adipose Percent, and Reproduction on PCB Concentrations and Profiles in an Extreme Fasting North Pacific Marine Mammal** 202
Sarah H. Peterson, Jason L. Hassrick, Anne Lafontaine, Jean-Pierre Thomé, Daniel E. Crocker, Cathy Debier, Daniel P. Costa

Chapter 20 **A Longitudinal Study of Steller Sea Lion Natality Rates in the Gulf of Alaska with Comparisons to Census Data** 215
John M. Maniscalco, Alan M. Springer, Pamela Parker, Milo D. Adkison

Permissions

List of Contributors

Index

PREFACE

This book attempts to study the vast field of marine mammals. The field is a study of marine or aquatic mammals that depends on ocean ecosystems and marine ecosystems for their survival. It has a compilation of topics that is varied in nature catering to the different needs of the readers. The book highlights the classification, evolution and the distribution of habitat of the marine mammals in detail. This text also elucidates strategies and tools to restore the marine mammals on the verge of extinction that are being used by scientists from all over the world. Through this book, we attempt to further enlighten the readers about the new concepts in this field.

The purpose of the book is to provide a glimpse into the dynamics and to present opinions and studies of some of the scientists engaged in the development of new ideas in the field from very different standpoints. This book will prove useful to students and researchers owing to its high content quality.

At the end, I would like to appreciate all the efforts made by the authors in completing their chapters professionally. I express my deepest gratitude to all of them for contributing to this book by sharing their valuable works. A special thanks to my family and friends for their constant support in this journey.

Editor

1

A Critical Time Window for Organismal Interactions in a Pelagic Ecosystem

Kelly J. Benoit-Bird[1]*, Margaret A. McManus[2]

1 College of Oceanic and Atmospheric Sciences, Oregon State University, Corvallis, Oregon, United States of America, **2** Department of Oceanography, University of Hawaii Manoa, Honolulu, Hawaii, USA

Abstract

To measure organismal coherence in a pelagic ecosystem, we used moored sensors to describe the vertical dynamics of each step in the food chain in shelf waters off the west shore of Oahu, Hawaii. Horizontally extensive, intense aggregations of phytoplankton, zooplankton, and micronekton exhibited strong diel patterns in abundance and vertical distribution, resulting in a highly variable potential for interaction amongst trophic levels. Only around dusk did zooplankton layers overlap with phytoplankton layers. Shortly after sunset, micronekton ascended from the deep, aggregating on the island's shelf. Short-lived departures in migration patterns were detected in depth, vertical distribution, density, and total abundance of micronekton when zooplankton layers were present with typical patterns resuming within one hour. Layers of zooplankton began to disappear within 20 minutes of the arrival of micronekton with no layers present after 50 minutes. The effects of zooplankton layers cascaded even further up the food chain, affecting many behaviors of dolphins observed at dusk including their depth, group size, and inter-individual spacing. As a result of these changes in behavior, during a 30-minute window just after dusk, the number of feeding events observed for each dolphin and consequently the feeding time for each individual more than doubled when zooplankton layers were present. Dusk is a critical period for interactions amongst species in this system from phytoplankton to top predators. Our observations that short time windows can drive the structure and function of a complex suite of organisms highlight the importance of explicitly adding a temporal dimension at a scale relevant to individual organisms to our descriptions of heterogeneity in ocean ecosystems.

Editor: Hans G. Dam, University of Connecticut, United States of America

Funding: This work was funded by the US Office of Naval Research, awards #N00014-08-1-1212 (MAM), and #N00014-08-1-1210 (KJBB); http://www.onr.navy.mil/. The funders had no role in study design, data collection and analysis, decision to publish, or preparation of the manuscript.

Competing Interests: The authors have declared that no competing interests exist.

* E-mail: kbenoit@coas.oregonstate.edu

Introduction

Interactions between predators and prey have shaped all life on earth, affecting individual behavior, species morphology, population dynamics, species diversity, community structure, and ecosystem function [1]. The probability of these interactions between trophic levels is largely a function of the density of predator and prey and the overlap of their populations. [2] The roles of ecological process have been shown to be scaled by the spatial context in which an organism is found [3]. In pelagic systems, there are four dimensions of overlap to consider – the x-y or horizontal plane, depth (z), and time (t). Much attention has been paid to the horizontal spatial relationships between predator and prey and their dependence on scale [4,5,6,7]. However, the challenge of accessing organisms beneath the surface of the ocean has made adding measurements of vertical distribution of both predator and prey simultaneously difficult [8,9,10].

The emphasis on spatial distribution in ecological interactions (and the development of the field of landscape ecology) was, in part, driven by seminal experimental and theoretical work in systems with a two-dimensional surface onto which one member of the dyad is tied physically [11,12,13] or computationally [14]. However, even a description of spatial coherence in three dimensions is not sufficient to understand the coincidence of organisms. The fourth, temporal dimension is particularly crucial in pelagic systems because even immobile organisms can be moved by currents and, unlike terrestrially influenced three-dimensional habitats [15,16], there are relatively few fixed features to provide cover or anchor behavior. Surveys of predator and prey densities implicitly assume a constant relationship within each observation, making it possible to observe correlations at climatological [17,18], annual [19], seasonal [20,21], or even daily [22] time scales. Knowledge of the behavioral state of observed predators can allow correlations at smaller time scales, sometimes even approaching a single foraging event [8,10,23,24]. However, it is typically difficult to connect the small scale observations of an individual predator to the larger scale observations of populations or communities necessary to understand the ecology of the system [25].

The complexities of resource heterogeneity in both space and time are critically important to ecosystems; they may underlie the timing of outbreaks and blooms, successional patterns, stability of predator-prey interactions, and coexistence of multiple competitors for the same resources [26]. However, adding temporal dynamics to our understanding of spatial variance has been identified as the most challenging task in ecosystem description and prediction [27]. To move beyond interpretation of temporally separated snapshots of a system, here, we use high-resolution moored sensors to describe the vertical dynamics of each component of Hawaii's nearshore pelagic ecosystem, a relatively well-described food web beginning with phytoplankton and culminating in spinner dolphins (*Stenella longirostris*). The distribution of biomass in each of these trophic levels is spatially heterogeneous, which we showed in previous analyses affects predator-prey interactions throughout the system [28]. In describing the importance of spatial pattern, however, our previous efforts did not measure true coherence between predators and prey as they focused on long-term temporal dynamics. By not addressing vertical position of aggregations of organisms in the water column or time variance at scales of less than a day, the mechanisms of these interactions remain obscured. Particularly as short time scales and vertical position are predicted to be important based on the strong diel patterns in both local biomass and vertical distribution observed in each step of the food chain (spinner dolphins [23]; zooplankton [29]; phytoplankton [30]; micronekton [31]). In this contribution, we use fine vertical (tens of cms) and temporal (one sample every second to every 15 minutes, depending on the instrument) resolution descriptions of the distribution of biomass in the food web to elucidate the interactions between predators and prey in both space and time.

Methods

Over two, three-week periods in the spring of 2009 and 2010, we used moored sensors complemented by shipboard surveys to measure the physical and biological properties of the shelf off the leeward coast of Oahu, Hawaii. This section of coastline runs north to south with little variation, and isobaths are roughly parallel to the shoreline. The study was conducted near the 25 m isobath from 20 April to 12 May 2009 and 10 April to 5 May 2010 in the area of 21° 30.5 N, 158° 14.2 W (Figure 1), roughly 1 km from the shoreline and 5 km inshore of the habitat that serves as the daytime habitat for mesopelagic animals of the deep-scattering layer [31].

Ethics Statement

Observations of spinner dolphins were conducted under a US National Marine Fisheries Service Permit 1000–1617. The area accessed was not privately owned or protected and no protected species were sampled. This effort complied with all relevant regulations.

Moorings

A moored autonomous profiler (the Seahorse; Brooke Ocean Technology) collected temperature, salinity, pressure (SeaBird SBE-19 CTD), dissolved oxygen (SBE-43), and chlorophyll fluorescence (WET Labs ECO-FLS) data every 30 minutes between the near-bottom and the near-surface with <1 cm vertical resolution.

Throughout each study period, an upward looking multi-frequency echosounder (ASL Environmental Acoustic Water Column Profiler: 200, 420, 740 kHz) collected acoustic backscatter data from zooplankton, micronekton, and dolphins once per second. The echosounder used a pulse length of 256 ms, had a 3-dB beamwidth of 7 degrees, and a vertical resolution of 1 cm. The echosounder was calibrated in two ways: in a seawater tank using a reference sphere [32] and at sea using both a reference target as well as a comparative approach with the calibrated shipboard echosounders.

Shipboard sampling

Ship-based sampling was conducted from the 9 m *Alyce C.* anchored near the moored sensors. Sampling of zooplankton, micronekton and dolphins with echosounders and phytoplankton, zooplankton, and micronekton with a high-resolution profiling package was carried out continuously for 24 hours during the full, new, and quarter moon phases in each year. Additional sampling conducted over eight shorter intervals in each year that were dispersed over each study period. These efforts were supplemented by periodic zooplankton net tows and phytoplankton bottle samples.

Echosounders

Downward-looking echosounders mounted 1 m below the surface (calibrated [32] split-beam Simrad EK 60 s at 38 kHz (12° beamwidth) and 70, 120, and 200 kHz (7° beamwidth) and

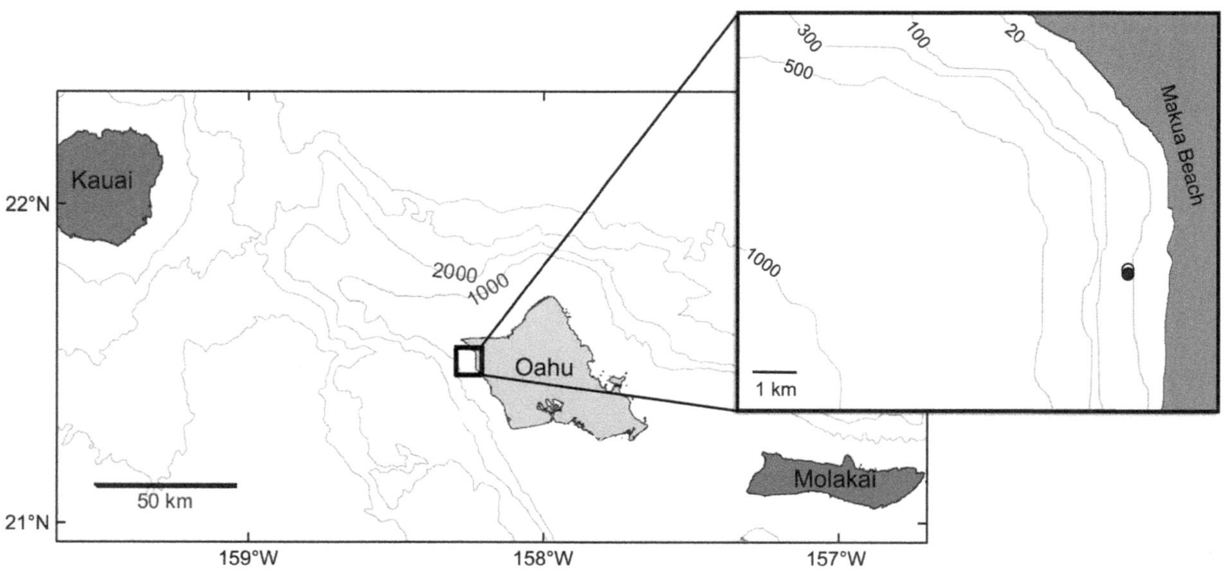

Figure 1. A map of the field site off leeward Oahu, Hawaii showing the location of the ship-based sampling (open circle) and the moored sensors located approximately 100 m away (filled circle), both on the 25 m isobaths.

Simrad ES60, 5° single beam at 710 kHz, each with an outgoing pulse length of 256 μs and a pulse rate of 8–10 Hz) were used to continuously measure the distribution and density of micronekton and zooplankton aggregations. A 200 kHz multibeam sonar (Kongsberg-Mesotech SM2000) was also deployed on the rigid pole. This system has a 120 degree by 1 degree field of view that was used to observe groups of foraging dolphins [23].

Profiler

The physical and biological characteristics of the water column were measured near-continuously with a profiling package during each sampling period. Profiles covered 1 m from the surface to 1 m above the bottom approximately every 4 minutes throughout each sampling period. To maximize the vertical resolution of the profiles the instrument package was ballasted to achieve an average descent rate of 10 cm s^{-1} and was decoupled from ship motion during descent by allowing cables to remain slack. The profiling package was equipped with a CTD (SeaBird 25; temperature, salinity, pressure), a dissolved oxygen sensor (SBE-43), and a fluorometer to measure chlorophyll a fluorescence (WETLabs WETStar). Data from CTD casts were low passed filtered and edited for loops before the raw variables were converted to variables of interest using factory calibrations.

The profiling package also had a Tracor acoustic profiling system (TAPS) which uses acoustical scattering at six frequencies (265, 420, 700, 1100, 1850 and 3000 kHz) to quantitatively estimate zooplankton abundance in size classes [33]. Volume scattering strength profiles were averaged into 0.2 m vertical bins and transformed to estimates of zooplankton biovolume in equivalent spherical diameter (ESD) classes via a constrained, non-linear, least-squares algorithm [34,35,36], that employed a simple spherical model; a choice guided by the body forms observed in net tows. These estimates of biovolume were converted to estimates of density by dividing the biovolume at a given size by the volume of a spherical animal of that ESD. Large individual targets (e.g. micronekton) were rarely sampled due to the small sample volume of TAPS (about 3 l). After initial processing, TAPS profiles were analyzed for the presence of discrete layers, as well as vertically integrated to provide an estimate of total zooplankton abundance.

To obtain information on the taxonomic composition, numerical density, and size of micronekton, the profiler was equipped with a two-camera low-light video system with infrared illumination. Still views were extracted every 0.25 m from each of the two cameras and analyzed following Benoit-Bird and Au [31].

Net tows

Periodically throughout shipboard sampling efforts, vertical net tows were conducted with a 0.5 m opening/closing, 200 μm mesh net equipped with a pressure sensor with real time communication (Simrad PI32) and flow meter modified to spin only on the upcast (General Oceanics). These tows were used to provide information on the identity and density of zooplankton in identified features and more generally throughout the water column. When an acoustic feature of interest was identified during shipboard sampling, the feature was discretely sampled at least once. Each targeted samples was followed by a sample integrated from 1 m above the bottom to the surface. Water-column integrated samples were also taken at regular intervals when no features of interest were identified. Samples were preserved in a buffered 5% formalin solution in seawater for later analysis following the methods described in Benoit-Bird et al.[29].

Data Analysis

Echosounder data

All echosounder data were analyzed using Myriax's Echoview software. For shipboard echosounders, all echoes from solitary individual targets, that is targets at densities lower than one per sampling volume were identified as large individual fish or marine mammals and removed from the data. In moored data, intense, large targets that were present at all frequencies were identified as marine mammals [23] and removed from the data set. The remaining volume scattering data from both the shipboard and moored sensors were thresholded at a value of −80 dB and integrated in 10 second horizontal by 0.30 m vertical bins. The volume scattering in each bin was then compared across all frequencies. Those bins that had volume scattering that did not increase by more than 2 dB with any increasing step in frequency were classified as micronektonic animals [37]. Of the remaining bins, those that had the scattering at least 4 dB higher at the two highest frequencies used than the lowest were classified as zooplankton [38,39]. These two classifications accounted for 95% of all intervals for both echosounder systems.

Using only data classified as zooplankton, the total scattering at 200 and 710 kHz for the shipboard echosounders and 420 and 740 kHz for the moored echosounder were integrated over the entire water column in 10 s bins. Using only data classified as micronekton, raw 200 kHz volume scattering from both the moored and shipboard echosounders was integrated into 10 second by 0.5 m bins. Numerical density of micronektonic animals was calculated from the 200 kHz volume scattering data, determined to be the most accurate frequency for this estimate [37], using echo energy integration. Estimates of individual target strength were made by combining laboratory target strength measurements with animal size and identity information from the video system on the high-resolution profiler averaged across all recordings made at the same time of day and depth as the echosoundings.

Identification of biotic layers

Both fluorescence and acoustic scattering were found in discrete, persistent features or layers. Layers in all data sets were identified following the general approach detailed in Benoit-Bird et al. [40] and summarized here. Layers in scattering consistent with micronekton were identified in the 200 kHz data from both moored and shipboard echosounders, while layers in scattering consistent with zooplankton were analyzed at all frequencies 200 kHz and above. A running, 5 m vertical median was taken for each profile which was an individual cast of TAPS or the fluorometer or 10 s averaged classified echosounder data (micronekton or zooplankton, analyzed separately) to define the local background in acoustic scattering or fluorescence. If the maximum measurement exceeded this background by 1.25 times for at least two consecutive casts or 30 s of echosounder data, the feature was considered a layer. The points at which scattering crossed below the running median were used to define the upper and lower edges of the layer. Scattering or fluorescence between these two points was integrated for each layer. Layer peak depth was defined as the point at which the layer reached a maximum value while median depth was the depth at which the integrated value within the layer was split in half. For plankton layers, thickness was calculated as the range of values within half the peak intensity of the layer, sometimes called the full width half maximum (FWHM), for comparison with previous studies. The average of these characteristics was calculated for each detected layer from the mooring data for statistical comparisons. The persistence of all zooplankton layers less than 5 m in thickness was characterized from the moored echosounder as the total time a layer was identified with no gaps in detection greater than 10 seconds. This analysis was not conducted on the time series of fluorescence data because of its low temporal resolution of one profile every 15 minutes.

Zooplankton effects on micronekton and dolphins

To test for potential effects on micronekton occurring before their arrival at the field site, an ANOVA was used to examine the effects of zooplankton layer presence on the timing of the arrival of micronekton at the site. To determine if the vertical distribution of the mesopelagic layer was affected by the presence of a zooplankton layer, the median depth, thickness, integrated abundance, and mean density of the layer of mesopelagic animals was measured on each night 30 min after sunset as well as 2 hours after sunset as reported by the US Naval Observatory. A multiple Analysis of Variance (MANOVA) was used to test for the effects of zooplankton layer presence on micronekton layer characteristics.

An ANOVA was used to determine the effect of zooplankton layers on spinner dolphin median depth measured 15–45 min after sunset. To examine other effects of zooplankton on spinner dolphin behavior. we examined data from 18 days of shipboard data that covered at least the 30 minutes before sunset and two hours after. On 11 of these days a zooplankton layer was present. Following the methods of Benoit-Bird and Au [23], we used the multibeam sonar system to count the number of dolphins within each foraging group, the average spacing between pairs of dolphins within each foraging group, and the average time each pair spent inside the circle of foraging dolphins, a measure of feeding time for all groups observed 15–45 minutes and 90–120 minutes after sunset. We used a MANOVA with sampling date as a covariate to determine if there was a significant effect of zooplankton layer presence on these measures of spinner dolphin behavior in each time interval. We counted the number of dolphin groups observed during each of these 30 minute intervals and used an ANOVA to determine if there was an effect of zooplankton layer presence on the number of dolphin groups observed.

Micronekton effects on zooplankton layers

To examine the impact of micronekton on zooplankton layers, scattering at 740 kHz from the moored echosounder was integrated over the thickness of detected zooplankton layers. Scattering within a layer was tracked in 5 minute intervals from 15 minutes before the first detection of micronektonic animals to 60 minutes after. The average value for each layer in the three, 5 minute intervals immediately before the arrival of micronekton was subtracted from each subsequent observation of the same feature to normalize the results.

Results

Fluorescent layers

A total of 206 discrete, fluorescent layers were identified from the moored data set. Of these, all but 5 of these were thinner than 5 m. A comparison of layers detected with the echosounder mooring and those detected with the profiler showed detection rates and vertical locations coincided. Fluorescence within these layers accounted for 18%–94% of the water column integrated fluorescence while layers covered 1–20% of the water column with the average thickness layer covering about 10% of the water column's depth. The fluorescence from phytoplankton layers was highly disproportionate to their vertical extent.

Fluorescent layers showed a diel pattern during both sampling years (Figure 2A), with more layers found at night than during the day and peaks in layer frequency at dawn and dusk. The depth of layers also showed a diel pattern, with the shallowest layers early in the morning and the deeper layers around dusk (Figure 3A). Note that too few layers were present between 0800 and 1600 to represent their vertical distribution with any confidence. Although we profiled continuously with the high-resolution shipboard profiler during the surveys, we only discuss layers occurring between 1600 hours and 0800 hours. During these evening and early morning hours, the average PAR in the upper 5 m of the water column was extremely low ($<200\ \mu E\ m^{-2}\ s^{-1}$); thus, we avoid non-photochemical quenching of fluorescence. For comparison, the maximum PAR was 1,469 $\mu E\ m^{-2}\ s^{-1}$ at 1230 hours on 3 May.

Zooplankton acoustic layers

A total of 372 layers with scattering characteristics consistent with zooplankton were identified from the moored data set. Of these, 361 (all but 11) were thinner than 5 m and only 4 identified layers were between 3 and 5 m in thickness. A comparison of layers detected by the moored echosounder and those detected by TAPS showed that detection rates and vertical locations were identical. Similarly, the same layers were detected using the 710 kHz shipboard echosounder and the 740 kHz moored echosounder. Zooplankton layers during daylight hours were detectable at all frequencies used by the echosounder mooring and all frequencies 120 kHz and above on the shipboard echosounders, however, at night when mesopelagic animals were abundant, zooplankton layers were not detectable at frequencies 200 kHz and below.

A criterion of 1.25 times the local background was used to define layers, however, in zooplankton layers detected in mooring data, the minimum intensity of scattering at 740 kHz was 2.65 times the local background, the maximum was approximately 8000 times the local background, and the average layer was about 1000 times as intense the local background. Acoustic scattering at 740 kHz from these layers accounted for between 31% and 100% of the scattering attributable to zooplankton in the water column while layers covered between 1 and 20% of the water column with the average thickness layer covering about 7% of the water column. These layers of zooplankton were intense and scattering from layers of zooplankton was highly disproportionate to their vertical extent.

The fraction of sampling days in which zooplankton layers were detected showed a diel pattern during both sampling years (Figure 2B), with more zooplankton layers detected during the day than at night. No layers of zooplankton were ever detected between 90 minutes and 3 hours after sunset. The persistence time of zooplankton layers also showed a diel pattern (Figure 2B) with layers increasing in persistence time throughout the day until they peak at dusk before disappearing entirely. Figure 3B shows how the depth of zooplankton layers changed throughout the day with layers generally found shallower during the day than at night. The depth distribution of layers during the day also had a much larger spread. Layer position was more consistent at night with the narrowest range of depth values found around dusk.

Zooplankton layers detected in mooring data showed a significant pattern in peak intensity relative to the local background as a function of layer thickness (Figure 4), which can be described by an exponential curve (Peak intensity $= 3.93.6\ e^{-1.17\ *\ \text{thickness}}$; $R^2 = 0.38$; $p<0.01$). There was no significant relationship between layer thickness and the total integrated scattering within the layer at 720 kHz, even when outliers and thicknesses with small sample sizes were removed ($R^2 = 0.02$, $p>>$

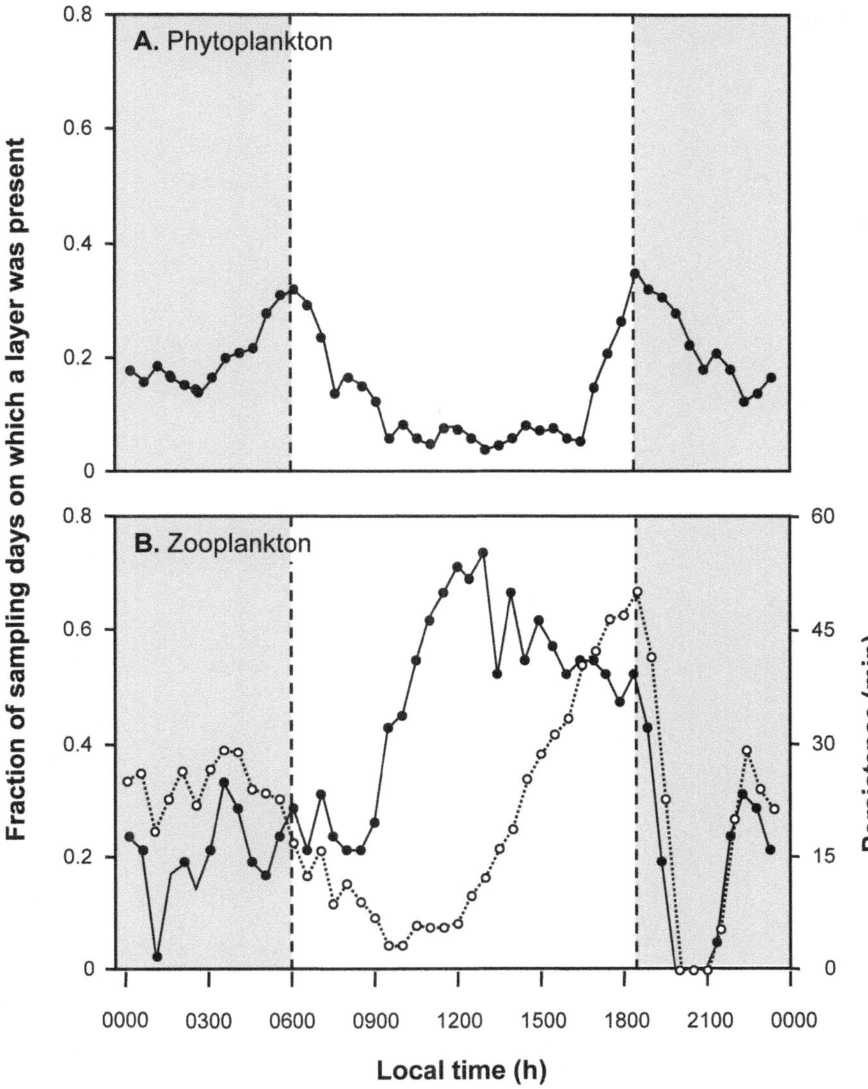

Figure 2. The frequency of occurrence of layers of phytoplankton (A) and zooplankton (B) as a function of local time detected using moored instruments across the two, 3-week studies is shown in the black symbols and solid lines. The y-axis shows the fraction of the 42 sampling days on which layers were detected in each time interval. Dusk and dawn are indicated by dashed lines separating the gray (night) from white (day) regions. Open circles and dashed lines in B indicate the average persistence, the total time a zooplankton layer was identified with no gaps in detection greater than 10 seconds, of zooplankton layers.

0.05). The median integrated scattering within each layer was about 250 $m^2 nmi^{-2}$ at 740 kHz, regardless of layer thickness.

Identification of zooplankton layer constituents

The frequency response ($S_{V\ 740\ kHz}$ - $S_{V\ 420\ kHz}$) of zooplankton layers detected by the mooring was bimodal with one mode centered at −2.5 dB and a larger mode at 5.5 dB (Figure 5). There were no layers with frequency response values between −1.6 and 2.7 dB. Layers with a positive frequency response were sampled using stratified net tows on 11 occasions. When these layers were found, net tows, both those that covered the entire water column and those that targeted acoustic features, were dominated in both number and biomass by calanoid copepods between 0.5 and 2.0 mm. The single most abundant group was *Clausocalanus* spp. A paired t-test showed that there was a significant increase in the density of copepods in net tows targeting the acoustic layers compared to those covering the entire water column ($t = 86.92$, $df = 10$, $p >> 0.05$).

The smaller mode in frequency response from the moored echosounder was accounted for by 31 layers, all detected in the second half of the 2009 field experiment. On five occasions, layers with stronger scattering at 420 kHz relative to 740 kHz were discretely sampled. These net tows were unlike those taken at any other time during the experiment as they were dominated both in number and biomass by hyperiid amphipods with equivalent spherical diameters between 2.6–3.7 mm. Vertically integrated tows of the entire water column taken at the same time showed no difference from stratified tows in the total number of amphipods caught (paired t-test: $t = 0.72$, $df = 4$, $p >> 0.05$). There was a significant increase in the total number of copepods in net tows covering the entire water column relative to the tows targeting just the layer (paired t-test: $t = 11.86$, $df = 4$ $p < 0.01$) but no increase in copepod density (paired t-test: $t = 1.20$, $df = 4$, $p >> 0.05$).

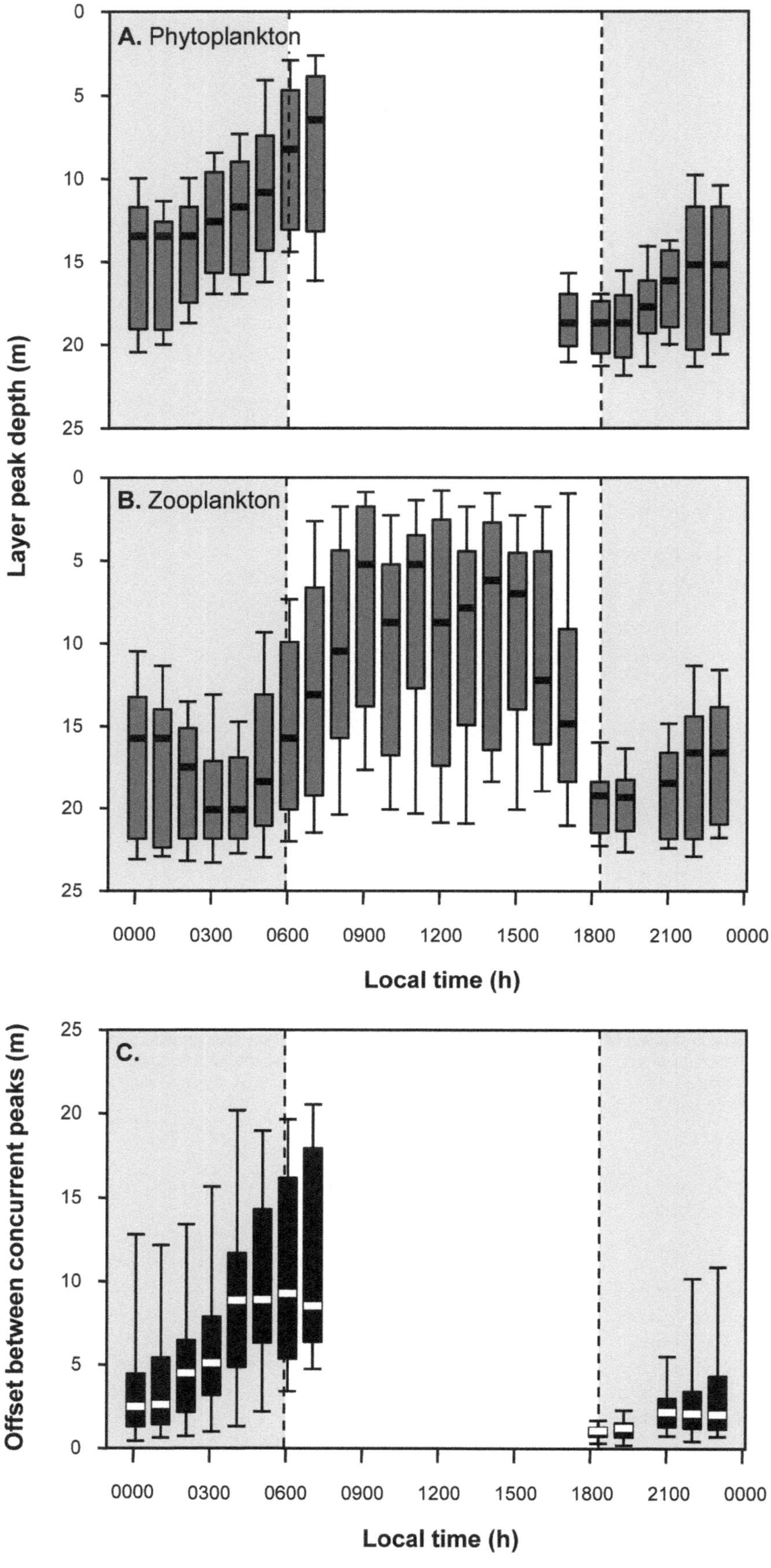
A. Phytoplankton
B. Zooplankton
C.
Layer peak depth (m)
Offset between concurrent peaks (m)
0
5
10
15
20
25
0000
0300
0600
0900
1200
1500
1800
2100
0000
Local time (h)

Figure 3. The peak depth of phytoplankton (A) and zooplankton layers (B) as well as the vertical offset between concurrently detected layer peaks (C) is shown as a function of local time. In each plot, heavy bars indicate the median of the distribution, the box one interquartile range, and the error bars show the 95% confidence interval.

During ship sampling, TAPS profiles were used to characterize the frequency response of scattering layers, which can provide information on the type and size of dominant scatterers. The frequency response of all layers was characteristic of small, fluid-like scatterers; generally increasing scattering strength with increasing frequency, sometimes reaching an asymptote at the highest frequencies with the presence of relatively small local nulls in scattering [41]. The average scattering responses for the two types of layers identified from the moored data are shown in Figure 6.

Overlap of plankton layers

The absolute value of the offset between the peaks of co-occurring fluorescent and acoustically scattering layers is shown in Figure 3C. Despite the apparent parallel movements of fluorescent layers and zooplankton layers, the median offset between paired layer peaks was less than 2.5 m only between dusk and 0100 h. Given the average thickness of each layer type (~2.5 m) and the fact that >90% of the layer's integrated fluorescence or acoustic scattering occurred within this thickness, this is the maximum separation that would allow consistent overlap between individuals within each layer type. Only at and immediately after dusk was the offset between layers consistently less than 2.5 m, with layers of zooplankton consistently deeper than phytoplankton layers (Figures 7 and 8).

Zooplankton effects on micronekton and spinner dolphins

The diel migration of micronekton into the shallow waters of Oahu's shelf has been well described previously [31,42]. The general patterns of micronekton during the time periods of this study matched those observations. For example, previous work has shown that nearly 100% of the micronekton found at night in these shallow waters were myctophids [31]. Both the camera system on the profiler and the frequency response of scattering from micronekton measured using the shipboard sampling indicated that held true during this study. However, while timing of arrival of the micronektonic layer at the 25 m site matched those observed previously and was not affected by layer presence (ANOVA: df = 1,25; F = 1.8; p = 0.41), the vertical location of the layer just after dusk showed more variation relative to previous studies that measured this time period with coarser temporal resolution [42]. The vertical distribution of micronekton at dusk on a night when a mode 2 zooplankton layer was detected is

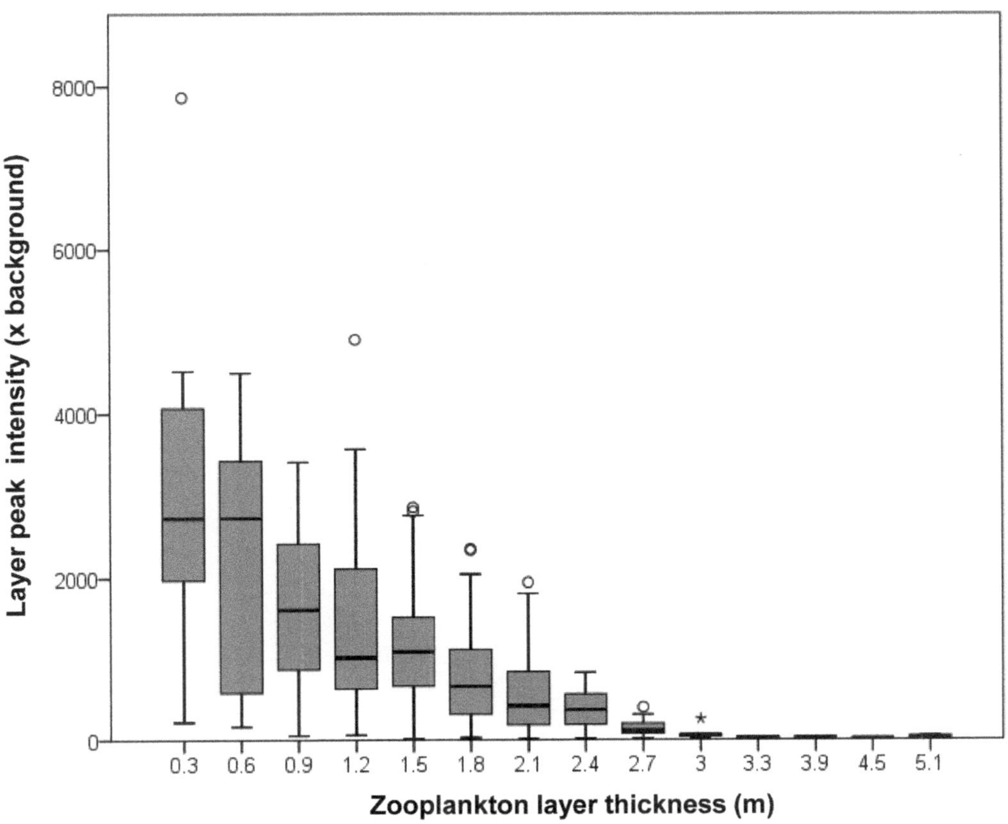

Figure 4. The peak intensity of layers detected at 740 kHz from the moored sensor is shown as a function of layer thickness. Heavy bars indicate the median of each distribution, the box one interquartile range, the error bars the 95% confidence interval, open circles outliers, and stars are those values that are least three box lengths from the median. All extreme outliers were mode 1 layers, those that had much stronger volume scattering at 420 than 740 kHz, revealed by net tows to be comprised of hyperrid amphipods rather than the copepods that made up the majority of layers.

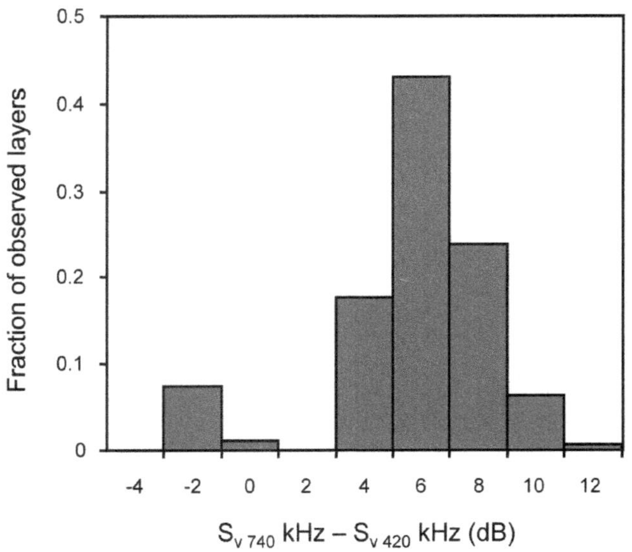

Figure 5. A histogram of the frequency response ($S_{v\ 740\ kHz}$ - $S_{v\ 420\ kHz}$) of layers detected by the moored echosounder. The y-axis shows the fraction of all observations accounted for by each 2 dB frequency response bin. There were no layers with frequency response values between −1.6 and 2.7 dB.

shown in Figure 8. The vertical distribution of micronekton at the same time on the next night when no zooplankton layer was detected is also shown. A multiple analysis of variance showed that the presence of zooplankton layers significantly affected the characteristics of micronekton layers measured just after dusk (df = 1,25; F = 68.3, p<0.001; Figure 9) but not later in the night (df = 1,25; F = 0.85, p = 0.91). Between subjects effects tests showed that there was a significant effect of layer presence on micronekton median depth (F = 24.3; p<0.005), micronekton layer thickness (F = 107.6; p<0.001), integrated abundance (F = 11.5; p<0.01), and micronekton density just after dusk (F = 46.9; p<0.001). A regression analysis showed a strong, linear relationship between the peak depth of the zooplankton layer and the peak depth of the micronekton layer observed on the same night, both 30 minutes after sunset (micronekton depth = 0.96 * zooplankton depth – 1.05; $r^2 = 0.83$; p<0.001).

The presence of zooplankton layers had a significant effect on the mean depth of dolphins observed in the 15–45 minutes after dusk (Mean dolphin depth: layer present = 19.6 m, absent = 12.3 m; ANOVA: df = 1,16; F = 17.13; p<0.005) but not during the 90–120 minutes after dusk (Mean dolphin depth = 10.28; ANOVA: df = 1,16, F = 1.7; p = 0.51). The presence of zooplankton layers had a significant effect on the number of groups of dolphins observed in the 15–45 minutes after dusk (Average number of groups: layers present = 4.8, absent = 2.6; ANOVA: F = 30.42; df = 1,16; p<0.001) but again, not during the later time period (Average number of groups = 3.8; ANOVA: F = 0.41; p = 0.88). The presence of layers but not sampling date significantly affected the measured characteristics of dolphin behavior just after dusk (MANOVA: df = 3,66; $F_{day} = 0.13$, p = 0.94; $F_{layers} = 100.8$, p<0.001) but neither was important later in the night (MANOVA: df = 3,66; $F_{day} = 1.21$, p = 0.71; $F_{layers} = 1.42$, p = 0.57). Between subjects effects tests for the time period just after dusk (df = 1,71) showed that there was a significant effect of layer presence on foraging dolphin group size (Mode group size: layers present = 20, absent = 24; F = 20.1; p< 0.001) and average inter-pair spacing within the group (Average inter-pair spacing: layers present = 3.76 m, absent = 2.77 m; F = 261.1; p<0.001 for both comparisons) but not on time for a feeding event by an individual pair (mean = 10.2; F = 0.21, p = 0.65). Observations that involved the entire feeding stage of a foraging event (e.g., from the onset of pairs moving into the circle to surfacing) showed no significant effect of layer presence on the duration of the feeding stage of foraging events (df = 1,26; F = 1.07, p = 0.31). As a result, while the duration of each pair of dolphins' individual feeding event did not increase, the number of feeding

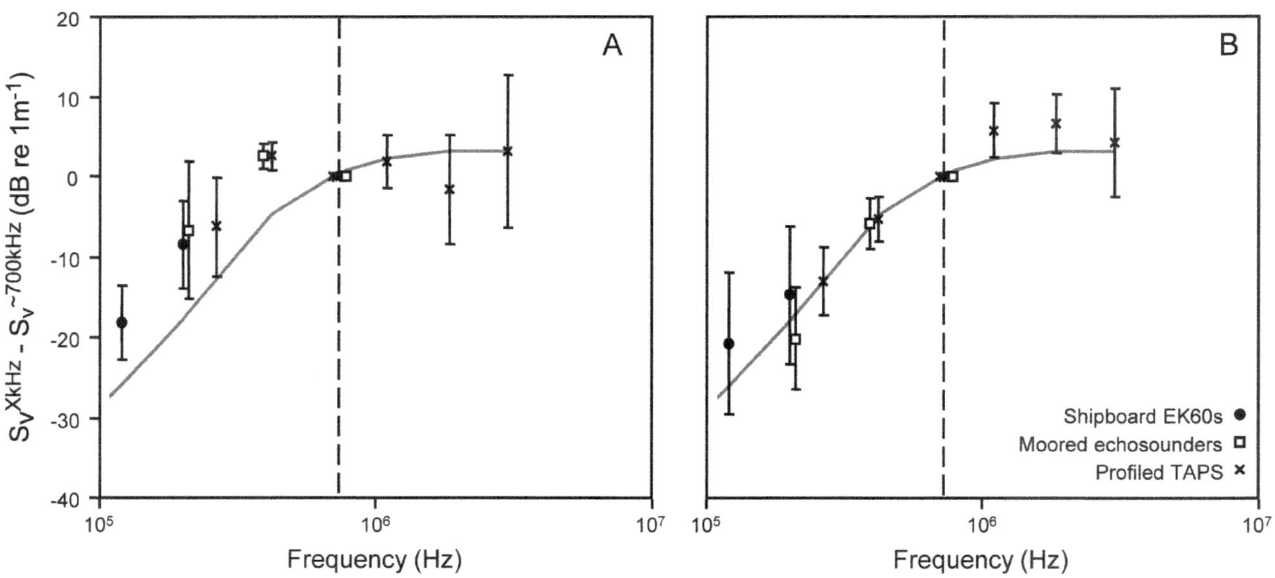

Figure 6. The mean volume scattering for each identified thin layer relative to that measured around 700 kHz (dashed line) is shown as a function of frequency for each of three acoustic instruments. Error bars show standard deviation. A) The frequency response of mode 1 layers, those that had stronger scattering at 420 kHz relative to 740 kHz in the moored data set. B) The frequency response of mode 2 layers. The gray line in each figure represents the scattering response expected for a fluid sphere of the median size of copepods measured from net tows.

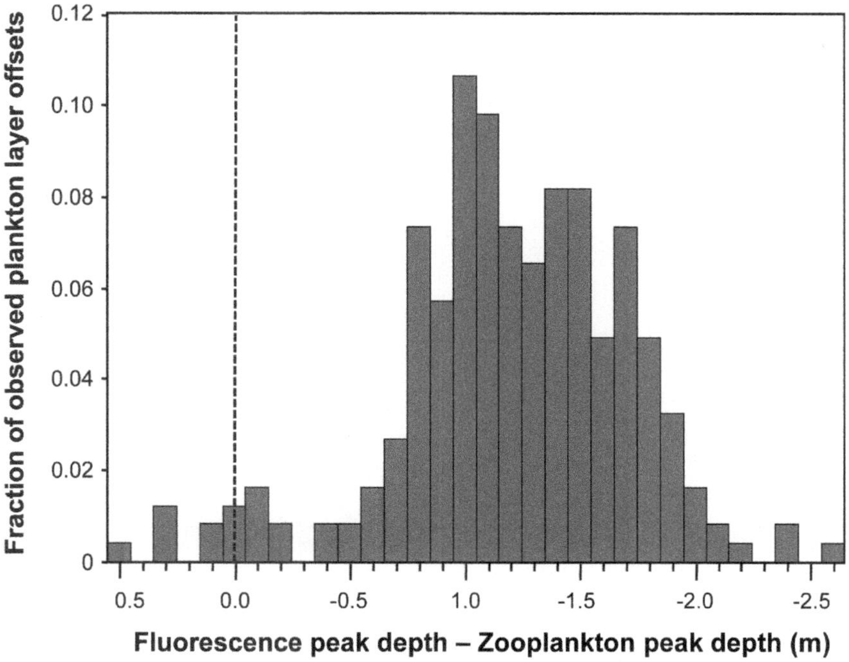

Figure 7. A histogram of the distance between the peaks of simultaneously detected fluorescence and zooplankton layers. The y-axis shows the fraction of all observations accounted for by data within each 0.1 m bin. Positive values indicate the zooplankton layer is shallower than the fluorescent feature while negative values mean the zooplankton layer is deeper in the water column than the fluorescent layer.

events was affected by the presence of zooplankton layers, increasing from an average of 2.5 when layers were not present to 5.1 when layers were present.

Micronekton effects on zooplankton layers

To examine the effect of micronekton on zooplankton layers, integrated scattering at 740 kHz within a layer was normalized to the integrated scattering over the 15 minutes before the first detection of micronekton and tracked in 5-minute intervals from 15 minutes before to 60 minutes after micronekton arrived. The average value for each layer in the three, 5 minute intervals immediately before the arrival of micronekton was subtracted from each subsequent observation of the same feature to normalize the results. The integrated scattering strength began to decrease within 15 minutes and layers began to be lost entirely 25 minutes after micronekton arrived at the mooring station (Figure 10).

Discussion

Our goal in this work was to add temporal dynamics to our growing understanding of spatial variance in ecosystem processes in Hawaii's nearshore pelagic environment. To do this, we used moored sensors to describe the fine-scale vertical dynamics of each component of system over a variety of timescales. By measuring true coherence between predators and prey in both space and time, we gain insight into the mechanisms underlying the interactions between these organisms. Considerable attention has been paid to micronekton aggregations [31,42,43], spinner dolphins [23], and more recently, phytoplankton aggregations [30,44] at this site. Our results for these components of the system are individually consistent with previous descriptions. Micronekton are present in the nearshore only at night where primarily myctophid fishes achieve considerable densities. Spinner dolphins track the migratory patterns of micronekton, foraging primarily in cooperative groups that herd patches of micronekton into even more intense aggregations. The overall concentrations of phytoplankton in nearshore waters are relatively low, consistent with other subtropical habitats. However, the regular appearance of intense (up 5 mg m^{-3}), thin (typically less than a few meters) layers dominated the fluorescence in the water column, accounting for up to 94% of the water column integrated value.

Zooplankton aggregations in this system have not been well characterized [29,45] and the critical links involving zooplankton are only beginning to be explored [28]. During our two, 3-week study periods, zooplankton layers were relatively common, occurring during at least some time during every day of the study. During most of the study, net tows showed that zooplankton layers were made up of copepods, as was the remainder of the water column. These dense layers reflected the general composition of the entire system. However, during the second half of 2009, net tows within layers were dominated by hyperiid amphipods while in the remainder of the water column, copepods still dominated by number and biovolume. These differences in layer composition were reflected in differences in the frequency response of acoustic backscatter with features made up of copepods having higher scattering at 740 kHz while features dominated by hyperiid amphipods showed higher scattering at 420 kHz. The unique zooplankton layers found only in one week in the second half of the 2009 experiment occurred during a period of uncommonly persistent low wind speeds and increased stratification.

Zooplankton within layers dominated the zooplankton biomass measured in the water column when layers were present. Despite covering only an average of 10% of the water column's depth, zooplankton in layers accounted for between 31 and 100% of the measured zooplankton biomass. Over the course of our two, 3-week studies, the average concentrations of copepods in the water

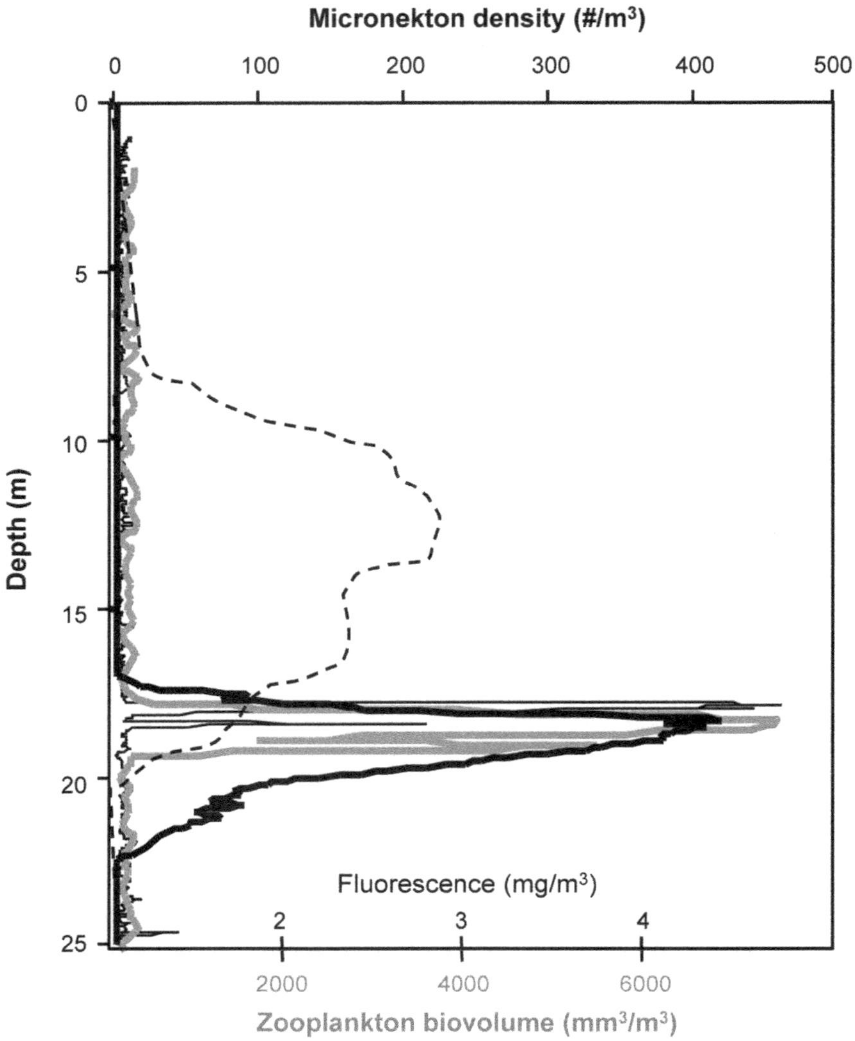

Figure 8. A sample profile of fluorescence (thin black line), zooplankton biovolume (gray line), and micronekton density (heavy black line) measured using shipboard sampling tools 30 minutes after sunset. The zooplankton layer shown was a mode 2 layer comprised of copepods. The vertical distribution of micronekton at the same time on the next night when no zooplankton layer was detected is shown as a dashed black line.

column were approximately 200 m^{-3} for a biovolume of 400 mm^2 m^{-3}. The average layer of zooplankton had a density of zooplankton at its peak of about 5,000 individuals m^{-3} while the strongest layer had a peak density of over 100,000 individuals m^{-3}. However, these densities covered a very narrow vertical range, typically less than 3 m so the areal density of a layer was often lower than its peak density. Despite large differences in the peak intensity of layers and layer thickness, the integrated scattering within a layer was relatively constant at 250 m^2 nmi^{-2} as thinner layers became more intense but did not contain more individual zooplankters. The origins of this biomass limit and its implications are not clear.

Zooplankton layers showed strong, predictable diel patterns in their occurrence and location in the water column. Zooplankton layers were most common during the afternoon, disappeared just after dusk, a pattern also observed in other sites in Hawaii [46], reappearing just before midnight. Zooplankton layers were found deep in the water column at night and an average of about 10 m shallower during the day. Around dusk, the abundant zooplankton layers showed a very narrow range of depths centered at 20 m, 5 m above the seafloor. This diel pattern is likely to be the result of behavioral dynamics of the zooplankters, perhaps as a response to convective overturning that occurs after the sun sets [30].

Layers of fluorescent phytoplankton also showed distinctive diel patterns with more layers observed at night than during the day with peaks in layer abundance at dawn and dusk. The peak in abundance at dusk is consistent with observations by McManus et al [30], whoalso observed a peak in abundance just before sunset. This peak is coincident with the time of day when thermal stratification is the strongest and most persistent [30], however, that stratification breaks down throughout the night due to convective overturn and thus cannot explain the somewhat smaller dawn peak in layer frequency. The depth of fluorescent layers of phytoplankton was restricted around dusk to just shallower than 20 m. Throughout the night, phytoplankton layers moved upwards through water column with an increasingly variable range of depths amongst them until they were within 7 m of the surface around dawn. During daylight hours, too few phytoplankton layers were identified to characterize their depth distribution. This limited number of detectable phytoplankton layers during the

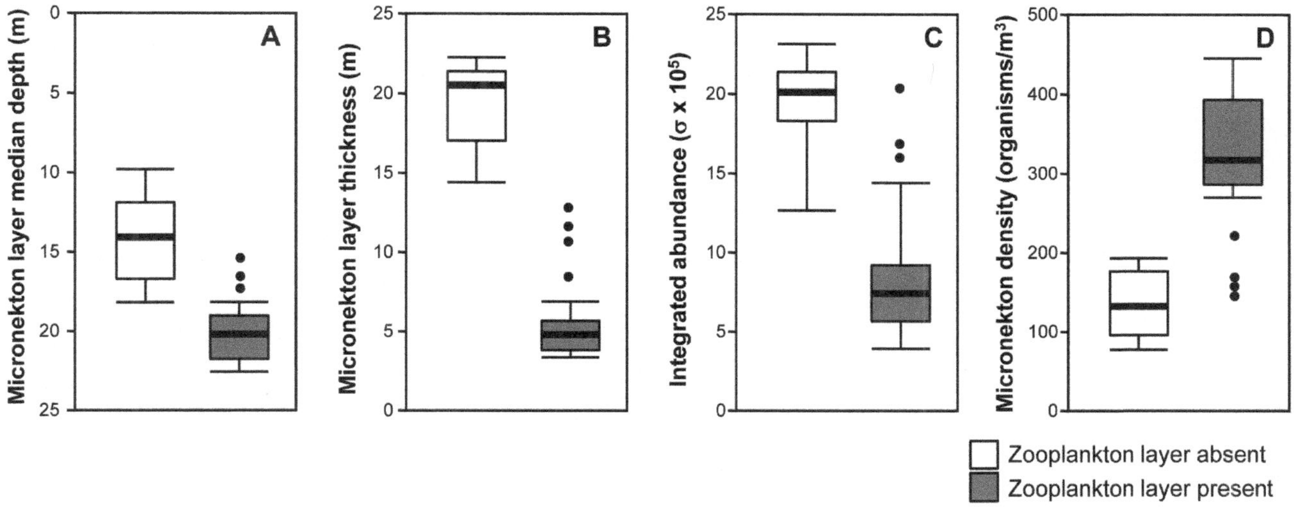

Figure 9. Micronekton layer characteristics 30 minutes after sunset when a zooplankton layer was absent (white boxes) and when a zooplankton layer was present (gray boxes). For all plots, the median of the distribution is shown by a heavy bar, the box represents one interquartile range, the error bars indicate the 95% confidence interval, and circles show outliers.

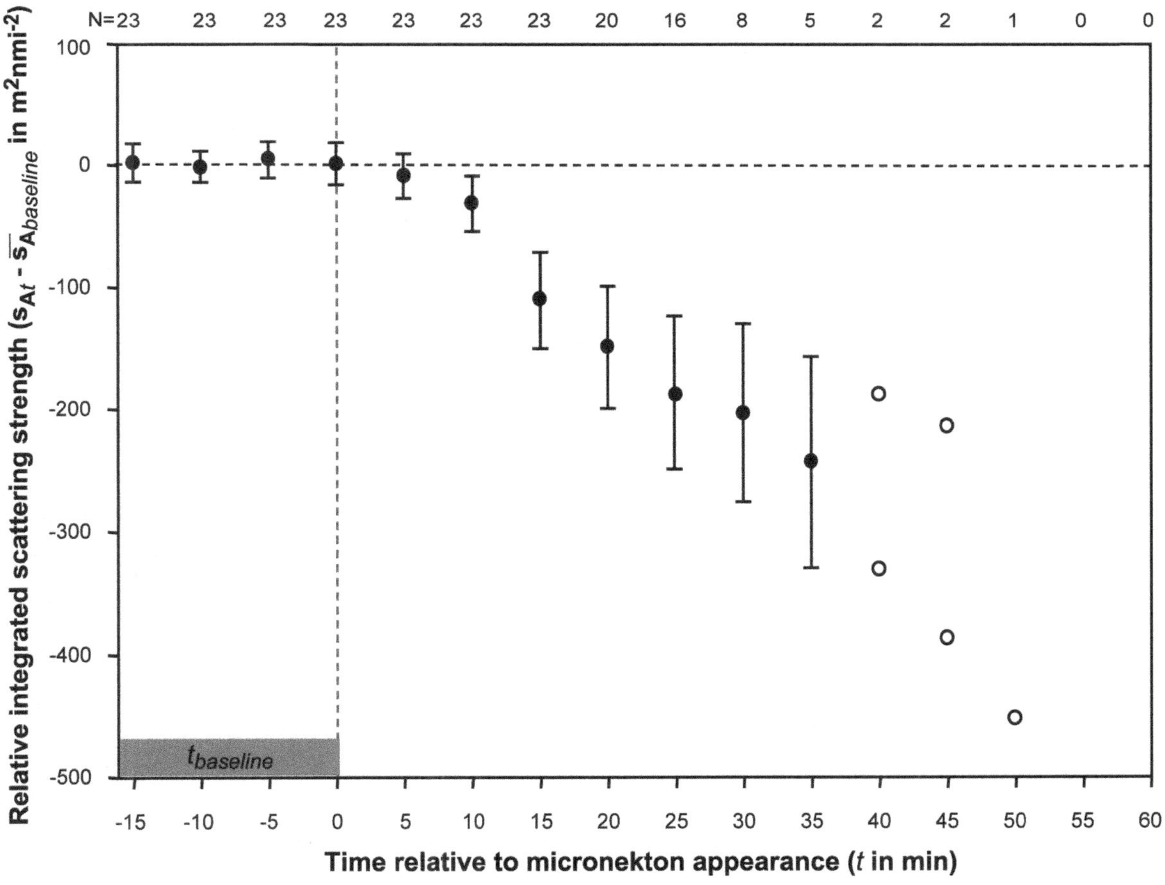

Figure 10. Changes in the biomass within zooplankton layers after the arrival of micronekton at the study site indicate rapid loss of these features each night. Scattering at 740 kHz integrated over the depth of each scattering layer was normalized to the average scattering for the same layer over a baseline defined as the three, 5-minute intervals before the arrival of micronekton. Mean changes relative to zero values are shown in 5 minute intervals (points) with error bars indicating 95% confidence intervals. When 2 or fewer layers were identified in a time interval, actual points are shown with open symbols. The number of layers in each time interval is indicated near the top of the plot.

day could be due to dispersion of phytoplankton during the day followed by reformation into layers at dusk, migration of layers to the surface where they would not be detected by the moored or shipboard profilers as has been observed in layers of motile phytoplankton in other habitats [47], or perhaps intense non-photochemical quenching of fluorescence during the day.

The strong diel patterns of layers of both phytoplankton and zooplankton result in a highly variable potential for interaction between the two groups. Layers of phytoplankton and zooplankton show different temporal trends in their occurrence but similar patterns in their vertical position in the water column. However, despite the similar patterns in vertical position, layers that occurred at the same time did not often occur within a few meters of each other vertically and the zooplankton layer was always deeper than the phytoplankton layer so there was no potential for interaction during vertical crossings of layers. A significant number of phytoplankton and zooplankton within layers co-occur in both three-dimensional space and time at night, however, only just after dusk did the organisms in these layers predictably overlap. This is consistent with observations of other copepod species which focus their feeding during dusk and/or dawn [48]. However, the pattern was quite different from that observed in Monterey Bay, California where layers of zooplankton overlapped with phytoplankton infrequently but predictably as a function of fraction of available phytoplankton in layers and not diurnal cycling [40]. Interestingly, in the observations here, all phytoplankton layers contained at least 18% of the total phytoplankton fluorescence in the water column, the threshold at which zooplankton layers actively responded to phytoplankton layers in the richer, temperate waters of Monterey Bay, indicating different mechanisms for layer co-occurrence.

While not evidence of consumption, overlap between predator and prey is necessary for foraging and thus only during this less than one hour each day could zooplankton be consuming fluorescent phytoplankton found in layers in this system. However, despite the predictable co-location of layers of phytoplankton and zooplankton at dusk, complete overlap of these layers did not occur at this time or any other. The peak of zooplankton layers was always deeper than the peak of fluorescent layers. At the time of consistent co-location of layers, throughout all observations the peak of zooplankton layers was approximately 1 m deeper than the phytoplankton layer occurring at the same time. As a result, only the upper ~40% of a layer of zooplankton interacts with the lower ~30% of the phytoplankton layer. These offsets could be the result of physical forcing affecting these different sized organisms differently, a foraging tactic of the predator attacking prey from beneath, a response of consumers to their own predators, or a balance of potentially opposing forces. Offsets between layers of different types of plankton have been observed but only rarely have small offsets been predictable [40,49,50]. Benoit-Bird et al [40] found that when fluorescent and acoustic layers were in close proximity, layer peaks were not coincident but zooplankton were nearly equally likely to be above the phytoplankton layer as below it. So, while small offsets between layers of plankton appear to be common, the factors driving those relationships likely vary amongst systems and species.

Dusk is also a critical period for other interactions within the system. Shortly after sunset, micronektonic organisms, primarily myctophid fishes, ascend from the deep, aggregating on the narrow island shelf to consume the relatively rich resources present near the island at night [29,31]. The phenology and typical vertical distributions of these organisms have been well documented [42]. We noticed dramatic departures from these typical patterns whenever dense layers of zooplankton were detected at dusk as illustrated in Figure 8. The upper edge of micronekton was not nearly as shallow as expected. This also affected the thickness of the layer as the bottom edge of the micronekton layer is constrained by the seafloor, resulting in an increase in the density of micronekton. The total abundance of micronekton within the layer, however, decreased as layers reached these high densities. The effects on micronekton were not observed to be as strong when zooplankton layers were made up of hyperiid amphipods as opposed to pelagic copepods. The outliers in Figure 9 were all measured from zooplankton layers with scattering at 420 kHz that exceeded that at 740 kHz, associated with large numbers of hyperiid amphipods in net tows through the layer. These outliers occurred in a single week during May of 2009, during a period of uncommonly persistent low wind speeds and increased stratification. A preference for copepods is consistent with the known diets of similar myctophid species in adjacent waters [51].

The presence of a dense, thin layer of zooplankton, particularly those made up of copepods, caused micronekton to abbreviate their vertical migration in these shelf waters. This change in behavior likely allows migrating micronekton to exploit the rich resource these layers represent. Net tows and acoustic sampling both reveal that each layer of zooplankton accounted for at least 30% of the total biomass of zooplankton in the water column, sometimes containing all of the water column's zooplankton biomass in the sampling location. It should not be surprising that zooplankton consumers that move vertically through such layers either stop moving when they reach the upper limit of these features or quickly move back into them, particularly if the original vertical movement was instigated as a quest for food as most assume nocturnal upward migration to be [52].

The effects of zooplankton layers on micronekton behavior were relatively short lived. Within an hour of their appearance, all micronekton were observed to resume their more typical vertical distribution. Within that same period of time, each zooplankton layer became undetectable as its integrated acoustic scattering gradually dropped (Figure 10). Layers of zooplankton began to disappear within 20 minutes of the arrival of midwater micronekton and all layers were undetectable after 50 minutes. The disappearance of these layers could be due to dispersion of individuals or to loss by consumption. After micronekton resumed their normal movement, layers of zooplankton reappeared beginning about 2 hours later. The reappearance of layers did not result in any measurable changes in micronekton despite their continued co-occurrence in the x-y plane and some vertical overlap between the lower edge of the micronekton layer and the zooplankton layer. The reappearance of zooplankton layers would suggest that dispersion plays some role in the disappearance of these features after dusk, however dramatically reduced peak intensities (e.g. 10x or more) indicate that consumption likely plays a role in their demise as well.

The short temporal overlap between micronekton and zooplankton layers might suggest a limited importance of these features in the energy acquisition of the micronekton that make a stop within them. However, simple calculations indicate that relative increases in foraging gains may indeed by significant. The density of zooplankton within these layers is at least two orders of magnitude higher than the average density of all zooplankton in the water column typical for the same time and location [29]. In fact, the biomass of zooplankton in the average layer exceeds by two-fold the average biomass of the entire water column in this site. If copepods are a preferred food source as indicated by the increased response to layers made up of copepods relative to those made up of amphipods, the relative local density difference is even more striking as zooplankton layers were not representative of the entire water column's zooplankton composition but rather an intensification only of the copepod community, increasing the ratio of preferred to non-preferred food resources which may increase foraging efficiency [53]. Combining estimates of biomass with Figure 10 shows that after about 30 minutes, the biomass of zooplankton in layers roughly equals the water column integrated average. If prey encounter rates for each myctophid are roughly related to the density of zooplankton in these two conditions, then those 30 minutes within the layer are equal to approximately 3.5 hours foraging under more typical conditions. If prey consumption is decreased proportionately with micronekton density, then the increased density of competitors during that time could reduce the benefit from a factor of 7 to a factor of ~3, a likely underestimation of the benefit given results that indicate higher foraging success rates on more densely aggregated prey [54,55,56]. However, when foraging time is limited by the day night cycle to about half of each 24-period, even a factor of 3 for this short post-dusk time window would result in a ~10% overall increase in foraging gains.

The presence of layers of zooplankton cascaded even further up the food, affecting the behavior of dolphins observed at dusk. On evenings when zooplankton layers were present, spinner dolphins were an average of 60% deeper in the water column, in groups that were 17% smaller, with spacing between pairs of dolphins 35% higher and were detected nearly twice as often relative to those evenings when a zooplankton layer was absent. Perhaps most importantly, the number of feeding events observed in each pair of dolphins and consequently the feeding time for each individual more than doubled when zooplankton layers were present. These changes only occurred in a 30 minute window just after dusk, the same time period where significant effects of layer presence were observed in micronekton. Some of the changes observed in dolphin behavior may involve costs (deeper diving), however the increase in the number of feeding events indicate a net gain as a result of zooplankton layers during the post-dusk time window. As spinner dolphin foraging is limited by feeding time rather than directly by the availability of food [23], the doubling in duration of feeding time over about 5% of their total foraging time each day could have significant impacts on an individual's fitness.

Previous analyses of data from the field efforts described here have shown the importance of the layers of organisms we observed in their interactions [28]. In part because of the intensity of these aggregations and strong, once per day patterns, correlations amongst trophic levels were robust to averaging over each day, a much longer time scale than the scale at which the interactions between organism types were occurring. However, this averaging obscured the details of these interactions. Here, we show that the interactions driving the overall patterns observed amongst trophic levels occur over the same very short time interval – within the hour following sunset each night. The pattern of spatial coincidence between predator and prey is an emergent outcome of a behavioral race between them [11,57]. In this case, the outcome we observed is the result of a number of interdependent races. The solutions for each step in the trophic level converged at a short, dynamic period of each day – dusk. Given the changes in physical processes [30] and the remarkable redistribution of the animals in the ecosystem [29,42] that occurs at this time, foraging and avoidance tradeoffs change quickly during this time, ultimately allowing a brief period of intense interaction between species. It has long been recognized that dawn and dusk "transition periods" are dynamic and often critical periods for a variety of species [58,59,60,61] and, as a result, for predator prey interactions [4]. However, dawn and dusk have often been explicitly excluded from analyses of diurnal patterns because of the challenges of obtaining sufficient samples to describe these short periods.

To understand the interactions between predators and prey in pelagic systems, we need concurrent data from each that is coincident and time and space. However, even under ideal conditions, we must recognize that studies that provide only snapshots in time are likely to miss the critical periods when interactions occur amongst species, leaving us with weak or inconsistent correlations between predator and prey, a common observation in pelagic ecosystems [5,6,62]. One approach that has proven helpful in dealing with this challenge for single predator-prey pairs is to utilize the identification of foraging behavior to focus sampling efforts [10]. Multi-trophic level interactions, however, present additional challenges. Because of the importance of place in Hawaii's slope associated ecosystem, a fixed network of sensors that could sample with relatively fine resolution in vertical space over time allowed us to identify a critical time window for predator-prey coherence throughout the food chain. Long-term, multifaceted arrays of ocean sensors are or will soon be coming online in a variety of pelagic systems. We are optimistic that these observational tools will provide opportunities to move beyond examination of temporally separated snapshots towards mechanistic descriptions of ecosystems. Space use decisions by predator and prey determine the spatial overlap between the two that affects encounter rates, predation rates and population and community dynamics [63]. These space use decisions are dynamic over a variety of time scales, particularly in the constantly moving fluid environment of the pelagic. While few studies of spatial ecology have explicitly included a temporal component at scales smaller than seasonal or larger than single behavioral interactions or simple day/night contrasts [22], our observations that short time windows can drive the structure and function of a complex suite of organisms highlight the importance of explicitly adding a temporal component to our descriptions of heterogeneity.

Acknowledgments

Chad Waluk, Jeff Sevadjian, Ross Timmerman, Gordon Walker, Linnaea Jasiuk, Jennifer Patterson, Donn Viviani, Amanda Timmerman, D. Bloedorm, Lindsey Benjamin, Rebecca Baltes, Gordon Walker, and Jon Whitney provided assistance in the field. Captains Joe Reich, Tom Swenarton, and Tony Brown provided logistical support. Tim Cowles, Mike Chapman of MECCO, and Deep-Sea Power & Light loaned equipment.

Author Contributions

Conceived and designed the experiments: KJBB MAM. Performed the experiments: KJBB MAM. Analyzed the data: KJBB MAM. Contributed reagents/materials/analysis tools: KJBB MAM. Wrote the paper: KJBB MAM.

References

1. Lima SL (1998) Nonlethal effects in the ecology of predator-prey interactions. BioScience 48: 25–34.
2. Williamson CE, Stoeckel ME, Schoeneck LJ (1989) Predation risk and the structure of freshwater zooplankton communities. Oecologia 79: 76–82.
3. Silvertown J, Holtier S, Johnson J, Dale P (1992) Cellular automaton models of interspecific competition for space—the effect of pattern on process. Journal of Ecology: 527–533.
4. Rose GA, Leggett WC (1990) The importance of scale to predator-prey spatial correlations: An example of Atlantic fishes. Ecol 71: 33–43.

5. Fauchald P (2009) Spatial interaction between seabirds and prey: review and synthesis. Mar Ecol Prog Ser 391: 139–151.
6. Russell RW, Hunt GL, Coyle KO, Cooney RT (1992) Foraging in a fractal environment: Spatial patterns in a marine predator-prey system. Landscape Ecology 7: 195–209.
7. Hunt G (1990) The pelagic distribution of marine birds in a heterogeneous environment. Polar Research 8: 43–54.
8. Benoit-Bird KJ, Kuletz K, Heppell S, Jones N, Hoover B (2011) Active acoustic examination of the diving behavior of murres foraging on patchy prey. Mar Ecol Prog Ser 443: 217–235.
9. Williamson CE, Stoeckel ME (1990) Estimating predation risk in zooplankton communities: the importance of vertical overlap. Hydrobiologia 198: 125–131.
10. Hazen EL, Friedlaender AS, Thompson MA, Ware CR, Weinrich MT, et al. (2009) Fine-scale prey aggregations and foraging ecology of humpback whales *Megaptera novaeangliae*. Mar Ecol Prog Ser 395: 75–89.
11. Sih A (1984) The behavioral response race between predator and prey. Am Nat 123: 143–150.
12. Bell AV, Rader RB, Peck SL, Sih A (2009) The positive effects of negative interactions: Can avoidance of competitors or predators increase resource sampling by prey? Theoretical Population Biology 76: 52–58.
13. Roughgarden J (1974) Population dynamics in a spatially varying environment: how population size" tracks" spatial variation in carrying capacity. Am Nat: 649–664.
14. Lima SL (2002) Putting predators back into behavioral predator-prey interactions. Trends Ecol Evol 17: 70–75.
15. Nachman G (1981) Temporal and spatial dynamics of an acarine predator-prey system. The Journal of Animal Ecology: 435–451.
16. Dupuch A, Dill LM, Magnan P (2009) Testing the effects of resource distribution and inherent habitat riskiness on simultaneous habitat selection by predators and prey. Anim Behav 78: 705–713.
17. Hunt GL, Stabeno PJ, Strom S, Napp JM (2008) Patterns of spatial and temporal variation in the marine ecosystem of the southeastern Bering Sea, with special reference to the Pribilof Domain. Deep Sea Research Part II: Topical Studies in Oceanography 55: 1919–1944.
18. Durant JlM, Hjermann DO, Ottersen G, Stenseth NC (2007) Climate and the match or mismatch between predator requirements and resource availability. Climate Research 33: 271.
19. Sigler MF, Kuletz KJ, Ressler PH, Friday NA, Wilson CD, et al. (2012) Marine predators and persistent prey in the southeast Bering Sea. Deep Sea Research Part II: Topical Studies in Oceanography 65–70: 292–303.
20. Womble JN, Sigler MF (2006) Seasonal availability of abundant, energy-rich prey influences the abundance and diet of a marine predator, the Steller sea lion Eumetopias jubatus. Mar Ecol Prog Ser 325: 281–293.
21. Similä T, Holst JC, Christensen I (1996) Occurrence and diet of killer whales in northern Norway: seasonal patterns relative to the distribution and abundance of Norwegian spring-spawning herring. Can J Fish Aquatic Sci 53: 769–779.
22. Hampton SE (2004) Habitat overlap of enemies: temporal patterns and the role of spatial complexity. Oecologia 138: 475–484.
23. Benoit-Bird KJ, Au WWL (2009) Cooperative prey herding by the pelagic dolphin, *Stenella longirostris*. J Acoust Soc Am 125: 125–137.
24. Sims DW, Witt MJ, Richardson AJ, Southall EJ, Metcalfe JD (2006) Encounter success of free-ranging marine predator movements across a dynamic prey landscape. Proceedings of the Royal Society B: Biological Sciences 273: 1195–1201.
25. Schneider DC (2001) The Rise of the Concept of Scale in Ecology: The concept of scale is evolving from verbal expression to quantitative expression. BioScience 51: 545–553.
26. Grunbaum D (2002) Predicting availability to consumers of spatially and temporally variable resources. Hydrobiologia 480: 175–191.
27. Horne JK, Schneider DC (1995) Spatial variance in ecology. Oikos 74: 18–26.
28. Benoit-Bird KJ, McManus MA (2012) Bottom-up regulation of a pelagic community through spatial aggregations. Biology Letters 8: 813–816.
29. Benoit-Bird KJ, Zirbel MJ, McManus MA (2008) Diel variation of zooplankton distributions in Hawaiian waters favors horizontal diel migration by midwater micronekton. Mar Ecol Prog Ser 367: 109–123.
30. McManus MA, Sevadjian JC, Benoit-Bird KJ, Cheriton OM, Timmerman AH, et al. (2012) Observations of Thin Layers in Coastal Hawaiian Waters. Estuaries Coasts: 1–9.
31. Benoit-Bird KJ, Au WWL (2006) Extreme diel horizontal migrations by a tropical nearshore resident micronekton community. Mar Ecol Prog Ser 319: 1–14.
32. Foote KG, Vestnes G, Maclennan DN, Simmonds EJ (1987) Calibration of acoustic instruments for fish density estimation: A practical guide. ICES Coop Res Rep 144: 57 pages.
33. Holliday DV, Pieper RE (1995) Bioacoustical oceanography at high frequencies. ICES J Mar Sci 52: 279–296.
34. Holliday DV (1977) Extracting biophysical information from the acoustic signals of marine organisms. In: Anderson NR, Zahuranec BJ, editors. Oceanic Sound Scattering Prediction. New York: Plenum. pp. 619–624.
35. Medwin H, Clay C (1997) Fundamentals of Acoustical Oceanography. San Diego: Academic Press.
36. MacLennan DN, Simmonds EJ (1992) Fisheries Acoustics. New York: Chapman and Hall.
37. Benoit-Bird KJ (2009) Effects of scattering layer composition, animal size, and numerical density on the frequency response of volume backscatter. ICES J Mar Sci 66: 582–593.
38. Korneliussen RJ (2000) Measurement and removal of echo integration noise. ICES J Mar Sci 57: 1204–1217.
39. Kang M, Furusawa M, Miyashita K (2002) Effective and accurate use of difference in mean volume backscattering strength to identify fish and plankton. ICES J Mar Sci 59: 794–804.
40. Benoit-Bird KJ, Moline MA, Waluk CM, Robbins IC (2010) Integrated measurements of acoustical and optical thin layers I: Vertical scales of association. Cont Shelf Res 30: 17–28.
41. Stanton TK, Chu D, Wiebe PH (1998) Sound scattering by several zooplankton groups: II. Scattering models. J Acoust Soc Am 103: 236–253.
42. Benoit-Bird KJ, Au WWL (2004) Diel migration dynamics of an island-associated sound-scattering layer. Deep-Sea Res 51: 707–719.
43. Benoit-Bird KJ, Au WWL (2003) Echo strength and density structure of Hawaiian mesopelagic boundary community patches. J Acoust Soc Am 114: 1888–1897.
44. Sevadjian J, McManus M, Benoit-Bird K, Selph K (2012) Shoreward advection of phytoplankton and vertical re-distribution of zooplankton by episodic near-bottom water pulses on an insular shelf: Oahu, Hawaii. Cont Shelf Res.
45. Benoit-Bird KJ, Shroyer EL, McManus MA (2013) A critical scale in plankton aggregations across coastal ecosystems. Geophysical Research Letters in press.
46. Sevadjian J, McManus M, Pawlak G (2010) Effects of physical structure and processes on thin zooplankton layers in Mamala Bay, Hawai'i. Mar Ecol Prog Ser 409: 95–106.
47. Sullivan J, Donaghay P, Rines J (2010) Coastal thin layer dynamics: Consequences to biology and optics. Cont Shelf Res 30: 50–65.
48. Mackas D, Bohrer R (1976) Fluorescence analysis of zooplankton gut contents and an investigation of diel feeding patterns. J Exp Mar Biol Ecol 25: 77–85.
49. Alldredge AL, Cowles TJ, MacIntyre S, Rines JEB, others a (2002) Occurrence and mechanism of formation of a dramatic thin layer of marine snow in a shallow Pacific fjord. Mar Ecol Prog Ser 233.
50. Sevadjian J, McManus M, Ryan J, Greer A, Cowen R, et al. (2014) Across-shore variability in plankton layering and abundance associated with physical forcing in Monterey Bay, California. Cont Shelf Res 72: 138–151.
51. Clarke TA (1980) Diets of fourteen species of vertically migrating mesopelagic fishes in Hawaiian waters. Fish Bull 78: 619–640.
52. Enright JT (1977) Diurnal vertical migration: Adaptive significance and timing. Limnol Oceanogr 22: 856–886.
53. Krebs JR (1978) Optimal foraging: Decision rules for predators. In: Krebs JR, Davies NB, editors. Behavioural Ecology, an Evolutionary Approach. Sunderland, MA: Sinauer. pp. 23–63.
54. Beyer JE (1995) Functional heterogeneity: Using the interrupted poisson process (IPP) model unit in addressing how food aggregation may affect fish ration. ICES Council Meeting Papers: 14.
55. Lasker R (1975) Field criteria for the survival of anchovy larvae: The relation between inshore chlorophyll maximum layers and successful first feeding. Fish Bull 73: 453–462.
56. Leising AW (2001) Copepod foraging in patchy habitats and thin layers using a 2-D individual-based model. Mar Ecol Prog Ser 216: 167–179.
57. Abrahams M, Townsend L (1993) Bioluminescence in dinoflagellates: A test of the burglar alarm hypothesis. Ecol 74: 258–260.
58. Ringelberg J (1995) Changes in light intensity and diel vertical migration: A comparison of marine and freshwater environments. J Mar Biol Assoc UK 75: 15–25.
59. Forward R (1988) Diel vertical migration: Zooplankton photobiology and behavior. Oceanog Mar Biol Annu Rev 26: 361–392.
60. Lobel P (1978) Diel, Lunar, and Seasonal Periodicity in the Reproductive Behavior of the Pomacanthid Fish, Centropyge potteri, and Some Other Reef Fishes in Hawaii. Pac Sci 32: 193–207.
61. Freon P, Gerlotto F, Soria M (1996) Diel variability of school structure with special reference to transition periods. ICES J Mar Sci 53: 459–464.
62. Goss C, Bone DG, Peck JM, Everson I, Murray GLJ, et al. (1997) Small-scale interactions between prions *Pachyptila spp.* and their zooplankton prey at an inshore site near Bird Island, South Georgia. Mar Ecol Prog Ser 154: 41–51.
63. Sih A (2005) Predator-prey space use as an emergent outcome of a behavioral response race. Ecology of predator-prey interactions. New York: Oxford. pp. 256–278.

2

Mortality of Inshore Marine Mammals in Eastern Australia Is Predicted by Freshwater Discharge and Air Temperature

Justin J. Meager*, Colin Limpus

Threatened Species Unit, Department of Environment and Heritage Protection, Dutton Park, Queensland, Australia

Abstract

Understanding environmental and climatic drivers of natural mortality of marine mammals is critical for managing populations effectively and for predicting responses to climate change. Here we use a 17-year dataset to demonstrate a clear relationship between environmental forcing and natural mortality of inshore marine mammals across a subtropical-tropical coastline spanning a latitudinal gradient of 13° (>2000 km of coastline). Peak mortality of inshore dolphins and dugongs followed sustained periods of elevated freshwater discharge (9 months) and low air temperature (3 months). At a regional scale, these results translated into a strong relationship between annual mortality and an index of El Niño-Southern Oscillation. The number of cyclones crossing the coastline had a comparatively weak effect on inshore marine mammal mortality, and only in the tropics. Natural mortality of offshore/migratory cetaceans was not predicted by freshwater discharge, but was related to lagged air temperature. These results represent the first quantitative link between environmental forcing and marine mammal mortality in the tropics, and form the basis of a predictive tool for managers to prepare responses to periods of elevated marine mammal mortality.

Editor: Vincent Laudet, Ecole Normale Supérieure de Lyon, France

Funding: This study was funded by the Queensland Department of Environment and Heritage Protection. The funders had no role in study design, data collection and analysis, decision to publish, or preparation of the manuscript.

Competing Interests: The authors have declared that no competing interests exist.

* E-mail: Justin.Meager@gmail.com mailto

Introduction

Many marine mammal species are conservation dependent, with low population growth rate and poor resilience to human impacts [1,2]. Managing anthropogenic mortality of marine mammals with tools such as precautionary reference points [3,4] depends on an understanding of the dynamics of natural mortality. Natural mortality is rarely constant in space and time-even for large, long-lived mammals [1,5]. The risk of disease, parasitism, starvation and predation varies with seasonal fluctuations and natural perturbations such as climatic phenomena and habitat disturbance [2,6–8]. Understanding the relationship between extrinsic environmental forcing and mortality in marine mammal populations is an important first step in predicting possible impacts of extreme weather events and climate change [7].

Australia is recognised as both a hotspot for marine mammal diversity [9] and as a region that experiences extreme climatic and rainfall variation [10]. Tropical and subtropical eastern Australia supports a globally significant population of the herbivorous dugong, *Dugong dugon* [2], and a diverse assemblage of resident, transient and migratory cetaceans [11,12]. Two species of tropical dolphins of special conservation concern are resident in the region, at least one of which is endemic to Australasia [13,14]. Climatic variation in eastern Australia is largely dominated by the El Niño-Southern Oscillation phenomenon [15,16], and often alternates between drought- and flood-dominated periods. Over the long term (i.e. since the early 1900s), air temperature and sea surface temperature (SST) in the region have significantly warmed and are predicted to continue to increase [17]. Exactly how warming temperatures will interact with ENSO is uncertain, but precipitation extremes and the frequency of severe weather events such as floods, storms and cyclones are predicted to increase into the future [17]. Such extreme weather events were exemplified by summer of 2010-11, when high sea-surface temperatures and a strong La Niña resulted in widespread and severe flooding in eastern Australia. A very large tropical cyclone then crossed the coast in northern Queensland (Cyclone Yasi) in February 2011. Several months later, high numbers of stranded and dead dugongs and dolphins were reported along the Queensland coast [18,19].

Episodic floods and cyclones have long been suspected to be key drivers of dugong mortality [20–22], but statistical evidence or predictive models have not been forthcoming because long-term data over a relevant spatial scale have not until recently been available. It is thought that the primary reason for elevated dugong mortalities following severe weather events is because of large-scale disturbance to seagrass pastures [20–22]. The effect of floods and cyclones on other marine mammals, such as inshore dolphins, has received much less attention, but impacts on prey resources for inshore dolphins are also well documented (i.e. fish, cephalopods and crustaceans: [23,24]). Further, the impacts of freshwater discharge on estuarine and coastal ecosystems are not limited to periods of flood; sustained periods of high or low freshwater discharge have a strong influence on seagrass pastures and prey assemblages [23,25].

Impacts of freshwater discharge and cyclones are likely to be most pronounced in shallow, coastal or estuarine-influenced areas that represent critical habitat for dugongs and inshore dolphins [26,27]. Conversely, the effects of freshwater discharge and cyclones should be less pronounced on species that do not feed in coastal waters or that are present in coastal waters for short time periods, such as offshore, transient or migratory cetacean species.

In the current paper, we test the prediction that freshwater discharge and cyclones have a strong impact on mortality of dugongs and resident inshore dolphins, but not on mortality of offshore/migratory cetaceans, by developing nonparametric models to empirically explore a 17-year dataset of marine mammal mortality spanning a latitudinal gradient of 13° (>2000 km of coastline). Because marine mammals are affected by both past and present conditions, we also examine how the timing and duration of environmental forcing influences marine mammal mortality.

Methods

Biological data

The Queensland Marine Wildlife Strandings and Mortality Program maintains records of sick, injured, incapacitated or dead marine wildlife in Queensland, Australia, in an online Oracle database (StrandNet). Records are obtained from government departments, community groups, businesses, environmental organisations, and the general public via a telephone hotline. Records lodged in StrandNet include information such as location, date, sex, life-history stage and carcass condition. All records are verified by experts. The probable cause of death is established through necropsies by veterinarians, examination of carcasses by trained staff or, in some cases, through photos and/or case histories. Where there is no evidence for a probable cause of the death, it is recorded as 'unknown'.

Marine mammal strandings and mortalities have been systematically recorded along the 'urban' coast of Queensland (Cairns to the Queensland-New South Wales border) since 1996. Within this area, coverage is most comprehensive in densely populated and highly trafficked regions, such as southern Queensland, and in areas regularly patrolled by marine park rangers. However, marine mammal strandings attract significant attention from the public and there are very few locations along the Queensland coast where beach-washed marine mammals are not reported. In the current study, we analyse relative spatio-temporal patterns of natural marine mammal mortality (i.e. by month and latitudinal block) rather than estimating true natural mortality, because the proportion of marine mammal carcasses which reach the shore is unknown (*sensu* [28]).

Data were exported from StrandNet on 21/08/2013 and were summed over a 1° gradient from the Queensland-New South Wales border (28°S) to northern Queensland (16°S) for each month from January 1996 to December 2012 (i.e. 13 latitudinal blocks). These data spanned in excess of 2000 km of coastline that included a range of marine systems, such as open-ocean beaches, protected bays, estuaries and coral reefs. Only cetacean and dugong mortalities that were either attributed to natural or unidentified causes are examined in this paper. No mass mortality events (e.g. mass strandings) occurred in the dataset.

Environmental predictors

Two climatic indicators: the Southern Oscillation Index (SOI, based on the difference in sea level pressures between Tahiti and Darwin, Australia); and SST anomaly (difference between annual SST and the long-term average) were obtained from the Australian Bureau of Meteorology at monthly intervals. From the Bureau of Meteorology, we also obtained monthly mean minimum air temperature (i.e. monthly mean of daily minimum) and monthly mean maximum air temperature (i.e. monthly mean of daily maximum) for the coastal station nearest to the centre of each latitudinal block. SST data from the wave-monitoring buoy network along the Queensland east coast were provided at 30-minute intervals by the Queensland Department of Science, Information Technology, Innovation and the Arts (DSITIA),

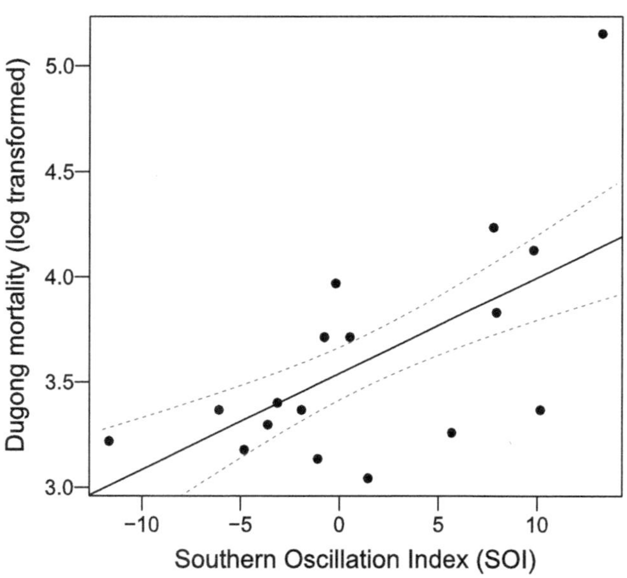

Figure 1. Relationship between annual natural mortality of dugongs and ENSO (SOI, annual average). The model was fit using robust regression (dotted red lines: 90% confidence intervals) and the outlier at the top of the graph is from 2011.

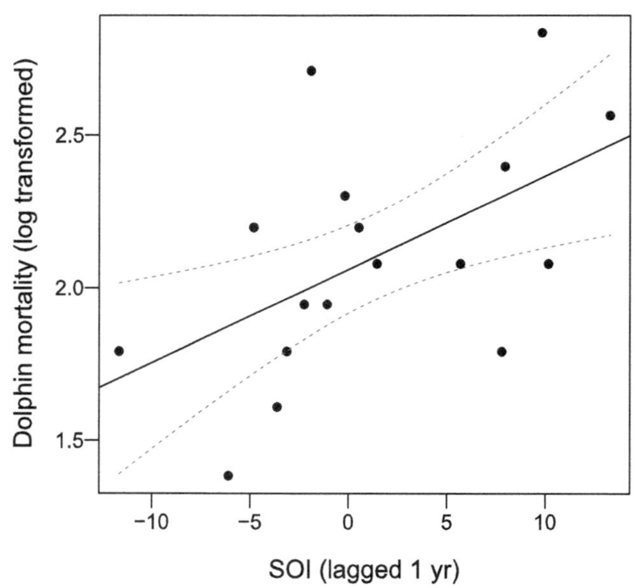

Figure 2. Relationship between annual natural mortality of inshore dolphins and ENSO (annual average SOI lagged by 1 year) (OLS regression; dotted red lines: 90% confidence intervals).

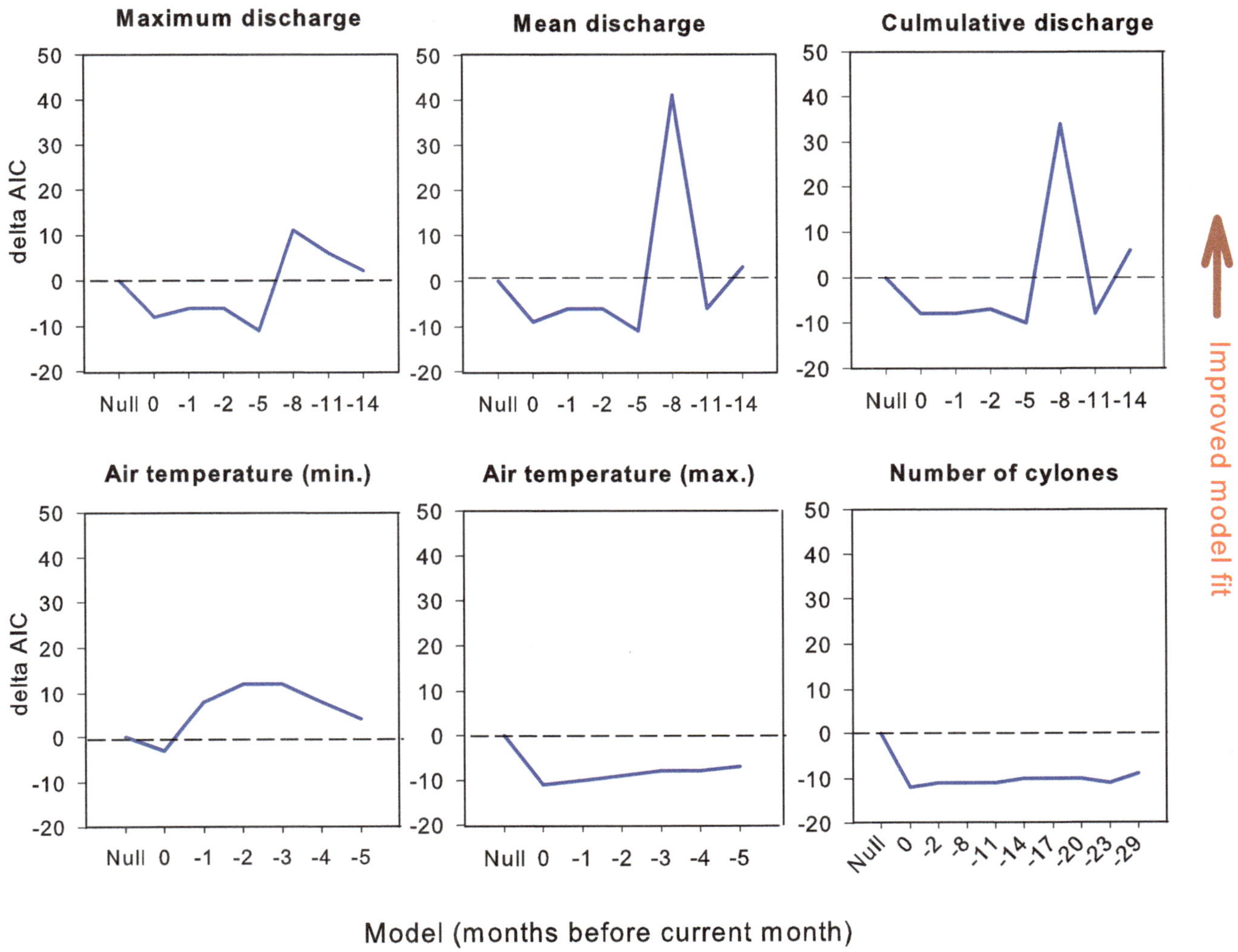

Figure 3. Fit of different additive models of dugong natural mortality as a function of lagged environmental predictors, at lags from zero (same month) up to 29 months (-29). In each case predictors are compared to the null model, which includes only the random effects and time-series component. Delta AIC represents the difference in predictive power between the null model and the model in question (higher values represent improved model fit).

under the Creative Commons Attribution 3.0 (CC BY) licence. Only the buoy (Datawell Waverider Buoy) located off North Stradbroke Island in the Moreton Bay region (27°29.285'S; 153° 37.931'E) had a time series with sufficient SST data for the purposes of the current analysis (January 1997 to December 2012).

Monthly maximum and mean freshwater discharge (m^3s^{-1}) data were downloaded from the Department of Natural Resources and Mines (http://watermonitoring.derm.qld.gov.au/host.htm [accessed June 30, 2013]), under the CC BY licence. Discharge data for the downstream gauging station of each drainage basin (33 in total) were binned to 1° latitudinal blocks, depending on where the point of seawards discharge was situated. Three monthly metrics of freshwater discharge were then calculated for each of the 13 latitudinal blocks: (1) peak discharge, which was the maximum rate of discharge across each latitudinal block in a given month; (2) mean discharge, which was monthly mean discharge averaged over each latitudinal block; and (3) cumulative mean discharge, which was the sum of monthly mean discharge across each latitudinal block.

Because marine mammals are influenced by both the past and present environment, predictor variables were lagged. Annual predictors (mean SST anomaly and mean SOI) were lagged at 1-yearly intervals to 5 years, i.e. a lag of 1 represented the previous year's environmental data tested against the present year's marine mammal mortality data. Monthly environmental predictors were lagged using moving window functions to examine both chronic and acute effects of environmental forcing in detail. Freshwater discharge was lagged at intervals of 0, 1, 3, 6, 9, 12 and 15 months (where a lag of 0 equalled the focal month for the response variable, and a lag of 1 equalled both the focal month and the previous month). *Lagged peak discharge* was calculated using a moving maximum function as the maximum rate of discharge over the time period. *Lagged cumulative discharge* was calculated using a moving cumulative function as the total cumulative mean discharge over the time period. *Lagged mean discharge* was calculated using a simple moving average (SMA) as the mean discharge over the time period. Air temperature was expected to affect marine mammals over shorter timescales than freshwater discharge and was therefore lagged at intervals of 0, 1, 2, 3, 4, 5 and 6 months using SMA functions. Cyclones were too infrequent to be lagged over shorter time periods and can have severe impacts on marine habitats that can take years to recover [29,30]. The number of

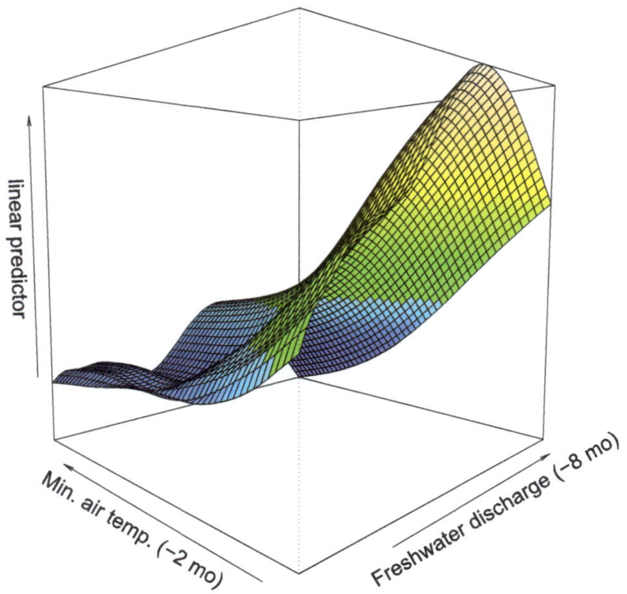

Figure 4. Final model of dugong natural mortality (linear predictor, $\log_e x + 1$) in response to mean minimum air temperature (°C, lagged by 2 months) and freshwater discharge (m^3s^{-1}, lagged by 8 months and square-root transformed).

cyclones crossing each latitude block was therefore lagged using moving cumulative functions at intervals of 9, 12, 15, 18, 21, 24 and 30 months. SST was lagged using a SMA function at intervals of 0, 1, 2, 3, 4, 5, 6, 7, 9 and 12 months.

Data analysis

Dugongs, as the only marine mammal herbivore, were analysed separately. Cetaceans were assigned to functional ecological groups, because individual species were too infrequent for detailed analysis. 'Inshore dolphins' included the three species of dolphins known to be resident in coastal waters: the 'inshore' Indo-Pacific bottlenose (*Tursiops aduncus*), Indo-Pacific humpback (*Sousa chinensis*) and Australian snubfin (*Orcaella heinsohni*) dolphins. 'Offshore/migratory cetaceans' included oceanic or neritic dolphins such as the 'offshore' bottlenose (*Tursiops truncatus*), short-beaked common (*Delphinus delphis*) and spinner (*Stenella longirostris*) dolphins, in addition to migratory species such as the humpback whale (*Megaptera novaeangliae*) and species occasionally present in coastal waters (e.g. pygmy sperm whale, *Kogia breviceps* and melon-headed whale, *Peponocephala electra*). Bottlenose dolphins not identified to species were not analysed because they could have been in either the inshore or the offshore groups.

We first examined the relationship between annual natural mortality and environmental predictors (SST anomaly and SOI) across the entire region. Environmental predictors were lagged at intervals of 1 to 5 years. Response variables were normalised using log transformation ($\log_e x + 1$) and relationships were assessed using linear regression and standard regression diagnostics.

We then examined the full region-by-month dataset with the modelling approach recommended by Zuur *et al.* [31]. Briefly, temporal autocorrelation plots, spatial variograms, residual diagnostics and Akaike's Information Criterion (AIC) of different mixed models were used to determine the optimal variance structure. This indicated significant seasonal autocorrelation (Figures S1–S3), and that the optimal variance structure included separate intercepts for each region and year. In effect, this approach accounted for heterogeneity in the marine mammal mortality-environmental relationship between regions and years. A cubic-spline function of month (coded from 1 to 12) was used to model seasonal autocorrelation [32]. This general mixed-model structure was used to compare the effects of different environmental predictors on marine mammal mortality. Because there was no *a priori* expectation of a linear relationship between marine mammal mortality and environmental predictors, a generalised additive mixed modelling (GAMM) approach was used (library gamm4 v 0.1-6 of R v 3.0.1, the R foundation for statistical computing). Dugong monthly mortality data were log-normally distributed and were hence normalised by $\log_e(x+1)$ transformation and analysed using a Gaussian error distribution. Inshore dolphin and offshore/migratory cetacean mortalities were modelled with a Poisson error distribution and a log-link function.

We analysed each set of lagged predictor variables separately by comparing the AIC of each model against the null model (i.e. the model including no environmental predictors). The best predictor variables were chosen based on (a) lowest AIC and (b) residual diagnostics. The final model was then chosen by comparing the AIC of different interactive and main effect models including the best predictor variables. Final models, including more than one predictor variable, were screened for collinearity (i.e. $|r| < 0.4$).

We then examined three marine mammal 'hotspots' separately (Townsville, Moreton Bay and Hervey Bay) using area-specific generalised additive models (GAM). The same general model structure for seasonal effects was used as above, but no random effects were included. The influence of SST was examined only for the Moreton Bay region between January 1997 and December 2012.

Results

Annual mortality and environmental forcing

Annual dugong mortalities were significantly associated with SOI (b = 0.052 ± 0.015; t = 3.35; r^2 = 0.417; df= 15; *p*-value = 0.0044). To account for the clear outlier of 2011 (Figure 1), the regression was refit with this value excluded (b = 0.0315 ± 0.013; t = 2.34; r^2 = 0.281; df= 14; *p*-value = 0.035) and using robust regression (b = 0.0456 ± 0.016; Library MASS, Huber's M-estimator; Figure 1). The relationship between annual dugong mortalities and SOI was not as strong for lagged SOI, remaining significant for a 1-year lag (b = 0.048 ± 0.016; t = 2.98; r^2 = 0.372; df= 15; *p*-value = 0.009) but not for lags of 2, 3, 4 and 5 years (*p*-values from 0.14 to 0.83). There was no significant relationship between annual dugong mortalities and SST temperature anomaly, either for the same year for lags up to 5 years (*p*-values from 0.13 to 0.95).

Annual mortality of inshore dolphins was significantly associated with SOI lagged by 1 year (b = 0.030 ± 0.012; t = 2.50; r^2 = 0.29; df= 15; *p*-value = 0.025, Figure 2), but not for the same year or at lags of 2, 3, 4 and 5 years (*p*-values from 0.11 to 0.94). Offshore/migratory cetacean mortality was not significantly associated with SOI (*p*-values from to 0.12 to 0.96) for lags of up to 5 years. SST anomaly influenced neither annual inshore dolphin mortality (*p*-values from 0.08 to 0.92, lagged up to 5 years) nor annual mortality of offshore/migratory cetaceans (*p*-values from 0.08 to 0.92, lagged up to 5 years).

Monthly mortality and environmental forcing

The predictive power of GAMMs for monthly dugong mortality by region was not improved by including the number of cyclones crossing the coast or maximum air temperature (Figure 3).

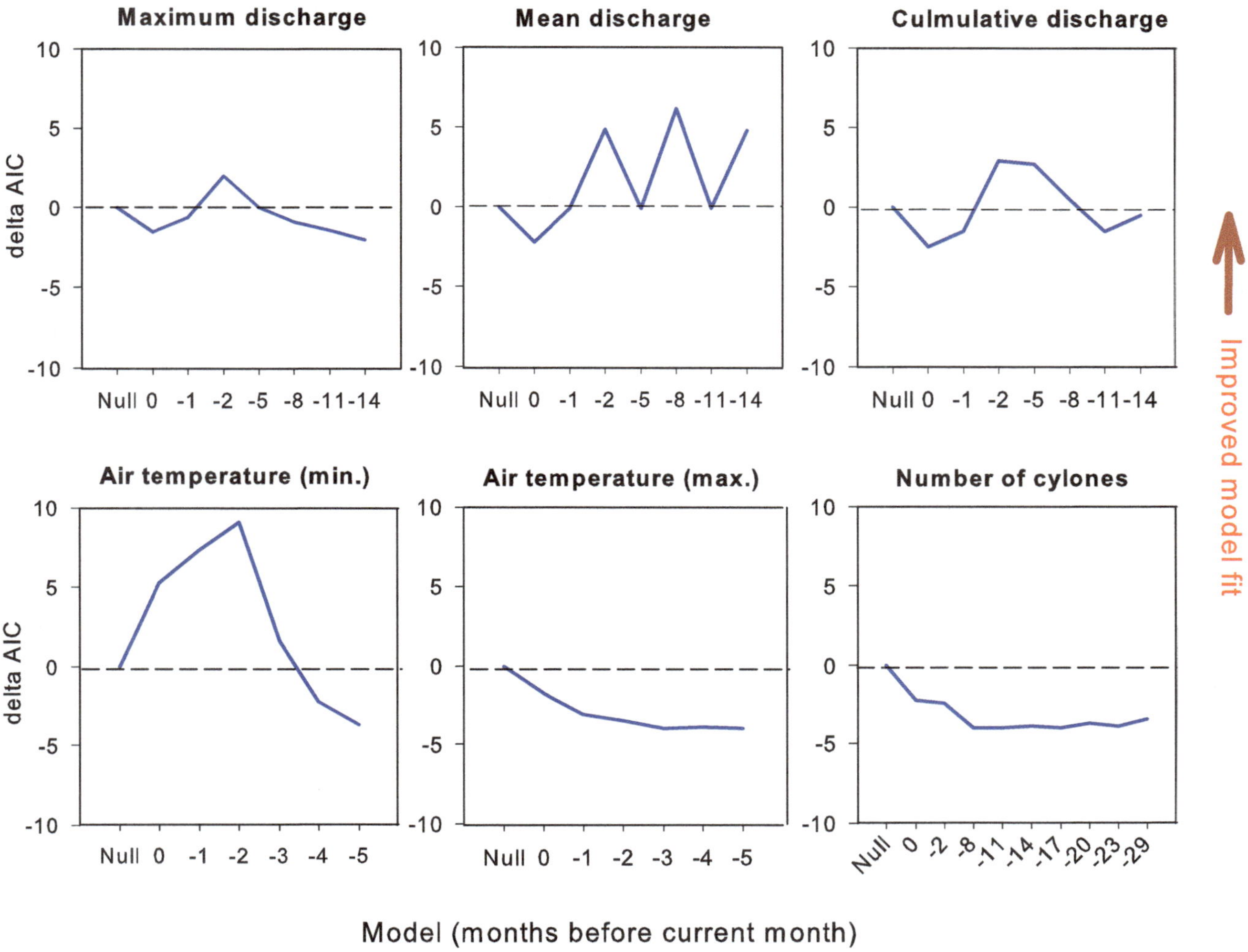

Figure 5. Fit of different additive models of inshore dolphin natural mortality as a function of environmental predictors at lags from zero (same month) up to 29 months (-29). Refer to the caption of Figure 3 for a full explanation.

However, model fit was improved by including minimum air temperature lagged 2 or 3 months (i.e. by a 3- to 4-month SMA window including the focal month). A strong improvement in the predictive power of models occurred with the inclusion of mean discharge lagged by 8 months. To a lesser extent, cumulative discharge and maximum discharge also improved model fit for the same lagged time period.

On the basis of lowest AIC and residual diagnostics, the final model included the main effects of minimum air temperature (lagged by 2 months) and mean discharge (lagged by 8 months), in addition to an interaction between lagged mean discharge and lagged minimum air temperature (Figure 4). Freshwater discharge was square-root transformed to reduce the influence of outliers (i.e. major floods in large catchments). This model predicted that on average, i.e. across latitudinal regions, elevated dugong mortalities occur after a 9-month period of elevated rainfall and a 3-month period of low air temperatures. The interaction accounted for the fact that elevated dugong mortality was more closely linked to sustained freshwater discharge when it also coincided with prolonged low air temperatures.

A very similar pattern was evident for inshore dolphin mortality. Again, neither the number of cyclones nor maximum air temperature improved the fit of the GAMM (Figure 5). Of the predictors of freshwater discharge, lags of 2 months improved GAMM fit in each case, but the most pronounced improvement occurred when mean discharge was lagged by 8 months. Minimum air temperature lagged by 2 months also resulted in a marked improvement in the fit of the model. Using the same model selection criteria as above, we arrived at a final model that included the main effects of mean freshwater discharge (lagged by 8 months, square-root transformed) and minimum air temperature (lagged by 2 months) (Figure 6).

In contrast, freshwater discharge did not improve the fit of GAMMs for offshore/migratory cetaceans (Figure 7). This was also true for the number of cyclones, but minimum air temperature again improved model fit. As for dugongs and inshore dolphins, minimum air temperature lagged by 2 months improved model fit. However, in the case of offshore cetaceans, maximum air temperature (lagged by 2 months) also improved model fit. Examination of model residuals indicated that the maximum air temperature models were strongly influenced by outliers (i.e. extreme temperatures). We therefore arrived at a final model that included only the linear effect of lagged minimum air temperature (Figure 8).

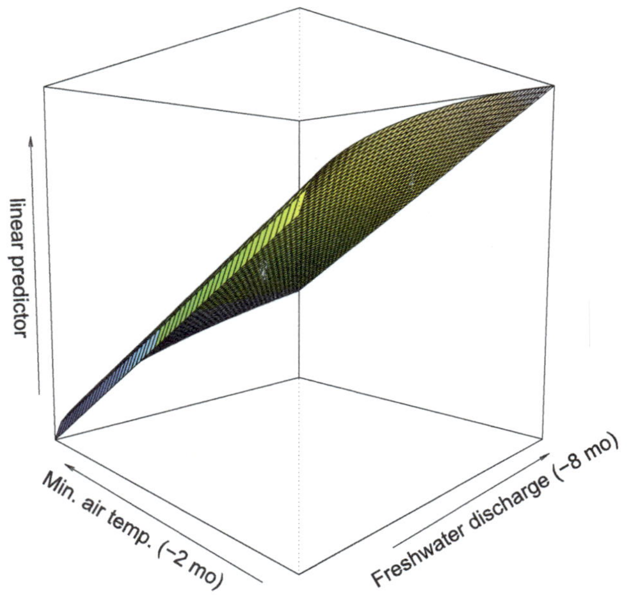

Figure 6. Final model of inshore dolphin natural mortality in response to mean minimum air temperature (°C, lagged by 2 months) and freshwater discharge (m^3s^{-1}, lagged by 8 months and square-root transformed). A Poisson error and log-link function were used for this relationship, the response surface is on the scale of the linear predictor.

Area specific models: dugongs

For the Moreton Bay region, a model including the main effects of minimum air temperature (lagged by 2 months) and mean discharge (lagged by 8 months) best predicted dugong mortality. SST had a similar effect on dugong mortality as minimum air temperature (lagged by 2 months). A GAM including SST explained approximately 19.6% of deviance compared to 19.7% explained by the GAM including minimum air temperature (lagged by 2 months). There was significant co-variance between air temperature and SST, especially for mean minimum air temperature. Correlation between air temperature and SST was the strongest for data lagged by 2 months (r = 0.955), as was also the case for maximum mean air temperature (r = 0.926), suggesting that air temperature lagged by 2 months is a good proxy for SST in the region.

For the Hervey Bay region, the model with the lowest AIC scores was slightly different to the overall region-by-month mixed model, and included the main effects of cumulative discharge (lagged by 8 months) and maximum mean air temperature (lagged by 4 months). The model including only the main effects of cumulative discharge (lagged by 8 months, square-root transformed) and month, explained approximately 22.6% of deviance, whereas the model including only the main effect of maximum mean air temperature (lagged by 4 months) explained 23.9% of deviance. No SST data were available for the Hervey Bay region.

In the Townsville area, the final area-specific model included similar terms as the regional model, although cumulative discharge (lagged by 8 months, 42% of deviance, AIC = 240.86) performed better than mean discharge (lagged by 8 months, 41.9% of deviance, AIC = 306.95). A model including minimum air temperature (lagged by 2 months) explained 25.2% of deviance. The final model explained 39.5% of deviance and included cumulative discharge (lagged by 8 months, square root transformed), month and minimum air temperature (lagged by 2 months).

There were insufficient data to include the effects of cyclones for either the Moreton Bay or Hervey Bay regions. In the Townsville area, there were only enough data to fit a model for the cumulative number of cyclones lagged by 29 months, which explained 19.1% of deviance.

The area between Coolangatta and Hervey Bay was analysed separately for inshore dolphins, because the data did not support a detailed analysis for each individual area. Again, on the basis of AIC and deviance, the best predictors were lagged mean discharge (8 months, 22.9% of deviance) and lagged air temperature, although in this case minimum air temperature lagged by 1 month was the best fit (18.2% of deviance). The final model included minimum air temperature (lagged by 1 month) and mean discharge (lagged 8 months, square-root transformed) and explained 26.5% of deviance.

Discussion

Freshwater discharge predicted mortality of inshore dolphins and dugongs in subtropical and tropical eastern Australia, and the relationship was the strongest for models that included lagged effects. The relationship was clear both across a very large spatial scale (>2000 km) and in empirical models specific to latitudinal regions, and underpinned correlations between inshore marine mammal mortality and SOI. As predicted, freshwater discharge was not an important driver of natural mortality in offshore or migratory cetaceans. To our knowledge, this is the first demonstration of a long-term relationship between freshwater discharge and mortality of inshore marine mammals in tropical or subtropical waters. Lagged air temperature also predicted mortality trends in dugongs, inshore dolphins and offshore/migratory marine mammals, although relationships were less consistent at a regional level.

Correlative spatio-temporal models are an important first step to understand and predict the timing and magnitude of mortality events. They are also a fundamental step towards generating testable hypotheses for the functional basis of observed relationships. To this end, the two most parsimonious functional explanations for the relationship between freshwater discharge and inshore dolphin and dugong mortality are (1) reduced food availability, and (2) direct impacts on health.

Strong support for the first hypothesis is available for dugongs. The density and species composition of seagrass pastures in inshore waters are closely related to freshwater discharge [33]. Elevated discharge reduces photosynthetic available radiation (PAR), exported sediment can smother seagrasses, and flood discharge can scour seagrass beds and seed banks [34]. Dugongs have a high metabolic demand and are thought to be impacted from food loss comparatively quickly [2]. The correlation between SOI (lagged by approximately 2 years) and nesting numbers of the only other large herbivorous marine species in the region, the green turtle, *Chelonia mydas*, also supports the contention that freshwater discharge can regulate populations of herbivores by limiting food availability [35,36]. Reduced food availability is also the simplest explanation for the observed link between inshore dolphin mortality and freshwater discharge. Given that freshwater discharge has a strong influence on the distribution, abundance and phenology of many species of fish and crustaceans in eastern Australia (e.g. [23,37] and other references herein), it would not be surprising to see these effects realised in apex predators.

The impact of sustained periods of elevated freshwater discharge (i.e. lagged mean discharge) on inshore marine

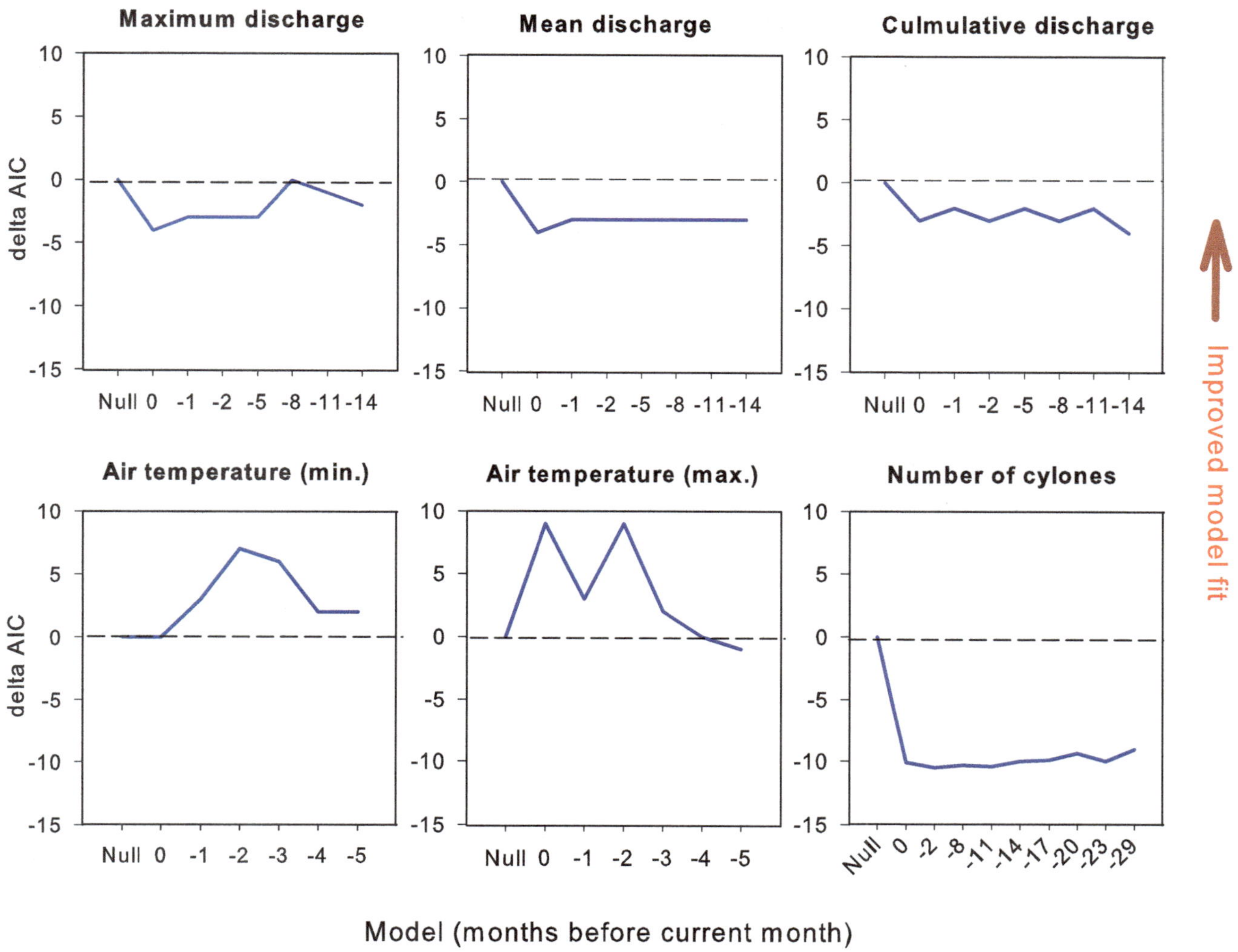

Figure 7. Fit of different additive models of offshore/migratory cetacean natural mortality as a function of environmental predictors at lags from zero (same month) up to 29 months (-29). Refer to the caption of Figure 3 for a full explanation.

mammals in our study was more pronounced than the effect of the magnitude of an individual flood event *per se* (i.e. lagged maximum discharge). Seagrass pastures of dugongs are known to be able to survive short-term periods of reduced PAR, but not sustained periods [38]. Similarly, while prey species of dolphins may be able to tolerate pulses of elevated discharge by moving to areas of higher salinity, sustained periods of rainfall can impose constraints on prey resources by limiting the availability of suitable habitat or by inhibiting post-larval recruitment [25,39,40].

Freshwater discharge may also impact on the health of marine mammals by increasing exposure to infectious pathogens such as toxoplasmosis and faecal coliform bacteria [41–43], or to contaminants with immunosuppressive effects [44]. Toxoplasmosis or toxoplasmosis antibodies have been observed in Indo-Pacific humpback dolphins [41] and dugongs [45] in Queensland. Elevated contaminant concentrations have also been reported in dugongs and inshore dolphins in the region [46–49], including levels of polychlorinated biphenyls where immunosuppression may occur based on dose-response relationships for other species [49].

Temperature also influences the health and physiology of marine mammals [50,51], in addition to impacting foraging pasture [52] and prey production [53]. In our study, sustained low air temperature was associated with mortality of dugongs, inshore dolphins and offshore/migratory cetaceans. The very close correspondence between air temperature and SST in the Moreton Bay region suggests that lagged air temperature may simply be a good proxy of SST, rather than an important driver of marine mammal mortality in its own right.

There are several lines of evidence to suggest that the health of dugongs is limited by cooler temperatures at the southern limits of their range. First, seasonal movements are correlated to temperature changes [2,54–56]. Second, dugongs seem to move further southwards along the east coast of Australia in warm years [57]. Finally, cold-stress syndrome has been described in dugongs in Moreton Bay [51]. Much less is known of the effects of temperature on tropical dolphins-most studies on the effects of temperature on mortality of cetaceans have focused on impacts in temperate regions [58,59]. Elsewhere in the subtropics, sustained cold temperature was thought to be a contributing factor to the high mortality rate of bottlenose dolphins (*T. truncatus*) observed in the northern Gulf of Mexico after the Deepwater Horizon oil spill [43].

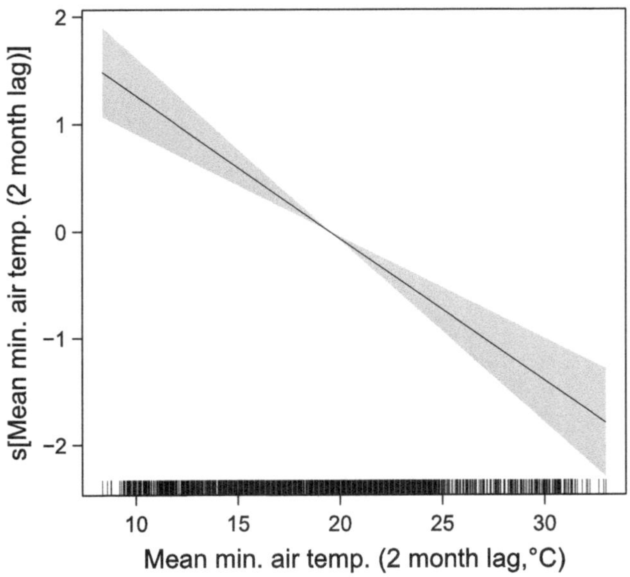

Figure 8. Relationship between offshore/migratory cetacean mortality and mean minimum air temperature (lagged by 2 months). The solid line is the smoother from the final GAMM model and the shaded area represents the 95% confidence intervals. The vertical black lines on the *x* axis denote individual values. A Poisson error and log-link function were used for this relationship, the response surface is on the scale of the linear predictor.

The fact that lagged minimum air temperature also explained mortality of dugongs in the tropics (i.e. Townsville region), where temperature is not expected to be cold enough to impair health, suggests that impacts on health may not be the only reason for the association between low temperature and mortality in our study. Other seasonal dynamics, such as the seasonal availability of seagrass pastures or prey, may have explained some of this association [60]. In the case of offshore/migratory cetaceans, the relationship between mortality and lagged air temperatures may be indicative of larger-scale climatic or oceanographic drivers, as the results of other recent studies imply [61,62].

Despite the fact that cyclones are known to cause major disturbance to coastal ecosystems [24,29], the number of cyclones crossing the coast did not improve the fit of the overall model, and only improved the fit of the Townsville regional model for dugong mortality. The main reason for this is that cyclones very rarely cross the Queensland coast in subtropical and southern areas, which meant that there were insufficient data for rigorous statistical analysis in the overall model.

The accuracy of predictive models of marine mammal mortality would be undoubtedly improved by including covariates such as population abundance. However, estimates of population abundance are still too inaccurate at a relevant spatial scale for this to be of value in our study area, even in the case of relatively well monitored species such as dugongs [63]. Even so, our study highlights the importance of incorporating past environmental forcing in developing predictive models for inshore marine mammal mortality, because the effects of temperature and freshwater discharge on inshore marine mammal mortality were not fully realised until they were lagged using moving average functions of 3 and 9 months, respectively.

Overall, our study provides an important basis for further investigation into the proximate basis of the functional relationship between marine mammal mortality and environmental forcing, ideally through detailed necropsies and dedicated surveys of population health. Further, our results illustrate how data collected from a marine mammal strandings network can be used in conjunction with routinely measured environmental and climatic variables to predict and prepare for periods of increased natural marine mammal mortality. Our study also highlights the need for further investigation into the influence of long-term trends of climate variability on marine mammals in tropical and subtropical regions. Current climate models do not have a strong degree of certainty in predicting how climate change will affect ENSO-related precipitation variability [64]. Yet, our results suggest that this aspect of climate may have a central role in the population dynamics of marine mammals in tropical-subtropical eastern Australia.

Supporting Information

Figure S1 Autocorrelation function (ASF) of dugong mortality. Lags are in years and mortalities were summed across regions. Values outside the dotted lines are significant (α = 0.05).

Figure S2 Autocorrelation function (ASF) of inshore dolphin mortality. Lags are in years and mortalities were summed across regions. Values outside the dotted lines are significant (α = 0.05).

Figure S3 Autocorrelation function (ASF) of offshore/ migratory cetacean mortality. Lags are in years and mortalities were summed across regions. Values outside the dotted lines are significant (α = 0.05).

Figure S4 Monthly dugong natural mortalities and environmental predictors (mean monthly freshwater discharge, lagged by 8 months; and mean minimum air temperature, lagged by 2 months) in the Moreton Bay region (27°S latitudinal block).

Figure S5 Monthly dugong natural mortalities and environmental predictors (mean monthly freshwater discharge, lagged by 8 months; and mean minimum air temperature, lagged by 2 months) in the Hervey Bay region (25°S latitudinal block).

Figure S6 Monthly dugong natural mortalities and environmental predictors (mean monthly freshwater discharge, lagged by 8 months; and mean minimum air temperature, lagged by 2 months) in the Townsville region (19°S latitudinal block).

Figure S7 Inshore dolphin natural mortalities and environmental predictors (mean monthly freshwater discharge, lagged by 8 months; and mean minimum air temperature, lagged by 2 months) in southern Queensland (24 to 28°S).

Acknowledgments

The authors would like to thank the Queensland Marine Wildlife Strandings and Mortality network and all contributors to the StrandNet database. We would also like to thank Daryl Metters of the Coastal Impact

Unit, DSITIA for providing the wave buoy data, and the peer-reviewers of the annual cetacean and dugong stranding reports.

Author Contributions

Conceived and designed the experiments: JJM CJL. Performed the experiments: JJM. Analyzed the data: JJM. Wrote the paper: JJM CJL.

References

1. Perrin WF, Würsig BG, Thewissen JGM (2002) Encyclopedia of marine mammals. San Diego: Academic Press.
2. Marsh H, O'Shea TJ, Reynolds JE (2011) Ecology and conservation of the Sirenia: dugongs and manatees. New York: Cambridge University Press.
3. Wade PR (1998) Calculating limits to the allowable human-caused mortality of cetaceans and pinnipeds. Mar Mamm Sci 14: 1–37.
4. Moore JE (2013) Management reference points to account for direct and indirect impacts of fishing on marine mammals. Mar Mamm Sci 29: 446–473.
5. Erb J, Boyce MS, Stenseth NC (2001) Population dynamics of large and small mammals. Oikos 92: 3–12.
6. Halvorsen KM, Keith EO (2008) Immunosuppression cascade in the Florida Manatee (*Trichechus manatus latirostris*). Aquat Mamm 34: 412–419.
7. Schumann N, Gales N, Harcourt R, Arnould J (2013) Impacts of climate change on Australian marine mammals. Aust J Zool 61: 146–159.
8. Yan C, Stenseth NC, Krebs CJ, Zhang Z (2013) Linking climate change to population cycles of hares and lynx. Global Change Biol 19: 3263–3271.
9. Pompa S, Ehrlich PR, Ceballos G (2011) Global distribution and conservation of marine mammals. Proc Natl Acad Sci USA 108: 13600–13605.
10. Erskine WD, Warner R (1998) Further assessment of flood-and drought-dominated regimes in south-eastern Australia. Aust Geogr 29: 257–261.
11. Paterson R (1986) A list of specimens of the order Cetacea in the Queensland Museum. Mem Queensl Mus 46: 631–645.
12. Chilvers BL, Lawler IR, Macknight F, Marsh H, Noad M, et al. (2005) Moreton Bay, Queensland, Australia: an example of the co-existence of significant marine mammal populations and large-scale coastal development. Biol Conserv 122: 559–571.
13. Beasley I, Arnold P, Heinsohn G (2002) Geographical variation in skull morphology of the irrawaddy dolphin, *Orcaella brevirostris* (Owen in Gray, 1866). Raffles Bull Zool: 15–34.
14. Mendez M, Jefferson TA, Kolokotronis S-O, Krützen M, Parra GJ, et al. (2013) Integrating multiple lines of evidence to better understand the evolutionary divergence of humpback dolphins along their entire distribution range: a new dolphin species in Australian waters? Mol Ecol 22: 5936–5948.
15. Allan R, Lindesay J, Parker D (1996) El Niño: Southern Oscillation and climatic variability. Collingwood, Australia: CSIRO Publishing.
16. Power S, Casey T, Folland C, Colman A, Mehta V (1999) Inter-decadal modulation of the impact of ENSO on Australia. Clim Dynam 15: 319–324.
17. Intergovernmental Panel on Climate Change (IPCC) (2007) Climate Change 2007-The Physical Science Basis: Working Group I Contribution To The Fourth Assessment Report Of The IPCC: Cambridge University Press.
18. Meager JJ, Limpus CJ (2012) Marine wildlife stranding and mortality database annual report 2011. I. Dugong. Conserv Tech Data Rep 2012: 1–30. Available: http://www.ehp.qld.gov.au/wildlife/caring-for-wildlife/strandnet-reports.html. Accessed 5 July 2013.
19. Meager JJ, Winter KM, Biddle TM, Limpus CJ (2012) Marine wildlife stranding and mortality database annual report 2008–2011. II. Cetacean and Pinniped. Conserv Tech Data Rep 2012: 1–76. Available: http://www.ehp.qld.gov.au/wildlife/caring-for-wildlife/strandnet-reports.html. Accessed 5 July 2013.
20. Preen A, Marsh H (1995) Response of dugongs to large-scale loss of seagrass from Hervey Bay, Queensland, Australia. Wildl Res 22: 507–519.
21. Gales N, McCauley RD, Lanyon J, Holley D (2004) Change in abundance of dugongs in Shark Bay, Ningaloo and Exmouth Gulf, Western Australia: Evidence for large-scale migration. Wildl Res 31: 283–290.
22. Marsh H, Kwan D (2008) Temporal variability in the life history and reproductive biology of female dugongs in Torres Strait: The likely role of sea grass dieback. Cont Shelf Res 28: 2152–2159.
23. Gillanders BM, Kingsford MJ (2002) Impact of changes in flow of freshwater on estuarine and open coastal habitats and the associated organisms. In: Gibson RN, Barnes M, Atkinson RJA, editors. Oceanography and Marine Biology. pp. 233–309.
24. Loneragan NR, Kangas M, Haywood MDE, Kenyon RA, Caputi N, et al. (2013) Impact of cyclones and aquatic macrophytes on recruitment and landings of tiger prawns *Penaeus esculentus* in Exmouth Gulf, Western Australia. Est Cstl Shelf Sci 127: 46–58.
25. Loneragan NR, Bunn SE (1999) River flows and estuarine ecosystems: implications for coastal fisheries from a review and a case study of the Logan River, southeast Queensland. Aust J Ecol 24: 431–441.
26. Parra GJ (2006) Resource partitioning in sympatric delphinids: Space use and habitat preferences of Australian snubfin and Indo-Pacific humpback dolphins. J Anim Ecol 75: 862–874.
27. Sheppard JK, Marsh H, Jones RE, Lawler IR (2010) Dugong habitat use in relation to seagrass nutrients, tides, and diel cycles. Mar Mamm Sci 26: 855–879.
28. Peltier H, Dabin W, Daniel P, Van Canneyt O, Dorémus G, et al. (2012) The significance of stranding data as indicators of cetacean populations at sea: Modelling the drift of cetacean carcasses. Ecol Indicators 18: 278–290.
29. Preen AR, Long WJ, Coles RG (1995) Flood and cyclone related loss, and partial recovery, of more than 1000 km^2 of seagrass in Hervey Bay, Queensland, Australia. Aquat Bot 52: 3–17.
30. Gardner TA, Cote IM, Gill JA, Grant A, Watkinson AR (2005) Hurricanes and Caribbean coral reefs: Impacts, recovery patterns, and role in long-term decline. Ecology 86: 174–184.
31. Zuur AF, Ieno EN, Walker N, Saveliev AA, Smith GM (2009) Mixed effects models and extensions in ecology with R. New York: Springer-Verlag.
32. Wood SN (2006) Generalized additive models: an introduction with R. Boca Raton, Fla.: Chapman & Hall/CRC.xvii392 s
33. Campbell SJ, McKenzie LJ (2004) Flood related loss and recovery of intertidal seagrass meadows in southern Queensland, Australia. Est Cstl Shelf Sci 60: 477–490.
34. Larkum AWD (1989) Biology of seagrasses: a treatise on the biology of seagrasses with special reference to the Australian region. Amsterdam: Elsevier.
35. Limpus C, Nicholls N (2000) ENSO regulation of Indo-Pacific green turtle populations. In: Hammer GL, N N, Mitchell C, editors. Applications of seasonal climate forecasting in agricultural and natural ecosystems. London: Kluwer Academic Publishers. pp. 399–408.
36. Limpus CJ, Nicholls N (1988) The Southern Oscillation regulates the annual numbers of green turtles (*Chelonia mydas*) breeding around northern Australia. Aust Wildl Res 15: 157–161.
37. Gillson J, Suthers I, Scandol J (2012) Effects of flood and drought events on multi-species, multi-method estuarine and coastal fisheries in eastern Australia. Fish Manage Ecol 19: 54–68.
38. Longstaff BJ, Dennison WC (1999) Seagrass survival during pulsed turbidity events: the effects of light deprivation on the seagrasses *Halodule pinifolia* and *Halophila ovalis*. Aquat Bot 65: 105–121.
39. Staples DJ, Vance DJ (1986) Emigration of juvenile banana prawns *Penaeus merguiensis* from a mangrove estuary and recruitment to offshore areas in the wet-dry tropics of the Gulf of Carpentaria, Australia. Mar Ecol Prog Ser 27: 239–252.
40. Meager JJ, Vance DJ, Loneragan NR, Williamson I (2003) Seasonal variation, and environmental influences on juvenile banana prawn (*Penaeus merguiensis*) abundance in a subtropical estuary (Logan River) of eastern Australia. Est Cstl Shelf Sci 57: 569–576.
41. Bowater RO, Norton J, Johnson S, Hill B, O'Donoghue P, et al. (2003) Toxoplasmosis in Indo-Pacific humpbacked dolphins (*Sousa chinensis*), from Queensland. Aust Vet J 81: 627–632.
42. Schaefer AM, Bossart GD, Mazzoil M, Fair PA, Reif JS (2011) Risk factors for colonization of *E. coli* in Atlantic bottlenose dolphins (*Tursiops truncatus*) in the Indian River lagoon, Florida. Journal of Environmental and Public Health 2011.
43. Carmichael RH, Graham WM, Aven A, Worthy G, Howden S (2012) Were multiple stressors a 'perfect storm'for northern Gulf of Mexico bottlenose dolphins (*Tursiops truncatus*) in 2011? PLoS One 7: e41155.
44. Van Bressem MF, Raga JA, Di Guardo G, Jepson PD, Duignan PJ, et al. (2009) Emerging infectious diseases in cetaceans worldwide and the possible role of environmental stressors. Dis Aquat Org 86: 143–157.
45. Limpus CJ, Currie KJ, Haines J (2004) Marine wildlife stranding and mortality database annual report 2003. I. Dugong. Conserv Tech Data Rep 2004: 1–38.
46. Haynes D, Carter S, Gaus C, Müller J, Dennison W (2005) Organochlorine and heavy metal concentrations in blubber and liver tissue collected from Queensland (Australia) dugong (*Dugong dugon*). Mar Pollut Bull 51: 361–369.
47. Hermanussen S, Matthews V, Päpke O, Limpus C, Gaus C (2008) Flame retardants (PBDEs) in marine turtles, dugongs and seafood from Queensland, Australia. Mar Pollut Bull 57: 409–418.
48. Vetter W, Scholz E, Gaus C, Muller JF, Haynes D (2001) Anthropogenic and natural organohalogen compounds in blubber of dolphins and dugongs (*Dugong dugon*) from northeastern Australia. Arch Environ Contam Toxicol 41: 221–231.
49. Cagnazzi D, Fossi MC, Parra GJ, Harrison PL, Maltese S, et al. (2013) Anthropogenic contaminants in Indo-Pacific humpback and Australian snubfin dolphins from the central and southern Great Barrier Reef. Environ Pollut 182: 490–494.
50. Wilson B, Arnold H, Bearzi G, Fortuna CM, Gaspar R, et al. (1999) Epidermal diseases in bottlenose dolphins: impacts of natural and anthropogenic factors. Proc R Soc B-Biol Sci 266: 1077–1083.
51. Owen HC, Flint M, Limpus CJ, Palmieri C, Mills PC (2013) Evidence of sirenian cold stress syndrome in dugongs *Dugong dugon* from southeast Queensland, Australia. Dis Aquat Org 103: 1–7.
52. Campbell SJ, McKenzie LJ, Kerville SP (2006) Photosynthetic responses of seven tropical seagrasses to elevated seawater temperature. J Exp Mar Biol Ecol 330: 455–468.
53. Hobday AJ, Poloczanska ES, Matear RJ, (eds) (2009) Implications of climate change for Australian fisheries and aquaculture. A preliminary assessment. Report to the Department of Climate Change, Canberra, Australia.

54. Anderson PK (1986) Dugongs of Shark Bay, Australia - seasonal migration, water temperature, and forage. National Geographic Research 2: 473–490.
55. Preen AR, Marsh H, Lawler IR, Prince RIT, Shepherd R (1997) Distribution and abundance of dugongs, turtles, dolphins and other Megafauna in Shark Bay, Ningaloo Reef and Exmouth Gulf, western Australia. Wildl Res 24: 185–208.
56. Holley DK, Lawler IR, Gales NJ (2006) Summer survey of dugong distribution and abundance in Shark Bay reveals additional key habitat area. Wildl Res 33: 243–250.
57. Allen S, Marsh H, Hodgson A (2004) Occurrence and conservation of the dugong (Sirenia: Dugongidae) in New South Wales. Proc Linn Soc NSW 125: 211–216.
58. MacLeod CD, Bannon SM, Pierce GJ, Schweder C, Learmonth JA, et al. (2005) Climate change and the cetacean community of north-west Scotland. Biol Conserv 124: 477–483.
59. Truchon M-H, L'Hérault V, Brêthes J-C, Galbraith PS, Harvey M, et al. (2013) Marine mammal strandings and environmental changes: a 15-year study in the St. Lawrence ecosystem. PLoS One 8: e59311.
60. Mellors JE, Marsh H, Coles RG (1993) Intraannual changes in seagrass standing crop, Green Island, northern Queensland. Aust J Mar Freshw Res 44: 33–41.
61. Evans K, Thresher R, Warneke R, Bradshaw CJA, Pook M, et al. (2005) Periodic variability in cetacean strandings: links to large-scale climate events. Biol Lett 1: 147–150.
62. Pierce GJ, Santos MB, Smeenk C, Saveliev A, Zuur AF (2007) Historical trends in the incidence of strandings of sperm whales (*Physeter macrocephalus*) on North Sea coasts: An association with positive temperature anomalies. Fish Res (Amst) 87: 219–228.
63. Hagihara R, Jones RE, Grech A, Lanyon JM, Sheppard JK, et al. (2014) Improving population estimates by quantifying diving and surfacing patterns: A dugong example. Mar Mamm Sci 30: 348–366.
64. Intergovernmental Panel on Climate Change (IPCC) (in review) Climate Change 2013-The Physical Science Basis: Working Group I Contribution To The Fourth Assessment Report Of The IPCC: Cambridge University Press.

3

Recovery Trends in Marine Mammal Populations

Anna M. Magera[1]*, Joanna E. Mills Flemming[2], Kristin Kaschner[3,4], Line B. Christensen[5], Heike K. Lotze[1]

1 Department of Biology, Dalhousie University, Halifax, Nova Scotia, Canada, **2** Department of Mathematics and Statistics, Dalhousie University, Halifax, Nova Scotia, Canada, **3** Department of Biometry and Environmental System Analysis, University of Freiburg, Freiburg, Germany, **4** CESAB (Centre de Synthèse et d'Analyse sur la Biodiversité), Immeuble Henri Poincaré, Domaine du Petit Arbois, Aix-en-Provence, France, **5** Fisheries Centre, University of British Columbia, Vancouver, British Columbia, Canada

Abstract

Marine mammals have greatly benefitted from a shift from resource exploitation towards conservation. Often lauded as symbols of conservation success, some marine mammal populations have shown remarkable recoveries after severe depletions. Others have remained at low abundance levels, continued to decline, or become extinct or extirpated. Here we provide a quantitative assessment of (1) publicly available population-level abundance data for marine mammals worldwide, (2) abundance trends and recovery status, and (3) historic population decline and recent recovery. We compiled 182 population abundance time series for 47 species and identified major data gaps. In order to compare across the largest possible set of time series with varying data quality, quantity and frequency, we considered an increase in population abundance as evidence of recovery. Using robust log-linear regression over three generations, we were able to classify abundance trends for 92 spatially non-overlapping populations as Significantly Increasing (42%), Significantly Decreasing (10%), Non-Significant Change (28%) and Unknown (20%). Our results were comparable to IUCN classifications for equivalent species. Among different groupings, pinnipeds and other marine mammals (sirenians, polar bears and otters) showed the highest proportion of recovering populations, likely benefiting from relatively fast life histories and nearshore habitats that provided visibility and protective management measures. Recovery was less frequent among cetaceans, but more common in coastal than offshore populations. For marine mammals with available historical abundance estimates (n = 47), larger historical population declines were associated with low or variable recent recoveries so far. Overall, our results show that many formerly depleted marine mammal populations are recovering. However, data-deficient populations and those with decreasing and non-significant trends require attention. In particular, increased study of populations with major data gaps, including offshore small cetaceans, cryptic species, and marine mammals in low latitudes and developing nations, is needed to better understand the status of marine mammal populations worldwide.

Editor: Konstantinos I. Stergiou, Aristotle University of Thessaloniki, Greece

Funding: Financial support was provided by the Census of Marine Life's Future of Marine Animal Populations (FMAP) program (www.fmap.ca) with a grant to HKL, the Natural Sciences and Engineering Research Council (NSERC, http://www.nserc-crsng.gc.ca) with grants to HKL and JMF, and Dalhousie University (www.dal.ca). Financial support for data collection was provided by Pew Charitable Trusts (www.pewtrusts.org) through the Sea Around Us Project (www.seaaroundus.org) for populating the abundance databases provided by KK and LBC. The funders had no role in study design, data collection and analysis, decision to publish, or preparation of the manuscript.

Competing Interests: AMM currently works as a Fisheries Management Biologist with the Nunavut Wildlife Management Board. All other authors have declared that no competing interests exist. This does not alter the authors' adherence to all the PLOS ONE policies on sharing data and materials.

* E-mail: ammagera@gmail.com

Introduction

In the marine realm, mammals appear to be one of the taxa that have benefitted the most from a shift from resource exploitation toward wildlife conservation [1–4]. Marine mammals, a loose grouping of approximately 127 species, include cetaceans (whales, porpoises and dolphins), pinnipeds (true seals, fur seals and sea lions), as well as marine otters and sea otters, sirenians (manatees and dugongs) and polar bears [5]. Throughout history, humans have exploited and often depleted marine mammal populations [3,6–13]. Yet in the 20th century, substantial population declines led to relatively early and widespread reduction or cessation of commercial exploitation and implementation of conservation measures [1,2,4,14]. While several marine mammals have been held up as key conservation success stories (e.g. eastern North Pacific gray whales, *Eschrichtius robustus* [15,16] and sea otter, *Enhydra lutris,* populations [17]), not all marine mammal populations have recovered from earlier exploitation-driven declines. However, with data indicating positive abundance trend estimates for various populations, we now have the opportunity to not only evaluate threat status, decline, and extinction risk in marine mammals, but also increase and recovery.

Threats to marine mammals are numerous and have changed over time. Historically, marine mammals have been prized sources of meat, oil, fur, baleen and ivory [6,7,9,11]. They have also been captured for display in aquariums, culled when declared nuisances, used for bait, and indirectly exploited as bycatch [18–24]. Numerous marine mammal species were reduced to very low abundances by or during the 1900 s [2,7,25], and some almost to extinction, such as Northern elephant seals (*Mirounga angustirostris*) [26] and Guadalupe fur seals (*Arctocephalus townsendi*) [27]. Certain populations were regionally extirpated, including the sea otter throughout most of its range [17], the Atlantic gray whale, and the walrus (*Odobenus rosmarus*) in parts of the Northwest Atlantic [1]. A handful of species have become globally extinct, including the Steller's sea cow (*Hydrodamalis gigas*), sea mink (*Neovison macrodon*),

Caribbean monk seal (*Monachus tropicalis*), and Japanese sea lion (*Zalophus japonicus*) [5,26]. Among the survivors, however, some substantial population recoveries have occurred, for example for gray whales in the eastern North Pacific, multiple populations of humpback whales (*Megaptera novaeangliae*), southern right whales (*Eubalaena australis*) [28], sea otters [29], northern elephant seals [26], grey seals (*Halichoerus grypus*) in the UK [30] and Northwest Atlantic [31], and numerous fur seal species [27]. Yet direct and indirect exploitation, as well as disease, competition for prey, habitat degradation from coastal development and dams, ship traffic, offshore oil and gas exploration, pollution (chemical, physical and acoustic), and climate change continue to impact marine mammal populations today [23,24,32–34].

Although it is a commonly used term, "recovery" can have many definitions in different management and conservation contexts [25,35]. What definition of recovery is chosen depends on the goals of the study and can alter the conclusions. Generally, recovery is taken to be "a return to a normal state of health... or strength" [36]. In analyses of wild animal populations, however, we often do not know the "normal state" of a population [25]. Pre-commercial exploitation abundance or carrying capacity (K) can be used as a reference point, with the increase of a population towards such a reference point indicating recovery [6,25,37]. However, neither pre-exploitation nor K estimates exist for many species that lack records or data on past catches, traded products (e.g. oil, fur), scientific surveys, genetics, life history, or population structure. Furthermore, there is often debate as to whether K estimates should refer to pre-exploitation or current ecosystem conditions [6,38]. Therefore, a basic and practical approach in many data-limited (in terms of time span, number, frequency, and precision of data points) cases has been to view any significant abundance increase as evidence of a recovering population and at least partial recovery [25,39]. This is the definition we used in this study as it allowed us to compare across a maximal number of marine mammal population abundance time series with variable data quality.

Management bodies often judge recovery with respect to a proportion of K or pre-exploitation size. The U.S. Marine Mammal Protection Act specifies management for an "optimal sustainable population" level, which is defined by the U.S. National Marine Fisheries Services (NMFS) as "a population level between carrying capacity and the population size at maximum net productivity" [37]. An optimal sustainable population level for marine mammals is thought to be between 50–85% of K [37], but generally 60% is used [15], and the International Whaling Commission assumes 60% is the level at which whale populations are most productive [28]. In the absence of either K or pre-exploitation abundance, management bodies may use maximum observed population levels, based on survey data for example, as a reference point. Such is the case with Atlantic seal populations in Canada [40]. However, in cases where reports of substantial declines predate quantitative records, maximum observed population levels would not represent pre-exploitation levels, and could result in the underestimation of declines and the overestimation of recoveries. Additional criteria, such as targets for demographic features (e.g. ratios of juveniles to adults, or males to females), social dynamics, or ecological functions [41] may be relevant to evaluating recovery depending on the situation. Recovery may also be measured over different time periods, such as an entire time series of data, a set time period (e.g. 50 years, or 1950–2000), or with respect to the species' life history (e.g. three generations is commonly used by the International Union for Conservation of Nature (IUCN) [42]).

Despite the strong conservation focus on marine mammal management in many parts of the world, abundance data are limited and subject to high uncertainty. Reliable, effort-corrected catch data are absent for many populations, and marine mammals are notoriously difficult to survey accurately for abundance. Current population designations and/or distributions may not match historical records, leading to speculation that some populations have changed their distributions over time [43,44], and complicating assessments of long-term trends. Longer time series with historical population estimates are lacking for many populations that are or were not commercially valuable [2]. Many marine mammal species have elusive behaviors (e.g. extensive migrations or deep diving for prolonged time periods) which, combined with their widespread occurrences in remote areas, represent high logistical and economic challenges for monitoring their population statuses [15,45–47]. With the exception of land-breeding pinnipeds, accurate estimation of abundance trends over relatively short time periods is difficult in many cases [46], and abundance time series often encompass irregular survey intervals. Existing variability in survey coverage and methodology greatly hampers the assessment of population trends [47]. However, data availability has improved greatly with the population monitoring and modeling that have accompanied increased management and conservation efforts internationally and domestically since the 1970 s.

Although marine mammal abundance data are collected and assessed by several organizations for management and conservation purposes (e.g. IUCN, U.S. marine mammal stock assessments), a quantitative global synthesis of marine mammal recoveries on a population level has not been previously attempted. Populations within the same species may show different abundance trends, and population-level abundance monitoring has become increasingly valued. Thus, the objectives of our study were to compile publicly available population-level abundance time series for marine mammals around the world, assess population trends, and classify recovery. Where possible, we also aimed to quantify the relationship between historical population decline and subsequent recovery. Our overall goal was to enhance our understanding of recovery in formerly exploited marine mammals and marine species overall.

Materials and Methods

Data Compilation

We expanded a database of marine mammal abundance estimates collected by Kaschner (2004) [48] and Christensen (2006) [11]. We collected population-level abundance data from around the world from publicly available published journal articles, online government documents, stock assessment reports, and recovery plans. Major sources included: Fisheries and Oceans Canada (DFO), Committee on the Status of Endangered Wildlife in Canada (COSEWIC), U.S. National Oceanic and Atmospheric Administration (NOAA) and NMFS technical and administrative reports, U.S. Marine Mammal Stock Assessment Reports, St. Andrew's Sea Mammal Research Unit reports, Australian and New Zealand government documents, International Whaling Commission documents, and numerous published primary sources (Table S1). We collected data up to 2008 or the next most recent data point. We also collected information on generation times, methods of data collection and abundance estimation, and data reliability. For abundance trend analysis, we limited our attention to time series with at least three estimates. Although a population is generally described as a group of interbreeding organisms of the same species in a defined area [49], for the purpose of this study we chose population abundance data according to consistently defined areas described in the source literature. Thus, we did not exclusively adhere to population or stock definitions from monitoring or regulatory agencies, but the population designations often matched.

Population abundance data came from dedicated and opportunistic aerial, land-based and ship-based surveys, extrapolated total population or pup-count data, photo-identification and mark-recapture models, genetic diversity analysis, combined totals from the literature, models that relied heavily on bycatch, catch or catch-per-unit-effort data, various other model-derived estimates from these aforementioned types of data (including age- or stage-based, simple regression, Bayesian or state-space models), and in some cases unknown or unstated methods (Table S1). When year ranges were provided for abundance estimates, we used the mid-point of the range.

Error is present in virtually all marine mammal abundance data, but it may originate from different sources. For example, catch, bycatch, or product data could be subject to intentional or unintentional misreporting. In turn, abundance estimation from inaccurate data may provide faulty estimates of K or the intrinsic rate of increase (r_{max}), both of which may vary over time [6]. Historical abundance estimates from DNA have been questioned because of uncertainty over changes in mutation rates, appropriate designation of particular populations, and migration, all of which can affect population estimates [38]. Survey data are affected by biases such as avoidance or attraction behavior of the species surveyed, and estimates may also be skewed because of detectability issues, particularly for more cryptic species. Analysis methods have been developed, however, to correct survey data for unobserved animals that may be at sea or are otherwise not visible [45,50,51].

We collected any reported error information (e.g. coefficient of variation, confidence intervals, standard error, or standard deviation). Where available, we gathered pre-exploitation abundance estimates and K estimates. In cases where multiple historical estimates were found (e.g. catch vs. genetic data) we recorded them all. Not all data sources reported error, so we used a system to incorporate quantitative error information (where available) with qualitative information about the data source for each data point and its reliability into abundance trend estimation (Abundance Confidence ID (ACID), Text S1).

Data Analysis – Trend Overview

We used a general definition of recovering populations being those that displayed an increase in abundance. Our main objectives were to gather abundance data and statistical evidence so as to broadly classify populations as having either increasing or decreasing trends, rather than attempting to estimate and describe complex population trajectories. In order to do so we estimated a trend for each population over the three most recent generations. This corresponded with the criteria used by the IUCN for assessing population decline [42]. In cases where we did not have data for the entire three-generation time period, we used the available time period with a required minimum of ten years instead (i.e. the minimum time period used by the IUCN for assessing population declines [52]) (Table S2).

Trend Analysis

Given available data and our aim, we used robust regression to estimate population trends. Robust regression is a powerful tool for fitting linear models that simultaneously identifies and down weights outlying data points [53,54]. Specifically we used the lmRob command in the robust library in R (which makes available an S-estimator [55] with a high breakdown point). We performed a full exploratory analysis using simple and robust linear regression, as well as log-linear robust regression (i.e. with a log transformation of the abundance values) and with and without weighting to include abundance data error information. We fit simple and robust regressions to both regular abundance data (i.e. number of individuals) and standardized abundance data (obtained by subtracting the mean from each data point and dividing by the standard deviation). The standardization facilitated comparison of regression results amongst different populations. We decided to use robust linear and robust log-linear weighted regressions as they best reflected the data. The complete R code is available in Text S2.

Results of fitting the linear and log-linear robust weighted regressions were then used to classify each population as Significantly Increasing, Significantly Decreasing, Non-Significant Change, or Unknown. To start, any populations with insufficient data (i.e. <3 data points over three generations or minimum 10 years) were deemed Unknown. Those remaining were classified according to both the direction and significance of their abundance trend. Specifically, Significantly Increasing and Significantly Decreasing populations had positive or negative slope estimates, respectively, and 95% confidence intervals that did not include zero. Populations with 95% confidence intervals that did include zero were classified as showing Non-Significant Change. Our classifications of population trends allowed for an easily interpretable summary of patterns in the marine mammal population abundance data.

Trend Comparisons

We examined both the population trend estimates (and the ensuing classifications) for marine mammals overall, as well as for notable taxonomic divisions (cetaceans, pinnipeds, other marine mammals), main habitat type according to Jefferson et al. (2008) [5] (for cetaceans only: coastal, offshore, or both), and other relevant sub-groupings (e.g. mysticetes or baleen whales, odontocetes or toothed whales, a sub-group of just dolphins and porpoises, otariids or eared seals, phocids or true seals). The groupings were not necessarily mutually exclusive. For example, a pantropical spotted dolphin (*Stenella attenuata*) population would be included in the overall marine mammal category, as well as the cetacean, odontocete, and dolphin and porpoise sub-groupings. The groupings for each population are included in Table S2. Many of the populations for which time series data exist overlapped spatially or were nested in each other's range, thus likely representing a similar set of individual animals. To avoid such double counting, we grouped populations into the largest and smallest non-overlapping or non-nested areas and analyzed both sets separately. These groupings by large and small areas are also listed in Table S2.

Comparison with IUCN Data

We compared our population trend classifications (Significantly Increasing, Significantly Decreasing, Non-Significant Change, Unknown) to those available through the IUCN Red List (Increasing, Decreasing, Stable, Unknown) (www.iucnredlist.org, Table S3). Since our data were mostly collected at a population level, and the IUCN mostly works with species level, we first summarized our data at a species level to facilitate comparison. For each species, we selected from our database those populations that comprised the majority of the overall species abundance. In cases of multiple non-nested and approximately equal-sized populations, we took the most frequently occurring trend classification of the populations of the species. If this did not exist (e.g. as for three populations with different trend classifications), we listed the population classification for that particular species as Unknown.

Historical Declines and Recent Increases

With the available data, we also had the opportunity to examine another facet of marine mammal recovery. For some populations (n = 47), we had at least one historical estimate, as well as a minimum population estimate and a current population estimate that was above the minimum estimate. Thus, these populations demonstrated both a historical decline (decrease) and evidence of recent recovery (increase) (Table S4). We were interested in whether the magnitude of recent recovery was related to the magnitude of historical decline. Therefore, the historical abundance estimate(s) was compared to the population minimum and the most recent abundance estimates to estimate the magnitude of decline and the magnitude of recovery with respect to the historical population size. For the populations for which there existed multiple historical abundance estimates, we used the mean of the declines and recoveries relative to these historical estimates. In addition to the magnitude of recovery, we were also interested in comparing the magnitude of historical population declines to the rates of recovery (i.e. the slopes of our above regression analysis). Unfortunately, only a small subset of the populations with historical abundance estimates had enough data over three generations to derive a rate of recovery, which provided too small a sample size from which to derive meaningful patterns.

Results

State of Available Marine Mammal Population Data

Overall, we were able to compile 182 population abundance time series for 47 species of marine mammals (i.e. 37% of the 127 currently recognized marine mammal species) for which there were at least three abundance estimates over three generations or at least ten years. Originally, we had compiled 198 population abundance time series; however, some time series corresponded to the same population, representing both pup count and regular count (i.e. entire population) data. We used the regular data unless they were much sparser than the pup count data. With these duplicate population indices removed, we had 182 population abundance time series in total.

A breakdown of species by taxonomic groups and different sub-groupings represented in this study compared to all marine mammals is depicted in Figure 1. Groups with high representation (>50% of known species represented) were the baleen whales (mysticetes) and pinnipeds. Groups with low representation in our data (<50% of known species represented) included the sirenians and cetaceans overall, especially toothed whales (odontocetes) within the cetaceans grouping, and dolphins and porpoises within the toothed whales grouping. No species of beaked whales (n = 21 species globally) or river dolphins (n = 4 species globally) were included. It was difficult to obtain time series that met our criteria for many smaller cetacean species (notably porpoises, beaked whales and river dolphins), Antarctic true seals, and sirenians. However, we were able to obtain three-generation abundance data (rather than the 10 years of data) for 95 of the 182 populations.

Marine Mammal Population Trends

Population trends were estimated for 182 non-duplicated populations from 47 species (see Figure 2 for examples, Figure 3 for population trend classifications, Figure S1 for plots for all populations, and Table S5 for all regression results). The models that best fit the available data were found to be those based on scaled abundance data and estimated via robust log-linear regression (weighted by ACID). We compared the R-squared values for both the robust linear and log-linear models, and found that robust log-linear models provided a better fit to the data for 78 of the populations, while robust linear models were better for 50 of the populations. There was 1 population for which both models produced a similar result, and 33 populations with Unknown trends. However, the resulting trend classifications were largely the same, with the two models producing different results for only 20 populations. In these cases, the robust log-linear model provided a better fit for 13 populations, and for the other 7 populations it found no evidence of trend (i.e. Non-Significant Change) as opposed to a Significantly Increasing or Significantly Decreasing trend provided by the better fitting robust linear model. Since the robust log-linear model best described the data in the majority of cases, we reported these results in the paper. However, plots and results for both the robust linear and log-linear regressions are included in the supplementary materials (Figure S1 and Table S5).

In terms of marine mammal population trends for the largest non-overlapping areas (n = 92 populations), 42% were Significantly Increasing, 10% were Significantly Decreasing, 28% showed Non-Significant Change, and 20% were deemed Unknown (see 'All' in Figure 3). Since some populations were nested within each other and consequently not independent, we analyzed trends for populations chosen by the smallest (n = 135) and the largest (n = 92) non-overlapping areas. Results did not differ substantially so we report results by the largest non-overlapping areas. The number of other marine mammal populations (polar bears, otters and sirenians) in the analysis was low (n = 7), with sea otters comprising the majority (n = 4) of the sample of other marine mammals.

Some populations did have flat slopes (i.e. very close to zero) and small confidence intervals that might suggest stable population abundance, such as the New Zealand sea lion (*Phocarctos hookeri*) (Sandy Bay) and the Southern elephant seal (*Mirounga leonina*) (Kerguelen Isles) (Figure S1). However, they did not have significant trends and rather than include an additional trend classification category, they were grouped with other populations that had larger confidence intervals and/or more variable data as displaying Non-Significant Change.

When comparing different categories of marine mammals (Figure 3), our results indicate that proportionally more sirenian, polar bear and sea otter populations (i.e. "Other", 71%) and pinnipeds (50%) were Significantly Increasing, than marine mammals overall (42%) or cetaceans (31%). For the pinnipeds, eared seals (58%) showed more Significantly Increasing populations compared to true seals (44%). Among the cetaceans, all taxonomic groups showed relatively low numbers of Significantly Increasing populations (39% baleen whales, 23% all toothed whales, 28% just dolphins and porpoises). Primarily coastal cetaceans, however, had proportionally more Significantly Increasing populations (58%) than primarily offshore cetaceans (13%). Overall, pinnipeds (and especially eared seals), other marine mammals, and coastal cetaceans showed the highest proportions of Significantly Increasing populations. Toothed whales, and dolphins and porpoises, had the highest proportions of Significantly Decreasing populations. All categories of cetaceans, except for coastal cetaceans, had approximately one third or more Non-Significant trends. Offshore cetaceans and baleen whales also had over a third of populations with Unknown trends over three generations.

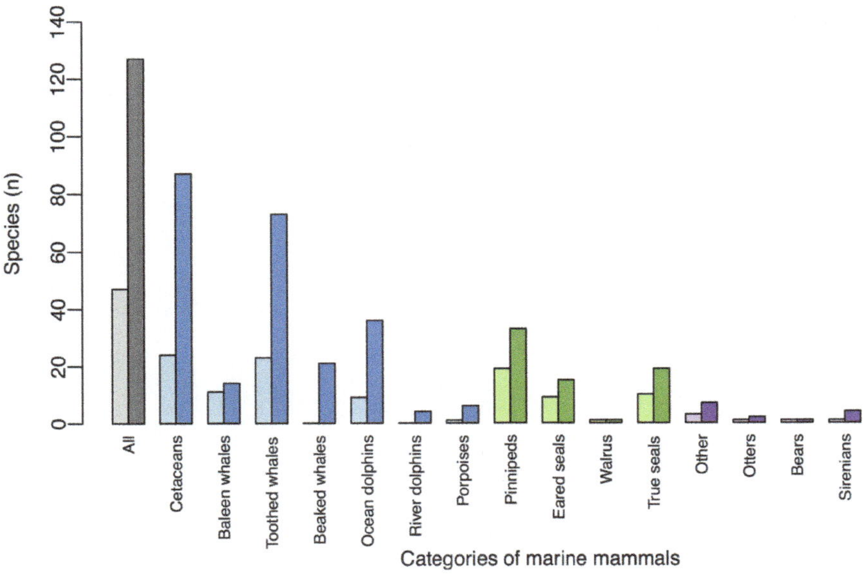

Figure 1. Species represented in population-level data in this study compared to marine mammal species overall. This study = light bars, marine mammals species overall = dark bars. Colors represent main taxonomic groups and their relevant sub-groupings: all marine mammals (grey), cetaceans (blue), pinnipeds (green), and other marine mammals (marine and sea otters, polar bears and sirenians) (purple).

Comparison with IUCN Data

The IUCN lists trend classifications for 127 marine mammal species, while we had data for populations from 47 species. We had enough population-level data to allow for comparison of 27 of the 47 species to the IUCN classifications (Figure 4, Table S3). Most trend classifications were similar, with approximately 37–41% of species showing Increasing trends, 11–19% Decreasing, 7% Non-Significant or 11% Stable (these were two categories that did not match between the IUCN classifications and the ones in our study), and 37% Unknown (Figure 4). However, compared to all 127 mammals in the IUCN database, our data showed a higher proportion of Increasing populations and lower proportions of Decreasing and Unknown populations.

Historical Declines and Recent Increases

Among all non-duplicated populations in our database (n = 182), we had 47 non-nested populations that included a historical, minimum and recent population estimate (Figure 5, Table S4). Historical abundance declines and recent recoveries, as a proportion of the historical abundance estimate(s), ranged from virtually zero to over 100%, with some population recoveries exceeding the best available historical population

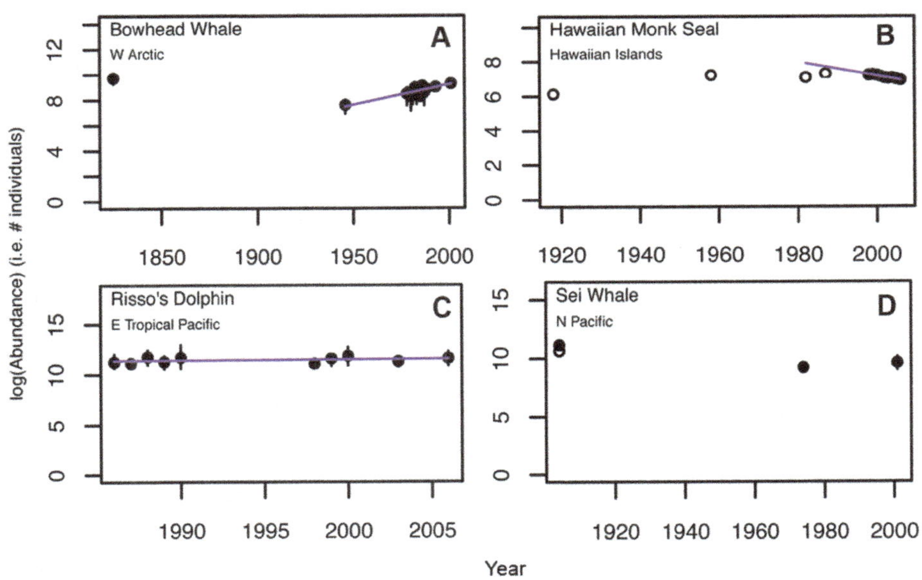

Figure 2. Examples of Significant Increase (A), Significant Decrease (B), Non-Significant Change (C) and Unknown (D) population abundance trends. Robust weighted log-linear regression line is depicted over three generations or at least ten years. Solid points = abundance estimates with reported error (95% confidence intervals). Open points = abundance estimates without reported error.

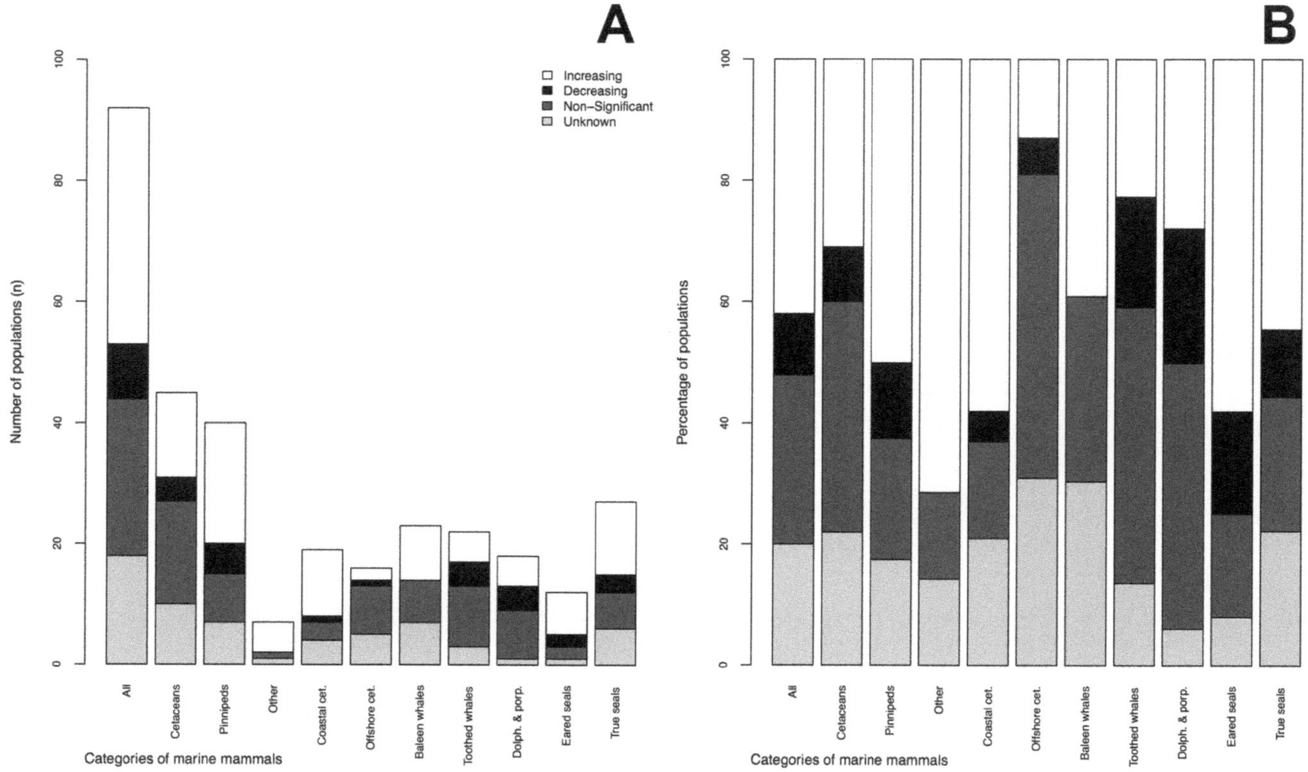

Figure 3. Trend classification for marine mammal populations by different categories as numbers (A) and proportions (B). Summary of results from robust weighted log-linear regressions for 92 (non-nested, including the largest possible areas) marine mammal populations. "Other" includes sirenians, polar bears and sea otters.

estimates we could find. One should note that populations with smaller historical declines (e.g. to 90% of historic levels) could

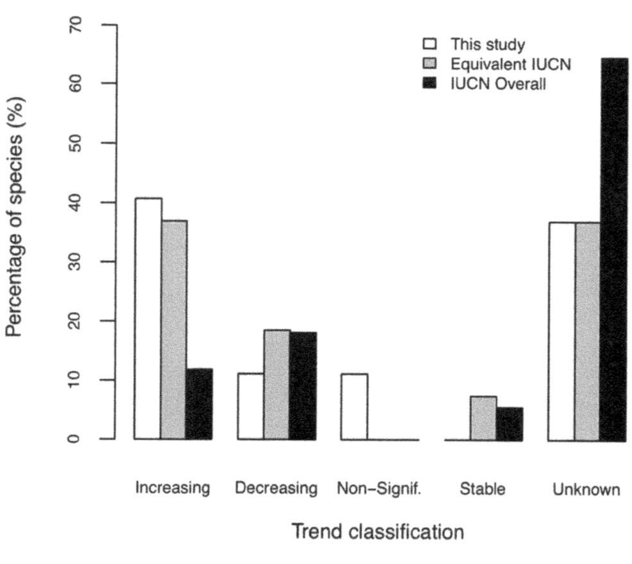

Figure 4. Comparison of marine mammal population trend classification results from this study with IUCN classifications. Percentage of marine mammals at the species level in abundance trend categories from this study (n = 27), the equivalent IUCN species (n = 27), and all marine mammal species assessed by IUCN (n = 127). The IUCN does not have a "Non-Significant" category as in this study, but it does have a "Stable" category not used in our results.

only increase over a smaller range (e.g. between 90 and >100%), while those with large historical declines (e.g. to 10% of historic levels) could increase over a much larger range (e.g. between 10 and >100%). Overall, however, there were many populations with very large historical declines >90% that have so far shown only small recent increases (clustered in the lower right corner of Figure 5A on or slightly above the diagonal line). Yet there was also considerable variation in recoveries among populations with declines >80%, and five populations with very high declines to <10% of historical levels also showed very high recoveries to >90% of historical levels. These included South African fur seals (South Africa and Namibia), harbour seals (Washington coast, Oregon), and humpback whales (North Atlantic and North Pacific).

On average, all marine mammal populations declined by 71% and have so far recovered to 61% of their historical abundance (Figure 5B). Recovery responses were generally more variable within each group than declines, and pinnipeds (true seals and eared seals) showed the most variation. Cetaceans that spend the majority of their time in coastal areas instead of offshore waters and other marine mammals (in this case, n = 1 sea otter population) showed the greatest mean declines (93% and 96% respectively), and coastal cetaceans the lowest recoveries (38%). The two groupings that had the lowest mean declines, just dolphins and porpoises (41%) and all toothed whales (49%), showed the highest mean recoveries (89% and 78% respectively).

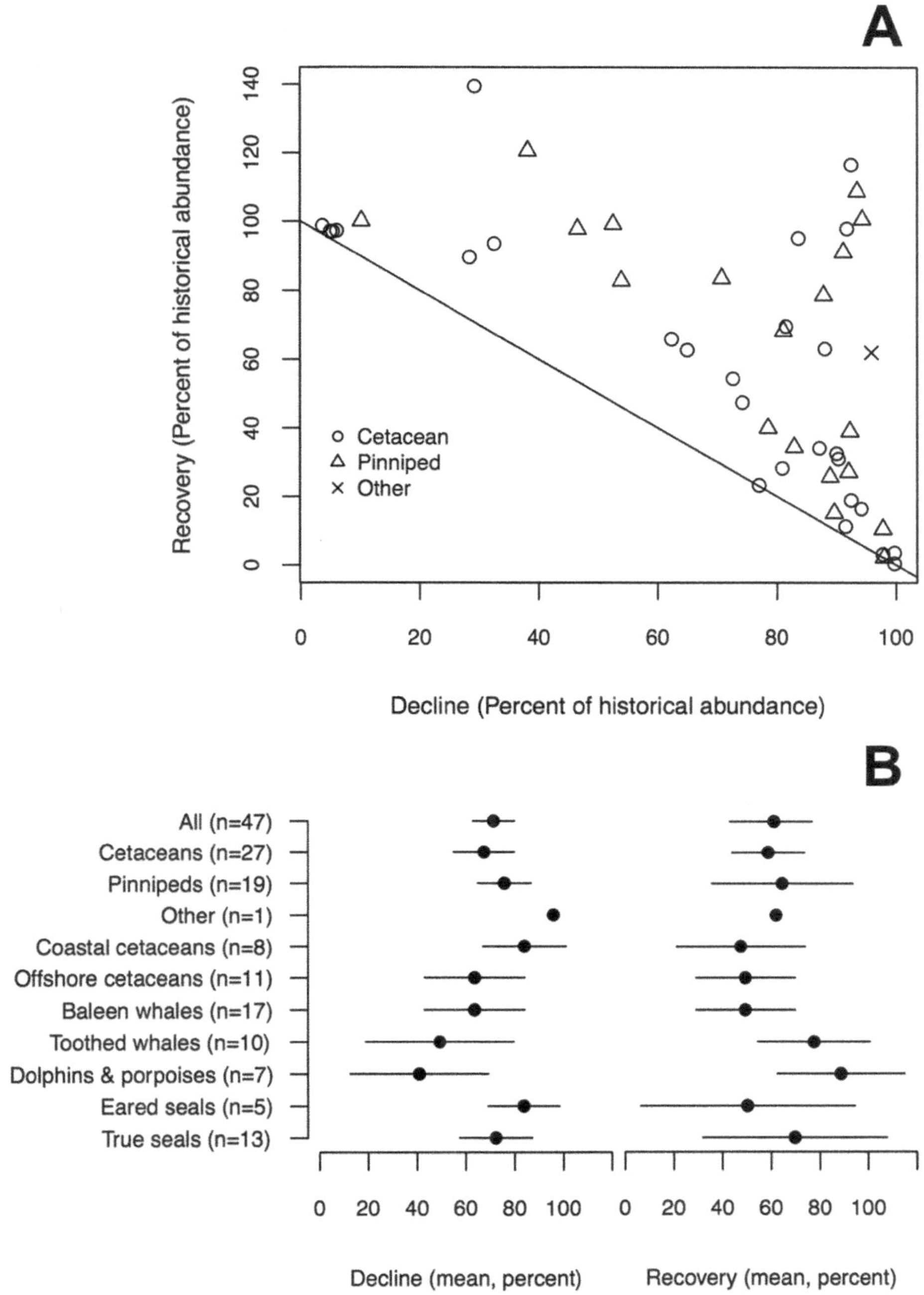

Figure 5. Historical decline and recent recovery relative to historical level for 47 non-nested marine mammal populations. Proportional decline and recovery in abundance for individual populations (A), and average decline and recovery across different categories of marine mammals (B). Note that in (A) that recovery must be equal to or greater than decline for a population (i.e. in the area above the diagonal line).

Discussion

The aim of our study was to perform a quantitative assessment of recovery in marine mammal populations worldwide using the best publicly available abundance data. Overall, our results suggest that more marine mammal populations are recovering than not, especially among pinnipeds, other marine mammals (e.g. sirenians, polar bears and otters) and coastal cetaceans, but a large proportion (43%) of populations have non-significant or unknown trends mainly due to high variability or data scarcity. Our study also suggests that, in general, more historically depleted populations have so far shown either low or variable recovery. These results highlight the many recovery successes while also drawing attention to those populations needing more attention in terms of monitoring, management and conservation.

Available Marine Mammal Population Data

Notwithstanding that population level abundance data are limited in many ways for marine mammals, we successfully compiled a substantial number of time series that represented 37% of marine mammal species worldwide. These time series provided insight into data availability and quality at the population level, as well as valuable information pertaining to recoveries. Any additional data could easily be included in our database for further investigations.

Populations with better abundance time series were found to share common characteristics. They typically had either present or past commercial or cultural value, as is the case with many large whales and pinnipeds, or they were iconic or charismatic species such as killer whales (*Orcinus orca*) or common bottlenose dolphins (*Tursiops truncatus*). Moreover, species with good data were generally easier to monitor because of some combination of factors that made them accessible and visible. These factors included aspects of their behavior, such as the regular use of haul-out and breeding areas by numerous pinniped species [46], consistent coastal migration routes as for the gray whale on the west coast of North America [56], or long times spent at the water surface, as with North Atlantic right whales (*Eubalaena glacialis*) [57]. Habitat or body size characteristics may have also contributed to better time series information: abundance data collection may be easier for animals in accessible coastal areas, with smaller, well-known ranges, with large body size (e.g. great whales), or with individually identifying markings (e.g. killer whales) [3,26,46].

Definite gaps in our knowledge of marine mammal abundance trends still exist, namely for beaked whales, river dolphins, sirenians and Antarctic true seal populations. Beaked whales are typically pelagic deep-divers that spend very little time at the surface where there can be detected. Moreover, they have large ranges and low densities [46]. Relative lack of commercial value and recognition in the public sphere may also have contributed to the lack of monitoring and data for these species, although short (typically <10 years), sparse, recent time series [46,47] or estimates [58] do exist for some populations. Time series data will likely improve due to interest in the susceptibility of beaked whales to acoustic disturbance, especially from seismic and naval sonar testing [46,59]. Many river dolphins are similarly cryptic, live at low densities, or do not gather in social groups. A lack of conservation and monitoring plans with standardized population and habitat assessment techniques has limited data availability for them [23]. Although intensely studied, West Indian or Florida manatees (*Trichechus manatus*) are difficult to observe and reliable abundance estimates are challenging to obtain [60]. The lack of management has also contributed to a lack of time series data for Amazonian (*T. inunguis*) and West African manatees (*T. senegalensis*) [61,62]. Despite intensive survey efforts in many areas, low densities and large ranges have inhibited reliable abundance estimates for dugongs (*Dugong dugon*) [63]. Many Antarctic true seal populations have relatively recent population estimates and are thought to have healthy populations, but they lack longer time series or historical estimates due to the inaccessibility of their remote breeding habitats until recently [24]. Abundance monitoring of Antarctic species may become more important for assessing and managing the effects of climate change and expanding Antarctic fisheries [64,65,66].

The time spans for which data were available varied between different groupings of marine mammals. Three generations of data were available for 52% of populations, while shorter time series were typically available for dolphins, porpoises, small whales, polar bears (*Ursus maritimus*), and sirenians. These data gaps are not surprising considering that many of these populations were not heavily commercially exploited, and thus abundance records likely only began recently with management and monitoring. Populations of great whales, Pacific dolphin species, northern true seals, belugas (*Delphinapterus leucas*) and narwhals (*Monodon monoceros*) made up the majority of the populations with historical abundance estimates. Historical population estimates for commercially valuable great whales typically came from either (1) catch data or product trade records, or (2) back-casting or genetic analysis techniques [6,67,68]. Numerous dolphin species found in Pacific U.S. waters were impacted by the tuna fishery that was established in the 1960 s [69,70], and they are a management concern under the U.S. Marine Mammal Protection Act. Certain northern marine mammal species (northern true seals, walrus, polar bear, beluga, narwhal, and bowhead whales) are still exploited for subsistence use by aboriginal groups, and in certain cases commercially or for sport hunts. They have been subject to study and management, although survey data collection can be logistically challenging (e.g. [9,40,71–74]). Commercial pinniped hunts exist in Canada, Greenland, Namibia, Norway and Russia for several true and fur seal populations [40,75]. With the lack of historical catch or abundance data for many populations, the development of genetic techniques [67,68] or habitat availability analyses [76] for estimating historic population size may provide more insight into pre-exploitation estimates for some populations in coming years.

We also found geographical biases in the data, which was typically from North America, Europe (especially northern Europe), and to a lesser extent Australia, New Zealand, Japan, South America and southern Africa. This bias is generally mimicked in global marine fish abundance datasets (e.g.[77–79]) and the distribution of worldwide line-transect cetacean survey effort [80]. A recent study by Kaschner et al. found that the Eastern Tropical Pacific has been the most surveyed region in the world via aerial or ship-based line-transects for cetaceans since the 1980 s [80]. The geographical biases in the data also reflect the availability of financial and logistical resources for monitoring and assessment in richer nations. With the increasing interest in the value of global biodiversity, monitoring may expand to other more data-poor areas. Population abundance monitoring is important for both conservation (e.g. recovery plans) and sustainable extractive management to assess trends and reference points for conservation goals and management targets [25,35,77].

Marine Mammal Population Trends

Our analysis provides a general overview of recovery trends across marine mammal populations for which data was available. Amalgamating data from numerous different sources posed challenges, but we chose analytical techniques and recovery definitions accordingly. Using a robust regression over three generations allowed us to estimate in a statistically sound manner the dominant recent abundance trends for the largest number of marine mammal populations, as scaled to life history and comparable to IUCN methods [52]. In a few cases, historical estimates were included in the 3-generation regression timespan, but the robust regression generally downweighted high leverage points (i.e. abundances that are far from the mean). The analysis of historical decline and recent recovery, however, allowed us to incorporate populations with longer time spans and historical abundance estimates, but insufficient data over three generations, into the study. Our method for identifying population trends and those populations showing signs of increase or recovery – i.e. those showing a statistically significant increase in abundance, robustly estimated using scaled data – also worked well for the challenges we faced, including non-uniform data time-spans and different life histories. Only in a few cases with short and/or highly variable time series did this method not capture what by sight appeared to be the actual trend of the data. In other analyses, depending on the goals, more specific definitions of recovery may be appropriate [25]. In addition, the sample size of other marine mammal populations in this analysis was low (n = 7) and further investigation of this category is recommended as more population data becomes available, particularly for species other than sea otters.

Overall, 42% of populations were Significantly Increasing, 10% were Significantly Decreasing, 28% showed Non-Significant Change, and 20% were deemed Unknown. The somewhat large proportion of Non-Significant and Unknown trends points to the difficulty of studying abundance trends in these animals and estimating their population size with accuracy, stressing the need for better monitoring efforts for certain groups. Improvements in increasingly used techniques such as acoustic monitoring, tagging, photo-identification and mark-recapture, as well as computerized database and analysis technologies, modeling, and data sharing among organizations, including incorporating local community knowledge [81], may improve data quality and quantity.

Despite the data gaps, those populations with sufficient data showed some interesting patterns. Half of the pinniped populations were Significantly Increasing. Their fast life history characteristics (e.g. shorter generation times) may have helped promote recovery. Management and conservation efforts, in terms of limiting direct exploitation, bycatch and trade, as well as either the isolation or protection of important haul-out or breeding habitats, likely also contributed to recoveries in some populations [24].

Coastal cetaceans were also recovering relatively well, possibly because of their early exploitation and subsequent relatively early management, conservation, and attention in the public sphere. Large, coastal cetaceans were often the first species commercially hunted in an area because of their economic value and relatively easy access (e.g. bowhead [82], North Atlantic right [82], gray [56], and humpback whales [82]). International concern over steep declines in numbers by the 1900 s for most populations led to some of the first multilateral conservation agreements, protection from international trade, domestic exploitation bans or regulations, habitat protection, and recovery planning [3,25,26,83]. Numerous marine mammals have also become endearing symbols of the environmental movement [84].

Less visible or charismatic species, such as more predominantly offshore or smaller cetaceans, may have suffered in terms of population recovery for a few reasons. They were often exploited after the depletion of more easily accessible coastal species. Later onset of management or lack of directed management may have also been a contributing factor.

Some toothed cetaceans, such as sperm, pilot (*Globicephala* species) and killer whales, also have highly developed social structures that may be important to survival. As a result they may be more heavily impacted by the effects of selected removal and small population size [85,86]. Critical factors which may determine population recovery have been investigated, and appear to support the assertion that primarily coastal marine mammals had a higher probability of recovering than offshore populations, as did earlier maturing populations and pinnipeds (Anna M. Magera, MSc thesis, Dalhousie University, 2011).

Comparison with IUCN Data

In order to verify our population-level trends, we compared our results with IUCN assessments for those species for which we had the majority of the species' abundance data at a population level. This included 27 species, approximately 21% of all marine mammal species worldwide. For these, we did have good agreement with the equivalent IUCN trend determination for the same species, which strengthens confidence in our results. However, we clearly had more Increasing, and fewer Decreasing and Unknown species (populations) compared to the global IUCN assessment of all 127 marine mammals (Figure 4). This is primarily attributable to the absence of data that met our criteria for many rare or difficult to monitor species, which resulted in them being excluded from the study (see Methods), combined with the large percentage (58%) of species classified as data-deficient by the IUCN.

Historical Declines and Recent Increases

Many populations with available time series did not have pre-exploitation or K estimates of historical population size. However, we were able to obtain estimates of historical population size and thus compare the magnitude of historical declines and recent recoveries for 47 of the non-nested populations. As shown for marine fish populations [39,87], smaller historical population declines were generally associated with more successful recent recoveries. In turn, populations with very large historical declines (>90%) in fish [39,87] and many marine mammals (this study) have shown low magnitudes of recovery so far, with some notable exceptions. The large proportion of populations with >60% to >90% declines in this study highlights the substantial historical declines in many marine mammal populations [2].

We also found differences in declines and recoveries between different types of marine mammals, with the smaller, typically less commercially valuable toothed whales, and especially porpoises and dolphins showing the smallest historical declines and greatest recent recoveries. In turn, easily accessible coastal cetaceans and other marine mammals showed the largest historical declines. Sethi et al. (2010) note that in commercial fisheries development is typically driven by which taxa will produce the most profit at the minimal cost [88]. Similar explanations certainly exist for the history of whaling and sealing [7,13]. Recovery among marine mammal populations showed quite high variability. However, on average eared seals, baleen whales, as well as offshore and coastal cetaceans seemed to show the smallest recoveries, likely because of heavy historical exploitation (eared seals, baleen and coastal cetaceans) or lack of management (offshore cetaceans) [7,13,24].

Other studies have quantified historical declines and, in part, recoveries in marine mammals. Christensen (2006) estimated historical baselines for exploited cetaceans and pinnipeds and found a cumulative decline of 22% (range: 0–62%) in numbers between 1800 and 2001. The largest declines were in the great whales (64%, range: 40–79%) during periods of increased catches [11], which is comparable to the declines we estimated for the baleen whales. Christensen's overall marine mammal declines, however, presented a smaller decrease in terms of overall numbers than calculated in our study, likely because Christensen [11] relied mainly on the use of catch data-driven models and limited her assessment of cumulative marine mammal declines to 1800–2001. By this time some populations (e.g. North Atlantic right whales) had already been substantially depleted. A second recent study of historical baselines for large marine animal populations (not limited to a specific period) estimated a historical decline of approximately 96% for pinnipeds, otters and sirenians, and a recovery to approximately 25% of historical abundance [2]. Whales declined by approximately 82% and recovered to approximately 32% of historical levels. The pinniped, sea otter and sirenian category was not directly comparable to any of our categories, but for pinnipeds, our study presented smaller declines and larger recoveries. This may have been because the study included populations that did not exhibit any recovery, including extirpated populations, while our examination of n = 47 populations did not. The whale category only included great whale populations (mainly baleen) and was comparable to our baleen whale decline and recovery results [2].

Although there was relatively high variability in population recoveries after large historical declines in our study, many populations showed minimal population recovery following very high declines despite many decades of protection (e.g. right and bowhead whales since the 1930 s), thus possibly indicating Allee effects [89]. Only five populations out of 18 that underwent very large declines to <10% of the historical population abundance also showed very high (>90%) recoveries. This reflected a similar finding in certain fast-growing clupeid populations, which recovered to levels that were not observed in any other types of fish after declines of similar magnitude [39,89]. Possible explanations include a faster life history and relatively early age at maturity (4–5.5 years) that allowed for more rapid population increase as well as a longer time since exploitation was halted and better protection [26,90–96]. Other possible explanations might be that historical population estimates underestimated true pre-commercial exploitation population size, or ecosystem conditions changed in these areas to favor substantial population growth and larger abundances.

Conclusion

Despite sparse data for many species and regions, lack of historical abundance estimates and often large error associated with available data, we were able to assess the recovery status for 182 marine mammal populations worldwide. Overall, 42% of populations for which we have good abundance time series are recovering from former depletions, especially pinnipeds, coastal cetaceans and other marine mammals (i.e. polar bears, otters and sirenians). Offshore cetaceans, all toothed whales, and just dolphins and porpoises showed relatively few recovering populations. On a species level, our results were comparable to assessments performed by IUCN, suggesting that our robust weighted regression over three generations is a useful and appropriate method for estimating general population trends. However, compared to all marine mammal species assessed by IUCN, our data over-represented recovering populations due to a lack of data for data-poor populations, most notably the sirenians, river dolphins and beaked whales. We also found that populations with smaller historical population declines were more likely to show stronger population recoveries in more recent times, while those that have been extensively depleted showed more variability in their recovery success, but tended to have smaller recoveries. This synthesis and compiled database are useful tools for other researchers interested in marine mammal population trends as well as decision makers to guide management and conservation efforts towards non-recovering populations. Moreover, our results stress the need for enhanced and increasingly innovative monitoring of offshore and cryptic marine mammals and those in low-latitude and developing nations that have not been extensively studied. Available and reliable abundance data are critical for better management and conservation, and will help to produce more complete and accurate trend estimates and recovery assessments in the future.

Supporting Information

Figure S1 Marine mammal population abundances over time and trends over three generations for robust log-linear (A) and robust linear (B) regressions. Species and population areas are described in the upper left hand corner of each plot (n = 198 populations with duplicates, n = 182 populations without duplicate regular and pup count data). Population robust regression trend classification (long-linear (A) or linear (B)) is indicated in the upper right hand corner: I = Significantly Increasing, D = Significantly Decreasing, NS = Non-Significant Change, NA = Unknown. Solid lines = robust regression weighted by Abundance Confidence ID (ACID). Solid points = abundance data with quantitative error information (95% confidence interval bars). Empty points = abundance data with no stated quantitative error information. Black points = regular data that was collected from the entire population. Grey points = indicate pup count data.

Table S1 Abundance data source information for study populations. For each population, denoted by a numeric area code (Population Area ID) and area description, abundance data sources are listed along with the data collection and/or additional analysis methods used to obtain the population abundance estimates. Abundance Confidence ID (ACID) provides an uncertainty rank (1 to 6, 1 = lowest, 6 = highest). CPUE = catch per unit effort.

Table S2 Overview of population time series: area descriptions, generation times, time spans, and data types. All time references are in years. Time series refer to population abundance time series. List of abbreviations used in table: Species Type = general species taxon – cetacean, pinniped or other (sirenian, otter or polar bear). Regular Data Type = non-pup count data. Pup Count Data Type = only pup count data. N, S, E, W = North, South, East, West. Include Small/Large Area = was the population included if the smallest or largest non-nested populations were considered for the species? (1 = yes, 0 = no). Dominant Habitat = dominant habitat type. Sub-type = - sub-type of marine mammal within species type. Dolphin or Porpoise? = whether the population is a type of a dolphin or porpoise.

Table S3 Abundance trends for 127 marine mammal species listed by the IUCN (2008). Species Type = general species taxon – cetacean, pinniped or other (sirenian, otter or polar bear). Data? = whether or not population level data for the species is included in this study. IUCN Trend = species trend listed on the IUCN Red List of Threatened Species (2008). IUCN typically describes abundance trends on a species level, while this study examines population level trends. Note, IUCN does not have a "Non-Significant" category as in this study, but does have a "Stable" category not used in our results.

Table S4 Historical population declines and recent recoveries with respect to historical level. For populations (n = 47) with historical, minimum and recent abundance estimates. Decline and recovery percentages are mean values for populations where more than one historical estimate is present. In some cases, recent abundance values are higher than historical population estimates, leading to recovery values of greater than 100%.

Table S5 Results from scaled robust log-linear and robust linear regressions for marine mammal populations (n = 198). The regressions were weighted by Abundance Confidence ID (ACID) over three generation times. Duplicate pup count/regular populations included in blue text. These will not be used in further analyses, resulting in 182 remaining populations (out of 198). Species Type = general species taxon (cetacean, pinniped or other (sirenian, otter or polar bear)). N, S, E, W = North, South, East, West. Include Small/Large Area = was the population included if the smallest or largest non-nested populations were considered for the species. Regression & Data Type = *log* lmRob or lmRob and Regular (i.e. non-pup count data) or Pup Count (i.e. only pup count data). Coefficient or slope = the scaled rate of population growth (i.e. change in abundance over time). SE = standard error. CI95 = 95% confidence interval. tval = t-value. pval = p-value. Increasing_Sig = significantly increasing abundance trend. Increasing_NS = non-significant increasing abundance trend. Decreasing_Sig = significantly decreasing abundance trend. Decreasing_NS = non-significant decreasing abundance trend. NA indicates a lack of sufficient data over the three-generation time period necessary to perform a robust regression analysis. Dominant Habitat = dominant habitat type. Sub-type = sub-type of marine mammal within species type. Dolphin or Porpoise? = whether the population is a type of a dolphin or porpoise.

Text S1 Abundance confidence ID (ACID) system: explanation and verification.

Text S2 R Statistical Software code for analyzing marine mammal population abundance trends.

Acknowledgments

We would like to thank W. Blanchard and S. Anderson for statistical and graphing advice, and J. Hutchings, H. Whitehead, T. Gerrodette, T. Branch and an anonymous reviewer for valuable comments and discussions. We would also like to thank all the researchers, collaborators and research assistants who contributed to the abundance database, in particular J. Beblow and K. Willis Ladel.

Author Contributions

Conceived and designed the experiments: AMM JMF HKL. Analyzed the data: AMM JMF HKL. Contributed reagents/materials/analysis tools: AMM KK LBC. Wrote the paper: AMM JMF KK LBC HKL.

References

1. Lotze HK, Milewski I (2004) Two centuries of multiple human impacts and successive changes in a North Atlantic food web. Ecological Applications 14: 1428–1447.
2. Lotze HK, Worm B (2009) Historical baselines for large marine animals. Trends in Ecology & Evolution 24: 254–262.
3. Aron W (1988) The commons revisited - thoughts on marine mammal management. Coastal Management 16: 99–110.
4. Adams WM (2004) Against extinction - the story of conservation. London: Earthscan. 311 p.
5. Jefferson TA, Webber MA, Pitman RL (2008) Marine mammals of the world: A comprehensive guide to their identification. LondonUK: Elsevier. 573 p.
6. Baker CS, Clapham PJ (2004) Modelling the past and future of whales and whaling. Trends in Ecology & Evolution 19: 365–371.
7. Lotze HK, Lenihan HS, Bourque BJ, Bradbury RH, Cooke RG, et al. (2006) Depletion, degradation, and recovery potential of estuaries and coastal seas. Science 312: 1806–1809.
8. Colten RH (2002) Prehistoric marine mammal hunting in context: Two western North American examples. International Journal of Osteoarchaeology 12: 12–22.
9. Reeves RR (2002) The origins and character of 'aboriginal subsistence' whaling: a global review. Mammal Review 32: 71–106.
10. Smith I (2005) Retreat and resilience: fur seals and human settlement in New Zealand. In: Monks G, editor. The Exploitation and Cultural Importance of Sea Mammals: Oxbow Books. 6–18.
11. Christensen LB (2006) Marine mammal populations: Reconstructing historical abundance at the global scale. Vancouver: University of British Columbia Fisheries Centre. 167 p.
12. Pauly D, Watson R, Alder J (2005) Global trends in world fisheries: impacts on marine ecosystems and food security. Philosophical Transactions of the Royal Society B: Biological Sciences 360: 5–12.
13. Reeves RR, Smith TD (2006) A Taxonomy of world whaling operations and eras. In: Estes JA, DeMaster DP, Doak DF, Williams TM, BrownellJr RL, editors. Whales, whaling, and ocean ecosystems. Los Angeles and London: University of California Press. 82–101.
14. Lotze HK (2007) Rise and fall of fishing and marine resource use in the Wadden Sea, southern North Sea. Fisheries Research 87: 208–218.
15. Gerber LR, DeMaster DP, Roberts SP (2000) Measuring success in conservation. American Scientist 88: 316–324.
16. Angliss RP, Allen BM (2009) Alaska marine mammal stock assessments, 2008. U.S. Department of Commerce, National Oceanic and Atmospheric Administration, National Marine Fisheries Service, Alaska Fisheries Science Center. 269 p.
17. Sea Otter Recovery Team (2007) Recovery strategy for the sea otter (*Enhydra lutris*) in Canada. Vancouver, B. C.: Fisheries and Oceans Canada. vii +56 p.
18. Read AJ (2008) The looming crisis: interactions between marine mammals and fisheries. Journal of Mammalogy 89: 541–548.
19. Bigg MA (1988) Status of the Steller Sea Lion, *Eumetopias jubatus*, in Canada. Canadian Field Naturalist 102: 315–336.
20. Reijnders PJH (1994) Historical population size of the harbour seal, *Phoca vitulina*, in the Delta area, SW Netherlands. Hydrobiologia: 557–560.
21. Kasuya T (2007) Japanese whaling and other cetacean fisheries. Environmental Science & Pollution Research 14: 39–48.
22. Ward EJ, Holmes EE, Balcomb KC (2009) Quantifying the effects of prey abundance on killer whale reproduction. Journal of Applied Ecology 46: 632–640.
23. Reeves RR, Smith BD, Crespo EA, Notarbartolo di Sciara G (2003) Dolphins, whales and porpoises: 2002–2010 conservation action plan for the world's cetaceans Gland, Switzerland and Cambridge, UK: IUCN. 147 p.
24. Reijnders P, Brasseur S, vanderToorn J, vanderWolf P, Boyd I, et al. (1993) Seals, fur seals, sea lions and walrus. GlandSwitzerland: IUCN. 98 p.
25. Lotze HK, Coll M, Magera AM, Ward-Paige C, Airoldi L (2011) Recovery of marine animal populations and ecosystems. Trends in Ecology & Evolution 26: 595–605.
26. Gerber LR (1998) Seeking a rational approach to setting conservation priorities for marine mammals. Integrative Biology 1: 90–98.
27. Gerber LR, Hilborn R (2001) Catastrophic events and recovery from low densities in populations of otariids: implications for risk of extinction. Mammal Review 31: 131–150.
28. Best PB (1993) Increase rates in severely depleted stocks of baleen whales. ICES Journal of Marine Science 50: 169–186.
29. Nichol LM, Watson JC, Ellis GM, Ford JKB (2005) An assessment of abundance and growth of the sea otter population (*Enhydra lutris*) in British Columbia. NanaimoBC: Fisheries and Oceans Canada. 22 p.
30. Thomas L, Harwood J (2008) Estimating the size of the UK grey seal population between 1984 and 2007. St. AndrewsUK: NERC Sea Mammal Research Unit and Centre for Research into Ecological and Environmental Modelling, University of St Andrews. 18 p.
31. Thomas L, Hammill MO, Bowen WD (2007) Estimated size of the Northwest Atlantic grey seal population 1977–2007. In: Fisheries and Oceans Canada, editor: Canadian Science Advisory Secretariat. 2007/082. 31 p.
32. Schipper J, Chanson JS, Chiozza F, Cox NA, Hoffmann M, et al. (2008) The status of the world's land and marine mammals: Diversity, threat, and knowledge. Science 322: 225–230.
33. Kappel CV (2005) Losing pieces of the puzzle: threats to marine, estuarine, and diadromous species. Frontiers in Ecology and the Environment 3: 275–282.
34. Davidson AD, Boyer AG, Kim H, Pompa-Mansilla S, Hamilton MJ, et al. (2012) Drivers and hotspots of extinction risk in marine mammals. Proceedings of the National Academy of Sciences of the United States of America 109: 3395–3400.
35. Hutchings JA, Butchart SHM, Collen B, Schwartz MK, Waples RS (2012) Red flags: correlates of impaired species recovery. Trends in Ecology & Evolution 27: 542–546.
36. Oxford Dictionaries (2010) Recovery. Oxford Dictionaries: Oxford University Press.
37. Gerrodette T, Demaster DP (1990) Quantitative determination of optimum sustainable population level. Marine Mammal Science 6: 1–16.
38. Freeman MMR (2008) Challenges in assessing cetacean population recovery and conservation status. Endangered Species Research 6: 173–184.
39. Hutchings JA (2000) Collapse and recovery of marine fishes. Nature 406: 882–885.
40. Fisheries and Oceans Canada (2011) 2011–2015 Integrated Fisheries Management Plan for Atlantic Seals. 34 p.
41. Sanderson EW (2006) How many animals do we want to save? The many ways of setting population target levels for conservation. Bioscience 56: 911–922.
42. IUCN (2001) IUCN Red List categories and criteria: Version 3.1. GlandSwitzer-Switzerland and Cambridge, UK: IUCN Species Survival Commission. 30 p.
43. Kaschner K, Watson R, Trites AW, Pauly D (2006) Mapping world-wide distributions of marine mammal species using a relative environmental suitability (RES) model. Marine Ecology Progress Series 316: 285–310.
44. Whitehead H (2002) Estimates of the current global population size and historical trajectory for sperm whales. Marine Ecology Progress Series 242: 295–304.

45. Amstrup SC, Laake JL, Manly BFJ, MacDonald LL, Robertson DG (1999) Preface. In: Garner GW, Amstrup SC, Laake JL, Manly BFJ, Macdonald LL et al., editors. Marine mammals survey and assessment methods. Rotterdam: AA Balkema. pp. XI–XII.
46. Taylor BL, Martinez M, Gerrodette T, Barlow J, Hrovat YN (2007) Lessons from monitoring trends in abundance of marine mammals. Marine Mammal Science 23: 157–175.
47. Jewell R, Thomas L, Harris CM, Kaschner K, Wiff R, et al. (2012) Global analysis of cetacean line-transect surveys: detecting trends in cetacean density. Marine Ecology Progress Series 453: 227–240.
48. Kaschner K (2004) Modelling and Mapping Resource Overlap between Marine Mammals and Fisheries on a Global Scale. Vancouver: University of British Columbia. 241 p.
49. Gotelli NJ (1995) A Primer in Ecology. SunderlandMassachusetts: Sinauer Associates, Inc. 206 p.
50. Berta A, Sumich JL, Kovacs KM (2006) Population structure and dynamics. Marine Mammals - Evolutionary Biology. Second Ed. London: Academic Press. 416–455.
51. Forney KA, Wade PR (2006) Worldwide distribution and abundance of killer whales. In: Estes JA, Demaster DP, Doak DF, Williams TM, BrownellJr RL, editors. whales, whaling, and ocean ecosystems. Los Angeles and London: University of California Press. 145–162.
52. IUCN Standards and Petitions Subcommittee (2010) Guidelines for using the IUCN Red List categories and criteria. Version 8.0. IUCN. 85 p.
53. Wright DB, London K (2009) Modern regression techniques using R. London: Sage. viii +204 p.
54. Wilkinson L, Task Force on Statistical Inference (1999) Statistical methods in psychology journals - guidelines and explanations. American Psychologist 54: 594–604.
55. Wang J, Zamar R, Marazzi A, Yohai V, Salibian-Barrera M, et al. (2009) Package 'robust'. CRAN.
56. Swartz SL, Taylor BL, Rugh DJ (2006) Gray whale *Eschrichtius robustus* population and stock identity. Mammal Review 1: 66–84.
57. NMFS (2005) Recovery plan for the North Atlantic right whale (*Eubalaena glacialis*). Silver SpringMD: National Marine Fisheries Service. 137 p.
58. Whitehead H, Hooker SK (2012) Uncertain status of the northern bottlenose whale Hyperoodon ampullatus: population fragmentation, legacy of whaling and current threats. Endangered Species Research 19: 47–61.
59. Jepson PD, Arbelo M, Deaville R, Patterson IAP, Castro P, et al. (2003) Gas-bubble lesions in stranded cetaceans - Was sonar responsible for a spate of whale deaths after an Atlantic military exercise? Nature 425: 575–576.
60. Lefebvre LW, Reynolds III JE, Ragen TJ, Langtimm CA, Valade JA (2009) Manatee population ecology and management workshop proceedings. In: Lefebvre LW, Reynolds III JE, Ragen TJ, Langtimm CA, Valade JA, editors; April 2002; Gainsville, Florida. U.S. Geological Survey.
61. Reeves RR, Leatherwood S, Jefferson TA, Curry BE, Henningsen T (1996) Amazonian manatees, *Trichechus inunguis*, in Peru: Distribution, exploitation, and conservation status. Interciencia 21: 246–254.
62. Silva MA, Araujo A (2001) Distribution and current status of the West African manatee (*Trichechus senegalensis*) in Guinea-Bissau. Marine Mammal Science 17: 418–424.
63. Marsh H, Penrose H, Eros C, Hughes J (2002) Dugong: status report and action plans for countries and territories. UNEP. 172 p.
64. Southwell C, Paxton CGM, Borchers D, Boveng P, De La Mare W (2008) Taking account of dependent species in management of the Southern Ocean krill fishery: estimating crabeater seal abundance off east Antarctica. Journal of Applied Ecology 45: 622–631.
65. Lake S, Burton H, Barker R, Hindell M (2008) Annual reproductive rates of Weddell seals in eastern Antarctica from 1973 to 2000. Marine Ecology Progress Series 366: 259–270.
66. Forcada J, Trathan P, Boveng PL, Boyd IL, Burns JM, et al. (2012) Responses of Antarctic pack-ice seals to environmental change and increasing krill fishing. Biological Conservation 149: 40–50.
67. Roman J, Palumbi SR (2003) Whales before whaling in the North Atlantic. Science 301: 508–510.
68. Alter SE, Rynes E, Palumbi SR (2007) DNA evidence for historic population size and past ecosystem impacts of gray whales. Proceedings of the National Academy of Sciences of the United States of America 104: 15162–15167.
69. Gerrodette T, Watters GM, Perryman W, Ballance L (2008) Estimates of 2006 Dolphin abundance in the Eastern Tropical Pacific, with revised estimates from 1986–2003. U.S. Department of Commerce, National Oceanic and Atmospheric Administration, National Marine Fisheries Service. 43 p.
70. Wade PR, Watters GM, Gerrodette T, Reilly SB (2007) Depletion of spotted and spinner dolphins in the eastern tropical Pacific: modeling hypotheses for their lack of recovery. Marine Ecology Progress Series 343: 1–14.
71. COSEWIC (2004) COSEWIC Assessment and update status report on the beluga whale *Delphinapterus leucas* in Canada. Ottawa: Committee on the Status of Endangered Wildlife in Canada. ix +71 p.
72. COSEWIC (2005) COSEWIC Assessment and update status report on the narwhal *Monodon monoceros* in Canada. Ottawa: Committee on the Status of Endangered Wildlife in Canada. vii +50 p.
73. COSEWIC (2008) COSEWIC Assessment and update status report on the polar bear *Ursus maritimus* in Canada. Ottawa: Committee on the Status of Endangered Wildlife in Canada. vii +75 p.
74. COSEWIC (2006) COSEWIC Assessment and update status report on the Atlantic walrus *Odobenus rosmarus rosmarus* in Canada. Ottawa: Committee on the Status of Endangered Wildlife in Canada. viii +65 p.
75. Kirkman SP, Lavigne DM (2010) Assessing the hunting practices of Namibia's commercial seal hunt South African Journal of Science 106: 1–3.
76. Pyenson ND, Lindberg (2011) What happened to gray whales during the Pleistocene? The ecological impact of sea-level change on benthic feeding areas in the North Pacific Ocean. PLoS One 6: e21295.
77. Worm B, Hilborn R, Baum JK, Branch TA, Collie JS, et al. (2009) Rebuilding global fisheries. Science 325: 578–585.
78. Hutchings JA, Minto C, Ricard D, Baum JK, Jensen OP (2010) Trends in the abundance of marine fishes. Canadian Journal of Fisheries and Aquatic Sciences 67: 1205–1210.
79. Reynolds JD, Dulvy NK, Goodwin NB, Hutchings JA (2005) Biology of extinction risk in marine fishes. Proceedings of the Royal Society B: Biological Sciences 272: 2337–2344.
80. Kaschner K, Jewell R, Williams R, Harris CM (2012) Global coverage of cetacean line-transect surveys: status quo, gaps and future challenges. PLoS One 7: e44075.
81. Dale A, Armitage D (2011) Marine mammal co-management in Canada's Arctic: Knowledge co-production for learning and adaptive capacity. Marine Policy 35: 440–449.
82. Clapham P, Young SB, Brownell RL (1999) Baleen whales: conservation issues and the status of the most endangered populations. Mammal Reviews 29: 35–60.
83. Bhargava M (2005) Of otters and orcas: marine mammals and legal regimes in the North Pacific. Ecology Law Quarterly 32: 939–988.
84. Lavigne DM, Scheffer VB, Kellert SR (1999) The Evolution of North American attitudes toward marine mammals. In: J. R. Twiss Jr., Reeves RR, editors. Conservation and management of marine mammals. Washington and London: Smithsonian Institution Press. 10–47.
85. Whitehead H, Rendell L, Osborne RW, Wursig B (2004) Culture and conservation of non-humans with reference to whales and dolphins: review and new directions. Biological Conservation 120: 427–437.
86. Wade PR, Reeves RR (unpublished) Social and behavioural factors in cetacean responses to over-exploitation: Are odontocetes less 'resilient' than mysticetes?
87. Hutchings JA (2001) Influence of population decline, fishing, and spawner variability on the recovery of marine fishes. Journal of Fish Biology 59: 306–322.
88. Sethi SA, Branch TA, Watson R (2010) Global fishery development patterns are driven by profit but not trophic level. Proceedings of the National Academy of Sciences of the United States of America 107: 12163–12167.
89. Hutchings JA, Reynolds JD (2004) Marine fish population collapses: Consequences for recovery and extinction risk. Bioscience 54: 297–309.
90. Boyd IL (2002) Antarctic marine mammals. In: Perrin WF, Wursig B, Thewissen JGM, editors. Encyclopedia of Marine Mammals. San Diego: Academic Press.
91. Hoezel AR, editor (2002) Marine mammal biology: an evolutionary approach. Oxford: Blackwell Publishing.
92. Whitehead H, Mann J (2000) Female reproductive strategies of cetaceans: Life histories and calf care. In: Mann J, Connor RC, Tyack PL, Whitehead H, editors. Cetacean Societies. Chicago and London: University of Chicago Press. 219–246.
93. Brown RF, Wright BE, Riemer SD, Laake J (2005) Trends in abundance and current status of harbor seals in Oregon: 1977–2003. Marine Mammal Science 21: 657–670.
94. Perry SL, DeMaster DP, Silber GK (1999) Special issue: The great whales: History and status of six species listed as endangered under the U.S. Endangered Species Act of 1973. Marine Fisheries Review 61: 1–74.
95. David J, van Sittert L (2008) A reconstruction of the Cape (South African) fur seal harvest 1653–1899 and a comparison with the 20th-century harvest. South African Journal of Science 104: 107–110.
96. Jeffries S, Huber H, Calambokidis J, Laake J (2003) Trends and status of harbor seals in Washington state: 1978–1999. Journal of Wildlife Management 67: 207–218.

4

The Sun, Moon, Wind, and Biological Imperative–Shaping Contrasting Wintertime Migration and Foraging Strategies of Adult Male and Female Northern Fur Seals (*Callorhinus ursinus*)

Jeremy T. Sterling[1]*, Alan M. Springer[2], Sara J. Iverson[3], Shawn P. Johnson[4], Noel A. Pelland[5], Devin S. Johnson[1], Mary-Anne Lea[6], Nicholas A. Bond[7]

1 National Marine Mammal Laboratory, Alaska Fisheries Science Center, National Marine Fisheries Service, NOAA, Seattle, WA, United States of America, **2** Institute of Marine Science, University of Alaska Fairbanks, Fairbanks, AK, United States of America, **3** Department of Biology, Dalhousie University, Halifax, Nova Scotia, Canada, **4** Johnson Veterinary Service, The Marine Mammal Center, Sausalito, CA, United States of America, **5** School of Oceanography, University of Washington, Seattle, WA, United States of America, **6** Institute for Marine and Antarctic Studies, University of Tasmania, Hobart, Australia, **7** University of Washington/JISAO, Seattle, WA, United States of America

Abstract

Adult male and female northern fur seals (*Callorhinus ursinus*) are sexually segregated in different regions of the North Pacific Ocean and Bering Sea during their winter migration. Explanations for this involve interplay between physiology, predator-prey dynamics, and ecosystem characteristics, however possible mechanisms lack empirical support. To investigate factors influencing the winter ecology of both sexes, we deployed five satellite-linked conductivity, temperature, and depth data loggers on adult males, and six satellite-linked depth data loggers and four satellite transmitters on adult females from St. Paul Island (Bering Sea, Alaska, USA) in October 2009. Males and females migrated to different regions of the North Pacific Ocean: males wintered in the Bering Sea and northern North Pacific Ocean, while females migrated to the Gulf of Alaska and California Current. Horizontal and vertical movement behaviors of both sexes were influenced by wind speed, season, light (sun and moon), and the ecosystem they occupied, although the expression of the behaviors differed between sexes. Male dive depths were aligned with the depth of the mixed layer during daylight periods and we suspect this was the case for females upon their arrival to the California Current. We suggest that females, because of their smaller size and physiological limitations, must avoid severe winters typical of the northern North Pacific Ocean and Bering Sea and migrate long distances to areas of more benign environmental conditions and where prey is shallower and more accessible. In contrast, males can better tolerate often extreme winter ocean conditions and exploit prey at depth because of their greater size and physiological capabilities. We believe these contrasting winter behaviors 1) are a consequence of evolutionary selection for large size in males, important to the acquisition and defense of territories against rivals during the breeding season, and 2) ease environmental/physiological constraints imposed on smaller females.

Editor: Andreas Fahlman, Texas A&M University-Corpus Christi, United States of America

Funding: Principal financial support for this study was provided by NOAA's Undersea Research Program (http://www.nurp.noaa.gov). Additional crucial support was provided by the National Marine Fisheries Service (http://www.nmfs.noaa.gov), Alaska Sea Life Center (http://www.alaskasealife.org), and the TEMPEST project funded by NOAA's National Cooperative Research Program (http://www.st.nmfs.noaa.gov/cooperative-research/index). The funders had no role in study design, data collection and analysis, decision to publish, or preparation of the manuscript.

Competing Interests: Shawn Johnson was an employee of Johnson Veterinary Service at the time of the study. There are no patents, products in development or marked products to declare. The authors have declared that no competing interests exist and this does not alter our adherence to all the PLOS ONE policies on sharing data and materials.

* E-mail: Jeremy.Sterling@noaa.gov

Introduction

Body size commonly affects animal behavior and, in many pinniped species, sexual dimorphism results in habitat segregation and reduced intraspecific competition for resources [1–4]. Sexual selection or ecological divergence or both may explain the differences in size between males and females [3]. Larger size generally requires greater resources to meet energy requirements, but in pinnipeds it also allows for a greater capacity to dive longer, exploit resources at deeper depths, and endure harsher environmental conditions [5–7]. However, the necessity to dive deeper for food depends on the depth of preferred prey, which is affected by prey behaviors and a variety of environmental factors such as bathymetric and oceanographic gradients, high wind speeds due to winter storms, lunar periodicity, the day-night cycle, and seasonality in production [8–21].

While size predicts aerobic capacity and dive performance across pinniped species, factors affecting local foraging conditions also can cause inter- and intraspecific differences in diving and movement behavior [7,22–24]. For example, Antarctic and New Zealand fur seals (*Arctocephalus gazella* and *A. forsteri*) and northern and southern elephant seals (*Mirounga angustirostris* and *M. leonina*) show differences between sexes in the location and depth of

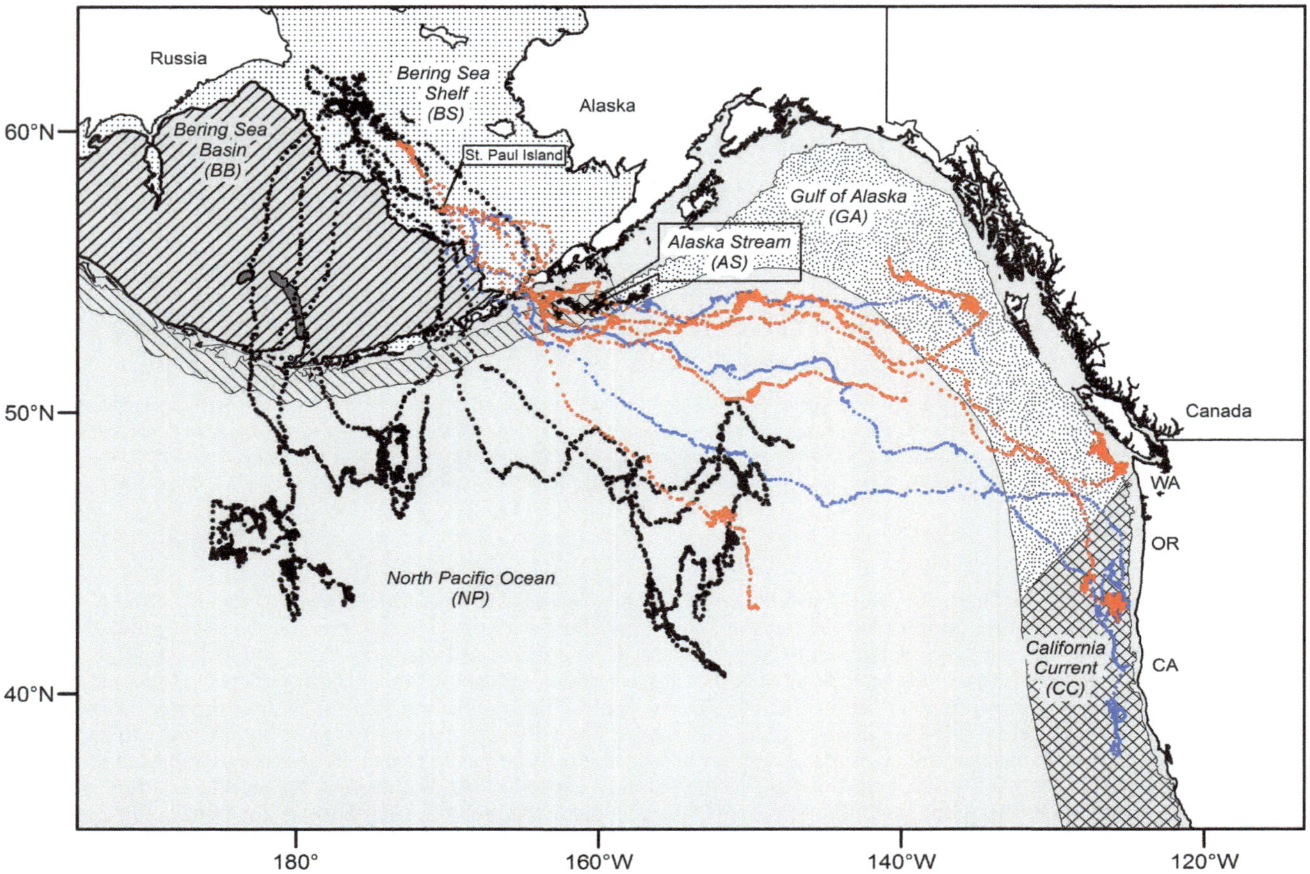

Figure 1. Adult male and female northern fur seal migratory routes. Map shows Large Marine Ecosystem delineations and all 6 h location estimates for five adult males (the black dots), six adult females with diving data (the red dots), and four adult females without diving data (the blue dots).

foraging [1,2,22,25]. Adult males of both species of fur seals are much larger, dive deeper, and forage in different habitats than their female counterparts [1,2,26]. However, in both northern and southern elephant seals, body mass predicts mean dive durations, but not mean dive depths. And, contrary to predictions based on the size differential between males and females, northern elephant seal females in some cases display longer mean dive durations than the much larger adult males [23,24]. These characteristics are explained by their foraging behavior along migration routes–adult male northern elephant seals migrate to and along the continental margins of western North America and undertake benthic dives to similar depths as females foraging farther offshore, which, in contrast, access the deep scattering layer while in pelagic waters [23,25]. Generally for both fur seals and elephant seals, larger size allows for longer, and in some cases, deeper dives; however, the interplay between predator, prey, and habitat varies and, as noted by Staniland and Robinson [2], the manner in which their behavior is expressed (i.e., dive durations, depths, and foraging routes) is dependent on the local environment and its effect on prey resources.

In polygynous species like elephant seals and fur seals, the large size of males confers advantages in fasting and holding territories to defend females against rivals during the breeding season, and thus strengthens males' reproductive potential [11]. Adult male northern fur seals (NFS; *Callorhinus ursinus*) may weigh as much as 350 kg, much more than adult females (30–50 kg), and have not been observed migrating into the same regions of the North Pacific Ocean [11,16,27–32]. Generally, the assumption has been that contrasting physiological and dive capabilities, due to sexual size dimorphism, combined with differences in prey distribution and abundance among ecosystems, lead to the divergent wintering areas. However, there has been limited empirical evidence to support this assumption.

Given that variability in seal movement and diving can be species, size, and habitat specific, acquiring spatial and temporal biological and physical measurements relevant at individual and group scales is crucial to aid our understanding of broader scale observations of intraspecific, interannual, and decadal variability in behavior, and ultimately of population dynamics. In this study, we examined the migratory behavior of adult female and male NFS from St. Paul Island (Bering Sea, Alaska, USA), where overall abundance has declined by 70% in the past few decades for uncertain reasons [33]. Previous studies summarized winter distribution and dive behavior for females [16] and winter distribution only for males [32]. Those studies were not conducted simultaneously, leaving open the possibility of interannual physical and/or biological oceanographic variability as an explanation for the contrasting winter behavior between sexes. We aimed to confirm and explain the apparent differences in migration patterns by employing remotely sensed physical environmental fields and data collected by animal-borne sensors integrated into satellite transmitters. Based on the previous studies of NFS and other pinnipeds, we predicted that adult males and females would migrate to different regions of the North Pacific Ocean. We

Table 1. Summary of capture information, satellite transmission durations, and Aleutian pass usage of adult male and female northern fur seals.

ID	Sex	Capture site on St. Paul Is.	Satellite transmitter	Dive data	Mass (kg)	Capture date	Migration departure date	Last day of transmission	Days post capture to migration	Total migrating days	Aleutian passes traveled through
662	F	Polovina Cliffs	SPLASH	x	33.6	2-Oct-09	11-Nov-09	13-Mar-10	40.5	122.0	Unimak Pass
665	F	Polovina Cliffs	KiwiSat 202		37.8	5-Oct-09	23-Nov-09	23-Jan-10	49.5	60.5	Unimak Pass
670	F	Polovina Cliffs	SPLASH	x	35.6	2-Oct-09	5-Nov-09	23-Feb-10	34.5	110.0	Unimak Pass
671	F	Polovina Cliffs	SPLASH	x	44.0	3-Oct-09	12-Nov-09	22-Dec-09	40.3	40.0	Unimak Pass
672	F	Polovina Cliffs	SPLASH	x	28.4	3-Oct-09	14-Nov-09	28-Dec-09	42.0	44.3	Unimak Pass
673	F	Polovina Cliffs	SPLASH	x	32.4	3-Oct-09	30-Oct-09	12-Dec-09	27.8	42.5	Unimak Pass
674	F	Polovina Cliffs	SPLASH	x	45.4	3-Oct-09	13-Nov-09	4-Mar-10	41.5	110.5	Unimak Pass
675	F	Polovina Cliffs	KiwiSat 202		27.0	5-Oct-09	14-Nov-09	15-Nov-09	40.0	1.0	n/a
676	F	Polovina Cliffs	KiwiSat 202		38.0	5-Oct-09	12-Nov-09	11-Mar-10	38.0	119.0	Akutan Pass
677	F	Polovina Cliffs	KiwiSat 202		37.6	7-Oct-09	15-Oct-09	3-Mar-10	8.0	139.0	Unimak Pass
678	M	Zapadni Sands	SRDL	x	149.1	22-Oct-09	25-Oct-09	31-Jan-10	3.5	98.0	Unimak Pass
679	M	Zapadni Sands	SRDL	x	152.9	22-Oct-09	8-Nov-09	22-May-10	17.5	194.9	Buldir – Kiska Pass, Amukta Pass, Tanaga Pass
680	M	Zapadni Sands	SRDL	x	114.2	23-Oct-09	25-Oct-09	9-Apr-10	2.8	166.0	Amchitka Pass
681	M	Reef	SRDL	x	118.7	23-Oct-09	6-Nov-09	25-Dec-09	14.3	49.0	Yunaska Is. Pass
682	M	Zapadni Sands	SRDL	x	221.8	24-Oct-09	11-Nov-09	19-May-10	18.5	189.4	Samalga Pass

F = female; M = male.

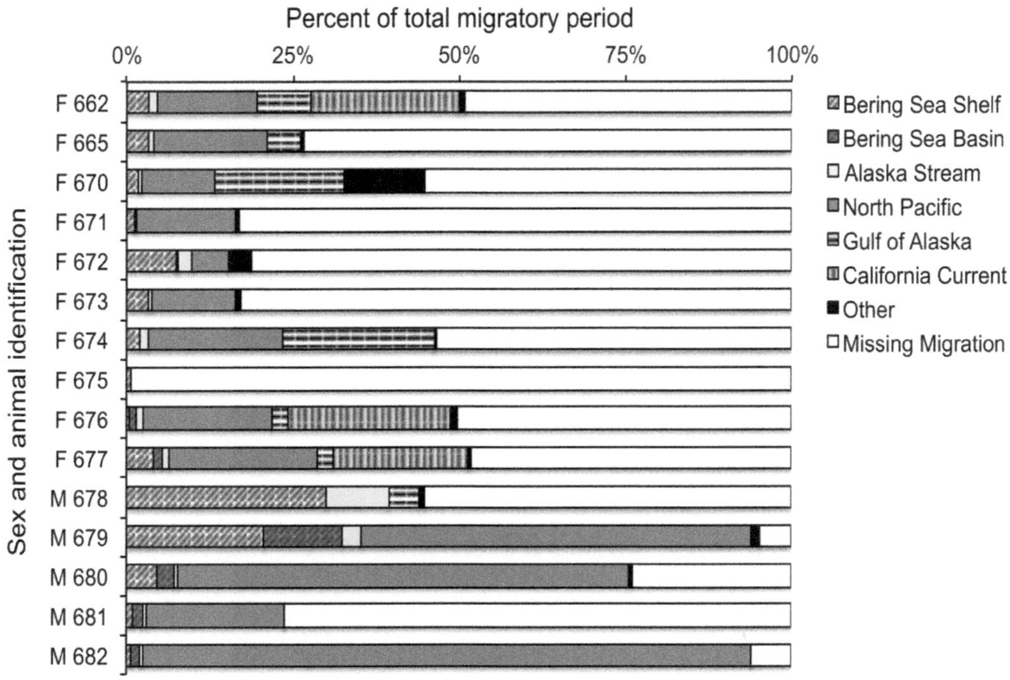

Figure 2. Percent of the estimated total adult male and female northern fur seal migratory period spent in each ecosystem. To make this calculation, we assumed return dates of 1 June for males (M) and 10 July for females (F).

hypothesized that size dimorphism and local environmental conditions would explain intraspecific differences in diving and movement patterns and migration routes. We compared movement and diving parameters between the sexes to determine whether size predicts behavior and whether behavior is modulated by the habitat variables light (sun and moon), wind speed, and mixed-layer depth within and between ecosystems of the North Pacific Ocean and Bering Sea. As explained below, each of these variables has been shown to affect the movement and foraging behavior of fur seals and other pinnipeds and the geographic and diel vertical distribution of prey fields.

Methods

Ethics Statement

All work was conducted in accordance with and under the authority of the United States Marine Mammal Protection Act (National Marine Fisheries Service, NMFS Permits 782–1708 and 14328). The Marine Mammal Protection Act was established in 1972 requiring all research conducted on marine mammals in the United States be done under the authority of federal permits issued by either NMFS or U.S. Fish and Wildlife Service (USFWS). All applications for a permit to conduct research on marine mammals have gone through a four-stage review process that includes: 1) agency review (either NMFS or USFWS); 2) a public notice and review period; 3) review and recommendation from the Scientific Advisers to the U.S. Marine Mammal Commission; and 4) a final action by the reviewing agency. All capture and handling activities described in this manuscript have gone through and been approved by this process. Additionally, the University of Alaska Fairbanks Institutional Animal Care and Use Committee approved the capture and handling methods used on adult male NFS (Protocol 09–50). However, at the time when adult female NFS were captured by NMFS there was no additional requirement for review of these procedures by an institutional review board or ethics committee. In 2010, a NMFS Institutional Animal Care and Use Committee was established for the Alaska Fisheries and Northwest Fisheries science centers and the capture and handling protocols described here were reviewed and approved by this committee.

Device Deployment

We captured ten adult female and five adult male NFS on St. Paul Island (57.2° N, −172.2°W; Fig. 1) during October 2009 (Table 1). Females were captured, physically restrained, administered gas anesthesia (isoflurane and oxygen), weighed, and an Argos transmitting KiwiSat 202 satellite transmitter (Sirtrack LTD, Havelock North, New Zealand) or a "SPLASH" satellite-dive recorder (Wildlife Computers, Redmond, Washington, USA) was glued to the fur above the shoulder blades using 5-minute epoxy. Males were immobilized with an intramuscular injection (IM) using a jab stick or dart containing the drug combination Telazol (0.77–1.14 mg kg^{-1}) and Medetomidine (0.03–0.04 mg kg^{-1}). Isoflurane and oxygen were administered via mask in four males, and with an endotracheal intubation in one male, to ensure adequate procedural time to weigh and attach satellite-linked conductivity, temperature, depth satellite-relayed data loggers (SRDLs) developed by the Sea Mammal Research Unit at University of St. Andrews, Scotland. Following these procedures a drug reversal combination of Atipamezole (0.14–0.26 mg kg^{-1}) and Flumazenil (0.002–0.005 mg kg^{-1}) was given IM and each seal was monitored to complete recovery.

All three satellite-transmitter types provided daily location estimates, while the two types of dive recorders (SPLASH and SRDL) stored and transmitted summary dive information. SPLASH satellite-dive recorders sampled depth every second and stored all dives ≥2 m (dive depth and duration) in 6 h blocks as histogram distributions starting at 0000, 0600, 1200, 1800 UTC [12]. Fourteen histogram dive depth bins (2, 4, 6, 10, 20, 34, 50,

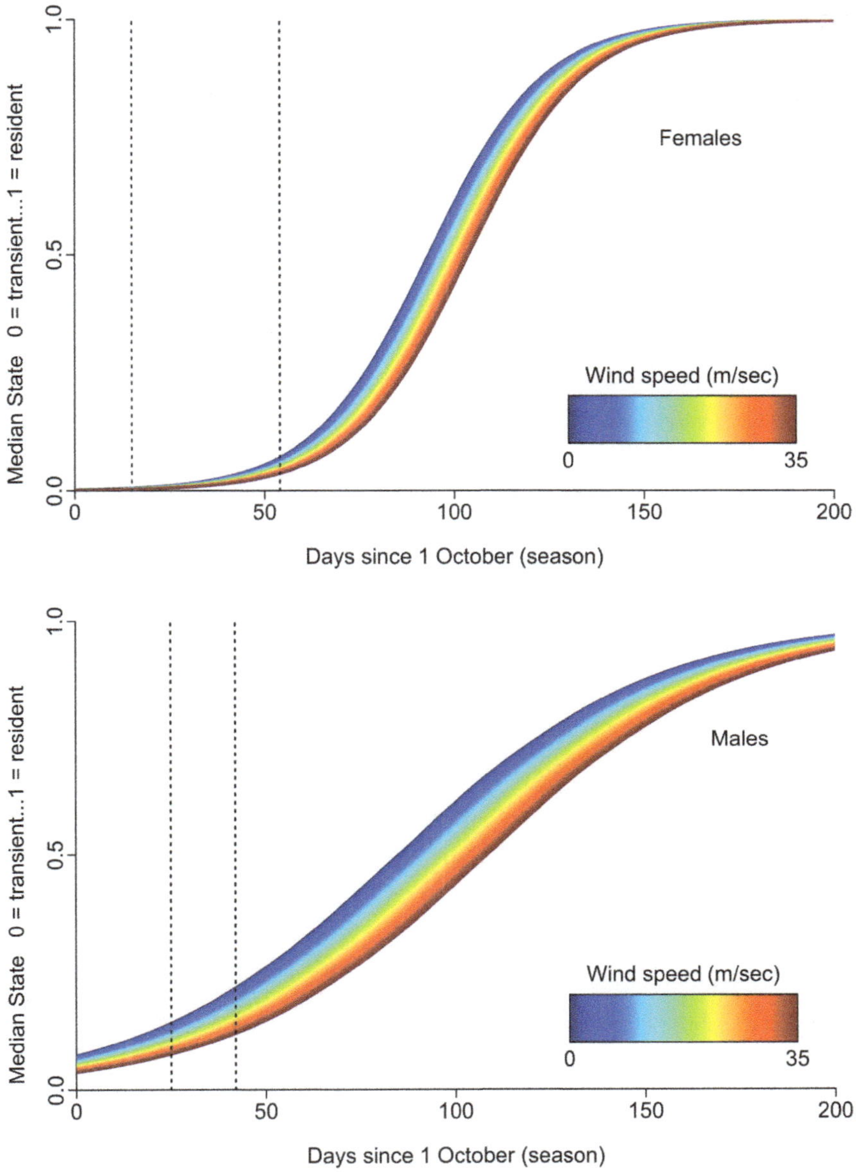

Figure 3. The effect of wind and season on adult male and female northern fur seal behavioral states. These figures were constructed using the linear mixed-effects model coefficients from Table 3 depicting days since 1 October (season) and wind speed. The vertical dashed lines indicate the first and last migration departure date of the seals in this study.

74, 100, 124, 150, 174, 200, >200 m) and dive duration bins (15, 30, 60, 90, 120, 150, 180, 210, 240, 300, 330, 360, >360 s) were defined prior to tag deployment. Each dive depth represented the maximum depth recorded during a dive, while each dive duration represented the total time of the entire dive, beginning and ending at 2 m depth. Maximum dive depth and duration were recorded and then stored in the corresponding predefined histogram bin. SRDLs sampled depth every second and recorded dives that were >6 m for at least 8 seconds. Like the SPLASH recorders, dive summaries were stored every 6 h at 0000, 0600, 1200, and 1800 UTC. However, unlike SPLASH recorders, SRDLs computed onboard and transmitted the average and standard deviation of dive depths, maximum of the dive depths, average and standard deviation of the dive durations, maximum of the dive durations, and the total number of dives for each 6 h time block.

Dive Data Processing and Analyses

Satellite-Relayed data loggers (Males). Satellite-relayed data loggers provided several measures of dive and environmental characteristics. We used average dive depth and dive duration, maximum dive depth and dive duration, and the number of dives for each 6 h block, as well as all conductivity, temperature, and depth (CTD) profiles. The CTD profiles were calculated from the deepest dive in each 6 h block as long as the dive exceeded ~4 m. Once the deepest dive was detected, the SRDL switched into sampling mode on the ascent portion of the dive by sampling temperature, conductivity, and depth at 1 Hz until the seal reached the surfaced. Several quality control measures were applied to the profile before 17 representative depth points (8 predefined, 7 inflection, minimum, and maximum) were selected for transmission [34]. Our goal for the CTD data was to determine the mixed-layer depth (MLD) for each 6 h block. To do

Table 2. Top three linear mixed-effects models selected using Akaike Information Criterion (AIC) for each adult male and female northern fur seal response variable examined.

GLMM model	K	AIC	ΔAIC
(1) Behavioral state (n = 5946)			
(a) WS+sex+S+LME (NP, AS, BB, BS, CC, GA)+sex*S	13	16449.15	
(b) WS+sex+S+LME (NP, AS, BB, BS, CC, GA) + S*LME (NP, AS, BB, BS, CC, GA)	17	16451.39	2.24
(c) WS+sex+S+sex*S	8	16451.88	2.73
(2) F average dive depth (n = 893)			
(a) D+M+LME (NP, AS, BS, CC, GA)+D*M+LME (NP, AS, BS, CC, GA)*M + LME (NP, AS, BS, CC, GA)*D	19	1496.00	
(b) D+M+LME (NP, AS, BS, CC, GA)+D*M+LME (NP, AS, BS, CC, GA)*M	15	1498.30	2.30
(c) D+M+S+LME (NP, AS, BS, CC, GA)+D*M	16	1500.07	4.07
(3) M average dive depth (n = 1473)			
(a) D+M +S+LME (NP, AS, BB, BS)+D*MLD+M*D+M*LME (NP, AS, BB, BS)	15	2508.14	
(b) D+S+LME (NP, AS, BB, BS)+D*MLD+M*LME (NP, AS, BB, BS)	14	2508.63	0.49
(c) D+M+LME (NP, AS, BB, BS)+D*MLD+M*D+M*LME (NP, AS, BB, BS)	14	2510.79	2.65
(4) F number of dives (n = 893)			
(a) D+LME (NP, AS, BS, CC, GA)+LME (NP, AS, BS, CC, GA)*D	13	2862.60	
(b) D+LME (NP, AS, BS, CC, GA)+LME (NP, AS, BS, CC, GA)*D+D*M	14	2864.58	1.98
(c) D+M+LME (NP, AS, BS, CC, GA)+LME (NP, AS, BS, CC, GA)*D+D*M	15	2866.48	3.88
(5) M number of dives (n = 1473)			
(a) D+LME (NP, AS, BB, BS)+LME (NP, AS, BB, BS)*D+D*MLD	12	4299.75	
(b) D+M+LME (NP, AS, BB, BS)+LME (NP, AS, BB, BS)*D+D*MLD	13	4301.13	1.38
(c) D+M+LME (NP, AS, BB, BS)+LME (NP, AS, BB, BS)*D+D*MLD+D*M	14	4303.12	3.37

F = female; M = male; K = number of parameters; NP = North Pacific Ocean; AS = Alaska Stream; BB = Bering Sea Basin; BS = Bering Sea Shelf; CC = California Current; GA = Gulf of Alaska; WS = wind speed (m sec^{-1}); S = season or days from 1 October; D = proportion of daylight in each 6 h period; M = fraction of the moon illuminated; MLD = mixed-layer depth.

so, we employed the algorithm described by Kara et al. [35], which uses a seasonally-optimized density difference criterion as a practical definition of the thickness of the well-mixed layer near the ocean surface. We chose this algorithm because it was calibrated using climatological data obtained from CTD casts along Canada's Department of Fisheries and Oceans "Line P", in the northeast Pacific, in a region proximate to the winter foraging grounds of NFS [36,37]. Starting from just below the surface, the Kara et al. [35] algorithm scans successively deeper samples from a given cast to find the vertical extent of the surface mixed layer of uniform density (σ_t). The MLD was then defined as the depth at which density rose by a specified increment ($\Delta\sigma_t$) above the density at the deepest sample in the mixed layer.

To identify the mixed layer for a given dive, the algorithm searched from 10 m for the first observed depth interval between samples n and n+1 (described as depth h_n to depth h_{n+1}) in which

Table 3. Best linear mixed-effects model results and estimated coefficients for the effects on adult male and female northern fur seal behavioral state.

Behavioral state	Estimate	*SE*	*df*	*t*	*P*
Intercept	−6.15	1.09	5923	−5.63	**<0.0001**
WS	−0.02	0.00	5923	−5.88	**<0.0001**
sex (male)	3.60	1.46	13	2.47	**0.0284**
S	0.07	0.01	5923	5.47	**<0.0001**
AS	−0.71	0.23	5923	−3.07	**0.0021**
BB	−0.74	0.39	5923	−1.91	0.0561
BS	−0.91	0.34	5923	−2.67	**0.0075**
CC	−0.76	0.59	5923	−1.29	0.1966
GA	−0.41	0.28	5923	−1.45	0.1463
sex*S	−0.04	0.01	5923	−3.04	**0.0024**

The reference levels are females and North Pacific Ocean. *P*-values in boldface are significant at $P \leq 0.05$. AS = Alaska Stream; BB = Bering Sea Basin; BS = Bering Sea Shelf; CC = California Current; GA = Gulf of Alaska; WS = wind speed (m sec^{-1}); S = season or days from 1 October.

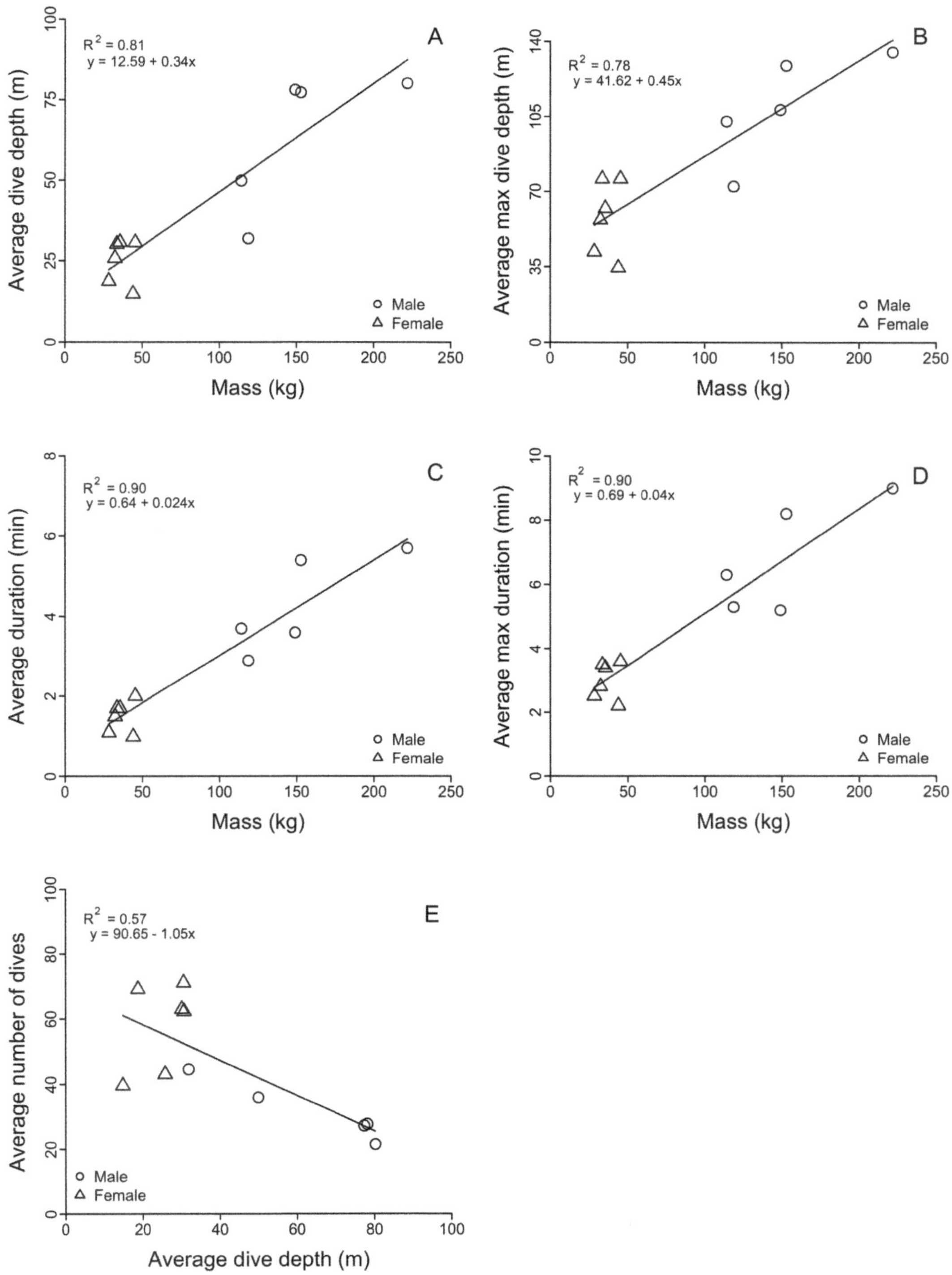

Figure 4. Relationships between adult male and female northern fur seal mass and dive behaviors. Significant linear regression relationships were found between fur seal mass and their average dive depth (A), average maximum dive depth (B), average dive duration (C), and average maximum dive duration (D). The relationship between average fur seal dive depths and the average number of dives (E) was also significant.

the density increase between the two samples was greater than 10% of $\Delta\sigma_t$. The density increment $\Delta\sigma_t$ was defined as the density increase that would result from reducing the temperature at h_n by the seasonally-prescribed value ΔT. In keeping with the values found by Kara et al. [35] to produce the best results along Line P, we used $\Delta T = 1.0°C$ in winter (January-March), $\Delta T = 0.8°C$ in spring (April-June), $\Delta T = 0.2°C$ in summer (July-September), and $\Delta T = 0.8°C$ in fall (October-December). Once a pair of samples meeting the above criteria was found, the density at h_n was then defined as the reference density ($\sigma_{t,\ ref}$) and the depth at which observed density exceeded $\sigma_{t,\ ref}+\Delta\sigma_t$ was the MLD.

Additionally, we employed various quality control measures to eliminate dives with erroneous density spikes or other outliers. Since the Kara et al. [35] algorithm assumes a minimum MLD of

Table 4. Average dive depth and the percentage of dives of adult male and female northern fur seals summarized for each Large Marine Ecosystem (LME) and grouped by daylight periods.

		Mean dive depth in meters ± SD				Percentage of dives (number of dives)				
LME	sex	day	>50% day	>50% night	night	day	>50% day	>50% night	night	Total Dives
BB	F									0
	M	117.4±54.0	115.7±24.1		39.6±40.8	62% (966)	5% (84)	0% (0)	32% (503)	1,553
BS	F	42.4±20.6	24.6±15.8		27.1±23.4	9% (205)	17% (378)	0% (8)	74% (1,679)	2,270
	M	78±45.0	74.8±50.0		62.4±36.7	19% (1,240)	9% (583)	0% (0)	73% (4,874)	6,697
NP	F	40.3±26.0	27.5±22.2	22.1±17.3	23.5±23.0	2% (584)	18% (4,749)	32% (8,254)	47% (12,101)	25,688
	M	90±40.2	64.6±32.6	46.7±30.7	24.8±18.3	63% (24,436)	12% (4,781)	13% (5,179)	12% (4,696)	39,092
AS	F	21.1±16.4	18.9±8.5		15.6±6.6	4% (170)	17% (697)	0% (0)	79% (3,205)	4,072
	M	158.7±72.0	169.3±58.2		31.9±33.7	51% (711)	32% (450)	0% (0)	17% (231)	1,392
CC	F	37.1±19.5	28.5±13.1	25.3±13.4	22.7±10.6	33% (1,323)	46% (1,857)	10% (391)	11% (437)	4,008
	M									0
GA	F	33.1±41.8	31.4±15.4	39.6±22.9	29.7±22.5	0% (31)	17% (3,115)	41% (7,460)	42% (7,559)	18,165
	M									0

NP = North Pacific Ocean; AS = Alaska Stream; BB = Bering Sea Basin; BS = Bering Sea Shelf; CC = California Current; GA = Gulf of Alaska; F = female; M = male.

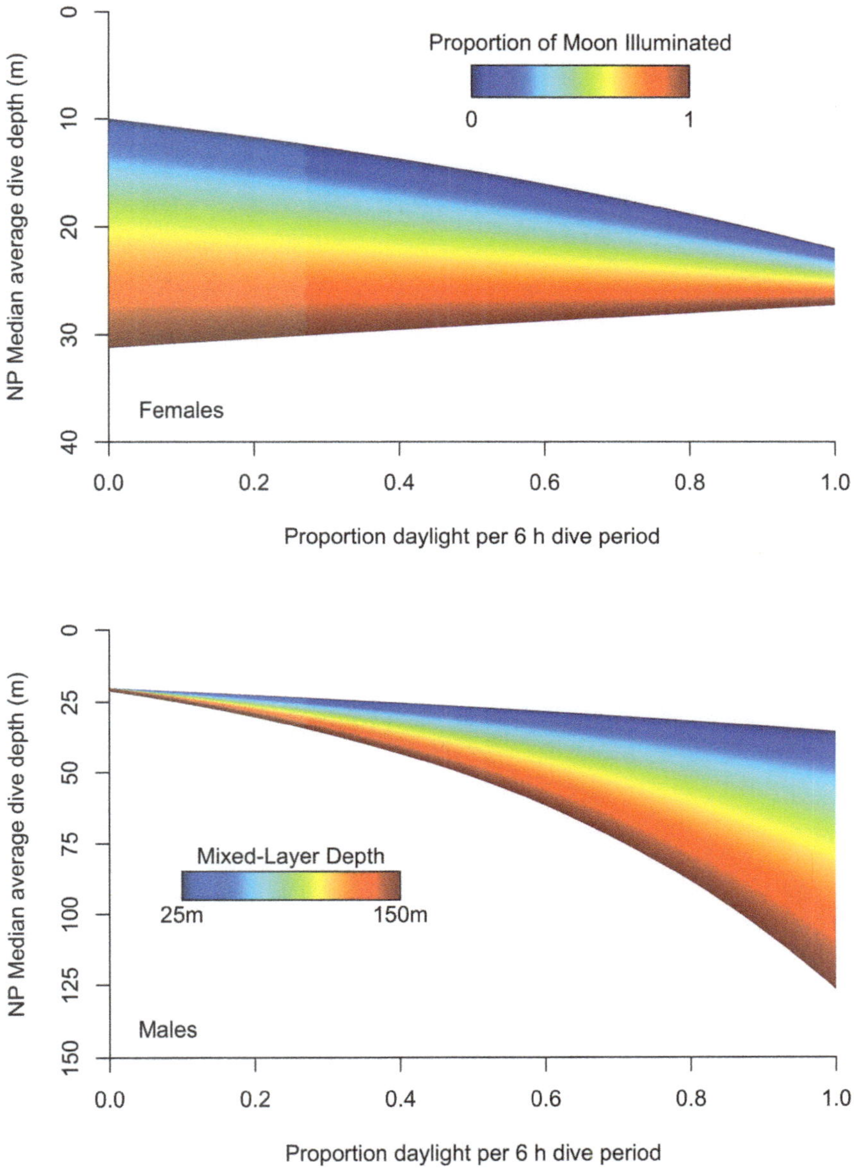

Figure 5. Dive response of adult male and female northern fur seals to light (moon and sun) and mixed-layer depth. These figures were constructed using the linear mixed-effects model coefficients from Table 5. The top figure shows the effect of moonlight and sunlight on female dive behavior while the bottom figure shows the effect of sunlight and mixed-layer depth on male dive behavior. Both plots represent adult diving in the North Pacific Ocean (NP) ecosystem.

10 m, for dives with CTD measurements whose maximum depth did not exceed 10 m, it was assumed that the maximum dive depth was shallower than the MLD. For dives that detected large density decreases or unstable regions in the top of the water column we used a secondary method in which the 10 m density was used as a reference rather than the well-mixed layer base. For dives with weak density stratification, where the algorithm could not find a MLD from the sample-pair method by the depth at which density increases by $2\Delta\sigma_t$ from the 10 m reference value, it again defaulted to the secondary method. If neither method was successful, it was presumed the maximum depth for that dive was shallower than the MLD.

SPLASH (Females). Dive data were processed using Wildlife Computers' WC-DAP program (V.2.0.29) and then filtered, quality controlled for spurious outliers, and analyzed using methods similar to those outlined by Lea et al. [12]. Duplicate histogram records were removed, only dives >6 m were analyzed in order to be consistent with male SRDL dive depth summaries, and dive durations >15 s were selected to reduce the effect of non-foraging behavior. The average dive depth and average dive duration were computed using each dive depth and dive duration histogram distribution and applying the same formula and methods described in Lea et al. [12].

Modeling Movement

Satellite tag transmission intervals were set to receive multiple daily location estimates [38]. The SRDLs were programmed to transmit all hours between noon and 10:00 UTC the next day; SPLASH tags were set to transmit between 0200–0600 and 1400–1800 UTC; and all four KiwiSat 202 s were set to transmit

Table 5. Best linear mixed-effects model results and estimated coefficients for the effects on adult male and female northern fur seal average dive depths.

Female Dive Depth	**Estimate**	*SE*	*df*	*t*	*P*
Intercept	2.314	0.092	872	25.044	**<0.0001**
D	0.782	0.115	872	6.781	**<0.0001**
M	1.122	0.124	872	9.071	**<0.0001**
AS	0.298	0.147	872	2.025	**0.0432**
BS	0.611	0.159	872	3.843	**0.0001**
CC	0.158	0.208	872	0.758	0.4487
GA	0.711	0.111	872	6.396	**<0.0001**
D*M	−0.918	0.155	872	−5.929	**<0.0001**
M*AS	−0.903	0.321	872	−2.811	**0.0051**
M*BS	−0.995	0.260	872	−3.832	**0.0001**
M*CC	−0.502	0.264	872	−1.900	0.0578
M*GA	−0.882	0.167	872	−5.273	**<0.0001**
D*AS	−0.320	0.185	872	−1.727	0.0845
D*BS	0.161	0.170	872	0.947	0.3441
D*CC	0.357	0.189	872	1.891	0.0589
D*GA	0.071	0.136	872	0.523	0.6009
Male Dive Depth	**Estimate**	*SE*	*df*	*t*	*P*
Intercept	3.031	0.156	1457	19.424	**<0.0001**
D	0.290	0.110	1457	2.635	**0.0085**
M	0.218	0.092	1457	2.369	**0.0179**
S	−0.001	0.001	1457	−2.158	**0.0311**
AS	1.374	0.182	1457	7.553	**<0.0001**
BB	0.167	0.154	1457	1.084	0.2785
BS	0.353	0.112	1457	3.149	**0.0017**
M*AS	−0.657	0.276	1457	−2.379	**0.0175**
M*BB	0.196	0.259	1457	0.759	0.4479
M*BS	0.616	0.152	1457	4.050	**0.0001**
D*M	−0.155	0.099	1457	−1.564	0.1180
D*MLD	0.010	0.001	1457	10.440	**<0.0001**

Females and males were examined separately and the reference level is the North Pacific Ocean. *P*-values in boldface are significant at $P \leq 0.05$. AS = Alaska Stream; BB = Bering Sea Basin; BS = Bering Sea Shelf; CC = California Current; GA = Gulf of Alaska; S = season or days from 1 October; D = proportion of daylight in each 6 h period; M = fraction of the moon illuminated; MLD = mixed-layer depth.

between 0030–0600, with one tag additionally transmitting between 1400–1500 and another between 1900–2350 UTC. Within these transmission intervals, Argos location estimates occur at irregularly spaced time intervals with some observation error. Our study objectives were to combine seal behavioral response variables, both horizontal movement and dive behavior, with remotely sensed fields and animal-borne CTD measurements. Seal dive behavior, animal-borne CTD measurements, and derived wind speeds and directions all summarized measurements taken in 6 h time blocks (see below). Thus, we needed to estimate locations every 6 h from the Argos location data set and quantify the seal movement behavior. To do this, we fitted a switching state-space model (SSSM) for each seal to account for the Argos observation error, create animal location estimates every 6 h, and identify movement behavior as either resident (area restricted or foraging movements) or transient (fast, somewhat linear or directed movements) [39–43]. Switching state space models were estimated with Markov Chain Monte Carlo estimation methods using WinBUGS and implemented through R statistical software and the R2WinBUGS package (http://www.r-project.org).

Predictors of Movement and Diving Behavior

Our hypothesis explaining adult male and female spatial segregation during their winter migration includes the confluence of NFS biology–the evolutionary effects of the mating system on sexual dimorphism [11]–and behavioral and environmental effects on NFS prey distribution. Specifically, we predicted deeper and longer dives for animals with greater mass, allowing the larger males to forage in regions where biophysical processes and behavioral adaptations deepen prey fields. Factors known to affect NFS potential prey include light (both sun and moon), season, and habitat [12,16,28,44], and to lesser extents the depth of the surface mixed layer [8] and winter storms [13]. Spatial and temporal alignment between NFS behavior (movement and diving) and light

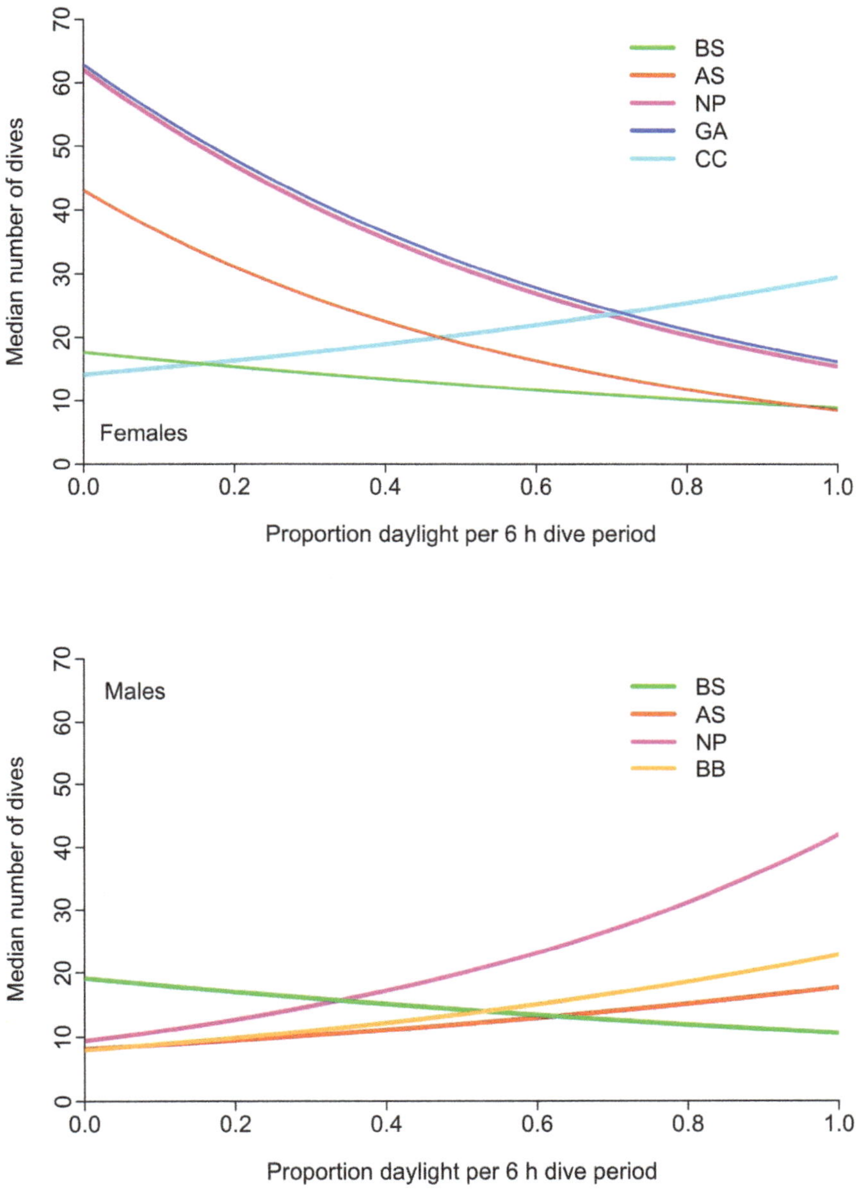

Figure 6. The effect of daylight on the number of dives of adult male and female northern fur seals. Each line represents an ecosystem: NP = North Pacific Ocean, AS = Alaska Stream, BB = Bering Sea Basin, BS = Bering Sea Shelf, CC = California Current, GA = Gulf of Alaska (see Fig. 1). In most cases, male dives (bottom) increased with increased proportion of light in each 6 h period. This response was opposite of adult females (top), which dived more during the night; a notable exception occurred in the CC ecosystem where one female exhibited an increase in the number of dives per 6 h period with increasing daylight.

(sun and moon), wind speed and direction, surface MLD, ecosystem (defined and discussed as Large Marine Ecosystems, LMEs, see below), and season was accomplished by the following methods.

Proportion daylight. The proportion of daylight in each 6 h period was calculated using custom coded functions in R that utilized the National Oceanic and Atmospheric Administration Sunrise/Sunset and Solar Position Calculators (http://www.srrb.noaa.gov/highlights/sunrise/sunrise.html). Nautical dawn and dusk (solar elevation equal to 12° below horizon) were calculated using each seal's estimated latitude and longitude at the midpoint of the 6 h block, and proportion daylight was defined as the amount of the 6 h block within daylight hours.

Lunar fraction. The lunar fraction is the illuminated area of the moon's apparent disk divided by the total area of the disk. Calculations of the lunar fractions were extracted from the United States Naval Observatory website (http://aa.usno.navy.mil/data/docs/MoonFraction.php) and temporally aligned with all 6 h periods containing summaries of dive behavior.

Wind speed and direction. National Centers for Environmental Prediction and National Center for Atmospheric Research (NCEP-NCAR) Reanalysis 2 data distributed by the NOAA/OAR/ESRL PSD, Boulder, Colorado, USA (http://www.esrl.noaa.gov/psd/) provided sub-daily analysis estimates of 10 m u (east-west) and v (north-south) wind velocities at 0000, 0600, 1200, and 1800 UTC on a 2.5° latitude × 2.5° longitude global grid. Seal location estimates from the SSSM were estimated at the same

Table 6. Best linear mixed-effects model results and estimated coefficients for the effects on adult male and female northern fur seal average number of dives.

Female number of dives	Estimate	*SE*	*df*	*t*	*P*
Intercept	4.128	0.121	878	34.057	**<0.0001**
D	−1.394	0.211	878	−6.613	**<0.0001**
AS	−0.365	0.225	878	−1.621	0.1054
BS	−1.262	0.231	878	−5.457	**<0.0001**
CC	−1.483	0.297	878	−4.998	**<0.0001**
GA	0.013	0.155	878	0.084	0.9333
D*AS	−0.230	0.413	878	−0.556	0.5783
D*BS	0.710	0.389	878	1.826	0.0683
D*CC	2.135	0.432	878	4.939	**<0.0001**
D*GA	0.036	0.323	878	0.113	0.9104
Male number of dives	**Estimate**	*SE*	*df*	*t*	*P*
Intercept	2.248	0.202	1460	11.152	**<0.0001**
D	1.498	0.204	1460	7.340	**<0.0001**
AS	−0.147	0.298	1460	−0.493	0.6222
BB	−0.173	0.198	1460	−0.874	0.3823
BS	0.700	0.125	1460	5.612	**<0.0001**
D*AS	−0.724	0.372	1460	−1.949	0.0515
D*BB	−0.446	0.273	1460	−1.630	0.1034
D*BS	−2.088	0.181	1460	−11.527	**<0.0001**
D*MLD	0.000	0.002	1460	−0.076	0.9393

Females and males were examined separately and the reference level is the North Pacific Ocean. *P*-values in boldface are significant at $P \leq 0.05$. AS = Alaska Stream; BB = Bering Sea Basin; BS = Bering Sea Shelf; CC = California Current; GA = Gulf of Alaska; D = proportion of daylight in each 6 h period; MLD = mixed-layer depth.

sub-daily time intervals and spatially and temporally aligned with *u* and *v* velocities. The *u* and *v* velocities were then converted to wind speed (m s^{-1}) and wind direction. The validity of atmospheric reanalyses for characterizing the winds of the North Pacific has been previously assessed. Using wind data from moored buoys that were not available for assimilation in the NCEP-NCAR Reanalysis, [45] found complex correlation coefficients between the measured and synthetic winds of about 0.9, with minimal systematic biases.

Mixed-Layer depth (Males). Mixed-layer depth calculations (see SRDL above) were temporally aligned with each seal location estimate. If more than one CTD profile and subsequent MLD calculation temporally corresponded to a single 6 h dive summary, then the average MLD was computed. Seventy-three percent of the 6 h dive summary bins had a corresponding MLD calculation. For some of the 6 h dive summaries without MLD calculations it was appropriate to estimate its depth by linear interpolation, particularly if the time gap was small and/or the variability in the MLD was low prior to and after the missing MLD calculations. This was a subjective process that increased paired MLD calculations and 6 h dive summaries to 86% of the total number of 6 h dive summaries.

Oceanic habitats. During their seasonal migration, NFS traverse marine ecosystems characterized by distinct hydrographic and topographic domains, productivity, and upper trophic level assemblages [16,28,30,46]. We determined the physical extent and boundaries of these marine ecosystems by modifying LMEs of the World map (http://www.lme.noaa.gov) [47]. Modifications incorporated continental shelf (≤1000 m) and oceanic ecosystems, including the addition of an Alaska Stream ecosystem, adjusting the East Bering Sea LME to lie north of the Aleutian Islands while further distinguishing Bering Sea shelf (≤1000 m) and basin (> 1000 m) ecosystems, and adding the North Pacific Ocean ecosystem (Fig. 1). The LMEs we used in the statistical analyses were the Bering Sea Shelf (BS), Bering Sea Basin (BB), Alaska Stream (AS), North Pacific Ocean (NP), Gulf of Alaska (GA), and California Current (CC). References to the Bering Sea refer to both the BB and BS ecosystems.

Season. Days since 1 October (referred to as season) were used to estimate any temporal signal in the data for each 6 h estimated location.

Statistical Analyses

Mass and dive behavior. Linear models were used to investigate the relationship between NFS mass and dive behavior. Mass was used as a predictor variable for both male and female average dive depths and durations, as well as the average maximum dive depths and durations. In addition, we investigated the relationship between the average numbers of dives with respect to male and female average dive depths.

Movement and dive behavior. Biophysical processes responsible for shaping the distribution, depth, and density of NFS preferred prey fields may also influence the horizontal movement and vertical diving behavior of individuals. Northern fur seal movement and diving behavior were simplified into three response variables – SSSM estimated movement behavioral state (hereinafter referred to as behavioral state), average dive depth every 6 h, and the total number of dives every 6 h. The SSSM provided an

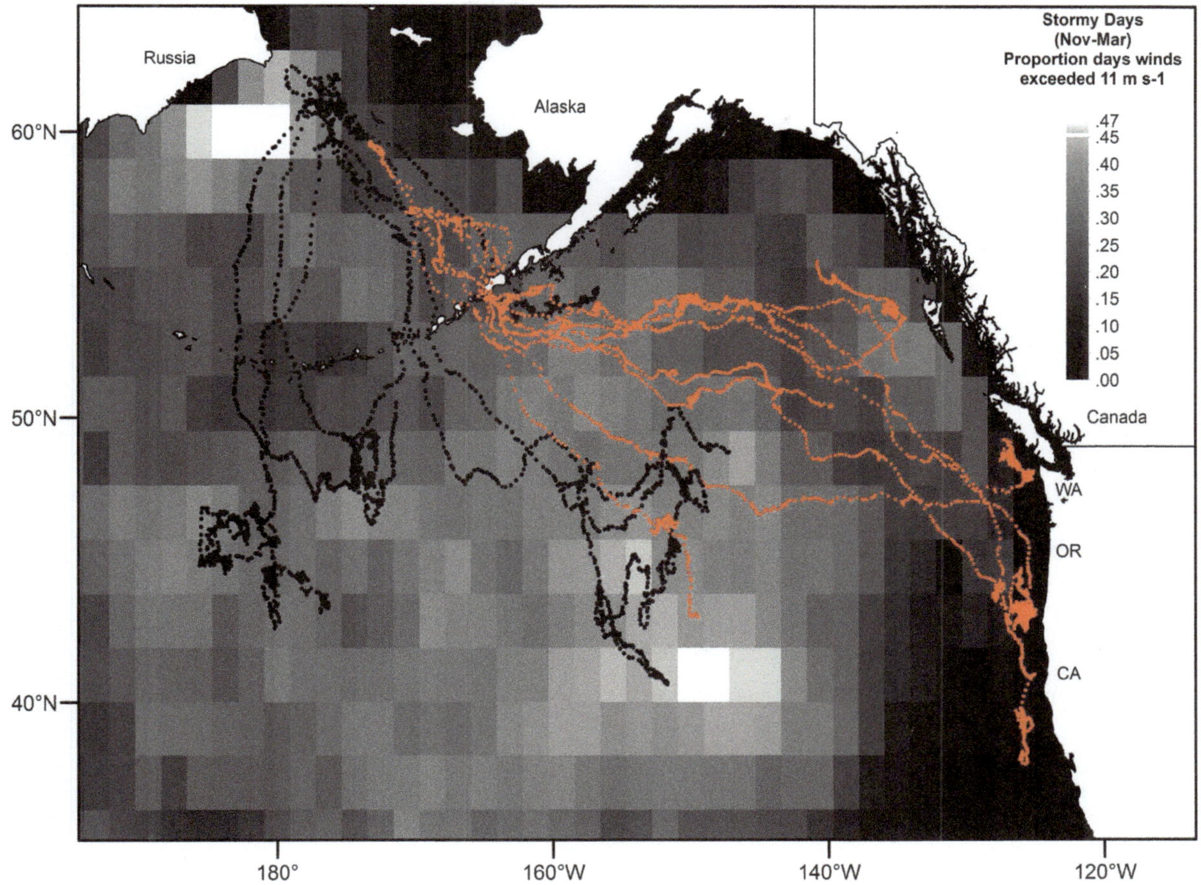

Figure 7. Spatial distribution of stormy days in relation to adult male and female northern fur seal migratory tracks. Stormy days, defined as the proportion of days wind speeds were greater than 11 m s^{-1} [53] during the November-March time period, were compared to male (the black dots) and female (the red dots) migratory tracks. Males remained in the stormy regions of the North Pacific Ocean, while females traveled to regions experiencing fewer stormy days along the continental margins of North America.

index of movement behavior scaled continuously between 0 and 1, with 0 indicating transient or fast moving linear movements and 1 indicating area-restricted resident or foraging movements [41]. To assist with normality we added 0.0001 to values equal to 0 and subtracted 0.0001 from values equal to 1 prior to logit transforming all SSSM behavioral state values [40]. Both the average dive depth and total number of dives every 6 h were log transformed.

Linear mixed-effects models were used to assess the relationship between movement and dive behavior with respect to light (sun and moon), wind speed, ecosystem, season, sex, and mixed-layer depth (adult males only). Models were built and computed by using the nlme V3.1-103 package [48] in the R statistical software package V2.14.1. In all models, individual seal was used as a random effect, an AR(1) autocorrelation structure within each seal was assumed, and several model combinations of the main effects and interaction terms were contrasted, ranked, and selected using Akaike's Information Criterion. Specifically, the response variable SSSM behavioral state was assessed with respect to wind speed, sex, season, and ecosystem. Linear mixed-effects models used for examining seal dive behavior were built separately for adult males and females due to the addition of the mixed-layer depth predictor variable to adult male diving response variables. We assessed the relationship between the average dive depth and number of dives in each 6 h period with respect to lunar fraction, proportion daylight in each 6 h dive period, ecosystem, season, and the mixed-layer depth (males only).

Qualitative Assessment of Winter Winds and Mesoscale Variability

Finally, changes in North Pacific Ocean wind patterns–specifically the increase in storm frequency and intensity [49,50] coupled with mesoscale oceanographic variability [51]–motivated a broad-scale seasonal qualitative assessment of spatial variation of high wind speeds and NFS eddy interactions. While we did not directly address the question of how the prevailing high wind patterns affected NFS population distribution as a whole, we discuss possible mechanisms to explain the distribution of adult males and females during the 2009/2010 winter and how high winds, mesoscale physical oceanographic variability, and NFS behavior are intertwined. For this, we employed two data sets to evaluate spatial wind speed and oceanographic variability over the entire NFS foraging range. NCEP-NCAR Reanalysis 2 10 m wind data were assessed for November through March by computing the proportion of days that wind speeds exceeded 11 m s^{-1} for each 2.5° latitude × 2.5° longitude grid cell–these were categorized as stormy days or high winds. For oceanographic variability we aligned fur seal locations with the weekly distribution of the eddy field as determined by Chelton et al. [51] and available online from http://cioss.coas.oregonstate.edu/

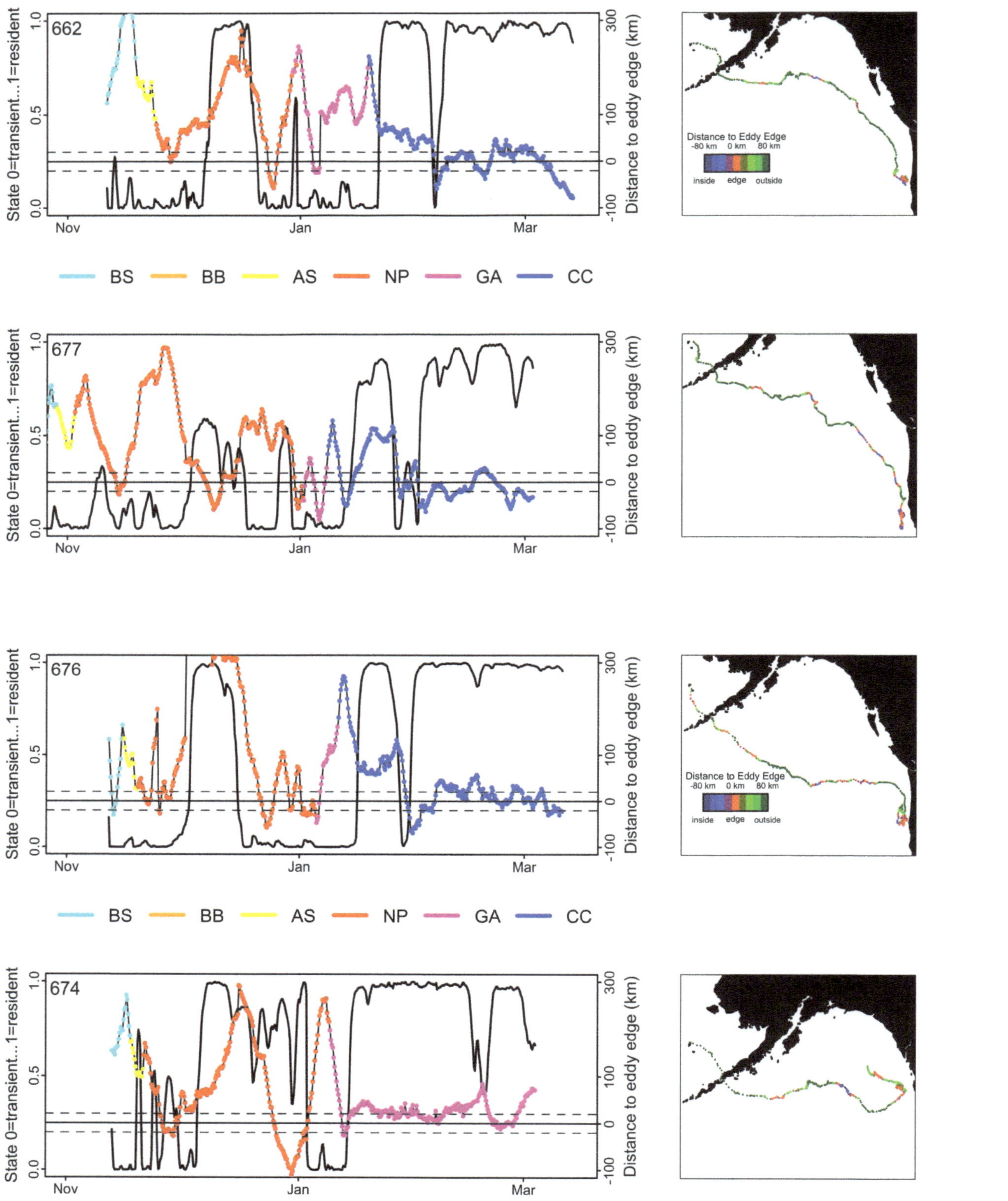

Figure 8. Relationship between adult female northern fur seal behavioral state, eddy distribution and ecosystem. Left panels: Behavioral states (solid black line; left y-axis) of four females compared to the distance of the closest eddy edge (colored dots; right y-axis) in each Large Marine Ecosystem traversed: NP = North Pacific Ocean, AS = Alaska Stream, BB = Bering Sea Basin, BS = Bering Sea Shelf, CC = California Current, GA = Gulf of Alaska (see Fig. 1). Right panels: female geographic distributions–color dots scale distance to the closest eddy edge. While females traversed several eddies during their winter migration, they appeared to dramatically shift their behavioral state to resident when aligned along eddy edges and once they arrived in the GA and CC ecosystems.

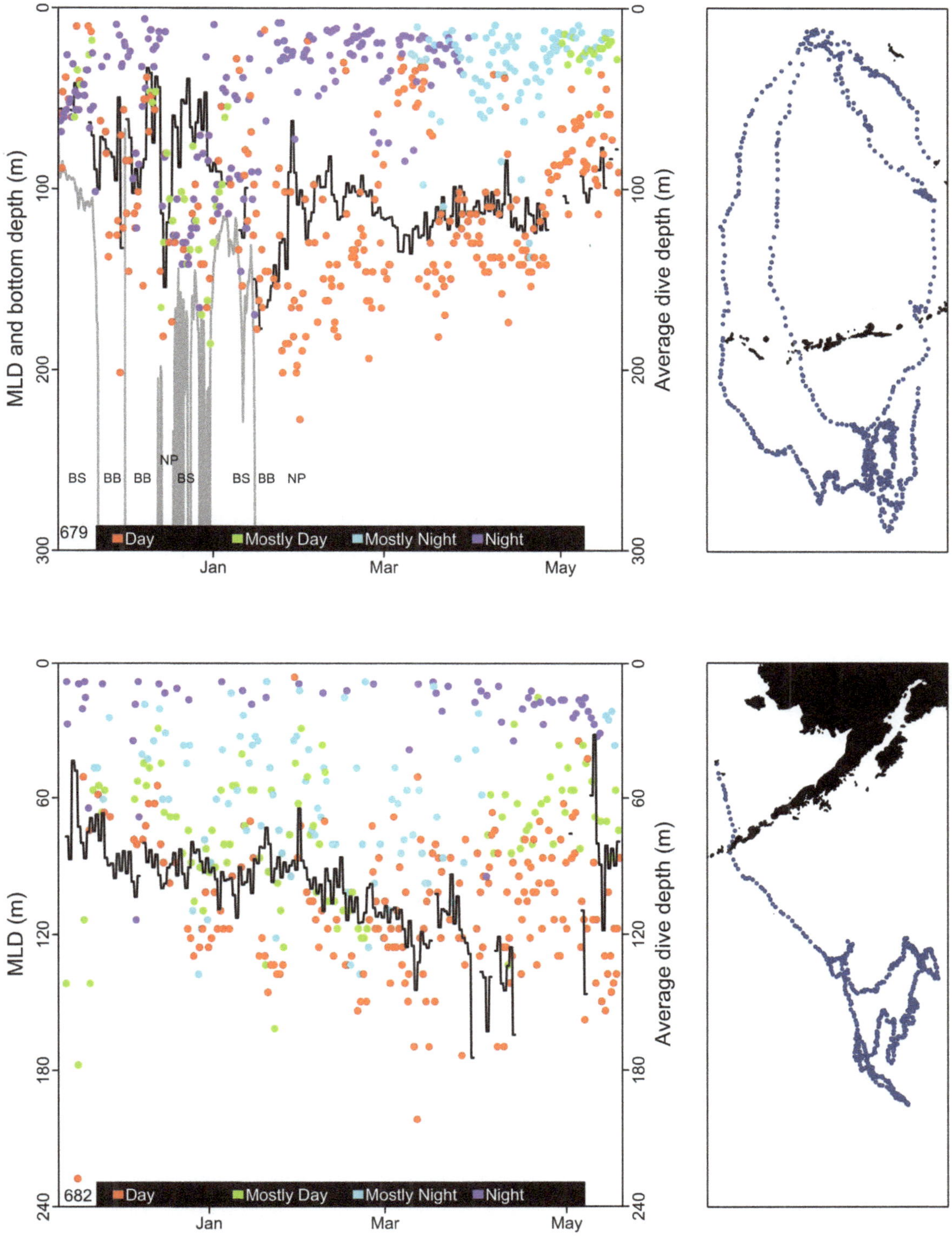

Figure 9. Adult male northern fur seal average dive depths in relation to the mixed-layer depth (MLD). Left panels: Average dive depth (colored dots) in relation to the MLD for males 679 and 682. The black line is the MLD calculated from the CTD measurements taken onboard the males, and the colored dots represent the proportion of light during each 6 h dive period grouped into four categories (Day, Mostly Day, Mostly Night, and Night). Male 679 (top panel) foraged in several Large Marine Ecosystems (LME) before spending the majority of his time in the North Pacific Ocean (NP) ecosystem. Each LME visited by male 679 is identified by a combination of the bottom depth (gray line) and the corresponding LME (NP = North Pacific Ocean, BB = Bering Sea Basin, BS = Bering Sea Shelf). Right panels: foraging locations of these males during the winter.

eddies/. This dataset covers the time period October 1992 to January 2011 and utilizes Version 3 of the AVISO Reference Series gridded sea surface height product to identify eddy center latitudes and longitudes, eddy radius scale, rotation direction, and approximate eddy strengths at 7-day time steps. At each 7-day interval we calculated the distance between seals to the closest eddy edge by using multi-dimensional visualization software (Eonfusion, Myriax Pty Ltd, Hobart, Australia).

Results

All 15 satellite transmitters communicated for >40 d after deployment. One adult female's transmitter stopped working one day into her winter migration after recording 40 days of foraging trips (not used in analyses) to the north of St. Paul Island, and transmitters of three females ceased working early into their winter migration while transiting the NP. We estimated the total migration duration by assuming a return date to St. Paul Island of 1 June for males and 10 July for females [11,52]. Using these return dates, we estimated that 45–52% of the adult female migration was recorded for half of the females tagged (Fig. 2); the longest migratory tracking period was 139 d (Table 1). Three of the five male SRDL tags transmitted into the spring months, recording an estimated 76–95% of the total migratory period, with the longest tracking duration lasting 195 d. The other two male tags stopped transmitting after 49 d and 98 d into the migration (Table 1, Fig. 2).

Initial Dispersal

Initial dispersal from St. Paul Island occurred between 25 October and 23 November, with 10 seals departing within a 10 d period associated with high wind speeds over the island generated by two storms (mean wind speed = 12 m s^{-1}, maximum 6 h average = 22 m s^{-1}). Two females remained in the Bering Sea for 14 d and 18 d before exiting into the NP, while the other seven females traveled out of the Bering Sea in less than 8 d. One female traveled through Akutan Pass and the remaining eight females traveled through nearby Unimak Pass (Table 1 and Fig. 1).

Two males traveled directly south from St. Paul Island and exited the Bering Sea in less than 4 d. The other three spent 16 d, 23 d, and 66 d foraging in the Bering Sea prior to traveling south and into the NP. Only one male traveled through Unimak Pass, while the others exited through passes further to the west (Table 1 and Fig. 1). One male traveled through the Aleutian Islands on three occasions in different locations (between Buldir and Kiska islands, Amukta Pass, and Tanaga Pass) as a result of foraging alternately in the NP and Bering Sea.

Ecosystem Preference

Destination areas and ultimate overwintering ecosystems varied by sex (Figs. 1 and 2). Adult females transited large regions of the NP to forage in the GA and CC. Of the four females whose tags stopped transmitting early, one ceased working while in the BS, one ceased working shortly after entering the NP, one appeared to be heading south toward the Transition Zone Chlorophyll Front (TZCF) [15,16], and one was heading toward the GA. The remaining six females averaged 43±11.5 SD days to transit the NP before arriving in the GA or CC. Four of the five males remained in the NP and Bering Sea and did not enter the GA or CC; the other male, with the shortest transmission duration, exited the BS and foraged in the AS before contact was lost.

Effects on Behavioral State

Of the top three linear mixed-effects models examining adult male and female behavioral state (Table 2), the best model identified season, wind speed, ecosystem, sex, and the interaction between season and sex as important predictors (Table 3). Model results revealed that high wind speeds led to a more transient behavioral state (lower index), and as the season progressed both males and females moved toward a resident behavioral state (higher index). The slope of this seasonal progression differed between the sexes due to females reaching a resident behavioral state earlier in the season compared to males (Table 3, Fig. 3). In addition, the extremely low behavioral states (indicating fast linear movements) of females within the first two months of their migration are consistent with previous observations of adult females rapidly exiting the BS and transiting the AS and NP en route to the GA and CC ecosystems.

Mass Effects on Adult Male and Female Dive Depths and Dive Durations

Adult males dove deeper and for longer durations than adult females, and average dive depth (Table 4, Fig. 4A; $R^2 = 0.81$, df = 9, $P<0.001$), average maximum dive depth (Fig. 4B; $R^2 = 0.78$, df = 9, $P<0.001$), average dive duration (Fig. 4C; $R^2 = 0.90$, df = 9, $P<0.001$), and average maximum dive duration (Fig. 4D; $R^2 = 0.90$, df = 9, $P<0.001$) were all positively correlated with seal mass. However, larger, deeper diving males performed fewer dives per 6 h compared to their female counterparts (Fig. 4E; $R^2 = 0.57$, df = 9, $P = 0.007$).

Exogenous Effects on Dive Depth

Of the top three linear mixed-effects models examining adult female dive depths (Table 2), the best model identified daylight, moonlight, and variation between ecosystems and interactions between daylight, moonlight, and ecosystem as explanatory variables (Table 5). As the percentage of daylight in each 6 h period and the lunar fraction increased, female dive depths increased in the NP (Fig. 5 top). When contrasted to the NP, high lunar fraction in the BS, AS, and GA ecosystems had less of an effect on female dive depths, and dives which took place during the night with a new moon resulted in significantly deeper dives (Table 5). Adult males showed a similar response to increased sunlight levels by also diving deeper during the day and to moonlight levels by diving deeper at night on moonlit nights, particularly in the BS (Tables 4 and 5). As expected, with increasing daylight the lunar effect decreased in females but not significantly for males, which may reflect the day-night differences in preferred foraging times between the sexes (Table 4, see next section). For males, as the percent daylight in each 6 h period increased, the MLD became a more significant influence on dive depths, leading to correspondence between their dive depths and MLD during the daytime (Table 5, Fig. 5 bottom). In addition, male dive depths responded to seasonal changes in the MLD by slightly increasing in depth in winter and shoaling in spring.

Exogenous Effects on Dive Activity

Of the top three linear mixed-effects models examining number of dives per 6 h period in adult females and males (Table 2), the best model showed that females had a greater number of dives per 6 h during the night in four of five LMEs visited–the exception was the CC where the only female with a functioning dive recorder performed more dives in each 6 h period with increasing proportion of daylight (Tables 4 and 6, Fig. 6 top). These trends are consistent with those of females tagged in previous years with dive recorders (unpublished data) [16], and thus are not likely an artifact of the small sample size in this study but instead represent ecosystem dive response. In contrast to adult females foraging in the AS, NP, and GA, adult males increased the number of dives with an increased proportion of daylight in each 6 h period while foraging in the AS, NP, and BB (Table 6, Fig. 6 bottom). The exception for male seals was the BS, a neritic region limiting the depth of prey to 200 m or less. Unlike their dive behavior in pelagic ecosystems (NP, BB, and AS), males performed more dives at night (73%) in the BS, a similar dive pattern as adult females foraging in the BS (Table 4).

Winter Winds and Mesoscale Eddies

The spatial gradient from many to few stormy days across the range of the animals tracked in this study showed that adult males foraged in areas experiencing high wind speeds, consistently, throughout November-March (Fig. 7). Unlike males, adult females continued to migrate south and east to ecosystems experiencing a lower proportion of stormy days (Fig. 7), all the while transitioning from transient to resident behavioral states (Fig. 8 left panels). The transition coincided with their proximity to mesoscale eddies located in known energetic regions of the GA and CC (Fig. 8). Interestingly, these were not the first eddies each of these females encountered along their migratory route. For each of the females shown in Figure 8, encounters with mesoscale eddies were common, but consistently passed by when the encounters occurred in the NP and in areas experiencing high wind speeds (Figs. 7 and 8).

Discussion

The annual migration of adult NFS from St. Paul Island begins in late autumn (October-November) and ends in late spring and early summer (May-July) when animals return to the island to reproduce and forage in the Bering Sea. Seasonal changes in the ocean and atmosphere, in particular the strong northerly winds of autumn cyclones, influence the migratory season with early storms causing earlier onset of migration than when autumn storms begin later [31,52,53]–10 of the 15 seals in this study departed during periods of high wind speeds generated by two autumn cyclones passing by St. Paul Island. The migration route begins with adult females departing the rookeries, traveling south through passes in the eastern Aleutian Islands, then across the NP to the continental margin of northwestern North America or the TZCF, as shown in this study and previously by [15,16,29–31]. In contrast, adult males are distributed in an expansive region of the northern NP, Aleutian Islands, GA, and Bering Sea throughout the entire winter and spring months, as also shown in this study and by [29,31,32].

Adult Female Migratory Behavior

The females in this study exhibited typical migration patterns–five of six with longer lasting tags (>60 d) departed to the southeast, exited the Bering Sea through Unimak Pass, and continued south and east across the NP. The other female briefly swam northwest of St. Paul Island before reversing direction and exiting the BS. Three of the six females arrived at their destinations in the CC, a fourth nearby in the southern GA region, and the remaining two farther north in the GA. Their behavioral states across the NP were influenced by wind speed and season–females traveled more (transient state) and with the wind (not shown) during high winds and travelled less (resident state) during low to moderate winds (Fig. 3). The seasonal shift in movement behavior from transient to resident occurred upon the arrival of females in the GA and CC, regions previously identified as destination areas [16,31]. We believe the shift in movement reflected arrival in a more equitable foraging environment that was less disrupted by strong North Pacific gales (Fig. 7). This is important in its own right, but wind also is important to the vertical distribution of biomass in mesoscale eddies, which are well known to be pelagic oases important for energy transfer to higher trophic level species [13,54–56] and which are targeted by females once they reach the GA and CC (Fig. 8).

Mesoscale eddies are ubiquitous across the migration route of females from the BB to the GA and CC [51,57] yet they appear to have little effect on female behavioral state when transiting the NP. The differential influence of eddies on female behavior might be related to female migratory momentum across the NP when they are prone to remain in transit mode to their destinations in the GA and CC, and once there enter into feeding mode, particularly around the periphery of eddies. However, the behavioral responses, or lack of responses, of females to eddies might be caused by a broader, multi-trophic level dispersion or aggregation of food web species within the females' foraging depth range. In the case of high winds interacting with eddies, Mackas et al. [13] concluded that aggregation and retention of zooplankton were linked to behavior in the water column, and the most successful taxa migrated below the MLD to minimize exposure, thus avoiding wash out due to an eddy "spinning down" or decaying (also described by Bakun [58]), Ekman transport during strong wind events, or displacement caused by wind-driven inertial currents. The depth range of zooplankton predators, such as squid and myctophids, thus would reflect the upper depth range of their prey during diel vertical movements and would be more or less accessible to female fur seals depending on the depth. Females in this study passed through several eddies during their transit across the NP (Fig. 8), particularly in regions experiencing high winds (Fig. 7). However, upon their arrival in the GA and CC, regions that experienced lower wind speeds, females transitioned from a transitory to a more resident behavioral state in association with eddies (Fig. 8). Under the Mackas et al. [13] scenario, eddies in the stormy regions far offshore may have enhanced biological aggregations, but the vertical distribution of fur seal prey fields may be too deep during the day and too dispersed at night for successful foraging. In contrast, eddy and storm interactions were greatly diminished along the continental margins of western North America, potentially allowing for shallower and more accessible prey fields that lead to the significant shifts in female movement and dive behavior.

Females foraged predominantly at night in the upper 30 m of the water column when in the BS, NP, and GA. In contrast, the female with a dive satellite tag that travelled to the CC foraged less at night and more during the day after arriving there. Nocturnal foraging is undoubtedly related to the diel vertical distribution of prey, as suggested by others based on behaviors of known vertically migrating mesopelagic prey species such as myctophids and squids [16,44,59–62]. The contrasting behavior of the female in the CC also may be related to the availability of prey–common prey of NFS in the CC in the past consisted of Pacific whiting (*Merluccius productus*), northern anchovy (*Engraulis mordax*), Pacific herring (*Clupea pallasii*), rockfish (*Sebastes* spp.), and salmon (Salmonidae) [44,63], epipelagic species that are typically found at comparatively shallow depths both day and night. Prey distribution and fur seal behavior may be further influenced by the MLD, which in winter along the eastern boundary of the North Pacific Ocean can be as much as ~50 m shallower than in central and western regions (see http://www.afsc.noaa.gov/Quarterly/amj2012/AMJ12_Feature-V2.pdf) [64], making it and associated prey fields more accessible to females during the day.

Day-night diving depths support our interpretation of foraging times–that is, females tended to dive deepest during the day when prey, even in the CC, are deeper in the water column. An exception occurred in the GA, where the deepest diving was during twilight and dawn. Changing light levels at dusk and dawn signal diel vertically migrating prey to either ascend or descend in the water column. The deeper dives performed by females during these periods may reflect prey reaching water column depths within their physiological dive limits. Why this pattern was not observed in the NP or CC is unknown, but one plausible explanation is that the GA is a transitional ecosystem between the NP and CC, such that prey fields are completely inaccessible during the day in the NP, marginally accessible in the GA during twilight periods, and accessible throughout the day at shallower depths in the CC where females shift to more daytime dives.

As expected, female diving depth at night was commonly, but not always, modulated by the lunar cycle, with depths greatest during times of greatest illumination (Fig. 5, Table 5). Moonlight at night is perceived by vertically migrating prey species similarly to sunlight during the day, attenuating the vertical range of their movements and forcing predators to dive deeper to access them. The exceptions to this pattern occurred in the BS, AS, and GA where there was little to no relationship between moonlight and diving depth. With respect to the BS, this could simply be a function of their departure dates relative to the lunar cycle–most females departed closer to the new moon and exited the BS before the full moon returned. This, combined with few dives and a quick exit (Table 1), reflects the lack of foraging and thus the lack of influence of prey behavior on female behavior.

Although we obtained only a small amount of diving data from females relative to males in the NP, where we were able to determine the MLD from the male CTD records, female daytime dive depths were greater in January than in the two previous months in accordance with a deepening MLD. Data from all females tracked since 2002 [16] indicate that their daytime dive depths during fall-spring in the CC, and thus the vertical distribution of prey fields, are indeed influenced by the MLD and seasonally deepen then shoal with changing MLD (http://www.afsc.noaa.gov/Quarterly/amj2012/AMJ12_Feature-V2.pdf).

Adult Male Migratory Behavior

Movement patterns and wintering distribution of males contrasted sharply with those of females. Males exhibited a wide variety of behaviors, from remaining in the northwestern BS near the advancing ice front throughout most of December before moving south into the NP, to swimming back and forth between the BS, BB and NP, to departing directly south and into the NP. Though all males ended up in widely separate regions, none of them ended up as far to the east as the females. Like females, their behavioral state was significantly affected by season and by high wind speeds throughout the migration (Table 3, Fig. 3). However, both the intercept and slope with respect to behavioral state differed from that of females (Table 3). Males were generally in a less transient state initially and the slopes of the seasonal effects differed from females by showing a more gradual shift to resident behavior as the season progressed (Fig. 3). The difference in the slopes between males and females could be due to males remaining in the stormy regions of the NP throughout the winter (Fig. 7) and frequently contending with strong gales.

Unlike females, which foraged mostly at night, males foraged predominantly during the day, with the exception of those in the BS where 73% of the dives occurred at night (Fig. 6, Table 4). Diving behavior was affected only marginally by the lunar cycle and primarily in the BS–had they fed more at night instead of during the day in other LMEs the lunar effect might have been stronger (Table 5). Males in oceanic LMEs apparently responded to the diel vertical movement of prey and dove deeper during the day than at night (Table 4). The day-night difference in mean dive depth was least in the BS (16 m) and much greater in the NP (65 m), BB (78 m), and AS (127 m). The small day-night difference in the BS likely reflects the shallow water depths on the shelf (<200 m), the ability of males to easily reach the seafloor, and the vertical distribution of their prey. While in the BS, males foraged at depths that corresponded approximately with the MLD, but that may be coincidental due to the close proximity of the MLD and the seafloor and not because there was a functional relationship between the MLD and distribution of prey fields.

Diving depths of adult male NFS have not been reported previously. Boyd et al. [8] suggested that diving depths of adult male Antarctic fur seals, a species of similar size and life history, reflected the depth of the mixed layer. Males in all oceanic LME in this study commonly foraged to depths at and below the MLD (Fig. 9), particularly during the day. As with females in the CC (see http://www.afsc.noaa.gov/Quarterly/amj2012/AMJ12_Feature-V2.pdf), diving depths of males in the NP showed some correspondence with the seasonal deepening and shoaling of the MLD, which further suggests a functional relationship between the MLD and prey fields (Fig. 9). Male dives were correspondingly longer than female dives and males performed fewer dives per unit time (Table 4, Fig. 4).

Diets of adult males are not known for any season. Historically, adult females in winter in the NP consumed primarily squids [44], many of which are diel vertical migrants [59–62] and would be accessible to females foraging at night. Many dives of males went well below the MLD, and extended into the deep scattering layer (DSL) that, in winter habitat of males, lies at about 275–300 m depth [65–67]. Myctophids, particularly *Stenobrachius leucopsarus*, constitute a significant portion of the DSL in the NP [10] and thus would be accessible to male NFS during their deepest dives. Most squids, however, reside during the day at depths >350 m and below the presumed foraging range of males [2,8,26,62], which suggests either that males are not targeting them during the day or that more is to be learned about the daytime depth distribution of squids.

Intraspecific Competition, Differential Dietary Needs, and Predation Risk

Three factors that have been invoked to explain the differing behaviors between sexes of other species are reduced intraspecific competition, differential dietary needs due to energetic requirements, and predation risk [25]. Among pinnipeds, Breed et al. [4] believe that geographic segregation of foraging male and female gray seals (*Halichoerus grypus*) on the Scotian shelf reduces competition between them and thereby increases fitness for both sexes. For NFS, size dimorphism between the sexes is much greater compared to that of gray seals (e.g. in this study, the largest male was eight times greater in mass than the smallest female). The extreme size dimorphism and thus disparate physiological capabilities between males and females likely limits their dietary overlap and thus competition for resources. While sexual selection may be the ultimate cause of NFS size dimorphism, ecological divergence cannot be discounted. That is, adult males may gain further momentum to diverge in size with females because of the increased diving capability large size imparts, thus giving them better access to potentially abundant and energetically profitable prey resources at depth. Whichever the case, the geographic separation during the migratory period may reduce intraspecific competition for food, although there is presently no way to evaluate this possibility because of the lack of information on adult male diets in any season and location and on contemporary winter diets of adult females.

The differences in foraging strategies between sexes could be due to dietary differences owing to the physical size differential–that is, NFS males might prefer or need, because of their greater energy requirements, larger or more energy-rich prey than females

– prey that might undertake attenuated or no diel vertical migration. Their much larger mouths and teeth seemingly would allow males to subdue and consume much larger prey, but again, because we know nothing about male diets, we cannot currently evaluate these possibilities.

Lastly, predation pressure has been invoked as a driver of sexual segregation in other marine mammals. Le Boeuf et al. [25] postulated that female northern elephant seals migrate to distant offshore areas of the NP to reduce the risk of predation by great white sharks (*Carcharodon carcharias*) and mammal-eating (Bigg's) killer whales (*Orcinus orca*). The much larger males should be less susceptible to those predators in general, although proximity to both great white sharks and killer whales on their migratory routes likely increases their risk. However, if the prey resources they exploit along continental margins are more energetically profitable than those occurring in the open ocean, males may balance the risk of predation with the benefit of enhanced foraging. Another example is that of sperm whales [68,69]: females and calves typically remain in the open NP during the summer feeding season, ostensibly to avoid killer whales, but thereby sacrificing access to the abundant prey resources enjoyed by males that migrate to the Aleutian Archipelago, BB, and GA to feed each summer and where they apparently can tolerate the risk posed by abundant killer whales. In the case of NFS the distributional patterns are reversed, making such an explanation for segregation of the sexes unlikely, since in winter the females should be more exposed to sharks and killer whales than are males. Their small size and high density along the continental margin might exacerbate the risk of predation, but because of their (former) abundance the risk to any individual might be minimized. On the other hand, one of their principal potential predators, great white sharks, is likely not abundant in winter in the colder northern part of its overall range [70], and the other, mammal eating killer whales, may not be a threat either as they typically are found in inshore areas on the continental shelf [71].

Conclusions

Contrasting winter migration behaviors of adult male and female NFS have been documented for over a century and likely known for millennia [16,28–32,72]. For example, Aleut knowledge of differing ecosystem preferences of adult male and female NFS and effects of wind on movement patterns were reported by Captain Hooper [31] and his interpreter Peter Shainsnakoff after they interviewed 80 hunters in the Aleutian Islands during the fall of 1892. The results from our study confirm the traditional ecological knowledge told to Captain Hooper and extend this and more recent scientific knowledge, empirically, by showing the interplay between physiology, behavior, and environment.

We conclude that adult male NFS winter in the northern NP and Bering Sea because they can. They can because of the biological imperative and their life history strategy that evolutionarily led to the large size and strength necessary to hold territories against rival males while fasting during the breeding season, thus ensuring reproductive fitness. The strong positive correlations between mass and dive depth and duration across the size range of females and males reflects a gradient in the physiological dive capacity of these seals with increasing size. The vertical segregation of foraging habitat between females and males would thus be best explained by the physiological capacity of males to access prey to depths of at least 250–350 m during daytime. By remaining in the open ocean, they reduce whatever predation risk and competition that might exist along the continental margin. Moreover, their larger mass and smaller surface to volume ratio confers a thermal advantage over females, and thus males should be better able to tolerate colder seawater temperatures of northerly latitudes [53,73–75]. Finally, their strength may confer a further advantage in contending with storms and turbulent seas typical of winter at higher latitudes.

Females pursue a different strategy because they must. In the context of reproductive life history traits, females are not under the same evolutionary selective pressure for greater size because they do not compete for space or mates in rookeries. But their small size places them at a disadvantage relative to males in that they are physiologically less able to reach prey occurring at greater depths at high latitudes. Instead, they seek a more equable environment, principally along the continental margin of western North America, where storms are less frequent and intense and where their dive capabilities allow them access to a diversity of prey that are predictably abundant and apparently concentrated by physical oceanographic features, particularly eddies.

Acknowledgments

We are indebted to the St. Paul Island community and the Ecosystem Conservation Office of the St. Paul Tribal Council for support of our work. L. Logerwell, R. Ream, D. Costa and one anonymous reviewer provided helpful critiques of the manuscript. Crucial logistical and analytical support came from the National Marine Fisheries Service, Alaska Sea Life Center, and Eonfusion software. Field-work assistance was provided by J. Baker, B Caruso, D. Deghetto, B. Dickerson, T. Gelatt, J. Leon-Guerrero, H. Harris, J. Harris, D. Johnson, C. Kuhn, B. Page, M. Rutishauser, B. Sinclair, W. Testa, and M. Williams. The findings and conclusions in the paper are those of the authors and do not necessarily represent the view of the National Marine Fisheries Service. Reference to trade names does not imply endorsement by the National Marine Fisheries Service, NOAA.

Author Contributions

Conceived and designed the experiments: JTS AMS SJI. Performed the experiments: JTS AMS SJI SPJ. Analyzed the data: JTS NAP DSJ MAL NAB. Contributed reagents/materials/analysis tools: DSJ. Wrote the paper: JTS AMS SJI NAP MAL NAB.

References

1. Page B, McKenzie J, Sumner MD, Coyne M, Goldsworthy SD (2006) Spatial separation of foraging habitats among New Zealand fur seals. Marine Ecology Progress Series 323: 263–279.
2. Staniland IJ, Robinson SL (2008) Segregation between the sexes: Antarctic fur seals, *Arctocephalus gazella*, foraging at South Georgia. Animal Behaviour 75: 1581–1590.
3. Stewart BS (1997) Ontogeny of differential migration and sexual segregation in northern elephant seals. Journal of Mammalogy: 1101–1116.
4. Breed GA, Bowen WD, McMillan JI, Leonard ML (2006) Sexual segregation of seasonal foraging habitats in a non-migratory marine mammal. Proceedings of the Royal Society Biological Sciences Series B 273: 2319–2326.
5. Kooyman GL (1989) Diverse divers: physiology and behavior. Berlín: Springer-Verlag.
6. Weise MJ, Costa DP (2007) Total body oxygen stores and physiological diving capacity of California sea lions as a function of sex and age. Journal of Experimental Biology 210: 278–289.
7. Weise MJ, Harvey JT, Costa DP (2010) The role of body size in individual-based foraging strategies of a top marine predator. Ecology 91: 1004–1015.
8. Boyd IL, McCafferty DJ, Reid K, Taylor R, Walker TR (1998) Dispersal of male and female Antarctic fur seals (*Arctocephalus gazella*). Canadian Journal of Fisheries and Aquatic Sciences 55: 845–852.
9. Francis R, Hare S, Hollowed A, Wooster W (1998) Effects of interdecadal climate variability on the oceanic ecosystems of the NE Pacific. Fisheries Oceanography 7: 1–21.
10. Frost B, McCrone L (1979) Vertical distribution, diel vertical migration, and abundance of some mesopelagic fishes in the eastern subarctic Pacific Ocean in summer. Fishery Bulletin 76: 717–749.

11. Gentry RL (1998) Behavior and ecology of the northern fur seal. Princeton, N.J.: Princeton University Press. 392 p.
12. Lea MA, Johnson D, Melin S, Ream R, Gelatt T (2010) Diving ontogeny and lunar responses in a highly migratory mammal, the northern fur seal *Callorhinus ursinus*. Marine Ecology Progress Series 419: 233–247.
13. Mackas DL, Tsurumi M, Galbraith MD, Yelland DR (2005) Zooplankton distribution and dynamics in a North Pacific eddy of coastal origin: II. Mechanisms of eddy colonization by and retention of offshore species. Deep Sea Research Part II: Topical Studies in Oceanography 52: 1011–1035.
14. Polovina J, Mitchum G, Evans G (1995) Decadal and basin-scale variation in mixed layer depth and the impact on biological production in the Central and North Pacific, 1960–88. Deep Sea Research Part I: Oceanographic Research Papers 42: 1701–1716.
15. Polovina JJ, Howell E, Kobayashi DR, Seki MP (2001) The transition zone chlorophyll front, a dynamic global feature defining migration and forage habitat for marine resources. Progress in Oceanography 49: 469–483.
16. Ream RR, Sterling JT, Loughlin TR (2005) Oceanographic influences on northern fur seal migratory movements. Deep Sea Research Part II: Topical Studies in Oceanography 52: 823–843.
17. Schumacher JD. Regime shift theory: A review of changing environmental conditions in the Bering Sea and Eastern North Pacific Ocean; Proceedings of the Fifth North Pacific Rim Fisheries Conference 1999, Anchorage, Alaska.
18. Sterling JT (2009) Northern fur seal foraging behaviors, food webs, and interactions with oceanographic features in the eastern Bering Sea. Seattle: Unpublished doctoral dissertation, University of Washington. 219 p.
19. Trillmich F, Limberger D (1985) Drastic effects of El Nino on Galapagos pinnipeds. Oecologia 67: 19–22.
20. Watanabe H, Kubodera T, Moku M, Kawaguchi K (2006) Diel vertical migration of squid in the warm core ring and cold water masses in the transition region of the western North Pacific. Marine Ecology Progress Series 315: 187–197.
21. Villegas-Amtmann S, Simmons SE, Kuhn CE, Huckstadt LA, Costa DP (2011) Latitudinal range influences the seasonal variation in the foraging behavior of marine top predators. PloS ONE 6: e23166.
22. Halsey LG, Butler PJ, Blackburn TM (2006) A phylogenetic analysis of the allometry of diving. American Naturalist 167: 276–287.
23. Le Boeuf BJ (1994) Variation in the diving pattern of northern elephant seals with age, mass, sex, and reproductive condition. In: Le Boeuf BJ, Laws RM, editors. Elephant Seals: Population Ecology, Behavior, and Physiology. Berkeley, Los Angeles, London: University of California Press. 237–252.
24. Hindell MA, Lea MA, Morrice MG, MacMahon CR (2000) Metabolic limits on dive duration and swimming speed in the southern elephant seal *Mirounga leonina*. Physiological and Biochemical Zoology 73: 790–798.
25. Le Boeuf BJ, Crocker DE, Costa DP, Blackwell SB, Webb PM, et al. (2000) Foraging ecology of northern elephant seals. Ecological Monographs 70: 353–382.
26. Page B, McKenzie J, Goldsworthy SD (2005) Inter-sexual differences in New Zealand fur seal diving behaviour. Marine Ecology Progress Series 304: 249–264.
27. Scheffer VB, Wilke F (1953) Relative growth in the northern fur seal. Growth 17: 129–145.
28. Kajimura H (1984) Opportunistic feeding of the northern fur seal, *Callorhinus ursinus*, in the Eastern North Pacific Ocean and Eastern Bering Sea. Seattle: NOAA Technical Report NMFS SSRF-779. 49.
29. Bigg MA (1990) Migration of northern fur seals (*Callorhinus ursinus*) off western North America. Canadian Technical Report of Fisheries and Aquatic Sciences 1764.
30. Kenyon KW, Wilke F (1953) Migration of the northern fur seal, *Callorhinus ursinus*. Journal of Mammalogy 34: 86–98.
31. Hooper CL (1895) Report of Captain Hooper, dated November 21, 1892. Fur-Seal Arbitration. Washington, D. C.: Government Printing Office. 228–233.
32. Loughlin TR, Baba N, Robson BW (1999) Use of surface-current model and satellite telemetry to assess marine mammal movements in the Bering Sea. In: Loughlin TR, Ohtani K, editors. Dynamics of the Bering Sea. Fairbanks, AK: University of Alaska Sea Grant. 615–630.
33. Towell RG, Ream RR, Fowler CW, Sterling JT, Fritz LW, et al. (2008) Population assessment of northern fur seals on the Pribilof Islands, Alaska, 2006–2007. Seattle, WA: U.S. Department of Commerce. NOAA Technical Memorandum NMFS-AFSC-188 NOAA Technical Memorandum NMFS-AFSC-188. 8–25 p.
34. Boehme L, Lovell P, Biuw M, Roquet F, Nicholson J, et al. (2009) Technical Note: Animal-borne CTD-Satellite Relay Data Loggers for real-time oceanographic data collection. Ocean Science 5: 685–695.
35. Kara AB, Rochford PA, Hurlburt HE (2000) An optimal definition for ocean mixed layer depth. Journal of Geophysical Research, C: Oceans 105: 16803–16821.
36. Crawford W, Galbraith J, Bolingbroke N (2007) Line P ocean temperature and salinity, 1956–2005. Progress in Oceanography 75: 161–178.
37. Tabata S (1978) An evaluation of the quality of sea surface temperatures and salinities measured at Station P and Line P in the Northeast Pacific Ocean. Journal of Physical Oceanography 8: 970–986.
38. Argos (2006) Argos User's Manual CLS America Inc.
39. Block BA, Jonsen ID, Jorgensen SJ, Winship AJ, Shaffer SA, et al. (2011) Tracking apex marine predator movements in a dynamic ocean. Nature 475: 86–90.
40. Breed GA, Jonsen ID, Myers RA, Bowen WD, Leonard ML (2009) Sex-specific, seasonal foraging tactics of adult grey seals (*Halichoerus grypus*) revealed by state-space analysis. Ecology 90: 3209–3221.
41. Jonsen I, Myers R, Flemming J (2003) Meta-analysis of animal movement using state-space models. Ecology 84: 3055–3063.
42. Jonsen ID, Basson M, Bestley S, Bravington MV, Patterson TA, et al. (2013) State-space models for bio-loggers: A methodological road map. Deep Sea Research Part II: Topical Studies in Oceanography 88–89: 34–46.
43. Jonsen ID, Flenming JM, Myers RA (2005) Robust state-space modeling of animal movement data. Ecology 86: 2874–2880.
44. Perez MA, Bigg MA (1986) Diet of northern fur seals, *Callorhinus ursinus*, off Western North America. Fishery Bulletin 84: 957–971.
45. Ladd C, Bond NA (2002) Evaluation of the NCEP/NCAR reanalysis in the NE Pacific and the Bering Sea. Journal of Geophysical Research, C: Oceans 107: 3158.
46. Sherman K, Laughlin TL (1992) The large marine ecosystem (LME) concept and its application to regional marine resource management, 1–6 October 1990, Monaco: conference summary and recommendations: IUCN in collaboration with the United States National Oceanic and Atmospheric Administration and the Musée Océanographique de Monaco.
47. Sherman K, Aquarone M, Adams S (2007) Global Applications of the Large Marine Ecosystem Concept. NOAA Technical Memorandum NMFS-NE-208. 72 p.
48. Pinheiro J, Bates D, DebRoy S, Sarkar D (2008) Linear and nonlinear mixed effects models. The nlme Package v 31–89.
49. Allan JC, Komar PD (2006) Climate controls on US West Coast erosion processes. Journal of Coastal Research 22: 511–529.
50. Graham NE, Diaz HF (2001) Evidence for intensification of North Pacific winter cyclones since 1948. Bulletin of the American Meteorological Society 82: 1869–1893.
51. Chelton DB, Schlax MG, Samelson RM (2011) Global observations of nonlinear mesoscale eddies. Progress in Oceanography 91: 167–216.
52. Peterson RS (1965) The behavior of the northern fur seal [Ph.D. dissertation]: John Hopkins University. 214 p.
53. Lea MA, Johnson D, Ream R, Sterling J, Melin S, et al. (2009) Extreme weather events influence dispersal of naive northern fur seals. Biology Letters 5: 252–257.
54. Atwood E, Duffy-Anderson JT, Horne JK, Ladd C (2010) Influence of mesoscale eddies on ichthyoplankton assemblages in the Gulf of Alaska. Fisheries Oceanography 19: 493–507.
55. Godø OR, Samuelsen A, Macaulay GJ, Patel R, Hjøllo SS, et al. (2012) Mesoscale eddies are oases for higher trophic marine life. PLoS ONE 7: e30161.
56. Mackas DL, Galbraith MD (2002) Zooplankton distribution and dynamics in a North Pacific eddy of coastal origin: 1. Transport and loss of continental margin species. Journal of Oceanography 58: 725–738.
57. Schumacher JD, Stabeno PJ (1994) Ubiquitous eddies of the eastern Bering Sea and their coincidence with concentrations of larval pollock. Fisheries Oceanography 3: 182–190.
58. Bakun A (2006) Fronts and eddies as key structures in the habitat of marine fish larvae: opportunity, adaptive response and competitive advantage. Scientia Marina 70: 105–122.
59. Beamish RJ, Leask KD, Ivanov OA, Balanov AA, Orlov AM, et al. (1999) The ecology, distribution, and abundance of midwater fishes of the Subarctic Pacific gyres. Progress in Oceanography 43: 399–442.
60. Roper FE, Young RE (1975) Vertical distribution of pelagic cephalods. Washington D.C.: Smithsonian Institution Press. 51 p.
61. Pearcy WG, Krygier EE, Mesecar R, Ramsey F (1977) Vertical distribution and migration of oceanic micronekton off Oregon. Deep-Sea Research 24: 223–245.
62. Nesis KN (1997) Gonatid squids in the subarctic North Pacific: Ecology, biogeography, niche diversity and role in the ecosystem. Advances in Marine Biology 32: 243–324.
63. Spalding DJ (1964) Comparative feeding habits of the fur seal, sea lion, and harbour seal. Bulletins of the Fisheries Research Board of Canada 146: 52.
64. Suga T, Motoki K, Aoki Y, Macdonald AM (2004) The North Pacific climatology of winter mixed layer and mode waters. Journal of Physical Oceanography 34: 3–22.
65. McGowan JA, Williams PM (1973) Oceanic habitat differences in the North Pacific. Journal of Experimental Marine Biology and Ecology 12: 187–217.
66. Tont SA (1976) Deep scattering layers: patterns in the Pacific. California Cooperative Ocean Fisheries Investigations 18: 112–117.
67. Isaacs JD, Tont SA, Wick GL (1974) Deep scattering layers: vertical migration as a tactic for finding food. Deep Sea Research and Oceanographic Abstracts 21: 651–656.
68. Pitman RL, Ballance LT, Mesnick SI, Chivers SJ (2001) Killer whale predation on sperm whales: observations and implications. Marine Mammal Science 17: 494–507.
69. Whitehead H (2003) Sperm whales: social evolution in the ocean. Chicago: University of Chicago press. 456 p.
70. Martin RA (2004) Northerly distribution of white sharks, *Carcharodon carcharias*, in the eastern Pacific and relation to ENSO events. Marine Fisheries Review 66: 16–26.

71. Bigg DM, Ellis GM, Ford JKB, Balcomb KC (1987) Killer whales: a study of their identification, geneology, and natural history in British Columbia and Washington State. Nanaimo, BC: Phantom Press and Publishers. 79 p.
72. Jordan DS (1898) The fur seals and fur seal islands of the North Pacific Ocean; Jordan DS, editor. 4 Parts, 5 Volumes. Washington, DC: U.S. Government Printing Office.
73. Dalton AJM, Rosen DAS, Trites AW (In press) Broad thermal capacity facilitates the primarily pelagic existence of northern fur seals (*Callorhinus ursinus*). Marine Mammal Science.
74. Rosen DAS, Trites AW (In press) Thermal limites in young northern fur seals. Marine Mammal Science.
75. Rutishauser MR, Costa DP, Goebel ME, Williams TM (2004) Ecological implications of body composition and thermal capabilities in young Antarctic fur seals (*Arctocephalus gazella*). Physiological and Biochemical Zoology 77: 669–681.

5

Wintering Habitat Model for the North Atlantic Right Whale (*Eubalaena glacialis*) in the Southeastern United States

Timothy A. Gowan*, Joel G. Ortega-Ortiz¤

Florida Fish and Wildlife Conservation Commission, Fish and Wildlife Research Institute, St. Petersburg, Florida, United States of America

Abstract

The coastal waters off the southeastern United States (SEUS) are a primary wintering ground for the endangered North Atlantic right whale (*Eubalaena glacialis*), used by calving females along with other adult and juvenile whales. Management actions implemented in this area for the recovery of the right whale population rely on accurate habitat characterization and the ability to predict whale distribution over time. We developed a temporally dynamic habitat model to predict wintering right whale distribution in the SEUS using a generalized additive model framework and aerial survey data from 2003/2004 through 2012/2013. We built upon previous habitat models for right whales in the SEUS and include data from new aerial surveys that extend the spatial coverage of the analysis, particularly in the northern portion of this wintering ground. We summarized whale sightings, survey effort corrected for probability of whale detection, and environmental data at a semimonthly resolution. Consistent with previous studies, sea surface temperature (SST), water depth, and survey year were significant predictors of right whale relative abundance. Additionally, distance to shore, distance to the 22°C SST isotherm, and an interaction between time of year and latitude (to account for the latitudinal migration of whales) were also selected in the analysis presented here. Predictions from the model revealed that the location of preferred habitat differs within and between years in correspondence with variation in environmental conditions. Although cow-calf pairs were rarely sighted in the company of other whales, there was minimal evidence that the preferred habitat of cow-calf pairs was different than that of whale groups without calves at the scale of this study. The results of this updated habitat model can be used to inform management decisions for a migratory species in a dynamic oceanic environment.

Editor: Z. Daniel Deng, Pacific Northwest National Laboratory, United States of America

Funding: This research was funded by the National Marine Fisheries Service (NMFS) [grant P.O. Number 6WCA90016; http://www.nmfs.noaa.gov/] and the Save the Manatee Trust Fund [http://myfwc.com/research/manatee/trust-fund/]. Other than a contribution from NMFS to original survey protocols and survey lines design for the right whale early warning system, the funders had no role in the analysis, decision to publish, or preparation of the manuscript.

Competing Interests: The authors have declared that no competing interests exist.

* E-mail: tim.gowan@myfwc.com

¤ Current address: Cooperative Institute for Marine and Atmospheric Studies, Rosenstiel School of Marine and Atmospheric Science, University of Miami, Miami, Florida, United States of America

Introduction

The North Atlantic right whale (*Eubalaena glacialis*) is highly endangered, and although it has received protection from the Endangered Species Act of 1973 and the Marine Mammal Protection Act of 1972, the species remains well below its optimum sustainable population level [1]. Threats to the recovery of this species include entanglement in fishing gear and collisions with ships [2]. Effective management decisions, and assessing risk of injury and mortality from these threats, require knowledge of right whale habitat preferences to identify areas where whales are likely to occur.

Western North Atlantic right whales occupy nearshore habitats of North America, with their primary feeding grounds from spring through autumn off the coast of New England and eastern Canada. The coastal waters off the southeastern United States (SEUS), however, have been identified as their primary calving grounds, with cow-calf pairs and some juveniles and adults occupying the waters off Florida (FL) and Georgia (GA) during winter months (December–March) [3]. Several management measures have been established in the SEUS due to threats from human activities in this region, especially the risk of ship strikes from the high volume of shipping traffic associated with the ports of Jacksonville, FL, Fernandina, FL, Brunswick, GA, and Savannah, GA. Measures designed to reduce the likelihood of ship collisions with right whales in high-risk areas include a mandatory ship reporting system, seasonal management areas with ship speed restrictions, recommended transit lanes for large ships, and aerial surveys during the calving season (Figure 1). In an effort to enhance protection measures for North Atlantic right whales, aerial surveys in the SEUS have been supported since the early 1990s by a collection of agencies, including the National Oceanic and Atmospheric Administration National Marine Fisheries Service (NOAA Fisheries), the US Coast Guard, the US Navy, and the US Army Corps of Engineers. As part of that effort, a multiagency, coordinated survey network, termed the Early Warning System (EWS), has been in place since the 1993/1994 calving season [4]. The level of effort varied in the early surveys, but a core survey area off the Florida-Georgia border was regularly flown [5]. In 2003 EWS survey lines were modified as depicted in Figure 1 to provide consistent coverage of nearshore waters from Butler Beach, FL (29.76° N) to Sapelo Sound, GA (31.56° N). A primary objective of these surveys is to locate right whales in the area and relay this information to mariners; other objectives include monitoring population vital rates and human-related injuries with the aid of photo identification and characterizing whale habitat use and distribution. Additional surveys off South Carolina and northern Georgia (SC-GA) were implemented in the 2004/2005 calving season (Figure 1) with similar objectives, providing coverage from Sapelo Sound, GA (31.56° N) to North Myrtle Beach, SC (33.82° N).

Effective management actions rely on the ability to know where and when right whales are likely to occur, yet the boundaries of management areas are typically static in time and space, while whale distribution varies spatially and temporally in response to dynamic environmental conditions [6,7]. Additionally, right whale distribution changes seasonally as whales migrate between wintering grounds in the SEUS and feeding grounds at higher latitudes. If whale distribution can be modeled and predicted from environmental variables and behavioral patterns, management strategies can be developed that take into account system variability and uncertainty. Several predictive habitat models have been developed for the right whale wintering grounds in the SEUS [5,8,9,10]. However, these models did not analyze the most current survey data, including surveys with more consistent protocols (see Methods) and data from the recent SC-GA surveys, which have significantly expanded coverage in the northern section of the wintering ground. Accurate inferences from species distribution models are frequently limited to the extent (spatial, temporal and environmental) of the input data (*e.g.*, [11]), and development of these models, particularly for rare and endangered species, should be an iterative process validated and informed by newly available data [12,13]. Right whale abundance (sighting rates) observed during more recent surveys in the northern portion of the wintering ground was not consistent with predictions from previous models. Additionally, none of the previous right whale habitat models for the SEUS accounted for latitudinal migration patterns or for differences in the probability of whale detection due to variable platforms (type of plane) and survey conditions, as has been done in other studies (*e.g.*, [14]).

In this study, we characterized right whale habitat in the SEUS using generalized additive models (GAMs) to relate whale sightings from aerial surveys to static and dynamic environmental variables. We modeled the occurrence (presence-absence) and relative abundance of whales using a hurdle model. We analyzed recent survey data (from 2003/2004 through 2012/2013), with spatial coverage ranging from southern Florida through South Carolina. Our model was developed at a two-week temporal resolution, and it included survey effort, adjusted for survey platform and sea state, and inter- and intra-annual effects as covariates. We generated predictions from this model, and the results demonstrate that right whale distribution varies within and among years. Additionally, we compared environmental conditions for sightings with a calf present to those without calves to determine whether cow-calf pairs utilize habitat that is different from other whales that migrate to the SEUS.

Methods

Ethics Statement

Whale surveys were conducted under permits #0594-1467, #594-1759-00 and #15488 issued to the Georgia Department of Natural Resources and permits #655-1652 and #14233 issued to Scott Kraus (New England Aquarium) by NOAA Fisheries. All research protocols were reviewed by NOAA's Office of Protected Resources and complied with the Endangered Species Act and the Marine Mammal Protection Act.

Aerial Surveys

Aerial surveys for right whales have been conducted in the SEUS during the right whale calving season (December–March) since 1991, but effort intensity and spatial coverage have varied considerably among years outside of a core EWS area [5]. We analyzed survey data from 2003/2004 through 2012/2013 for this study because survey methods were standardized and the location of EWS transect lines was consistent during these seasons. Sightings and effort data for these surveys were provided by the North Atlantic Right Whale Consortium [15]. Since 2003/2004, EWS surveys have been flown along fixed track lines running east-west, spaced 3 nautical miles (5.56 km) apart, and extending 24–34 nautical miles (44–63 km) offshore (Figure 1). Surveys were conducted from a Cessna 337 Skymaster or a de Havilland DHC-6 Twin Otter aircraft with a target altitude of 1000 ft (305 m). EWS surveys were designed to be flown daily during the calving season and complete the defined survey track lines, dependent upon weather conditions and aircraft availability. The SC-GA surveys began in 2004/2005 using Skymaster aircraft and the same data-collection protocols as in EWS surveys but with dedicated track lines running east-west and spaced 3 nautical miles (5.56 km) apart in the southern portion of the survey area and running northwest-southeast and spaced approximately 4 nautical miles (7.52 km) apart in the northern portion (Figure 1). Additional coastal ("Florida nearshore," *sensu* [5]) surveys south of these track lines were flown in some years. Most effort during coastal surveys was within 10 nautical miles (18.5 km) off and parallel to the shoreline (generally in a north-south direction).

During aerial surveys, one observer on each side of the aircraft searched for right whales, and a computer program was used to automatically record geographic location (latitude, longitude, and altitude) obtained from the aircraft's GPS every 10–30 seconds. Environmental conditions (sea state, visibility, weather) were recorded at the beginning of the survey and updated as conditions changed. When a right whale was observed, the aircraft deviated from the designated track line to approach the sighting location. Whale locations were recorded as the GPS location when the aircraft flew directly over the whale. After recording the behavior and number of whales and obtaining photographic documentation, the aircraft returned to the designated track line to resume survey.

Survey Data Processing

Data recorded during each flight from all EWS, SC-GA, and coastal surveys were entered into a GIS (ArcGIS version 10.0, Esri Inc., Redlands, CA), and the equidistant Universal Transverse Mercator (UTM) projection was used for data analysis. We filtered survey data to only include portions considered "on-effort": sea state ≤3 (Beaufort scale), altitude ≤365 m, visibility ≥3.7 km, and flying along a designated track line (*i.e.*, not in transit or circling a whale). Whale sightings were considered on-effort if that was the survey status at the time of the initial sighting. We removed all verification survey sightings and their associated effort (*i.e.*, surveys conducted to locate and verify a reported whale sighting) and all duplicate sightings (*i.e.*, whales already sighted on the same survey, as verified by photo identification). Each segment of a flight path was buffered on both sides with an effective search width, according to survey platform and sea state, to estimate the searched area. To determine effective search widths, perpendicular sighting distances were calculated from survey track lines and

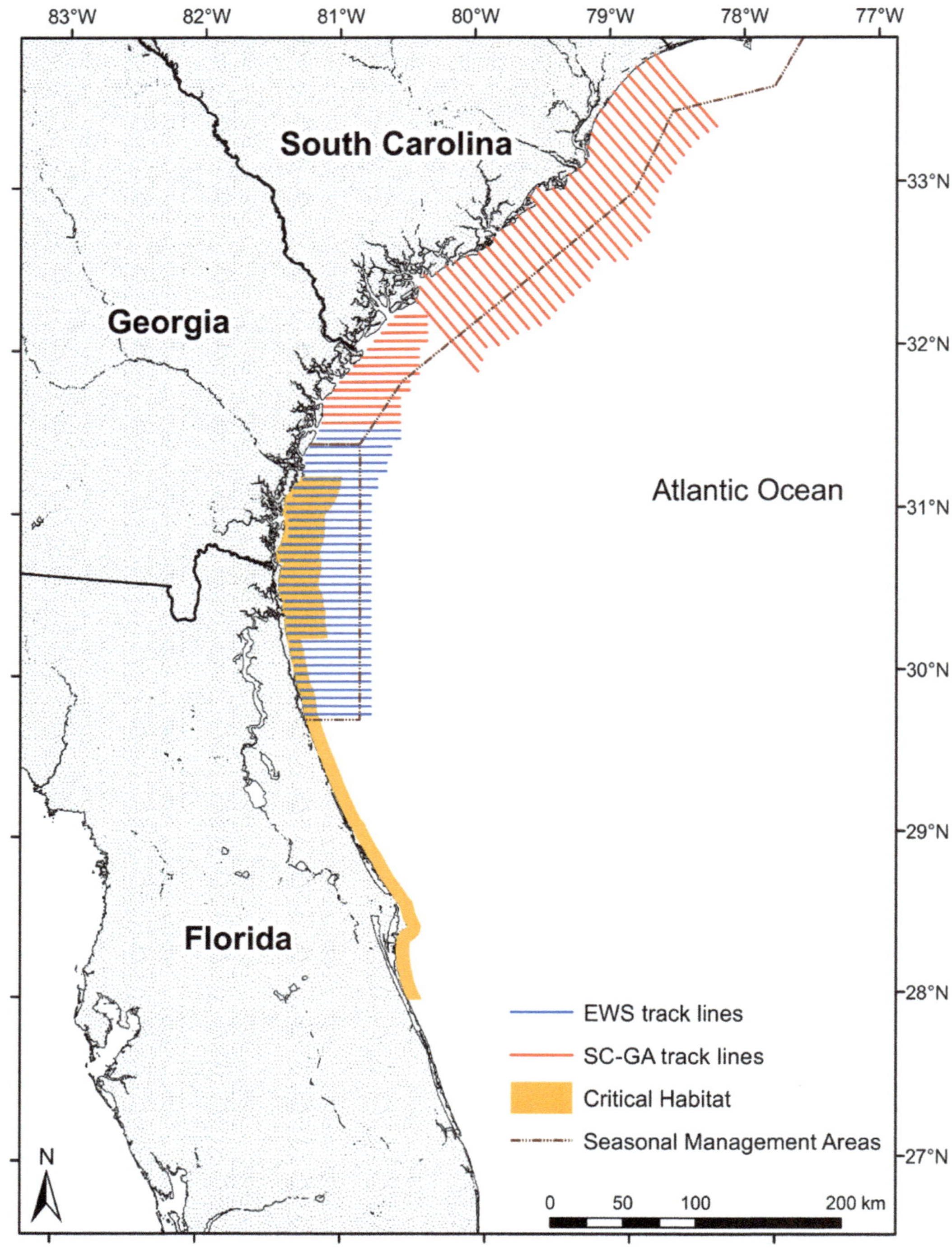

Figure 1. Aerial survey track lines for right whales in the southeastern United States. Transect lines are from the Early Warning System (EWS) and South Carolina/Georgia (SC-GA) survey areas. Critical habitat and seasonal management area boundaries are included for reference.

right whale sighting locations [16], and the multiple-covariate distance sampling (MCDS) engine in Distance 6.0 software was used to model detection probabilities [17]. Separate detection functions were created for each survey platform (Skymaster and Twin Otter aircrafts) using sea state as a covariate in MCDS, resulting in distinct effective search widths (range = 1.3–2.2 km) for each platform/sea state combination. Because the Skymaster has flat windows resulting in a blind spot beneath the aircraft, a section corresponding to 0.186 km (unpublished field data) on both sides of the flight path was removed from the estimated searched area for surveys from this platform. The Twin Otter had no blind spot because it has bubble windows that allowed observers to search directly under the aircraft.

A composite sampling grid for the study area was constructed to accommodate disparities in track line spacing and orientation, consisting of 5.56−×5.56-km cells oriented east-west in the south and 7.52−×7.52-km cells oriented northwest-southeast in the north (Figure 2). We overlaid the sampling grid onto the study area so the track lines bisected the grid cells, allowing all survey effort within a cell to be associated with a single track line. Because whale sightings were so rare, we aggregated all survey and environmental data into the grid cells at a semimonthly (approximately 2 week) temporal resolution–either the 1st through the 15th day of a month (A) or the 16th through the last day of a month (B)–to increase the number of records with whale presence, while maintaining a time frame with relatively stable environmental conditions. Thus each survey year consisted of eight semimonthly periods (December A through March B). Each cell summarized over a semimonth is hereafter considered a sampling unit. For each semimonth, the searched area (km^2) from all surveys during that period was summed within each grid cell and used as the measure of survey effort. Two sampling units (survey effort > 340 km^2) were identified as outliers on a Cleveland dotplot [18] and were excluded from analysis.

Environmental Data

Geographic locations (UTM easting and northing) were taken at the center point of each grid cell. Distance to shore was calculated as the distance between the center point of each cell and the high-resolution NOAA composite shoreline GIS layer (http://shoreline.noaa.gov). Bottom depth data were obtained from the 30-arc-second-resolution data set of the General Bathymetric Chart of the Oceans (http://www.gebco.net). Depth was calculated as the mean bottom depth of all values within each grid cell; cells with mean depth above sea level or deeper than 70 m (22 sampling units considered outliers) were excluded from analysis. Bottom slope (degrees) was derived from bathymetric data using Spatial Analyst in ArcGIS and was also summarized as the mean of values within each cell. Sea surface temperature (SST) was derived from Advanced Very High Resolution Radiometer (AVHRR) 1.47-km-resolution imagery from NOAA's CoastWatch data set (http://cwcaribbean.aoml.noaa.gov/data.html) for the East Coast South region. Within each semimonth all available daily images with minimal cloud cover were downloaded (mean = 8.1 images), image pixels with cloud interference were removed, and the semimonthly mean SST was calculated at each pixel. SST for a sampling unit was calculated as the mean of all semimonthly mean SST 1.47 km pixels within each cell. The potential environmental predictors mentioned above were chosen based on previous studies of right whales in the SEUS [5,8,9,10]. Additionally, the semimonthly mean SST data were used to estimate the location of the 22°C SST isotherm (Spatial Analyst in ArcGIS 10.0); this isotherm was chosen because 22°C is at the upper SST range for right whales in this region [5] and can be used as a proxy for the Gulf Stream boundary in winter [19]. Distance from this isotherm was calculated from the center point of each grid cell; distances for sampling units with SST >22°C were set as negative values, indicating that a cell was east or south of the isotherm.

Model Framework

GAMs were used to relate the number of right whale sightings to possible predictor variables. GAMs are extensions of generalized linear models (GLMs) that allow for smooth, nonlinear functions of predictor variables determined by observed data rather than by strict parametric relationships [20,21]. Like GLMs, GAMs use a specified error distribution for the response variable and a link function to relate the response variable to the predictor variables. Because the response variable, number of sighted whales, was overdispersed and zero-inflated due to the large number of sampling units (96%) with no sightings, we used a hurdle model [22]. A hurdle model consists of two steps: modeling presence-absence with a binomial distribution, and then modeling positive abundance, conditional on presence. Hurdle models are therefore useful for modeling zero-inflated data and allow the modeled process which determines presence to differ from the process which determines abundance [22]. We first used a quasibinomial distribution (to deal with excessive number of zeros) with a logit link to model presence-absence from all data. We then used a gamma distribution with a log link to model the number of whales from sampling units with whale sightings [23]. Predicted relative abundance can be calculated by multiplying the probability of occurrence, derived from the first model, by the expected number of whales, derived from the second model. Often, models of count data from wildlife surveys include effort as

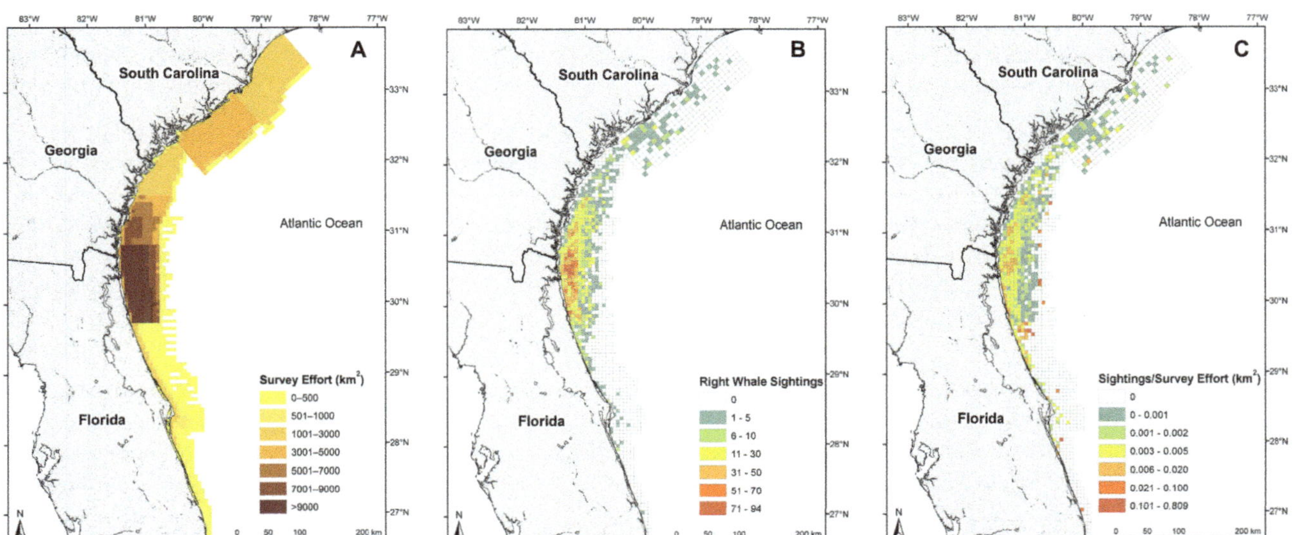

Figure 2. Aerial survey effort and right whale sightings in the southeastern United States. Values represent cumulative area surveyed (A), cumulative number of whales sighted (B), and number of whales sighted divided by area surveyed (C) per grid cell between December 2003 and March 2013 while observers were on-effort.

an offset term, assuming this variable as a linear predictor with a coefficient equal to one [21]; our data, however, did not meet this assumption (see Results), and we included effort as a smoothed covariate. GAMs were constructed using the R statistical software (version 3.0.1) with the mgcv package (version 1.7-24; [24]). The mgcv package determines the appropriate degrees of freedom (df) and smoothing for each predictor variable by minimizing the generalized cross validation (GCV) score [21]. To avoid overfitting by the model and to limit complexities of the smoothing function to an ecologically interpretable relationship, we set the basis dimension parameter to 3, thereby limiting the maximum df for each term to 2 (*e.g.*, [25,26]).

The response variable for the hurdle model was the total number of right whales sighted in each sampling unit. An individual whale could have been sighted multiple times within the same semimonth, and we did not correct for availability bias (*e.g.*, whales submerged at time of survey); therefore, our model is representative of the relative spatial distribution of whales rather than absolute abundance or density. Possible predictor variables included depth, distance to shore, SST, distance to the 22°C isotherm, and survey effort. Survey year was included as a factor variable to account for inter-annual variation in the total number of right whales in the SEUS, and the interaction between northing and semimonthly period was used to account for intra-annual shifts in distribution resulting from the timing of the whale migration. Easting was ultimately excluded as a predictor variable in favor of more informative environmental variables with which it was correlated. Slope was also excluded because all survey effort occurred on the continental shelf within a narrow range of values (0.01–1.29°), and it was consistently found to be the least significant term in a stepwise selection (see Table S1). Although collinearity was present among some of our explanatory variables, its impact was minimized by removing easting and slope to reduce variance inflation factors for the remaining predictors to <4.8 [27], employing penalized regression splines with shrinkage in the mgcv package to estimate smoothing functions [28,29], and limiting model predictions to the range of sampled data [29].

Model Selection and Validation

Model selection was accomplished with a forward stepwise selection procedure, using the following evaluation criteria: model GCV scores, percentage of deviance explained, and analysis of deviance tests. Starting with a null model, each term was added individually; the term resulting in the lowest GCV score was included in the next step. At each step, analysis of deviance was used to determine whether increasing model complexity with the addition of the selected term significantly improved the model. A term was not included in the final selected model if the confidence interval for the fitted response included zero for all observed values of the term or if the model at the previous step (which did not contain the term) had a lower model GCV score [24].

A five-fold cross-validation was used to evaluate each candidate model's performance in predicting novel data [30]. Model training was based on a random subset of 80% of the data and used to predict the validation subset (remaining 20%). Average squared prediction error (ASPE) was calculated from the predicted and observed number of whales sighted in the validation subset. This cross-validation was run five times, and the mean ASPE was calculated for each model to assist in model selection. Final specification of the selected best model used to estimate smoothing functions and create prediction maps was based on the complete data set.

Spatial autocorrelation in species distribution models can be problematic because it violates the assumption that residuals are independently distributed. We examined whether spatial autocorrelation was present by calculating Moran's I of deviance residuals of predicted relative abundance for all 80 semimonths (8 semimonths/year × 10 years) using the R-package spdep (version 0.5-56; [31]). Moran's I was calculated as a global statistic, considering all grid cells as neighbors, with an inverse distance weighting such that nearby grid cells exerted greater influence. Moran's I ranges from -1 (perfect negative correlation) to $+1$ (perfect positive correlation), with values near 0 indicating no spatial autocorrelation present.

Segregation of Calves

Studies of baleen whales, including southern right whales (*Eubalaena australis*), indicate that cows with calves may prefer environmental features different from those used by other demographic groups on wintering grounds [32,33], and it has been suggested that cow-calf pairs receive fitness benefits by segregating themselves from the harassment of other whales [34] or by avoiding areas with more predators [8]. We therefore compared environmental variables (depth, distance to shore, SST, distance to the 22°C isotherm, and UTM northing) at locations of sightings with calves present to those without calves. For this analysis, we overlaid exact sighting locations on the original environmental GIS layers (not the sampling grid) to maximize resolution for each variable. SST values were estimated only for sightings that occurred on days for which SST images were available and at locations free of cloud interference. Isotherms were constructed as stated above, based on semimonthly mean SST data. We excluded all sightings that were not on-effort or those with an estimated depth above sea level due to data precision. Sightings were classified based on the presence of a calf, pooled across years, semimonths, and levels of survey effort, and compared using a Mann-Whitney test. Due to the potential effect of migration timing, we tested for differences in UTM northing in each semimonthly period using a series of pair-wise comparisons with sequential Bonferroni correction [35].

Results

A total of 2191 surveys was flown in the EWS and SC-GA areas from December 2003 through March 2013, resulting in sightings of 3286 right whale groups and 7369 whales (not unique individuals, as some whales were resighted within and between seasons) while on-effort. The greatest concentration of both survey effort and sightings occurred in the EWS area near the coast of northern Florida and southern Georgia (Figure 2). A total of 56143 sampling units, containing 3104 sightings of 6953 whales, was retained for analysis after removing statistical outliers and those with missing SST values due to cloud interference.

Occurrence Model

For the presence-absence model, the GAM stepwise selection procedure yielded higher explained deviances, lower GCV scores, and lower ASPE values at each step, signifying that models had a better fit as complexity increased (Table 1, Table S1A). This finding was supported by analysis of deviance tests at each step; the more complex models were consistently identified as the most parsimonious despite having more parameters (Table S2A). The selected best model included seven predictor variables as significant (survey effort, SST, distance to the shoreline, bottom depth, interaction between semimonthly period and UTM northing, survey year, and distance to the 22°C SST isotherm) and explained 22.8% of the total deviance (Table 1). This model had the lowest ASPE values for all validation data sets,

Table 1. Summary of stepwise selection procedure for presence-absence model of right whales in the southeastern US.

Model	% Deviance	GCV	mean ASPE
null	0	0.3417	0.0392
s(SemiMonth:Northing)	10.3	0.3065	0.0377
s(SemiMonth:Northing)+s(DistToShore)	15.7	0.2882	0.0364
s(SemiMonth:Northing)+s(DistToShore)+Year	18.8	0.2777	0.0352
s(SemiMonth:Northing)+s(DistToShore)+Year+s(Effort)	20.9	0.2706	0.0345
s(SemiMonth:Northing)+s(DistToShore)+Year+s(Effort)+s(SST)	22.3	0.2658	0.0342
s(SemiMonth:Northing)+s(DistToShore)+Year+s(Effort)+s(SST)+s(Depth)	22.6	0.2647	0.0340
s(SemiMonth:Northing)+s(DistToShore)+Year+s(Effort)+s(SST)+s(Depth)+s(DistTo22Iso)	22.8	0.2642	0.0340

Predictor variables include interaction between semimonthly period and UTM northing, distance to the shoreline (DistToShore), survey year, survey effort, sea surface temperature (SST), bottom depth, and distance to the 22°C SST isotherm (DistTo22Iso). Smoothed covariates indentified by "s()". Evaluation criteria include the proportion of deviance explained, generalized cross validation score (GCV), and mean average squared prediction error (ASPE) from a five-fold cross-validation.

demonstrating its superior ability to predict the spatial distribution of right whales from novel data in the study area.

Smoothing functions for all terms in this model indicated nonlinear relationships and were consistent with the locations of sightings from the surveys. As expected, the probability of a right whale sighting increased with increasing survey effort but leveled off when effort was >250 km^2 (Figure 3A); this result supports our decision to model survey effort as a smoothed covariate rather than as an offset term. The SST for sampling units with a right whale present ranged from 9.0 to 22.6°C, with a mean (± SE) of 14.7°C (±0.04). The GAM predicted the highest probabilities between 12 and 16°C (Figure 3B). Right whales were more likely to be sighted close to shore, particularly within 25 km of the coast (Figure 3C), and at intermediate depths, between 10 and 25 m (Figure 3D). Generally, sightings occurred in cooler waters far from the 22°C isotherm, with only one sighting made south or east of this isotherm. However due to the proximity of the Gulf Stream to shore, especially in the southern portion of the study area, whales were limited by land to a maximum distance from the 22°C isotherm, and the response to this variable reached a plateau near this distance (Figure 3E). The interaction term between semimonth and northing was remarkably useful for capturing temporal shifts in distribution due to right whale migration in and out of the SEUS. Consistent with sighting data, the GAM predicted the highest probability of a whale sighting in January and February. Across all semimonthly periods, sighting rates were highest in intermediate northings of our study area, between latitudes of approximately 29.5 and 31.0°N. The interaction of these variables, however, demonstrated that whales occur in more northerly areas as they arrive at the beginning of the calving season in December, shift farther south in the middle of the season in January and February, and return north as they depart from the SEUS in March (Figure 3F). Survey year was also a significant predictor variable, as some years (*e.g.*, 2007/2008 and 2008/2009) had higher sighting rates than others (Figure 3G, Table S3).

Abundance Model

For the positive abundance model, the selected best model included six predictor variables as significant (SST, distance to the shoreline, distance to the 22°C SST isotherm, bottom depth, interaction between semimonthly period and UTM northing, and survey year) and explained 12.2% of the deviance for the presence-only data (Table 2). Adding survey effort increased the GCV score and did not significantly decrease the explained deviance for modeling the number of whales (Table S1B, Table S2B), so this term was excluded from the model.

The range of training data for the positive abundance model included only data from sampling units with whale sightings present, and the response for this model is the expected number of sighted whales per sampling unit, given whale presence. In general, smoothing functions in this model were similar to those in the presence-absence model, with more whales likely in sampling units with intermediate SST (Figure 4A), close to shore (Figure 4B), and in cooler waters far from the 22°C isotherm (Figure 4C). More whales were predicted, although with high uncertainty, near the upper limit of depth values with whales present (Figure 4D). Survey year was again significant, although the likelihood of multiple whales sighted in a sampling unit for a given year did not necessarily correspond with the overall sighting rate or probability of occurrence (Figure 4E, Table S3). Of the sampling units with whales present, multiple whales were more likely in sampling units at southerly northings and when whale densities were greatest, from late January through late February (Figure 4F). Smoothing functions for the positive abundance model generally had a higher standard error than those for the presence-absence model due in part to the smaller sample size.

Model Predictions

The final, combined hurdle model was successful at predicting the number of observed whales, as deviance residuals were overall close to zero (mean ± SE = −0.219±0.003). However, examination of the model residuals revealed that, in general, the model tended to overpredict when whales were not sighted (*i.e.*, predicting whale occurrence in sampling units with no whales observed) and underpredict when whales were sighted. For sampling units where no whales were sighted, the mean observed number of whales – the mean predicted number of whales = 0–0.106. For sampling units where whales were sighted, the mean observed number of whales – the mean predicted number of whales = 3.027–0.524. Excluding the last semimonthly period from March 2013, we found no evidence for spatial autocorrelation of the residuals during any of the semimonthly periods, as all Moran's I values approximated zero (mean = 0.030, range = −0.002 to 0.158). Moderate spatial autocorrelation was observed for late March 2013 (Moran's I = 0.376), although no sightings were made during this semimonth and all residuals were thus negative.

Using the selected hurdle model, we created hindcasts predicting the spatial distribution of whales for all semimonthly periods in the study. With the predict.gam function in the mgcv

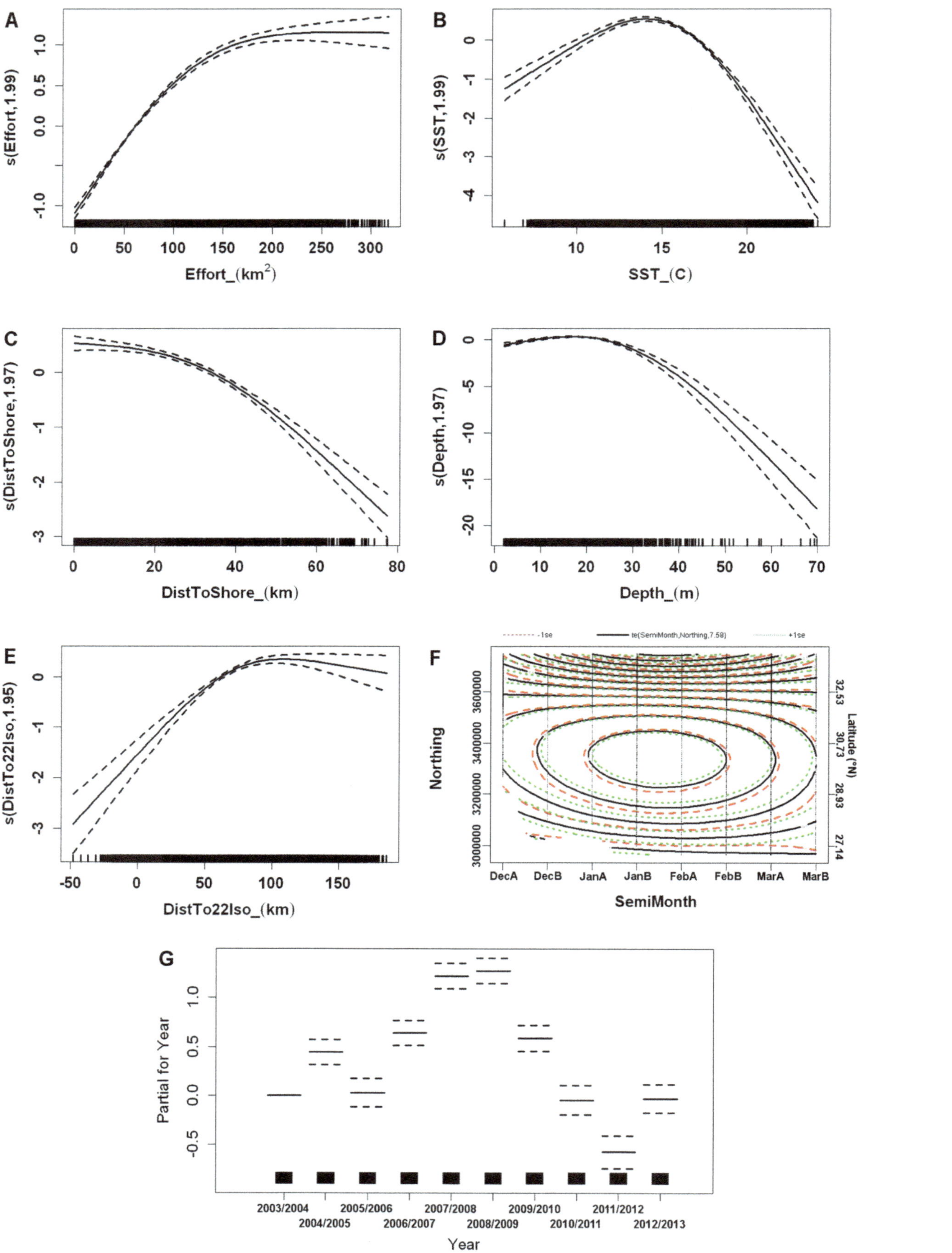
A
s(Effort,1.99)
Effort_(km²)
B
s(SST,1.99)
SST_(C)
C
s(DistToShore,1.97)
DistToShore_(km)
D
s(Depth,1.97)
Depth_(m)
E
s(DistTo22Iso,1.95)
DistTo22Iso_(km)
F
-1se
te(SemiMonth,Northing,7.58)
+1se
Northing
Latitude (°N)
32.53
30.73
28.93
27.14
DecA
DecB
JanA
JanB
FebA
FebB
MarA
MarB
SemiMonth
G
Partial for Year
2003/2004
2004/2005
2005/2006
2006/2007
2007/2008
2008/2009
2009/2010
2010/2011
2011/2012
2012/2013
Year

Figure 3. Smooth functions of predictor variables for presence-absence model of right whale sightings. Predictor variables include survey effort (A), sea surface temperature (B), distance to the shoreline (C), bottom depth (D), distance to the 22°C isotherm (E), interaction between semimonthly period and UTM northing (F), and partial effects for survey year (G). The y-axis is on the scale of the linear predictor, and dashed lines indicate ±2 standard errors. Estimated degrees of freedom for each smoothed variable are in parentheses on the y-axis; tick marks on the x-axis indicate all sampled values.

package, the observed values for predictor variables (except survey effort) in all grid cells were used to generate the predicted and standard error estimates for probability of a whale sighting and number of whales sighted from the first and second steps of the hurdle model, respectively. Predicted relative abundance was then calculated by multiplying the probability of occurrence by the expected number of sighted whales. For these predictions, survey effort was set constant for all grid cells (at 250 km^2, near which the smoothing function for the presence-absence model reached a plateau) to eliminate the confounding effect of variable survey effort on number of sightings. We did not extrapolate our results or predictions outside the range of our sampled data (*e.g.*, where depth was >70 m, the maximum value in our training data), nor did we make predictions for cells with missing data (*e.g.*, where SST data were lacking due to cloud interference). For illustrative purposes, we have included prediction maps for the first 15 days of each month for the 2009/2010 and 2011/2012 seasons (Figure 5). In general, sea surface temperatures were near average in December 2009 but colder than average for the remainder of the 2009/2010 season (Figure S1), and this season had intermediate sighting rates relative to other years in our data set (Figure 3G, Table S3). In contrast, the 2011/2012 season was warmer than average (Figure S1) and had low observed sighting rates relative to other years in our data set (Figure 3G, Table S3). Consistent with the smoothing functions and observed sightings from surveys, the greatest concentrations of predicted whale sightings for all semimonths occurred close to shore, in the relatively shallow and cooler waters west of the Gulf Stream (Figure 5). High predicted sighting rates were more common and distributed farther south in January and February than in December and March (Figure 5). Because 2009/2010 had a greater partial effect for the year term (Figure 3G), this season had higher predicted sighting rates overall compared to 2011/2012 (Figure 5). Notably, areas within the SEUS with the highest predicted number of whales were located farther south in a cold year (2009/2010; *e.g.*, Figure 5B) compared to a warm year (2011/2012; *e.g.*, Figure 5F). The presence-absence model seemed to drive the pattern for predicted relative abundance (Pearson correlation coefficient between predicted probability of presence and predicted relative abundance = 0.98); standard errors were generally larger for cells with high predicted probability of occurrence.

Segregation of Calves

Of the total 3286 sightings, 1344 (41%) were of unaccompanied cow-calf pairs and 1919 (58%) were sightings without calves (*i.e.*, juveniles, adult males, non-calving adult females, pregnant females, and females that had lost a calf that year). The remaining 1% of sightings was composed of a calf by itself (6), two cow-calf pairs in close association (4), or a cow-calf pair accompanied by other individuals (13). We were able to extract SST data for 459 sightings with a calf present and 675 sightings with a calf absent, and all other environmental data for 1349 sightings with a calf present and 1907 sightings with a calf absent. Sightings with a calf present were made in locations significantly shallower and closer to shore and tended to occur in warmer water than sightings without calves, although the differences were small and there was extensive overlap in the range of values between group types (Table 3). Distance to the 22°C isotherm did not differ significantly among groups (Table 3). UTM northing did not differ significantly between groups with or without calves for any semimonthly period except the first period in December, when sightings without calves occurred farther north ($p<0.01$; Figure 6).

Discussion

In agreement with previous studies, our results indicate that GAMs are a useful tool for relating cetacean distribution to environmental variables and predicting cetacean occurrence and relative density based upon those variables [26,36]. The development of a GAM, along with the ability to collect environmental data via remote sensing and to apply GIS processing techniques, permitted us to interpolate our results and estimate relative abundance of right whales, with associated standard error, in regions of our study area not sampled by aerial surveys. The flexibility of GAMs allowed right whale encounter rates to be modeled as a complex, nonlinear response to predictor variables, and a hurdle model approach allowed us to deal with zero-inflated data from surveys of a rare species summarized at a high spatial resolution. By using dynamic predictor variables and accounting

Table 2. Summary of stepwise selection procedure for positive abundance model of right whales in the southeastern US.

Model	% Deviance	GCV	mean ASPE
null	0	0.4534	8.284
s(SemiMonth:Northing)	6.5	0.4261	7.981
s(SemiMonth:Northing)+Year	10.2	0.4126	7.842
s(SemiMonth:Northing)+Year+s(SST)	11.2	0.4082	7.822
s(SemiMonth:Northing)+Year+s(SST)+s(DistTo22Iso)	11.7	0.4066	7.819
s(SemiMonth:Northing)+Year+s(SST)+s(DistTo22Iso)+s(DistToShore)	11.8	0.4065	7.820
s(SemiMonth:Northing)+Year+s(SST)+s(DistTo22Iso)+s(DistToShore)+s(Depth)	12.2	0.4054	7.807

Predictor variables and abbreviations same as in Table 1.

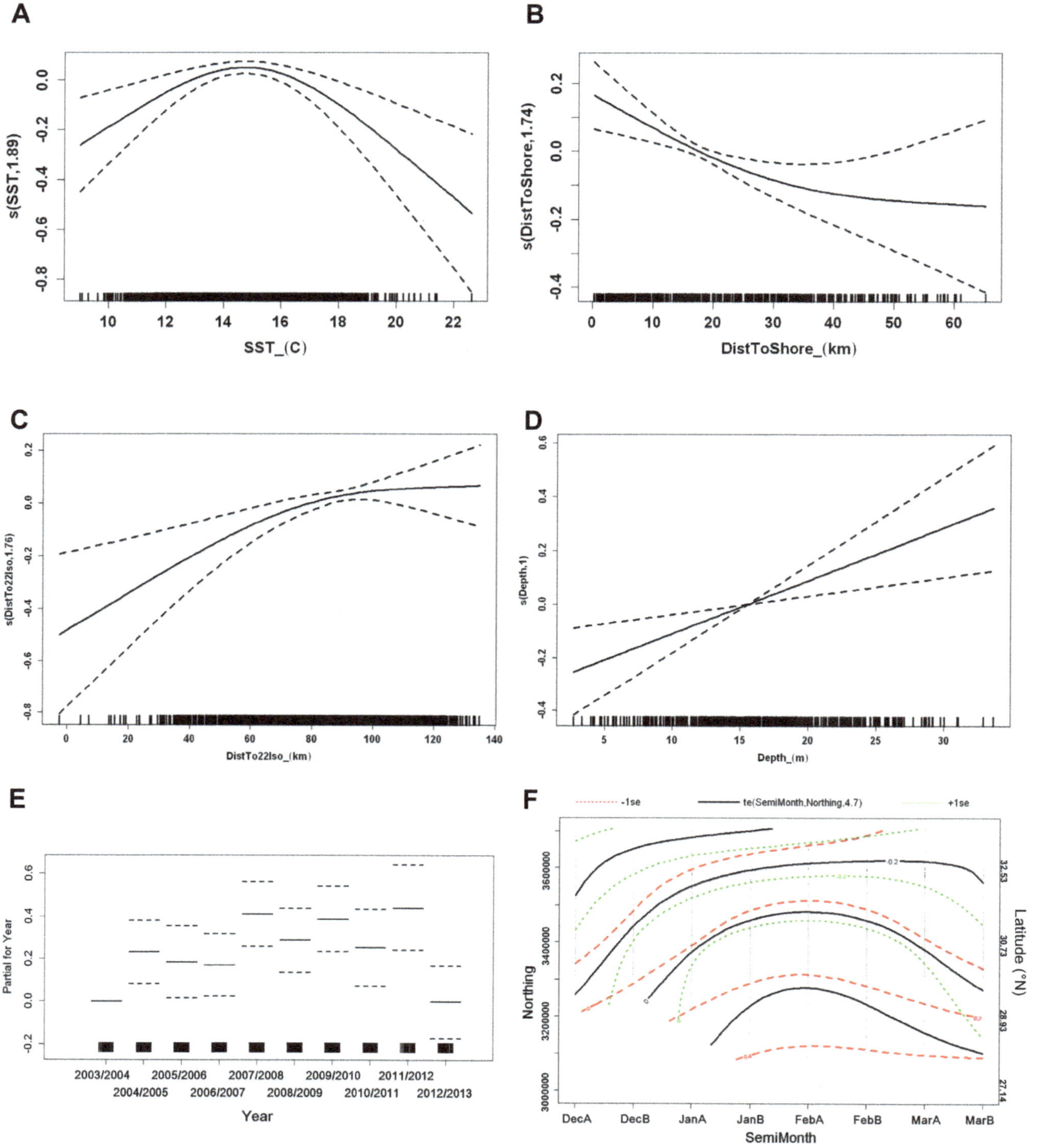

Figure 4. Smooth functions of predictor variables for positive abundance model of right whale sightings. Predictor variables and display same as in Figure 3.

for migration patterns, we were able to model changes in whale distribution through time.

Concordant with other studies in the SEUS [8,9,10], bottom depth and SST were significant predictors of right whale distribution; right whales were more likely to be sighted in waters 10 to 25 m deep and 12 to 16°C SST. Fonnesbeck *et al.* [9] and Keller *et al.* [10] also found year to be a significant factor variable related to differences in the total number of whales in the SEUS. Based on SST at sighting locations, Keller *et al.* [5,10] suggested that right whales likely avoid the Gulf Stream, although they did not include a term to represent Gulf Stream variability in their model (such as distance to the 22°C SST isotherm). Distance to

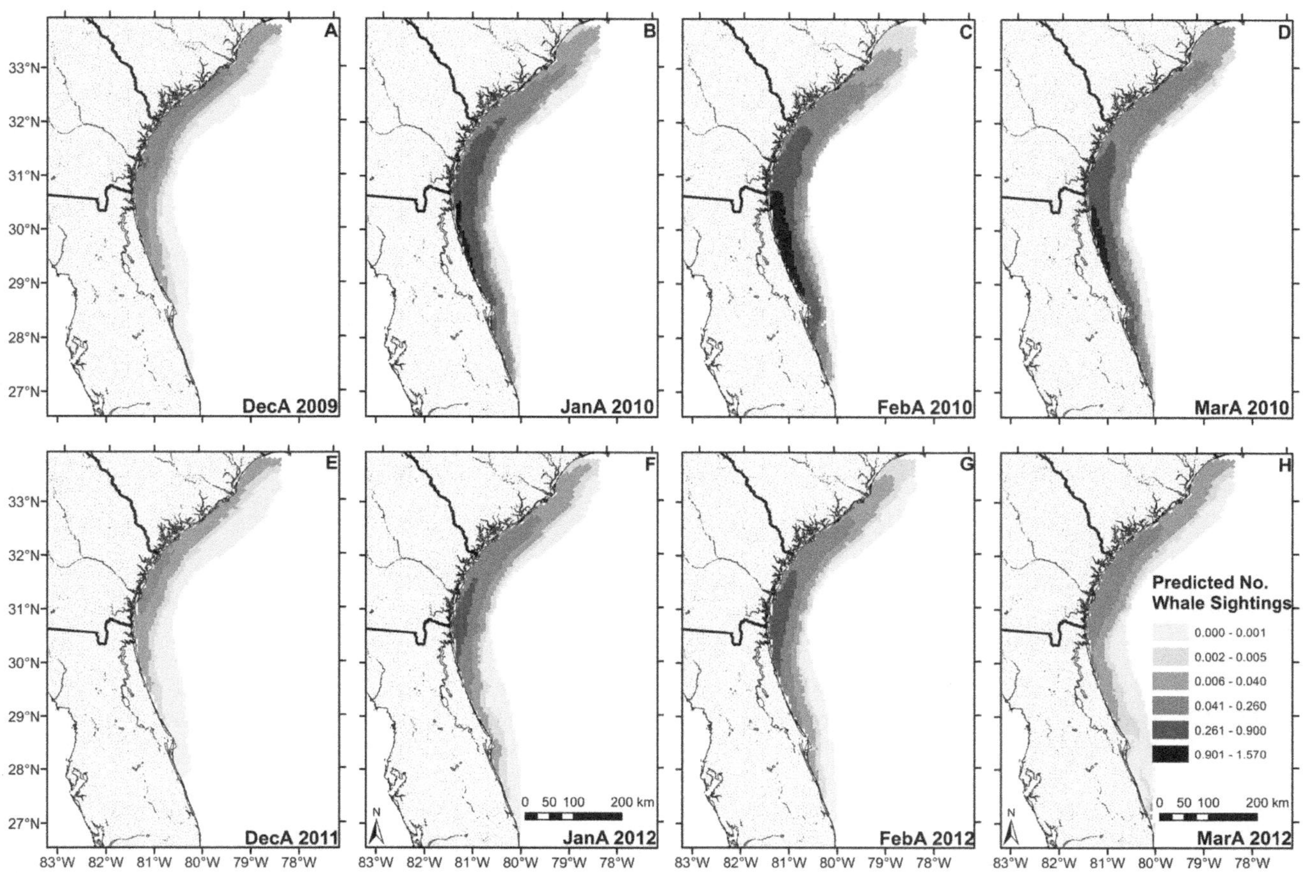

Figure 5. Predicted right whale relative abundance. Values represent predicted number of sighted right whales per grid cell (assuming uniform survey effort) during the 2009/2010 calving season (a relatively cold season with high sighting rates) for December 1–15 (A), January 1–15 (B), February 1–15 (C), and March 1–15 (D); and during the 2011/2012 calving season (a relatively warm season with low sighting rates) for December 1–15 (E), January 1–15 (F), February 1–15 (G), and March 1–15 (H).

shore was a significant predictor in our model, in contrast to these previous studies which either did not consider this term [8,9] or limited model complexity to four terms [10]. The range of preferred SST and avoidance of waters >22°C SST may be related to physiological optima and constraints on thermoregulation and growth, particularly in pregnant or nursing females and newborn calves [5,37,38]. It is less clear why whales were more common in shallow water close to shore, but it has been hypothesized that predator avoidance, weaker currents, and calm waters provided by these physical factors may be advantageous to calves and juveniles that are still developing swimming skills or more susceptible to predation [8,34]. Unfortunately, it will be difficult to adequately address these hypotheses unless predator abundance and wave/surface roughness data become available at finer temporal and spatial resolutions.

Our model also included the interaction between northing and semimonthly period to account for whale migration during the calving season. Both Good [8] and Keller *et al.* [10] predicted suitable habitat off northern Georgia and South Carolina throughout the winter. We, however, predict that these areas have relatively lower encounter rates compared to core-use areas off northern Florida and southern Georgia, particularly during the middle of the winter (Figure 5), congruent with sighting data (Figure 2). We believe that this difference in predictions is driven by the additional data with low sighting rates from the SC-GA surveys and by the inclusion of the latitude/semimonth interaction term in our model, which accounted for latitudinal shifts in whale distribution that were independent of SST variability. This is the first habitat model, to our knowledge, that explicitly considers a migration process in an area outside of the species' feeding grounds. Although the latitude/semimonth interaction term is only a proxy for the inherent processes that truly drive the migration timing, and it limits the application of the model to the area analyzed, it nevertheless improved the predictive capability of our model and is therefore useful for identifying areas in the SEUS with high whale encounter rates and predicting how these encounter rates change throughout a winter season. Moreover, the ecological factors that may influence the timing of whale migration into and within the SEUS (e.g., predation risk, calm seas, and historical distribution of con-specifics) are difficult to quantify and remain unknown. Although our model predicts that sightings are less common and whales likely spend less time in the area off northern Georgia and South Carolina, this does not mean that right whales do not use (and possibly give birth in) these areas. We hypothesize that, in most years, these more northerly areas are part of the right whale migration corridor [39,40]. Whales may display behaviors in these corridors that affect detectability (if they spend more time below the surface) and may have shorter residence times there than in the core wintering area. We therefore suggest that a separate model be created to characterize the right whale migration corridor when more data on migrating whales become available. Variability in sea surface roughness (not

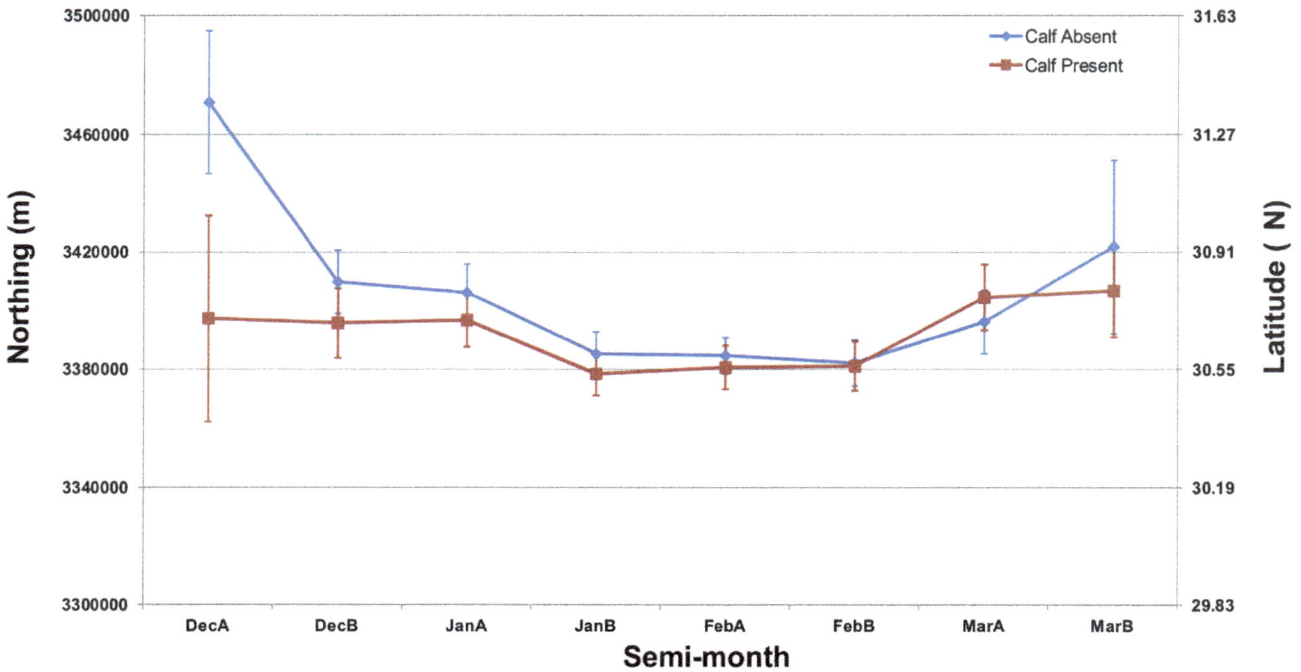

Figure 6. UTM northings for right whale sightings in the southeastern United States. Values indicate mean (±2 standard errors) UTM northing of sightings for each semimonthly period during all calving seasons. Sightings with a calf present displayed as red squares; sightings without a calf displayed as blue diamonds.

addressed in this study) and the occurrence of Gulf Stream meanders north of Charleston, SC, which generate onshore movements of unsuitably warm water [19,41], may yield a less stable habitat for whales in the northern areas of the SEUS. Nevertheless, we agree with Good [8] and Keller *et al.* [10] that in some years the core wintering area extends farther north than the current critical habitat boundary (*e.g.*, Figure 5G). Although extrapolating model results can help identify potential areas for future surveys [10,12], the different predictions among studies here highlight the importance of validating and updating models with additional data and may emphasize that caution is needed when applying model results beyond the spatial and temporal extent of training data [42].

Environmental variability among years leads to differences in both the total number of whales that migrate to the SEUS and the spatial distribution of whales once they arrive there (Figure 5, Keller *et al.* [10]). Variability in right whale spatial distribution among years seems to be driven by local conditions such as water temperature and location of the Gulf Stream, which vary in the SEUS among years and within a calving season (Figure S1). For example, whales are distributed farther south in relatively cold years (Figure 5), presumably to occupy areas of preferred SST. However, the number of whales that migrate to the SEUS in a given year and the amount of time they spend there are likely influenced by factors outside the SEUS [43,44]. Greene *et al.* [43] and Kenney [44] suggest that female right whale calving rates (and consequently the number of calving females that migrate to the SEUS each year) are determined by the number of females available to calve and by recovery times for reproductive females, ultimately influenced by food availability in summer foraging grounds. Encounter rates of whales will also be affected by the duration of whale residence in the SEUS, which may be a consequence of demographics (with calving females possibly having longer residence times than other whales [45]), energy reserves, changes in weather, or other, unknown factors. Although including year as a factor variable significantly improved our model fit and the ability to hindcast whale relative abundance for the years analyzed in our study, it precluded the ability to predict the magnitude of whale densities for years beyond our data set. Replacement of this term with a variable that describes the

Table 3. Mean (range) environmental conditions at locations of right whale sightings with a calf present or absent.

Variable	Calf Present	Calf Absent	z-score	*p*
Depth (m)	14.9 (3.0–34.0)	15.5 (2.0–33.0)	−2.95	0.003
DistToShore (km)	17.6 (0.2–62.1)	18.5 (0.1–66.2)	−2.11	0.035
SST (°C)	14.9 (9.9–21.3)	14.7 (8.9–22.9)	−1.84	0.066
DistTo22Iso (km)	86.2 (11.7–135.3)	85.7 (−3.9–132.3)	−0.44	0.664

Environmental variables include bottom depth, distance to the shoreline (DistToShore), sea surface temperature (SST), and distance to the 22°C SST isotherm (DistTo22Iso).

number and demography of right whales predicted to migrate to the wintering grounds each year (*e.g.*, [43]) would greatly improve the predictive capabilities of our model, particularly for forecasting the true value of whale encounter rates. Overall, the results of this study demonstrate that whale abundance and distribution vary within and between years. Recognizing and understanding this temporal variability can better inform the location and timing of management actions as whale distribution changes throughout a season, which would result in more effective risk mitigation and population monitoring of right whales.

While the final model explained the most variation in our dataset, overdispersion (partly caused by the high spatio-temporal resolution of our sampling units, the rarity of right whale sightings, and the potentially high number of false absences generated by aerial survey data) created challenges to achieving a good model fit. Indeed, it is common for GAMs modeling cetacean occurrence at a high spatial resolution, and non-normal models in general, to explain only a small proportion of the deviance in a dataset [26,36,46]. For example, our model frequently under-estimated the number of whales in grid cells with observed sightings, and this may be related to how the data were aggregated. Rather than predicting a high sighting probability in a single grid cell, the model smoothed predicted whale occurrence over several nearby cells with similar, preferred habitat conditions; in this regard, the model predictions may describe more accurately where a whale may be located over the span of two weeks (*i.e.*, temporarily occupying multiple nearby cells) than would an aerial survey sighting fixed in time and location. Conversely, the model tended to overpredict whale occurrence in areas where no whales were sighted. The small population size of right whales could have resulted in the absence of whales from suitable habitat [47]. Additionally, the lack of a sighting does not necessarily mean that whales were truly absent. Whales might not have been detected due to availability bias (*e.g.*, whales were submerged [48]), perception bias (*e.g.*, observer error [49]), or incomplete temporal coverage by surveys (*i.e.*, not flying at night, not flying during poor weather). In consideration of these limitations of aerial surveys and the confounding effect of unequal survey effort on the number of sightings, we believe that predictive habitat models can better characterize the distribution of whales than can sightings-per-unit-effort data (*e.g.*, Figure 2C). Rather than relying solely on costly surveys with imperfect detection probabilities and limited coverage, predictions from these models can be particularly helpful in making management decisions that require risk assessment based on expected whale distribution [50].

Whale groups with a calf present were sighted in waters that were shallower and closer to shore than were other whale groups. Even though shallow, nearshore locations in our study area generally had the coldest water, there was also a trend for groups with calves to be sighted in slightly warmer water. Sightings with a calf present occurred farther south during early December, although this finding is confounded by few sightings of calves early in the calving season and the fact that calves-of-the-year are not likely to be migrating from the north at this time. Studies of baleen whales on their wintering grounds have also found cow-calf pairs closer to shore and in shallower water than other demographic groups [32,33,51]. Differences in habitat preference may be a result of different energetic constraints or predation risks between demographic groups. Although our results were statistically significant and supported by other studies, the differences between groups were minimal, the range of values for environmental variables were very similar, and sightings of groups with calves compared with those without calves were not spatially segregated at the scale and spatial extent of our model. We therefore conclude that separate habitat models for cow-calf pairs and other demographic groups in the SEUS are not warranted and are impracticable at the scale used in this study, and management areas for one group will likely be equally effective for the other. Despite the large number of sightings in our data, however, it was extremely rare for a cow-calf pair to be sighted with another whale. Thus, at a fine scale, it appears that cow-calf pairs are segregated from other whales, but this pattern may be influenced more by social dynamics (*e.g.*, harassment avoidance; [34,52]) than by environmental features.

Our analysis presents an example of the use of static and dynamic environmental variables to predict species distribution and characterize habitat preferences of a migratory species in a dynamic seascape. Using temporally dynamic variables, such as SST, isotherms, and a seasonal migration index, we were able to predict semimonthly right whale distribution. Although our framework does not estimate absolute densities, it provides additional insights into how whale distribution changes over time compared to a single depiction of instantaneous density. This framework could be applied to other highly mobile migratory species in variable habitats that may benefit from dynamic management actions [6,53]. The model results corroborate that right whale distribution varies within and among years. We believe that the addition of data from a broader survey area relative to previous studies, and accounting for migration patterns, improved the model by predicting how whale distribution changes within a calving season and by separating core-use areas from short-term-use areas in the migration corridor. Hindcast predictions from this model can be used to assess risks to whales and effectiveness of management actions. Improvements of the model, including the addition of parameters for predicting the total number of whales that migrate to the wintering grounds [43], would allow for near-real-time forecasts of whale distribution that could be used to better inform management decisions. We encourage future evaluations of this and other habitat models as additional data, particularly from other predictors and from surveys in areas previously receiving little coverage, become available.

Supporting Information

Figure S1 Average daily sea surface temperatures for December through March. Data obtained from the National Data Buoy Center (http://www.ndbc.noaa.gov) at Gray's Reef (top panel, station 41008, 31.400°N 80.868°W) and at St. Augustine (bottom panel, station SAUF1, 29.857°N 81.265°W). Data include 2009/2010 (blue line), 2011/2012 (red line), long-term historical average with 95% confidence intervals (1991/1992 and 1997/1998–2012/2013 at Gray's Reef; 1988/1989–2001/2002 and 2004/2005–2011/2012 at St. Augustine), and long-term historical range.

Table S1 A. Summary of all models tested in stepwise selection procedure for presence-absence models of right whales in the southeastern United States. Predictor variables include interaction between semimonthly period and UTM northing, distance to the shoreline (DistToShore), survey year, survey effort, sea surface temperature (SST), bottom depth, distance to the 22°C SST isotherm (DistTo22Iso), and slope. Smoothed covariates indentified by "s()". Evaluation criteria include the proportion of deviance explained, generalized cross validation score (GCV), and mean average squared prediction error (ASPE) from a five-fold cross-validation. The best model at each step is in bold. B. Summary of all models tested in stepwise selection procedure for positive abundance models of right whales in the southeastern United

States. Predictor variables and abbreviations same as in Table S1A. The best model at each step is in bold.

Table S2 A. Results of analysis of deviance tests comparing presence-absence models at each step of the stepwise selection procedure. Models at each step refer to the respective best model (in bold) from Table S1A. Reductions in deviance, *F*-statistics, and *p*-values compare each model to the model in the previous step. B. Results of analysis of deviance tests comparing positive abundance models at each step of the stepwise selection procedure. Models at each step refer to the respective best model (in bold) from Table S1B. Reductions in deviance, *F*-statistics, and *p*-values compare each model to the model in the previous step.

Table S3 Number of observed sightings, number of observed whales (not unique individuals), total survey effort, and overall sighting rate (sightings/1000 km^2 surveyed) for each calving season from our data set.

Acknowledgments

We are grateful to the dedicated observers who conducted many hours of aerial surveys throughout the years and collected the data used in the analysis presented here. The North Atlantic Right Whale Consortium curates the aerial survey data provided to us by R. Kenney of the University of Rhode Island. We thank D. Dick, S. Cain, and A. Krzystan for their assistance with aerial survey data processing and GIS analyses. We also thank R. Muller, L. McEachron, L. Ward, B. Crowder, and two anonymous reviewers for their comments on this manuscript.

Author Contributions

Conceived and designed the experiments: JGO TAG. Performed the experiments: TAG JGO. Analyzed the data: TAG JGO. Contributed reagents/materials/analysis tools: TAG JGO. Wrote the paper: TAG JGO.

References

1. Waring GT, Josephson E, Maze-Foley K, Rosel PE, editors (2013) U.S. Atlantic and Gulf of Mexico marine mammal stock assessments - 2012. NOAA Technical Memorandum NMFS-NE-223: Available from: National Marine Fisheries Service, 166 Water Street, Woods Hole, MA 02543-1026. Available: http://www.nefsc.noaa.gov/nefsc/publications/. 319 p.
2. Knowlton AR, Kraus SD (2001) Mortality and serious injury of northern right whales (*Eubalaena glacialis*) in the western North Atlantic Ocean. J Cetacean Res Manag (Special Issue 2): 193–208.
3. Kraus SD, Rolland RM (2007) Right whales in the urban ocean. In: Kraus SD, Rolland RM, editors. The urban whale: North Atlantic right whales at the crossroads. Cambridge, MA: Harvard University Press. 1–38.
4. National Marine Fisheries Service (2005) Recovery plan for the North Atlantic right whale (*Eubalaena glacialis*). Silver Spring, MD: National Marine Fisheries Service, Office of Protected Resources. 137p.
5. Keller CA, Ward-Geiger LI, Brooks WB, Slay CK, Taylor CR, et al. (2006) North Atlantic right whale distribution in relation to sea-surface temperature in the southeastern United States calving grounds. Mar Mamm Sci 22: 426–445.
6. Redfern JV, Ferguson MC, Becker EA, Hyrenbach KD, Good C, et al. (2006) Techniques for cetacean-habitat modeling. Mar Ecol Prog Ser 310: 271–295.
7. Pendleton DE, Sullivan PJ, Brown MW, Cole TVN, Good CP, et al. (2012) Weekly predictions of North Atlantic right whale *Eubalaena glacialis* habitat reveal influence of prey abundance and seasonality of habitat preferences. Endanger Species Res 18: 147–161.
8. Good CP (2008) Spatial ecology of the North Atlantic right whale (*Eubalaena glacialis*) [Ph.D. dissertation]. Durham, NC: Duke University. 135 p.
9. Fonnesbeck CJ, Garrison LP, Ward-Geiger LI, Baumstark RD (2008) Bayesian hierarchichal model for evaluating the risk of vessel strikes on North Atlantic right whales in the SE United States. Endanger Species Res 6: 87–94.
10. Keller CA, Garrison L, Baumstark R, Ward-Geiger LI, Hines E (2012) Application of a habitat model to define calving habitat of the North Atlantic right whale in the southeastern United States. Endanger Species Res 18: 73–87.
11. Hamazaki T (2002) Spatiotemporal prediction models of cetacean habitats in the mid-western North Atlantic Ocean (from Cape Hatteras, North Carolina, U.S.A. to Nova Scotia, Canada). Mar Mamm Sci 18: 920–939.
12. Guisan A, Broennimann O, Engler R, Vust M, Yoccoz NG, et al. (2006) Using niche-based models to improve the sampling of rare species. Conserv Biol 20: 501–511.
13. Guisan A, Zimmermann NE (2000) Predictive habitat distribution models in ecology. Ecol Modell 135: 147–186.
14. Hedley S, Buckland S (2004) Spatial models for line transect sampling. J Agric Biol Environ Stat 9: 181–199.
15. North Atlantic Right Whale Consortium (2012) North Atlantic right whale consortium sightings database (New England Aquarium, Boston, MA, U.S.A.).
16. Buckland ST, Anderson DR, Burnham KP, Laake JL, Borchers DL, et al. (2001) Introduction to distance sampling: Estimating abundance of biological populations. Oxford: Oxford University Press. 432 p.
17. Thomas L, Buckland ST, Rexstad EA, Laake JL, Strindberg S, et al. (2010) Distance software: Design and analysis of distance sampling surveys for estimating population size. J Appl Ecol 47: 5–14.
18. Zuur AF, Ieno EN, Elphick CS (2010) A protocol for data exploration to avoid common statistical problems. Methods Ecol and Evol 1: 3–14.
19. Stegmann PM, Yoder JA (1996) Variability of sea-surface temperature in the South Atlantic Bight as observed from satellite: Implications for offshore-spawning fish. Cont Shelf Res 16: 843–861.
20. Hastie TJ, Tibshirani RJ (1990) Generalized additive models. Monographs on statistics and applied probability 43. London: Chapman & Hall. 335 p.
21. Wood SN (2006) Generalized additive models: An introduction with R. Boca Raton, FL: CRC Press.
22. Potts JM, Elith J (2006) Comparing species abundance models. Ecol Modell 199: 153–163.
23. Fauchald P, Skov H, Skern-Mauritzen M, Hausner VH, Johns D, et al. (2011) Scale-dependent response diversity of seabirds to prey in the North Sea. Ecology 92: 228–239.
24. Wood SN (2001) mgcv: GAMs and generalized ridge regression for R. R News 1: 20–25.
25. Austin M (2007) Species distribution models and ecological theory: A critical assessment and some possible new approaches. Ecol Modell 200: 1–19.
26. Best BD, Halpin PN, Read AJ, Fujioka E, Good CP, et al. (2012) Online cetacean habitat modeling system for the US East Coast and Gulf of Mexico. Endanger Species Res 18: 1–15.
27. Montgomery DC, Peck EA (1992) Introduction to linear regression analysis. New York: John Wiley. 527p.
28. Wood SN (2003) Thin plate regression splines. J R Stat Soc Series B Stat Methodol 65: 95–114.
29. Dormann CF, Elith J, Bacher S, Buchmann C, Carl G, et al. (2013) Collinearity: A review of methods to deal with it and a simulation study evaluating their performance. Ecography 36: 027–046.
30. Dormann CF, Purschke O, García Márquez JR, Lautenbach S, Schröder B (2008) Components of uncertainty in species distribution analysis: A case study of the great grey shrike. Ecology 89: 3371–3386.
31. Bivand R (2007) spdep. Spatial dependence: Weighting schemes, statistics and models. R Package Version 0.5–56.
32. Elwen SH, Best PB (2004) Environmental factors influencing the distribution of southern right whales (*Eubalaena australis*) on the south coast of South Africa I: Broad scale patterns. Mar Mamm Sci 20: 567–582.
33. Ersts PJ, Rosenbaum HC (2003) Habitat preference reflects social organization of humpback whales (*Megaptera novaeangliae*) on a wintering ground. J Zool 260: 337–345.
34. Elwen SH, Best PB (2004) Female southern right whales *Eubalanena australis*: Are there reproductive benefits associated with their coastal distribution off South Africa? Mar Ecol Prog Ser 269: 289–295.
35. Rice WR (1989) Analyzing tables of statistical tests. Evolution 43: 223–225.
36. Forney KA, Ferguson MC, Becker EA, Fiedler PC, Redfern JV, et al. (2012) Habitat-based spatial models of cetacean density in the eastern Pacific Ocean. Endanger Species Res 16: 113–133.
37. Brodie PF (1975) Cetacean energetics, an overview of intraspecific size variation. Ecology 56: 152–161.
38. Kshatriya M, Blake RW (1988) Theoretical model of migration energetics in the blue whale, *Balaenoptera musculus*. J Theor Biol 133: 479–498.
39. Knowlton A, Ring J, Russell B (2002) Right whale sightings and survey effort in the Mid-Atlantic region: Migratory corridor, time frame, and proximity to port entrances. A report submitted to the NMFS Ship Strike Working Group. Available: www.nero.noaa.gov/shipstrike/ssr/midatanticreportrFINAL.pdf.
40. Schick RS, Halpin PN, Read AJ, Slay CK, Kraus SD, et al. (2009) Striking the right balance in right whale conservation. Can J Fish Aquat Sci 66: 1399–1403.
41. Bane JM, Brooks DA (1979) Gulf Stream meanders along the continental margin from the Florida Straits to Cape Hatteras. Geophys Res Lett 6: 280–282.
42. Randin CF, Dirnböck T, Dullinger S, Zimmermann NE, Zappa M, et al. (2006) Are niche-based species distribution models transferable in space? J Biogeogr 33: 1689–1703.

43. Greene CH, Pershing AJ, Kenney RD, Jossi JW (2003) Impact of climate variability on the recovery of endangered North Atlantic right whales. Oceanography 16: 98–103.
44. Kenney RD (2007) Right whales and climate change: Facing the prospect of a greenhouse future. In: Kraus SD, Rolland RM, editors. The urban whale: North Atlantic right whales at the crossroads. Cambridge, MA: Harvard University Press. 436–459.
45. Burnell SR, Bryden MM (1997) Coastal residence periods and reproductive timing in southern right whales, *Eubalaena australis*. J Zool 241: 613–621.
46. Hosmer DW, Lemeshow S, Sturdivant RX (2013) Applied logistic regression. New York: John Wiley. 528p.
47. Pulliam HR (2000) On the relationship between niche and distribution. Ecol Lett 3: 349–361.
48. Hain JHW, Ellis SL, Kenney RD, Slay CK (1999) Sightability of right whales in coastal waters of the southeastern United States with implications for the aerial monitoring program. In: Garner GW, Amstrup SC, Manly BFJ, Laake JL, McDonald LL et al., editors. Marine mammal survey and assessment methods. Rotterdam, Netherlands: A.A. Balkema. 191–207.
49. Marsh H, Sinclair DF (1989) Correcting for visibility bias in strip transect aerial surveys of aquatic fauna. J Wildl Manage 53: 1017–1024.
50. Redfern JV, McKenna MF, Moore TJ, Calambokidis J, Deangelis ML, et al. (2013) Assessing the risk of ships striking large whales in marine spatial planning. Conserv Biol 27: 292–302.
51. Felix F, Haase B (2005) Distribution of humpback whales along the coast of Ecuador and management implications. J Cetacean Res Manag 7: 21–31.
52. Cartwright R, Sullivan M (2009) Associations with multiple male groups increase the energy expenditure of humpback whale (*Megaptera novaeangliae*) female and calf pairs on the breeding grounds. Behaviour 146: 1573–1600.
53. Bull JW, Suttle KB, Singh NJ, Milner-Gulland EJ (2013) Conservation when nothing stands still: Moving targets and biodiversity offsets. Front Ecol Environ 11: 203–210.

6

Probability of Detecting Marine Predator-Prey and Species Interactions Using Novel Hybrid Acoustic Transmitter-Receiver Tags

Laurie L. Baker[1]*, Ian D. Jonsen[1], Joanna E. Mills Flemming[2], Damian C. Lidgard[1], William D. Bowen[3], Sara J. Iverson[1], Dale M. Webber[4]

1 Department of Biology, Dalhousie University, Halifax, Nova Scotia, Canada, **2** Department of Mathematics and Statistics, Dalhousie University, Halifax, Nova Scotia, Canada, **3** Population Ecology Division, Bedford Institute of Oceanography, Dartmouth, Nova Scotia, Canada, **4** VEMCO Ltd., Halifax, Nova Scotia, Canada

Abstract

Understanding the nature of inter-specific and conspecific interactions in the ocean is challenging because direct observation is usually impossible. The development of dual transmitter/receivers, Vemco Mobile Transceivers (VMT), and satellite-linked (e.g. GPS) tags provides a unique opportunity to better understand between and within species interactions in space and time. Quantifying the uncertainty associated with detecting a tagged animal, particularly under varying field conditions, is vital for making accurate biological inferences when using VMTs. We evaluated the detection efficiency of VMTs deployed on grey seals, *Halichoerus grypus*, off Sable Island (NS, Canada) in relation to environmental characteristics and seal behaviour using generalized linear models (GLM) to explore both post-processed detection data and summarized raw VMT data. When considering only post-processed detection data, only about half of expected detections were recorded at best even when two VMT-tagged seals were estimated to be within 50–200 m of one another. At a separation of 400 m, only about 15% of expected detections were recorded. In contrast, when incomplete transmissions from the summarized raw data were also considered, the ratio of complete transmission to complete and incomplete transmissions was about 70% for distances ranging from 50–1000 m, with a minimum of around 40% at 600 m and a maximum of about 85% at 50 m. Distance between seals, wind stress, and depth were the most important predictors of detection efficiency. Access to the raw VMT data allowed us to focus on the physical and environmental factors that limit a transceiver's ability to resolve a transmitter's identity.

Editor: João Miguel Dias, University of Aveiro, Portugal

Funding: This study and a graduate student stipend to L. Baker were supported by a research network grant (NETGP 375118 – 08) from the Natural Sciences and Engineering Research Council of Canada (NSERC) for the Ocean Tracking Network; L. Baker was additionally supported by a Dalhousie Graduate Fellowship. Additional support was provided by NSERC Discovery Grants to J. Mills Flemming, D. Bowen and S. Iverson, and by the Department of Fisheries and Oceans Canada. The funders had no role in study design, data collection and analysis, decision to publish, or preparation of the manuscript.

Competing Interests: Dale Webber is affiliated with the commercial company (VEMCO, Ltd) that develop Vemco Mobile Transceivers (VMT). VEMCO Ltd. did not finance the project, and was not directly involved in the the study design and analysis of the paper. This affiliation does not alter the authors' adherence to PLOS ONE policies on sharing data and materials.

* E-mail: Laurie.Baker@dal.ca

Introduction

Electronic tracking and telemetry data have greatly improved our knowledge about the ecology of many marine species at the individual and population levels [1]. However, few studies have used these methods to investigate the nature of interactions between individual animals. Interactions among conspecifics and between species shape both social and ecosystem structures, and can affect population growth rates, distribution, diversity, and gene flow [2,3]. Studies of predator-prey, competitive and social interactions in marine species have largely been inferred from experiments [4], diet sampling [5], multi-species time series analyses [6,7], or direct observation [8]. These studies are often limited to accessible habitats (e.g. the intertidal, haul out sites) and may not provide insight at the individual level (e.g. time series analysis). Acoustic telemetry can overcome some of these shortcomings by providing information about interactions at the level of individuals from inaccessible marine environments, see Barnet et al. [9] and Barnet and Semmens [10] who simultaneously tracked predator and prey.

The deployment of dual transmitting and receiving acoustic Vemco Mobile Transceivers (VMT, www.vemco.com) and satellite-linked GPS tags or geolocation tags [11] on large marine vertebrates provides an opportunity to understand species interactions in space and time. The VMT is a hybrid acoustic tag, housing a 69 kHz coded transmitter and a 69 kHz monitoring receiver (similar to the VR2W). Whereas arrays of stationary acoustic receivers are often necessarily confined to continental shelf areas (e.g. [12]), the deployment of VMTs on marine animals provides the ability to extend detection ranges of conspecific and other marine species to biologically interesting regions that may be missed by fixed arrays. The dual transmitter and receiver capabilities of the VMT create a mobile receiving station by which non-surfacing acoustic-tagged organisms, such as fish, can be detected. With these data we have the capacity to better understand the role of predators in ecosystems and to improve our understanding of their interactions with commercial fish stocks and fish species of conservation concern.

To interpret interactions between two organisms we must accurately describe the interaction locations, duration, and frequency. At the most basic level, this relies on knowing whether or not a tagged organism is present. Quantifying the probability of detecting a tag if it is near a given receiver, particularly under changing field conditions, is vital for making accurate biological inferences when using these VMTs (e.g. Argos, [13]; geolocation, [14]). In general, the probability of detecting a transmitter depends on the distance the transmitter is from the receiver, the properties of the medium and transmission (e.g. sound frequency), and the presence of physical obstructions and noise [15]. Sound intensity attenuates with the square of the range according to geometric spreading of the sound in water [15]. Therefore the distance a transmission travels in the ocean depends strongly on the sound frequency of the signal and characteristics of the propagation medium (i.e. sea water composition). Detection probability can also be affected if parts of the transmission are masked by background noise or distorted (e.g. changes in transmission frequency).

Changes in detection efficiency may occur in response to changes in oceanographic and environmental conditions: wind stress [16], [17]; water column stratification [18], [19]; water density [20], [18]; bottom topography [21]. Detection efficiencies have been quantified using a range of approaches: boat based, diver based, fixed sentinel tags, fixed tag with receiver at set distances, post-analysis, single tag at different distance, etc. [22]. While these studies provide valuable data on detection ranges, they cannot fully describe conditions experienced off-shore, and therefore cannot be expected to assess the performance of the VMT when deployed on a free-ranging marine animal. Our case study is distinct from standard acoustic studies, where only the tag is in motion; in our case both the tag and receiver are in motion. The importance of understanding how a tagged marine animal's behaviour affects tag performance is therefore increased. Differences between VMTs may arise because some individuals spend a greater proportion of their time in noisier locations or near complex geomorphology, which may lead to more obstructed transmissions [23] than in other locations. Understanding these behavioural patterns and how they differ seasonally, by age, sex, and physiological state is of the utmost importance.

Pinnipeds are well suited for testing the performance of VMTs. Their frequent return to the surface provides highly accurate GPS locations. Grey seals (*Halichoerus grypus*) fitted with VMTs are known to interact frequently with each other [24], and exhibit high site fidelity, making them easy to recapture to retrieve archived data. Evaluating VMTs when deployed on grey seals provides an opportunity to assess the efficiency of VMTs under realistic behavioural and environmental conditions. Here, we define detection efficiency as how well VMTs are able to detect another VMT transmitter (i.e. with what probability) within a defined range.

We conducted two analyses of detection efficiency of VMTs deployed on grey seals using post-processed detection data (complete transmissions) and summarized raw VMT data (complete and incomplete transmissions), to explore the effect of environmental factors: wind stress, distance between VMTs, and temperature and depth gradients. The raw VMT data consists of a record of all acoustic pings (the smallest sound unit) recorded by the VMT receiver, and differs from the post-processed detection data in that it contains records of incomplete transmissions in addition to complete transmissions (confirmed detections) as well as pings from environmental and anthropogenic sources. Vemco provided us with summarized raw data for four VMTs consisting of acoustic pings classified by the time intervals between them and summed for each 10-minute period.

We evaluated the detection efficiency of VMTs, using calculated distances (based on GPS locations) between seals to generate a series of instances when detections are likely to have occurred. Access to the summarized raw VMT data allowed us to focus on the physical and environmental factors that limit a receiverability to resolve a transmitteridentity.

Materials and Methods

Ethics Statement

This research was conducted in accordance with guidelines for the use of animals in research [25] and of the Canadian Council on Animal Care. The research protocol for deployment of tags on grey seals was approved by the University Committee on Laboratory Animals, Dalhousie University's animal ethics committee (animal care protocol: 08–088) and the Department of Fisheries and Oceans, Canada (animal care permit: 10–65).

Study Site

The study was conducted between 8 September 2010 and 17 January 2011 on Sable Island, Nova Scotia, Canada (43°55, 60°00 and the Eastern Scotian Shelf in the northwest Atlantic Ocean (Figure 1). Sable Island is an important breeding site for grey seals [26] and the Eastern Scotian Shelf is an important foraging area [24,27].

Study Animals

Seventeen adult grey seals, *Halichoerus grypus* (Fabricius, 1791), selected from a pool of known-age adults were captured between 8 and 18 September 2010 on Sable Island and fitted with a VHF transmitter (164–165 MHz, www.atstrack.com), GPS satellite-linked tag (MK10-AF, www.wildlifecomputers.com) and a VMT according to the methods described in Lidgard et al. [24]. Briefly, the VHF and GPS tags were attached just below the neck to maximize the time the GPS tag spent above water where it could record the satellites in range. The VMT was attached to the lower back of the seal to increase the time the VMT spent in the water transmitting and receiving detections and to reduce electrical interference with the satellite tag. The GPS tag was programmed to collect light intensity, depth (m), and temperature (°C) every ten seconds and to record a GPS location every 15 minutes. GPS attempts were suspended when the unit was dry more than 20 minutes or when a location had been attained.

Peak sensitivities for hearing in phocids are between about 10 and 50 kHz with a high frequency limit of 100 kHz [28]. It is likely that seals could hear the 69 kHz VMT transmissions, given the power output of the transmitters (146–149 dB re 1 μPa SPL 1 m) [29]. However, we did not observe any differences in behaviour: seals in this study exhibited similar foraging and breeding patterns to seals previously tagged with satellite transmitters without an acoustic tag [27], [30], [31]. Ambient background noise, reflection and refraction of the signal, and habituation to the signal over time, make it unlikely that seals could localize other VMT tagged seals. Individuals were recaptured on Sable Island during the subsequent breeding season (December 2010 to January 2011) and their tags retrieved (median deployment period = 112 d range = 92–121 d).

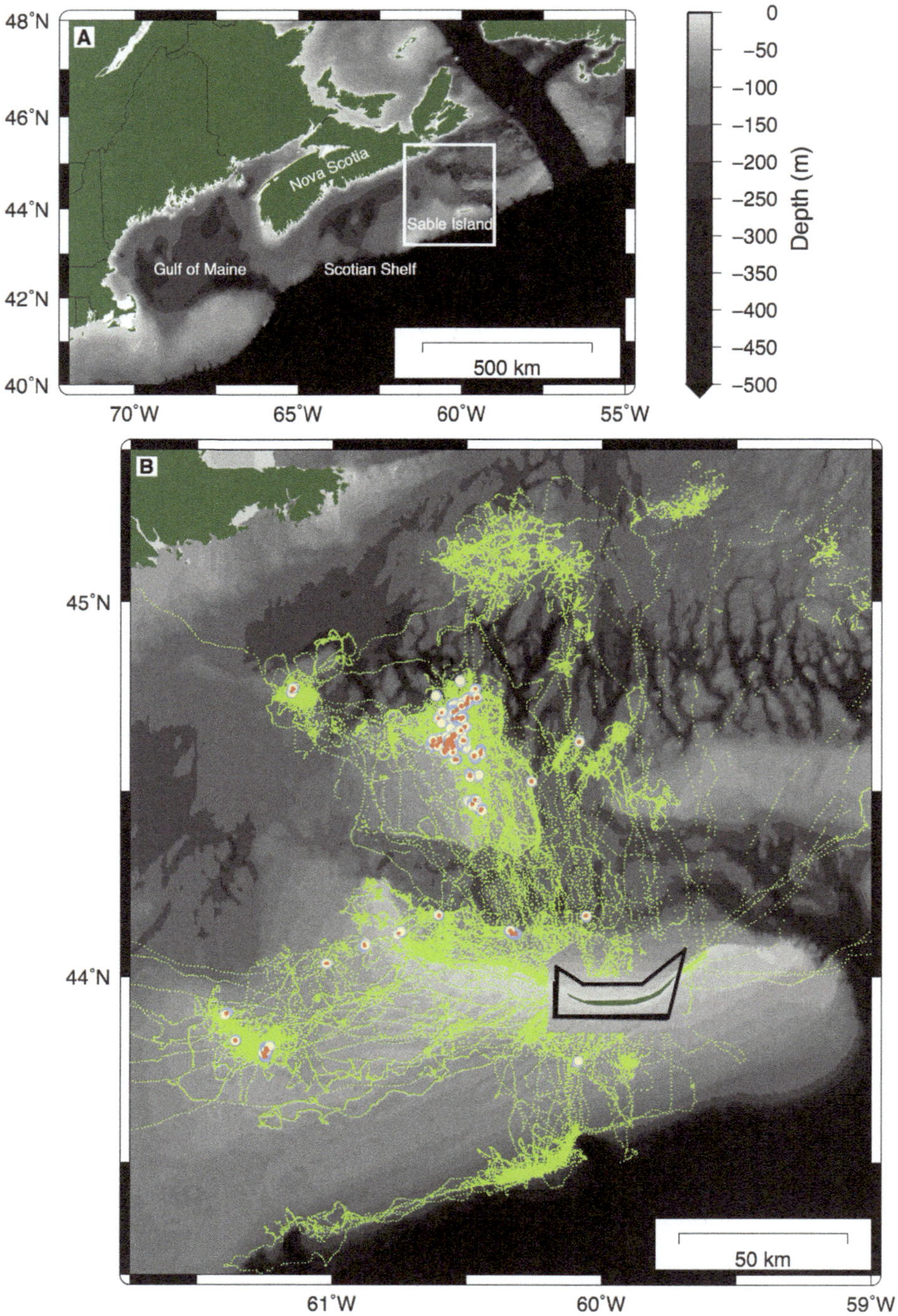

Figure 1. Study area and seal tracks. Nova Scotia and the Scotian Shelf (A) with the study area showing GPS tracks (green) and VMT expected (white) and observed (red) detections (B). The main shallow banks in the region are outlined with their 100 m isobaths (grey).

Post-processed Detection Data vs. Summarized Raw VMT Data

VMTs are coded transmitters, meaning they transmit a sequence of pings that form an acoustic code unique to each individual VMT. VMTs are programmed to transmit an acoustic code on an irregular schedule, every 60 to 180 seconds. During each code transmission the VMT turns off its receiver for approximately 3.5 s, to avoid receiving echos from its own transmission that could interfere with code validation, and records the date and time of the transmission. Each code transmission comprises a sequence of eight acoustic pings (acoustic code). Each acoustic code begins with a synchronisation interval (sync)– the time between the first two acoustic pings– that identifies the transmission format. The series of acoustic pings that follows creates the unique identification code; the interval between each of the eight acoustic pings creates the unique identification code (Figure 2). A checksum is applied to the entire acoustic code to identify the legitimacy of the transmission. Hereafter, we use the terms transmission and acoustic code synonymously.

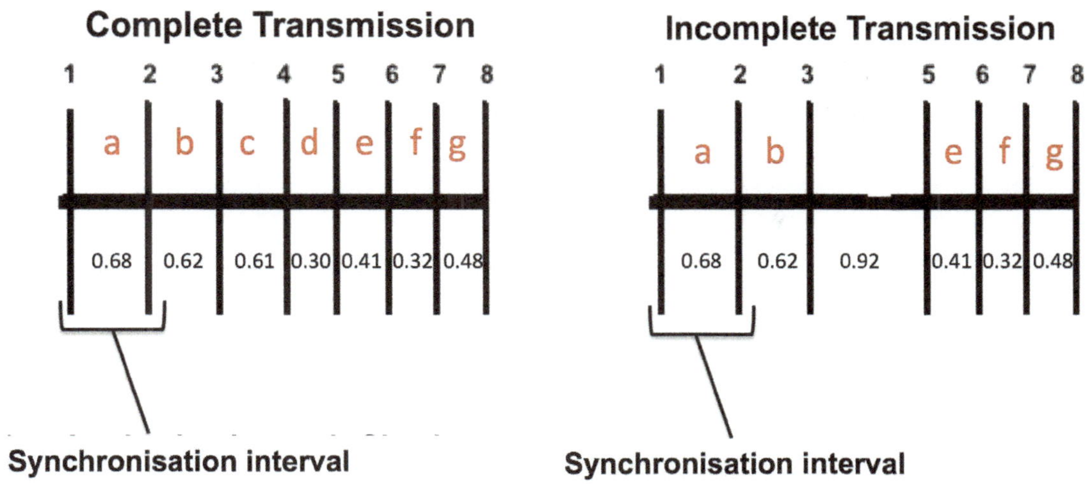

Figure 2. Complete vs. incomplete transmission. VMT transmissions comprise a series of 8 acoustic pings. Each oustic ping stringontains a synchronization interval (between the first two pings), used to identify acoustic-tag transmission format, followed by a series of pings unique to each individual tag. Intervals between 0.30–0.70 s correspond to consecutive pings. An interval between 0.70–1.50 s may indicate that one ping (of duration 0.01 s) is missing, e.g. time interval of 0.92 s in the incomplete transmission diagram. All 8 acoustic pings must be received for a detection to be recorded.

Post-processed detection data, available to all VMT users, comprises the complete received 69 kHz transmission– which may originate from a VMT or other 69 kHz transmitter– and a daily summary of the total number of acoustic pings, syncs, and rejected false detections. Received complete transmissions (detections), in VMT memory, comprise a date-time stamp and the identities of the transmitting and receiving acoustic tags. False detections are identified by VEMCO, using proprietary software, and removed from the dataset upon VMT retrieval. False detections may result from the collision of codes from other active transmitters that either generate a code that does not exist or an existing code that is known to be present elsewhere (e.g. tags deployed on freshwater species or on non-migratory species in other ocean basins).

The summarized raw VMT data is different from the post-processed detection data in that it includes all acoustic pings received by the transmitter, including those from incomplete transmissions. Acoustic pings may originate from a variety of sources such as other VMTs, acoustic transmitters and abiotic and biotic noise. Acoustics pings originating from VMTs and other VEMCO transmitters may be distinguished from background noise by the signature intervals between each ping in their acoustic codes (Table 1). VMTs are programmed such that consecutive acoustic pings in an acoustic code occur between 0.30 s and 0.70 s. Acoustic pings may also occur at intervals within 0.70 s and 1.50 s in cases where one or more acoustic pings in a code are missing (Figure 2). We therefore defined the range at which probable VMT pings occur as 0.30 s to 1.50 s. Acoustic pings occurring at intervals between 0.26 s and 0.30 s are thought to indicate possible echos, multipath transmissions, or transmission collisions. Acoustic pings occurring at intervals greater than 1.50 s are likely the result of environmental noise or are cases where VMTs are near their acoustic range limit.

Track Data and Expected vs. Observed Detections

We determined GPS locations by analyzing archival GPS data from each tag using software from the manufacturer. To be considered accurate, locations had to be acquired from >5 satellites with a residual error <30 m [32,33].

To link encounters between instrumented seals to locations interpolated at 3 min intervals from the seal tracks, clocks in the VMT and GPS tags were synchronized upon deployment and time corrected upon retrieval based on the respective clock drift calculated from GPS and VMT tags over the deployment time [24]. Distances between seals (m) were calculated from the 3-min interpolated locations.

Each seal's travel rate (m/s) was calculated using the original archival GPS location data. We matched these estimates to the respective transmitting and receiving VMTs using a date-time stamp. We assumed expected detections to occur every 180 s, based on tag specifications (every 60–180 s), when two VMTs encountered each other. We operationally defined an expected encounter as occurring when the VMTs were within 100–700 m of one another. We used 100 m as the lower limit of this range to avoid a decreased probability of detection, which may sometimes occur at close encounter ranges. We used 700 m as the upper limit of our range based on the manufacturers specifications and inspection of our detection data (Figure 3).

Despite being within range of VMTs that recorded data, two VMTs (66487, 66548) failed to record any detections, and one VMT (66494) was only recorded once by another VMT. Closer inspection of the seal tracks associated with these VMTs indicated they were spatially peripheral to the majority of the VMT-tagged seals, but still within range of certain known working VMTs. We

Table 1. Criteria used to determine ping origins.

Interval Length	Description
0.26–0.29 s	Possible echos or multipath transmissions
0.30–0.70 s	Interval range between **consecutive pings**
0.71–1.50 s	Interval range between **1 or more skipped pings**
>1.50 s	Spurious pings or 3 or more skipped pings

*Ping origins deduced from intervals between consecutive pings.

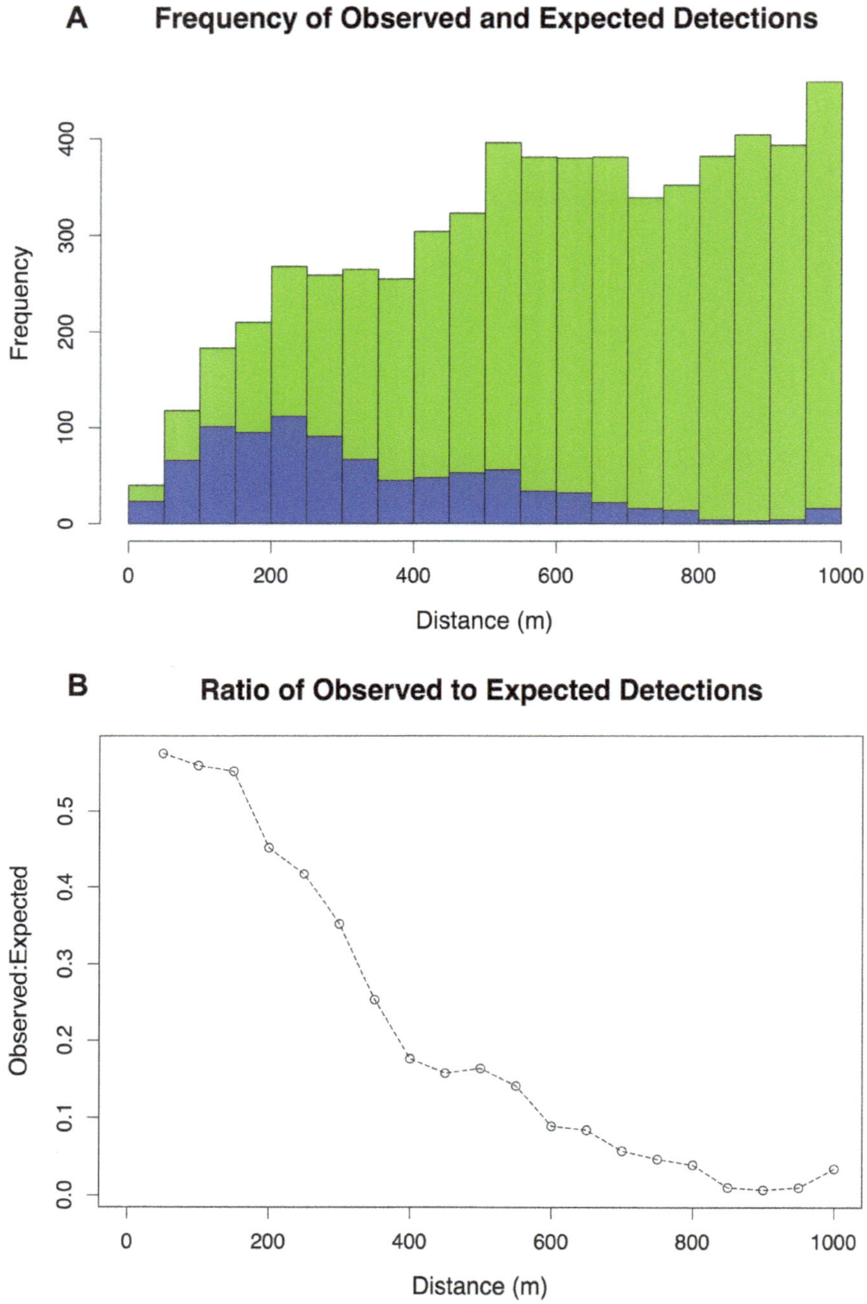

Figure 3. Density and ratio of detections. A. Density of observed (blue) and expected detections (green) with distance. B. Plot of the ratio of observed to expected detections.

excluded these non-functioning VMTs (66487, 66548, 66494). There were also confounding elements that could have affected the summarized raw VMT and post-processed detection data around the VMT deployment point, Sable Island. VMTs do not record signals out of the water; therefore it is important to exclude any periods the seal is out of the water from the analysis. Close to the island, it was difficult to determine if a VMT-tagged seal was out of the water if these durations were shorter than the wet-dry sensors on the GPS tag could detect. Furthermore, due to the shallow bathymetry and thus high noise disturbance around the island, we expected the capability of the VMT to record transmissions to be compromised. Thus, detection data around Sable Island were removed prior to analyses (see polygon outlined in Figure 1B).

Conversion Efficiency

Vemco provided summarized raw VMT data for four of the VMTs (66556, 66504, 66555, 66541). From these data we calculated the VMT conversion efficiency. Conversion efficiency was defined as the ratio of acoustic pings translated into detections (complete VMT transmissions) to those received (complete and incomplete VMT transmissions, Figure 2).

Statistical Model and Environmental Variables

We used a generalized linear model (GLM) with a negative binomial distribution to model VMT detection and conversion efficiency, where the response variable was the number of observed detections from new encounters in a 12 h period. New encounters were identified as detections (expected or observed) occurring when there was at least a 30 min interval between consecutive detections for a defined pair of seals. The number of expected detections in each 12 h period was included in the model as an offset term to account for the time VMT-tagged seals spent near each other.

Conversion efficiency was evaluated by modelling the number of acoustic pings from complete VMT transmissions (observed detections×8 pings), including the total number of pings from VMTs received (pings occurring at intervals between 0.3–1.5 s) in 10 min intervals as an offset.

Environmental Variables

Environmental variables were selected according to their relevance to sound propagation on the Scotian Shelf and their availability (Table 2). To avoid temporal and spatial scale mismatches, most variables were limited to those that we could collect from the MK 10-AF tags which sampled every 10 seconds and at the seal's exact location. Temperature (°C) and depth gradients (m) between the transmitting and receiving seals were included in the model to test for the effect of water stratification and density changes. The directional (positive or negative) difference in depth and temperature was included because the direction of signal travel with respect to the temperature or depth gradient affects sound transmission differently. Horizontal distance (m) was included in the model to represent detection range.

Wind stress (N/m^2) was included in the model to test the effect of increased noise and changes in the air-sea interface through the introduction of air bubbles. Wind stress (N/m^2) was calculated from hourly estimates of wind speed on Sable Island (Department of Fisheries and Oceans, Canada) in MATLAB (MathWorks, Inc.), using the function stresslp.m (package: air and sea) following Large and Pond [34]. We hypothesized that the effect of noise and/or air bubbles generated by wind stress would be greatest at the surface; we therefore tested for a possible interaction between wind stress (N/m^2) and the depth of the shallowest seal (m) in the model. Seal identity was included as a factor to account for variation in VMT performance and differences in seal behaviour and movement patterns. Travel rate (m/s) was included to describe the seal's horizontal movement rates.

Model Selection

Terms in the model were added and subtracted using forward and backward selection [35]. Variable selection was based on hypothesis testing (p-values) and by comparing the pseudo adjusted R^2 calculated from the residual and null deviance of the model. Residual diagnostics were examined to determine goodness of fit. To explore how sensitive the results were to the subsample distance range, we explored the data subset by distance ranging from 100–250 m, 100–400 m, and 100–700 m. This was done to control for varying amounts of time spent by seals at different distances from one another.

Results

All 17 deployed VMT and GPS tags were recovered from seals upon their return to Sable Island during the breeding season. GPS locations were acquired with a median of 9 satellites (<15 m residual error). A total of 1,168 detections were recorded, occurring at distances between 4 m and 1880 m (median = 320 m, mode = 250 m). Fewer detections occurred at both close range and and beyond 500 m. 60% of all detections occurred when the VMTs were within 500 m of one another (Figure 3A). We observed a decrease in the proportion of observed vs. expected detections with increased distance (Figure 3B). Only about half of the expected detections were recorded even when two VMT-tagged seals were estimated to be within 50–200 m. At a separation of 400 m, only about 15% of expected detections were recorded. The summarized raw VMT data provided a clearer picture of whether any part of a transmission was received with distance (Figure 4): the ratio of pings from complete transmission to pings from complete and incomplete transmissions fluctuated around 70%, with a minimum of around 40% at 600 m and a maximum of about 85% at 50 m (Figure 4).

Model 1: Expected and Observed Detections

The best model explained 35.7% of the variability in the detection efficiency. The probability of detection decreased with increasing distance between seals (−2.77, SE: 0.64), wind stress (−7.40, SE: 1.87), and depth of the shallowest seal (−0.03, SE: 0.01), (Figure 5).

Model 2: Conversion Efficiency

Wind stress (−1.59, SE: 0.35) and distance (−0.54, SE: 0.14) were both important predictors of conversion efficiency. Conversion efficiency decreased with increasing wind stress and increasing distance (Figure 6). Wind stress had the most significant effect on detection efficiency.

Sensitivity of Detection Efficiency to Distance Range

The results from each data subset were generally consistent with those of the main analyses. When encounters were defined at the 100–400 m range, results were consistent with the main analysis (100–700 m), but when encounters were defined at the 100–250 m range depth of the shallowest seal did not have a significant effect on detection efficiency. The signs and coefficients of model terms were conserved across distance ranges. The pseudo adjusted R^2 values were 19.5%, 28.1%, and 35.72% for the interval ranges: 100–250 m, 100–400 m, and 100–700 m respectively. These changes in explanatory power are likely the result of the increased influence of distance on decreases in detection efficiency.

Discussion

While it is relatively easy to ascertain if a tagged animal is present (true positive), it is more difficult to determine with certainty that it is absent (true negative) as it could be present but

Table 2. Environmental variables explored in VMT efficiency analyses.

Variable	Description
negtempdif	Directional temperature difference (±°C)
mindepth	Depth of the shallowest seal (m)
distance	Horizontal distance between seals (km)
negdepdif	Directional depth difference (±m)
travel rate	Travel rate of the receiving seal (m/s)

*Description of environmental variables tested in VMT efficiency analyses.

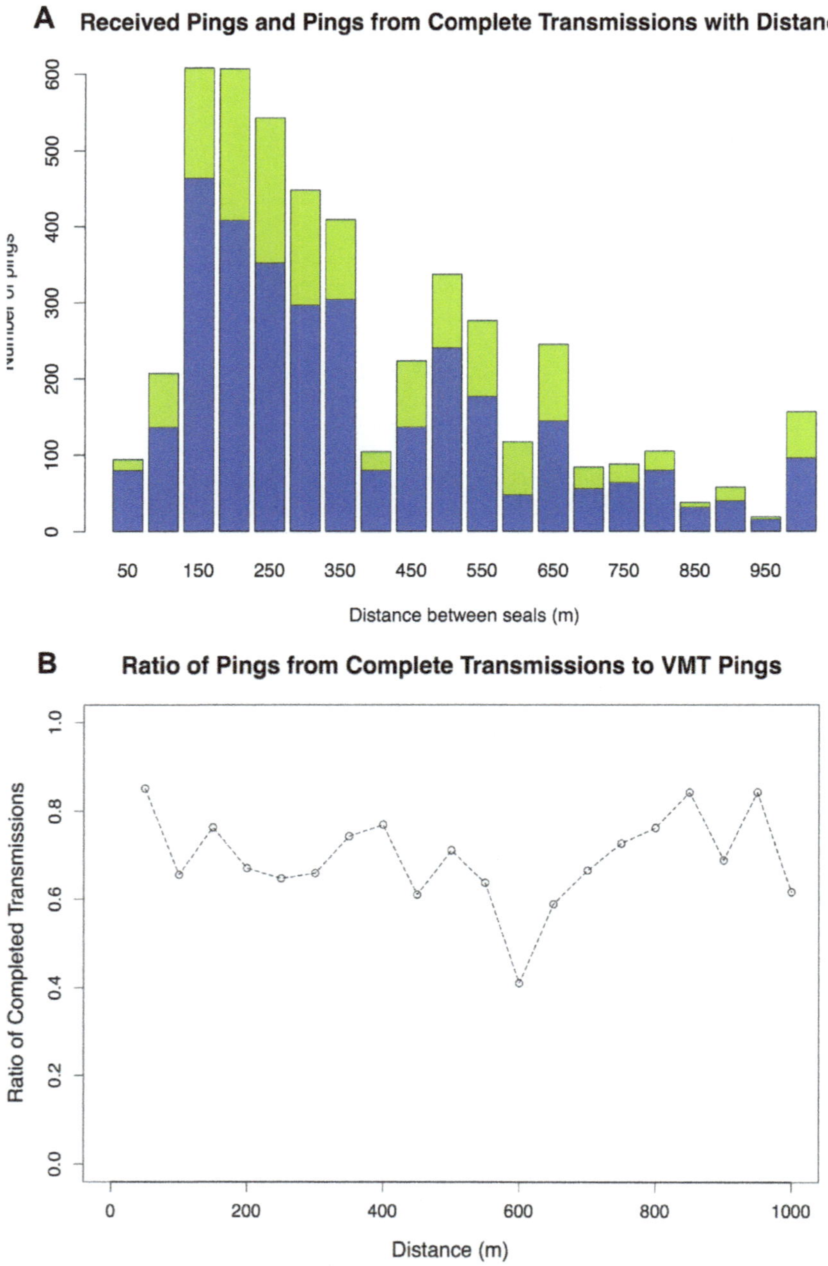

Figure 4. Density and ratio of VMT acoustic pings. A. Density of VMT acoustic pings received (green) and acoustic pings from VMT transmissions (blue) with distance. B. Plot of the ratio of pings from complete transmission to VMT pings received.

not detected (false negative). Quantifying the proportion of VMT transmissions that are not received and determining to what extent this is due to physical and environmental factors and the behaviour of the tagged animals, is vital to form accurate ecological conclusions from VMT data. Without an appreciation of these issues, these effects may lead to erroneous inferences.

We present one of the first studies to investigate the detection efficiency of acoustic VMT receivers deployed on marine animals and to analyze detection efficiency using summarized raw VMT data. Wind stress, depth of the shallowest seal, and distance between seals were significantly correlated with VMT performance. The summarized raw VMT data allowed us to determine the extent to which within-range VMTs are successfully detected and provided a clearer picture of whether any part of a VMT transmission is received. The ratio of VMT pings from complete transmissions to VMT pings received fluctuated around 70% with a minimum of around 40% at 600 m and a maximum of about 85% at 50 m. This shows a vast improvement when compared with at best 50% of expected detections received between 50–200 m, dropping to 15% at 400 m when using only the post-processed detection data. Examining conversion efficiency (the ratio of complete transmissions to all transmissions received) provides additional insight into VMT detection efficiency by focusing on factors that limit a transceiver's ability to resolve a transmitter's identity.

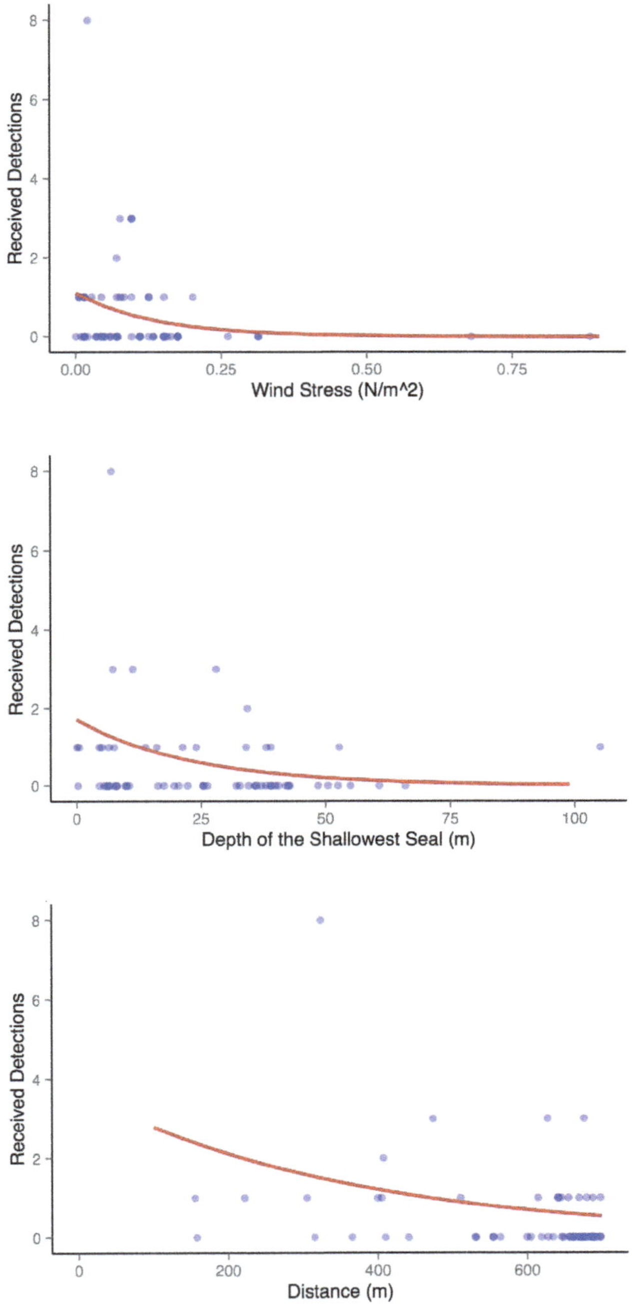

Figure 5. Factors affecting detection efficiency. The predicted effect on detection efficiency of the significant variables (red line): wind stress, minimum depth, and distance. Fitted values (observed detections offset by expected detections) as points. Points: dark blue indicates high intensity, light blue indicates low intensity.

To date, GPS tags provide the best location estimates for *in situ* studies of this nature. GPS locations were obtained with a small residual error (<15 m) [32], resulting in little uncertainty in the GPS locations and subsequently, little uncertainty in the actual detection distances observed. Therefore, although it is possible for the seals to be 60 m closer or further away than that reported, the chance of this occurring are low.

Environmental Factors Affecting VMT Performance

Distance between seals was a significant predictor of detection and conversion efficiency. In both cases, the probability of detection or conversion decreased with distance as expected. Detection range has long been identified as an important factor affecting the detection of acoustic tags [20]. Detection probability is hypothesized to decline proportionally to the decline in sound intensity, which is a combination of geometric and exponential decline due to sound spreading and attenuation resulting from water viscosity [15]. However, the exact shape of this relationship is unknown and modelling approaches vary. We were unable to resolve the shape of this relationship from our data due to its observational nature. However, results from our sensitivity analysis illustrate that the detection range, assumed *a priori*, did not affect the relationships observed.

We also observed a decrease in detection efficiency and conversion efficiency with increasing wind stress. Wind stress can introduce noise as well as air bubbles into the marine environment. Noise makes it difficult to distinguish the acoustic signal above the background noise and may result in failure to detect one or more of the pings. Air bubbles absorb a sound transmission because the acoustic signal has to pass between water and air. The absence of a significant interaction between wind stress and the depth of the shallowest seal suggests that the effect of wind stress on detection efficiency is not confined to surface waters. The observed decrease in detection efficiency with increasing depth may be indicative of sound attenuation occurring as a result of bathymetric effects [21].

Despite well established effects on sound transmission, we observed no effect of the propagation medium (temperature/depth gradients) on detection efficiency [15]. Sound propagation may be absorbed and deflected when traveling through density gradients (i.e., pycnocline). The coastal currents that transport source waters to the Scotian shelf exhibit strong seasonal cycles as well as significant interannual variability [36]. The Nova Scotia current reaches a peak velocity in winter, transporting low salinity and low temperature water originating in the Gulf of St. Lawrence [37] into the inshore waters. These forces generally result in a low salinity and low temperature signature inshore that is more pronounced during winter months [36]. Temperature and depth gradients are therefore more likely to affect detection efficiency after our deployment period (September–December) from January–March, than during our deployment period.

As animal-borne acoustic telemetry evolves beyond stationary receivers, it is unclear how factors such as the orientation of the VMT with respect to the animal or the size of the animal affect VMT performance. VMTs were placed on the lower back of the seal to maximize the time the VMT spent in the water receiving and transmitting signals. The sealbody might attenuate acoustic signals being transmitted to or received from a certain direction, regardless of VMT positioning. Although this effect has not been formally investigated, it would be extremely difficult to quantify *in situ*. A tri-axial accelerometer could be deployed to measure the sealspeed and VMT orientation, however, these devices also have limitations. Controlled experiments will be needed to investigate the influence of such factors on VMT performance. Other factors known to affect detection efficiency that were not included in our model are biotic and/or anthropogenic noise, e.g. [38], [39]. These, in addition to characteristics of the seals behaviour (e.g., the animalorientation during diving), may account for some of the unexplained variation in the model.

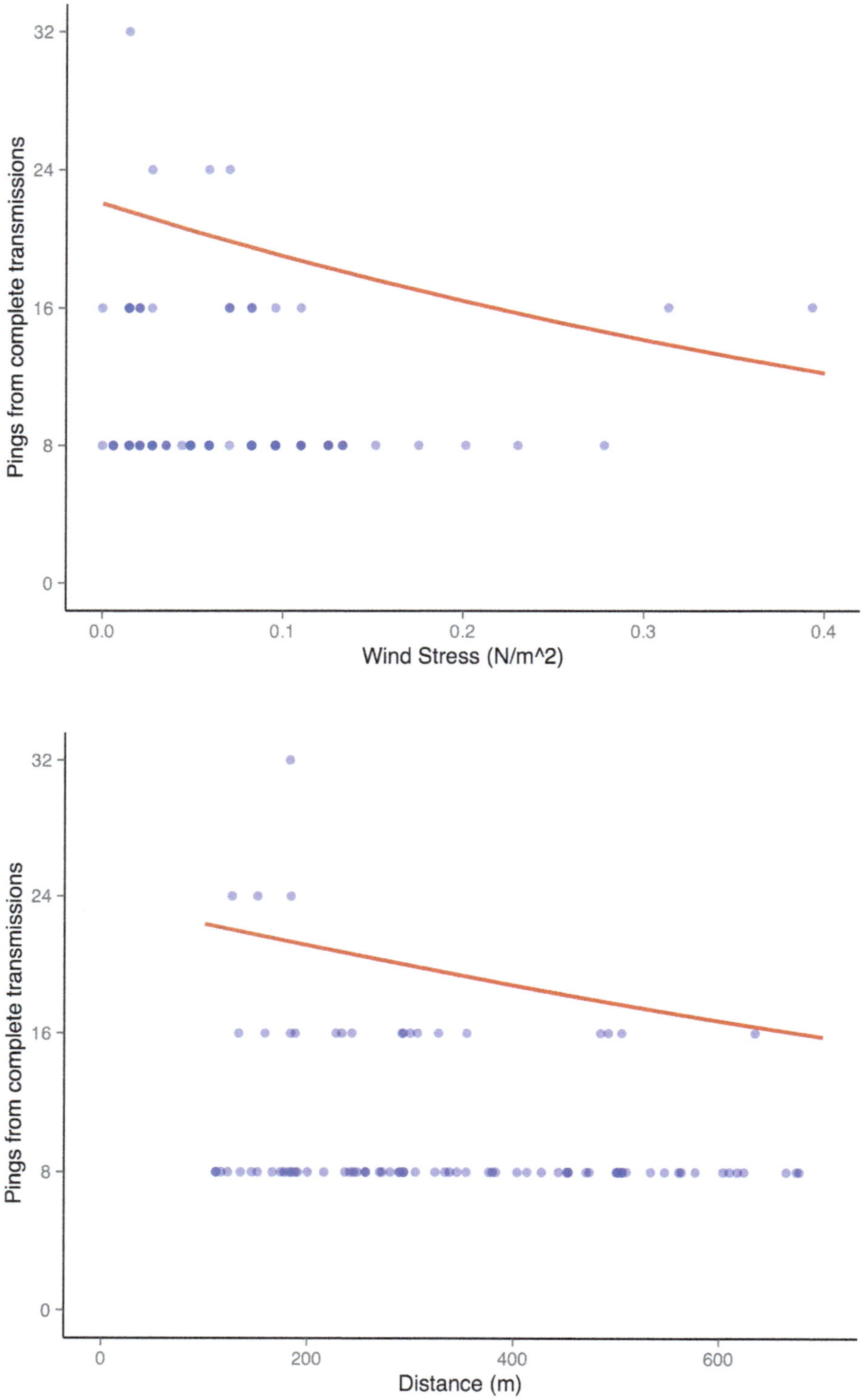

Figure 6. Factors affecting conversion efficiency. The predicted effect on conversion efficiency of the significant variables (red line): wind stress and distance. Fitted values (VMT acoustic pings from complete transmissions offset by total VMT acoustic pings received) as points. Points: dark blue indicates high intensity, light blue indicates low intensity.

VMT Engineering

To interpret interactions we must accurately define their location, duration, frequency, and confidently identify legitimate periods of silence (true negative), i.e. the absence of transmissions. For a detection to occur, the VMT receiver must be able to distinguish the acoustic signal from background noise. The background noise strength is dependent on weather and the fluid environment and other sources, including anthropogenic noise [15]. Distinguishing legitimate transmissions from background noise is an important component of measuring VMT performance. Simpfendorfer et al. [23] used syncs to estimate the volume of received incomplete and complete transmissions for a given period relative to the number of recorded transmissions; however, syncs are not precise. When tag transmissions collide, syncs can be created that are not from a tag transmission; consecutive pings from different tags may create a pseudo sync interval. The use of summarized raw VMT data addresses this shortcoming by utilizing aspects of the transmission that are less susceptible to false positives. With access to the summarized raw VMT data, users can examine the interval between consecutive pings to determine their origin and thus authenticity, i.e. whether the pings arose from echoes, multi-path collisions, environmental noise or are legitimate pings from a VMT.

Observational data in the ocean are often limited due to the technological, environmental, and physical challenges that accompany data collection. These constraints make it important to maximize what can be gleaned from such data. Currently, access to the summarized raw data is not routinely available. Wider access to data of this sort will provide users with an additional indicator of their tag's performance, and inform their analyses through the ability to identify false-negatives. In cases where the identity of the tagged individual is not pertinent, it may be sufficient to simply know that a seal was detected when part of a VMT transmission reached the VMT, even if we cannot account for the factors affecting the VMT transmission.

Without understanding the factors affecting detection efficiency, biological inferences regarding the prevalence and nature of species interactions via VMT/acoustic data will very likely be biased. For example, seasonal changes in environmental factors, that could reduce received transmissions, may be falsely attributed to seasonal changes in interaction rate. It is therefore vital that we account for changes in detection efficiency, as without this information, it is impossible to interpret what any given detection event represents.

Acknowledgments

We are very grateful to Suzanne Budge, Nell den Heyer, Susan Heaslip, Shelley Lang, Elizabeth Leadon, Jim McMillan, Sarah Wong, Rob Ronconi, and Sean Smith, for assistance in the field. S. Kessel, ML Bianchini, and one anonymous reviewer provided valuable feedback on the manuscript and improved its quality.

Author Contributions

Conceived and designed the experiments: DCL WDB IDJ SJI. Performed the experiments: DCL WDB. Analyzed the data: LLB IDJ JEMF. Contributed reagents/materials/analysis tools: DMW. Wrote the paper: LLB IDJ JEMF SJI WDB DCL DMW.

References

1. Cooke SJ, Hinch SG, Wikelski M, Andrews RD, Kuchel LJ, et al. (2004) Biotelemetry: a mechanistic approach to ecology. Trends in Ecology and Evolution 19: 334–343.
2. Whitehead H (2009) Socprog programs: analysing animal social structures. Behavioral Ecology and Sociobiology 63: 765–778.
3. Gorini L, Linnell JDC, May R, Panzacci M, Boitani L, et al. (2012) Habitat heterogeneity and mammalian predator-prey interactions. Mammal Review 42: 55–77.
4. Paine RT (1966) Food web complexity and species diversity. American Naturalist : 65–75.
5. Beck CA, Iverson SJ, Bowen WD, Blanchard W (2007) Sex differences in grey seal diet reect seasonal variation in foraging behaviour and reproductive expenditure: evidence from quantitative fatty acid signature analysis. Journal of Animal Ecology 76: 490–502.
6. Worm B, Myers RA (2003) Meta-analysis of cod-shrimp interactions reveals top-down control in oceanic food webs. Ecology 84: 162–173.
7. Lindegren M, Möllmann C, Nielsen A, Brander K, MacKenzie BR, et al. (2010) Ecological forecasting under climate change: the case of Baltic cod. Proceedings of the Royal Society B: Biological Sciences 277: 2121–2130.
8. Cantor M, Whitehead H (2013) The interplay between social networks and culture: theoretically and among whales and dolphins. Philosophical Transactions of the Royal Society B: Biological Sciences 368.
9. Barnett A, Abrantes KG, Stevens JD, Bruce BD, Semmens JM (2010) Fine-scale movements of the broadnose sevengill shark and its main prey, the gummy shark. PLOS ONE 5: e15464.
10. Barnett A, Semmens JM (2012) Sequential movement into coastal habitats and high spatial overlap of predator and prey suggest high predation pressure in protected areas. Oikos 121: 882–890.
11. O'Dor RK, Stokesbury M, Jackson GD (2007) Tracking marine species-taking the next steps. In: Australian Society for Fish Biology 2006 Workshop Proceedings. 6–12.
12. Cooke SJ (2012) Measuring the energetics and physiological status of wild fish using biotelemetry and biologging tools. In: AFS 142nd Annual Meeting. Afs.
13. Jonsen ID, Flemming JM, Myers RA (2005) Robust state-space modeling of animal movement data. Ecology 86: 2874–2880.
14. Winship AJ, Jorgensen SJ, Shaffer SA, Jonsen ID, Robinson PW, et al. (2012) State-space frame work for estimating measurement error from double-tagging telemetry experiments. Methods in Ecology and Evolution 3: 291–302.
15. Medwin H, Clay CS (1997) Fundamentals of acoustical oceanography. Academic Press.
16. How JR, de Lestang S (2012) Acoustic tracking: issues affecting design, analysis and interpretation of data from movement studies. Marine and Freshwater Research 63: 312–324.
17. Gjelland K, Hedger R (2013) Environmental influence on transmitter detection probability in biotelemetry: Developing a general model of acoustic transmission. Methods in Ecology and Evolution 3: 665–674.
18. Finstad B, Økland F, Thorstad EB, Bjørn PA, McKinley RS (2005) Migration of hatchery-reared Atlantic salmon and wild anadromous brown trout post-smolts in a Norwegian fjord system. Journal of Fish Biology 66: 86–96.
19. Singh L, Downey NJ, Roberts MJ, Webber DM, Smale MJ, et al. (2009) Design and calibration of an acoustic telemetry system subject to upwelling events. African Journal of Marine Science 31: 355–364.
20. Heupel MR, Semmens JM, Hobday AJ (2006) Automated acoustic tracking of aquatic animals: scales, design and deployment of listening station arrays. Marine and Freshwater Research 57: 1–13.
21. Kuperman WA, Lynch JF (2004) Shallow-water acoustics. Physics Today 57: 55–61.
22. Kessel ST, Cooke SJ, Heupel MR, Hussey NE, Simpfendorfer CA, et al. (2013) A review of detection range testing in aquatic passive acoustic telemetry studies. Reviews in Fish Biology and Fisheries: 1–20.
23. Simpfendorfer CA, Heupel MR, Collins AB (2008) Variation in the performance of 410 acoustic receivers and its implication for positioning algorithms in a riverine setting. Canadian Journal of Fisheries and Aquatic Sciences 65: 482–492.
24. Lidgard DC, Bowen WD, Jonsen ID, Iverson SJ (2012) Animal-borne acoustic transceivers reveal patterns of at-sea associations in an upper-trophic level predator. PLOS ONE 7: e48962.
25. Anon (2006) Guidelines for the treatment of animals in behavioural research and teaching. Animal Behaviour 71: 245–253.
26. Austin D, Bowen WD, McMillan JI, Iverson SJ (2006) Linking movement, diving, and habitat to foraging success in a large marine predator. Ecology 87: 3095–3108.
27. Breed GA, Bowen WD, McMillan JI, Leonard ML (2006) Sexual segregation of seasonal foraging habitats in a non-migratory marine mammal. Proceedings of the Royal Society B: Biological Sciences 273: 2319–2326.
28. Kastelein RA, Wensveen PJ, Hoek L, Verboom WC, Terhune JM (2009) Underwater detection of tonal signals between 0.125 and 100 kHz by harbor seals (*Phoca vitulina*). The Journal of the Acoustical Society of America 125: 1222–1229.
29. Bowles AE, Denes SL, Shane MA (2010) Acoustic characteristics of ultrasonic coded transmitters for fishery applications: could marine mammals hear them? The Journal of the Acoustical Society of America 128: 3223–3231.
30. Mellish JE, J IS, D BW (1999) Variation in milk production and lactation performance in grey seals and consequences for pup growth and weaning characteristics. Physiol Biochem Zool 72: 677–690.
31. Lidgard DC, Bowen WD, Jonsen ID, Iverson SJ (2005) State-dependent male mating tactics in the grey seal: the importance of body size. Behavioural Ecology 16: 542–549.
32. Bryant E (2007) 2D location accuracy statistics for Fastloc® cores running firmware versions 2.2 & 2.3. Wildtrack Telemetry Systems Ltd.
33. Hazel J (2009) Evaluation of fast-acquisition GPS in stationary tests and fine-scale tracking of green turtles. Journal of Experimental Marine Biology and Ecology 374: 58–68.

34. Large WG, Pond S (1981) Open ocean momentum flux measurements in moderate to strong winds. Journal of Physical Oceanography 11: 324–336.
35. Zuur AF, Ieno EN, Walker N, Saveliev AA, Smith GM (2009) Mixed effects models and extensions in ecology with R. Springer, 233–236.
36. Smith PC, Schwing FB (1991) Mean circulation and variability on the eastern Canadian continental shelf. Continental Shelf Research 11: 977–1012.
37. Sutcliffe Jr WH, Loucks RH, Drinkwater KF (1976) Coastal circulation and physical oceanography of the Scotian Shelf and the Gulf of Maine. Journal of the Fisheries Board of Canada 33: 98–115.
38. Voegeli FA, Pincock DG (1996) Overview of underwater acoustics as it applies to telemetry, volume 50. Taylor & Francis, 277–300.
39. Thorstad EB, Økland F, Finstad B (2000) Effects of telemetry transmitters on swimming performance of adult Atlantic salmon. Journal of Fish Biology 57: 531–535.

7

Habitat Fragmentation and Species Extirpation in Freshwater Ecosystems; Causes of Range Decline of the Indus River Dolphin (*Platanista gangetica minor*)

Gill T. Braulik[1,2,3]*, Masood Arshad[2], Uzma Noureen[2], Simon P. Northridge[1]

1 Sea Mammal Research Unit, Scottish Oceans Institute, University of St. Andrews, St. Andrews, Fife, United Kingdom, **2** World Wildlife Fund-Pakistan, Lahore, Pakistan, **3** Wildlife Conservation Society, Zanzibar, United Republic of Tanzania

Abstract

Habitat fragmentation of freshwater ecosystems is increasing rapidly, however the understanding of extinction debt and species decline in riverine habitat fragments lags behind that in other ecosystems. The mighty rivers that drain the Himalaya - the Ganges, Brahmaputra, Indus, Mekong and Yangtze - are amongst the world's most biodiverse freshwater ecosystems. Many hundreds of dams have been constructed, are under construction, or are planned on these rivers and large hydrological changes and losses of biodiversity have occurred and are expected to continue. This study examines the causes of range decline of the Indus dolphin, which inhabits one of the world's most modified rivers, to demonstrate how we may expect other vertebrate populations to respond as planned dams and water developments come into operation. The historical range of the Indus dolphin has been fragmented into 17 river sections by diversion dams; dolphin sighting and interview surveys show that river dolphins have been extirpated from ten river sections, they persist in 6, and are of unknown status in one section. Seven potential factors influencing the temporal and spatial pattern of decline were considered in three regression model sets. Low dry-season river discharge, due to water abstraction at irrigation barrages, was the principal factor that explained the dolphin's range decline, influencing 1) the spatial pattern of persistence, 2) the temporal pattern of subpopulation extirpation, and 3) the speed of extirpation after habitat fragmentation. Dolphins were more likely to persist in the core of the former range because water diversions are concentrated near the range periphery. Habitat fragmentation and degradation of the habitat were inextricably intertwined and in combination caused the catastrophic decline of the Indus dolphin.

Editor: Tom Gilbert, Natural History Museum of Denmark, Denmark

Funding: World Wildlife Fund-Pakistan -(http://www.wwfpak.org/) funded the interview surveys with support through the Pakistan Wetlands Project. The US Marine Mammal Commission (www.mmc.gov) provided funding to GB for reporting and analysis in grant number E4047595. The funders had no role in study design, data collection and analysis, decision to publish, or preparation of the manuscript.

Competing Interests: The authors have declared that no competing interests exist.

* Email: gillbraulik@downstream.vg

Introduction

Fresh waters are experiencing declines in biodiversity far greater than those in the most affected terrestrial ecosystems [1,2]. Dam construction has dramatically increased habitat fragmentation and degradation in freshwaters, which is likely to have incurred a large unredeemed extinction debt [3]. However, this debt is not yet well quantified or understood, as metapopulation ecology in freshwaters has lagged behind similar work in other habitats, such as tropical forests [4,5,6]. A fundamental, yet unanswered question for conservation biology is how rapidly freshwater species disappear from river fragments and which factors influence the extinction of freshwater species in habitat patches. We investigate this issue using the example of the highly fragmented Indus River system and its endemic, endangered freshwater cetacean.

The great rivers that drain the Himalaya are amongst the world's most biodiverse freshwater ecosystems, but they are increasingly under threat as the emerging nations of China, India, and Pakistan and the countries of Southeast Asia scramble to harness hydropower and provide water for expanding agrarian economies, in the midst of increasing water scarcity and climatic uncertainty [7,8,9]. Many hundreds of new dams and other water development projects are planned or under construction in the region, including mainstem and tributary mega-dams, run-of-the-river hydroelectric schemes, irrigation barrages, and inter-basin water transfers [10,11]. The Himalayan region will soon have the highest concentration of dams in the world [12]. The combined effects of these activities are predicted to cause rapid and escalating hydrological change and habitat fragmentation that will negatively impact riverine biodiversity and ecosystem services [9,13].

The freshwater cetaceans that inhabit the largest Himalayan rivers, the Indus, Ganges, Brahmaputra, Yangtze, Mekong, and Ayeyarwady, collectively form one of the world's most endangered groups of mammals, each listed as endangered or critically endangered on the IUCN RedList [14], and, in the case of the Yangtze River dolphin (*Lipotes vexillifer*), the species has probably been extinct since the mid-2000's [15]. River dolphins are iconic species that can serve as charismatic flagships for conservation of freshwater ecosystems; but they are poorly understood and increasingly threatened. This is exemplified by the demise of the Yangtze River dolphin which disappeared so quickly that a comprehensive evaluation of the causes of its decline was not conducted until after it was presumed extinct [16].

We examine the pattern, and causes of range decline of the Indus dolphin (*Platanista gangetica minor*), an obligate freshwater cetacean endemic to the Indus River system, which, after more than 150 years of barrage (gated-dam) construction and removal of water to feed irrigated agriculture, is one of the most fragmented and modified rivers in the world. The distribution of the Indus dolphin was carefully documented in the 1870's, (just prior to the first major barrage being constructed) and at that time the dolphin inhabited the entire lower Indus River system from the delta with the Indian Ocean, to the foothills of the Himalayas in what is now India and Pakistan [17]. By the early 1990's, Indus dolphins had undergone an 80% reduction in range, having been extirpated from the upper and lower reaches of the Indus and four of the largest tributaries [18]. They are now confined to five contiguous 'river sections' on the Indus mainstem in Pakistan, separated by barrages, and in the Beas River, in India (Fig. 1) [19,20]. Details of when dolphins were extirpated from different parts of their former range are vague, and the causes not clearly understood. However, the construction of twenty irrigation barrages between 1886 and 1971 (gated-dams used for water diversion) that fragmented the dolphins historical range into 17 river sections (numbered 1–17 on Fig. 1), and large-scale water abstraction for irrigation rendering many sections of river almost dry for many months, certainly played a role (see Fig. 2) [19,21].

In this paper we document the spatial and temporal dynamics of the Indus River dolphin range decline, and then use a series of regression models to determine the causes of the spatial pattern of decline, the timing of subpopulation extirpation, and the speed of subpopulation disappearance after habitat fragmentation. Greater understanding of how the Indus dolphin has responded to the presence of dams and water diversions within its habitat demonstrates how we may expect other vertebrate populations to respond as planned dams and water developments come into operation elsewhere. The results provide important and relevant insights into factors influencing species extinction in fragmented riverscapes.

Methods

Last Dolphin Sighting Date

The status of each of the six extant dolphin subpopulations is fairly well understood [19,22], but there is little information on when or why dolphins disappeared from the 11 river sections where dolphins are presumed extirpated (Fig. 1). In 2007, we compiled historical dolphin sighting dates and locations from areas in which dolphins are believed to have been extirpated, by conducting community interviews, and a review of historical literature. GB received clearance from the Pakistan Home Department for the survey, and as there was no appropriate ethics or review board in Pakistan to provide approval no other permits were required. Depending on location, dolphins are believed to have disappeared between 20 and 80 years ago, so the short structured interviews targeted elderly riverside inhabitants old enough to have seen dolphins in their lifetimes. The objective of the interviews was explained to potential interviewees and they provided their verbal, rather than written, consent to participate as the majority were illiterate. All interviews were anonymous. A calendar of significant local historical events was compiled to assist informants in recalling dates correctly. A last dolphin sighting date (LDSD) was allocated to each river section based on the most recent dolphin sighting that we identified. We did not attempt to identify the exact extinction date of each subpopulation [23], but used the LDSD as a general indicator of when dolphins disappeared [24]. Inexact dates were assigned to 5 year intervals, for example, if the date was early 1970s, a date of 1972 was assigned, if it was mid 1970s, 1975 was designated, and if it was late 1970s the date used was 1978.

Identifying the Causes of Range Decline

The following seven explanatory variables that may have contributed to the Indus dolphin range decline were determined for each of the 17 sections of the dolphin's former range:

1. *Fragmentation date* - The year that each river section was created; assigned as the date that the second of the two bounding barrages became operational. For the river section upstream of Harike barrage in India, the isolation date was assigned as the completion date of Hussainiwala barrage which is located only 30 km downstream of Harike.
2. *River length* - The number of river kilometres between two barrages.
3. *Proximity to range edge* - The distance along the river's course from the former dolphin distributional limit recorded by Anderson [17], to the barrage located closest to the range core.
4. *Size of river* - The mean annual discharge in Million Acre Feet reported for each river prior to implementation of the Indus Water Treaty in 1960 [25]. This illustrates original river size prior to large-scale water abstraction.
5. *Confluences* - The number of river confluences within each river section was included as an indicator of favourable habitat, as Indus dolphins occur with higher frequency at confluences [26].
6. *River slope* - The slope within each river section was calculated as the drop in elevation between the up and downstream barrages, or upstream range limit in the case of peripheral segments, divided by the length of river. Slope exerts a direct effect on flow velocity and sediment transport and therefore may influence dolphin habitat.
7. *Dry season river discharge* - River discharge data were obtained for all twelve barrages and two dams north of Guddu on the Indus River system in Pakistan for the period July 2008 to April 2013 (4 years 9 months). The daily discharge below Guddu and Kotri barrages was obtained from October 2010 to April 2013 (~2 ½ years), and discharge below Sukkur barrage was obtained from April 1994 to January 2000, and October 2010 to April 2013 (8 years 3 months). Occasional missing data were interpolated. It is very low flows that are likely to adversely impact dolphins, therefore median daily discharge during the dry season (1st October to 31st March) was determined using the years for which there was an entire dry season's data. Mean monthly discharge was available above and below Harike and Hussainiwala barrages in India from January 2009 to December 2011, and for these two river sections the median of the mean monthly dry season discharge was used. The number of years of data available differed according to barrage but the temporal discharge pattern was predictable and similar across years in each location.

Generalised Linear Models (GLMs) and a survival analysis were used with the seven explanatory variables described above as predictors of the continued presence of river dolphins in each of the 17 river sections. Generalised Additive Models were used in

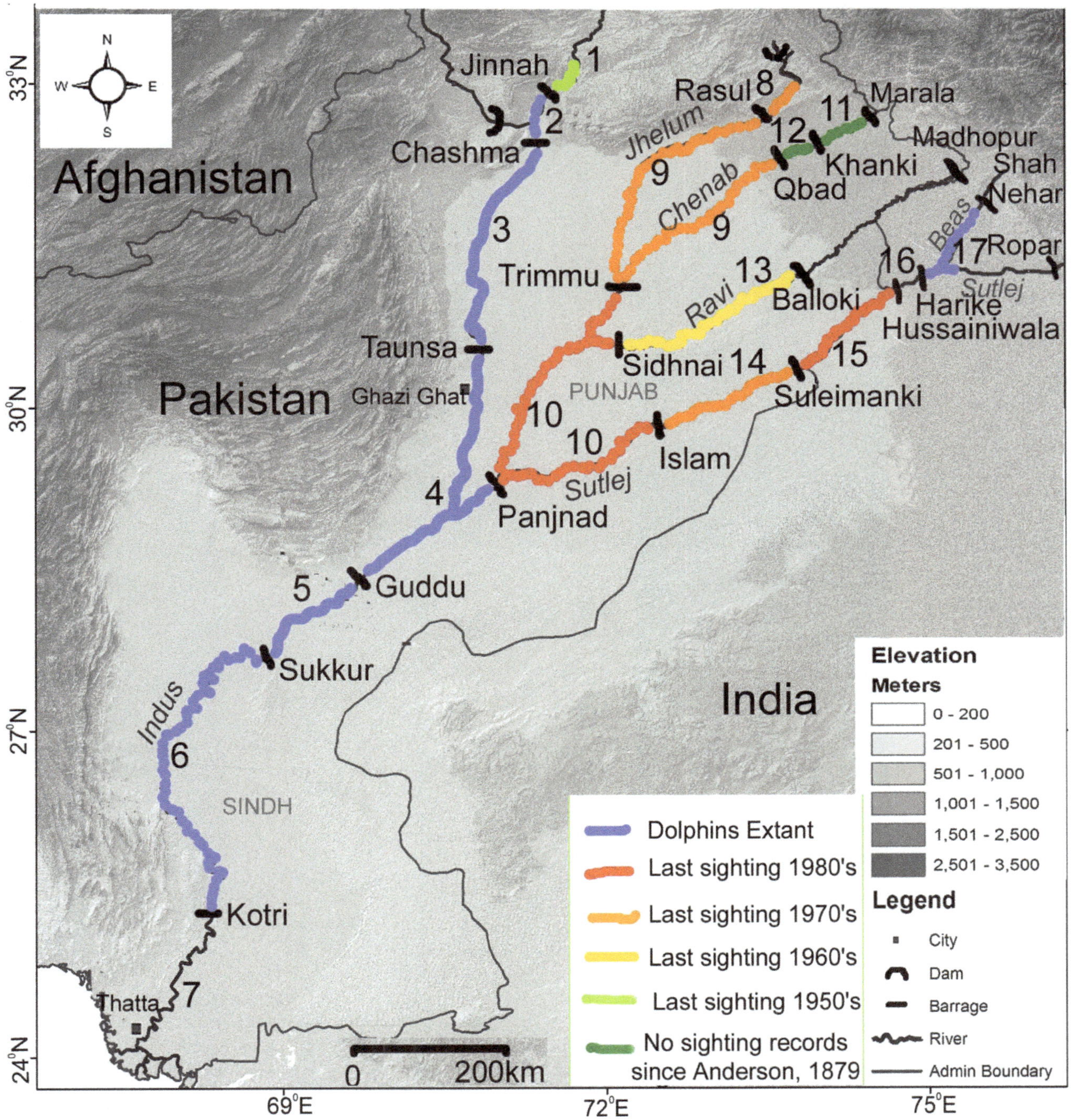

Figure 1. Map of the lower Indus River system. Rivers and barrages are named, and each river section is numbered and coloured according to whether river dolphins are extant, or the approximate date that they were extirpated (see Table A1 for details).

the initial data exploration to visually investigate whether the relationship between the predictor and explanatory variables was linear, and, which type of transformation could be used to best account for non-linearity. Three sets of models were developed, with objectives summarised below:

Spatial pattern of dolphin persistence. The objective of the first set of models was to identify which factors best explained the observed geographic pattern of range decline. The presence or absence of dolphins in each river section was modelled using a GLM and a binomial error distribution, with presently extant populations coded as 1 and extirpated populations coded as 0. The best fitting models were then used to predict the probability that dolphins are still present in the Harike-Hussainiwala river section on the India-Pakistan border (Fig. 1: no. 16), where dolphin presence is unknown.

Temporal pattern of decline. The second model set included only sections where dolphins have been extirpated and examined which factors influenced when dolphins disappeared. The number of years since dolphins were sighted (as of 2013) was

Figure 2. Aerial photograph of Sukkur Barrage. Image shows the seven canals diverting water out of the river, and demonstrates the dramatically reduced flows downstream (river flow direction right to left).

the response variable modelled using a GLM with a quasi-Poisson error distribution.

Time to extirpation. The third model set used a survival function to investigate which factors influenced the speed with which dolphin populations were extirpated following their isolation between barrages. We used the Kaplan–Meier estimate of survival which allows for the inclusions of censored data, in this instance allowing for the inclusion of river sections where dolphins have not yet been extirpated, as well as those where dolphins have disappeared. Each river section was qualified with a status assignment, where 1 = dolphins extirpated, and 0 = dolphins extant [27]. The time to extirpation was calculated as the number of years between the fragmentation date and either the LDSD if dolphins have been extirpated, or the current year, 2013, where they are still present. Time to extirpation and status together were the predictor variable (the Kaplan-Meier survivorship object) modelled using the 'survreg' function in the survival library of the program R [28]. Both an exponential and a Weibull error distribution were tested, and the Weibull distribution was selected as it provided a significantly better fit to the data (ΔAIC 15.17) [27].

All models were implemented using the program R 2.15.1 [28]. Logit, probit and cloglog link functions were included in global models and the logit function, which resulted in the best fit, applied. Variance Inflation Factors (VIFs) that demonstrate the degree of collinearity between variables were generated from the maximal models and collinear variables removed until VIF scores were less than five [27]. Three two-way interactions that described potentially meaningful relationships between variables (fragmentation date and dry season discharge, fragmentation date and river length, and river length and proximity to range edge) were included, as well as second and third order polynomials of significant variables. The binomial and survival models were simplified using backwards stepwise selection based on Akaike's Information Criteria (AIC). Quasi-Poisson models were selected on the basis of quasi-AIC (QAIC) scores, and non-significant terms were sequentially dropped based on their levels of significance. Models separated by at least two AIC/QAIC points were assumed to be significantly different [29]. Goodness of fit for the GLMs was measured by determining the proportion of the total deviance explained by the final model. Model plots were examined for non-normality of errors, statistical independence of observations, heteroscedasticity and influential points [30].

Results

Pattern of Range Decline

Historical dolphin sightings were obtained for all river sections formerly occupied by dolphins except for the area downstream of Kotri Barrage to the delta (Fig. 1: no. 7) and the stretch from Harike to Hussainiwala barrage (Fig. 1: no. 16) which is close to the India-Pakistan border. Our focus on retired fishermen and on areas from which dolphins have already been extirpated meant that the communities were forth-coming with information because it was not regarded by them as sensitive. However, there were very few elderly members of each fishing community and our pool of available informants was consequently small (n = 57). 79% of informants were, or had been, full-time commercial fishermen or contractors and the remainder were part-time subsistence fishermen. There was no significant difference in the age of informants interviewed at each barrage (GLM $p = 0.7$) but older individuals were significantly more likely to have seen dolphins than younger informants (GLM $p < 0.01$). We found no evidence that dolphins persist in any of the Indus tributaries in Pakistan. Of 17 sections of river, dolphins are extant in six, have been extirpated from ten, and in one border area that could not be surveyed (no. 16), dolphin presence or absence is unknown (Fig. 1).

Causes of Range Decline

For each river section, dolphin presence or absence, estimated LDSD, and the time to extirpation were compiled along with the physical characteristics and these data included in each of the models (Table S1). On the upper Chenab River (Fig1: no 11 & 12) the LDSD is that reported by Anderson in 1879. Time to extirpation was not calculated for these two sections reflecting the lack of recent sighting evidence and uncertainty of the extirpation date.

Where dolphins are still extant the median monthly dry season river discharge averaged 30,830 cusecs (ranged 7,224–47,040 cusecs), as compared to an average of 8,022 cusecs (range 0–38,000 cusecs) in locations from which dolphins have been extirpated. In general, sections of river where dolphins are still present were fragmented by barrages later, are further from the range periphery, are of longer length, have a shallower slope and greater dry season discharge than river sections where dolphins are no longer found.

Sixteen river sections, 6 where dolphins are extant and ten where they have been extirpated, were included in the spatial GLM models. The VIFs generated from the full model indicated that river discharge and slope were collinear. Slope was considered to be less important than discharge in explaining dolphin range decline because it has not changed substantially in hundreds of years, and it was therefore removed from further candidate models. The final model that best explained the observed spatial pattern in Indus dolphin range decline retained the explanatory variables dry season river discharge and distance from range edge (Table 1). The probability that an Indus dolphin subpopulation is still extant increases with increasing distance from the range edge, and with increasing dry season river discharge (Fig. 3). The final spatial model predicted only a 2.6% probability that dolphins are still extant in the Harike-Hussainiwala river section in India that was not surveyed for dolphins. This is not unexpected given that this section has very low dry season discharge and is near the periphery of the dolphin's range, both factors that increase the likelihood of subpopulation extirpation. Although the linear arrangement of river segments might suggest a lack of statistical independence, there was no clumping of residuals according to the geographic position of the river segments, and the independence of observations was further shown by the Durbin-Watson test of correlated errors ($p = 0.24$).

The variables that described the temporal pattern of Indus dolphin subpopulation extirpation were the same as those that influenced the spatial pattern of decline: dry season river discharge and distance from former range limit (Table 2). Within areas from which dolphins have disappeared, they were extirpated earlier from river sections where discharge was lower, and from those sections located near the periphery of the subspecies former range (Fig. 3).

In river sections where dolphins have been extirpated, the mean time from fragmentation to the LDSD was 50 years (SD = 23, range = 9–76). For river sections where dolphins are still extant, the mean time from subpopulation isolation to present (2013) was 57 years (SD = 15, range = 42–86). Thirteen river sections were included in the survival model (number 2, 10, 12 and 16 were excluded because of missing data) and the slope parameter was included. The final survival model retained four variables: median dry season river discharge, isolation date, length of river section and slope. Dolphin subpopulations were extirpated more quickly in sections with low dry season river discharge. Subpopulations persisted longer where the river slope is more gentle (e.g. in the lower reaches) and those that were isolated between barrages a long time ago persisted for longer than those in more recently

Table 1. Summary of spatial range decline model output.

Model	*AIC*	*Δ AIC*	*% explained Deviance*	*n*	*Deviance*						
					Q	**Range**	**Is. Date**	**L**	**Conf**	**Size**	**USD: Is. Date**
1	18.26	1.82	32.6	1	6.91	-					
2	**16.44**	**-**	**50.7**	**2**	**7.88**	**2.84**	-	-	-	-	-
3	17.72	1.28	54.1	3	7.88	2.84	0.72	-	-	-	-
4	19.67	3.23	54.3	4	7.88	2.84	0.68	0.08	-	-	-
5	21.64	5.20	54.4	5	7.88	2.84	0.42	0.31	0.07	-	-
6	23.64	7.20	54.5	6	7.88	2.84	0.42	0.31	0.07	-	0.001
7	25.16	8.72	56.7	7	7.88	2.84	0.79	0.31	0.07	0.0001	0.11

n = number of covariates, Is. Date = Isolation Date, L = Length of river section, Range = Distance from range edge, Size = River size, Conf = confluences, Q = River discharge, USD: Is. Date = Interaction between Isolation Date and River Discharge. Model in bold was the final selected model.

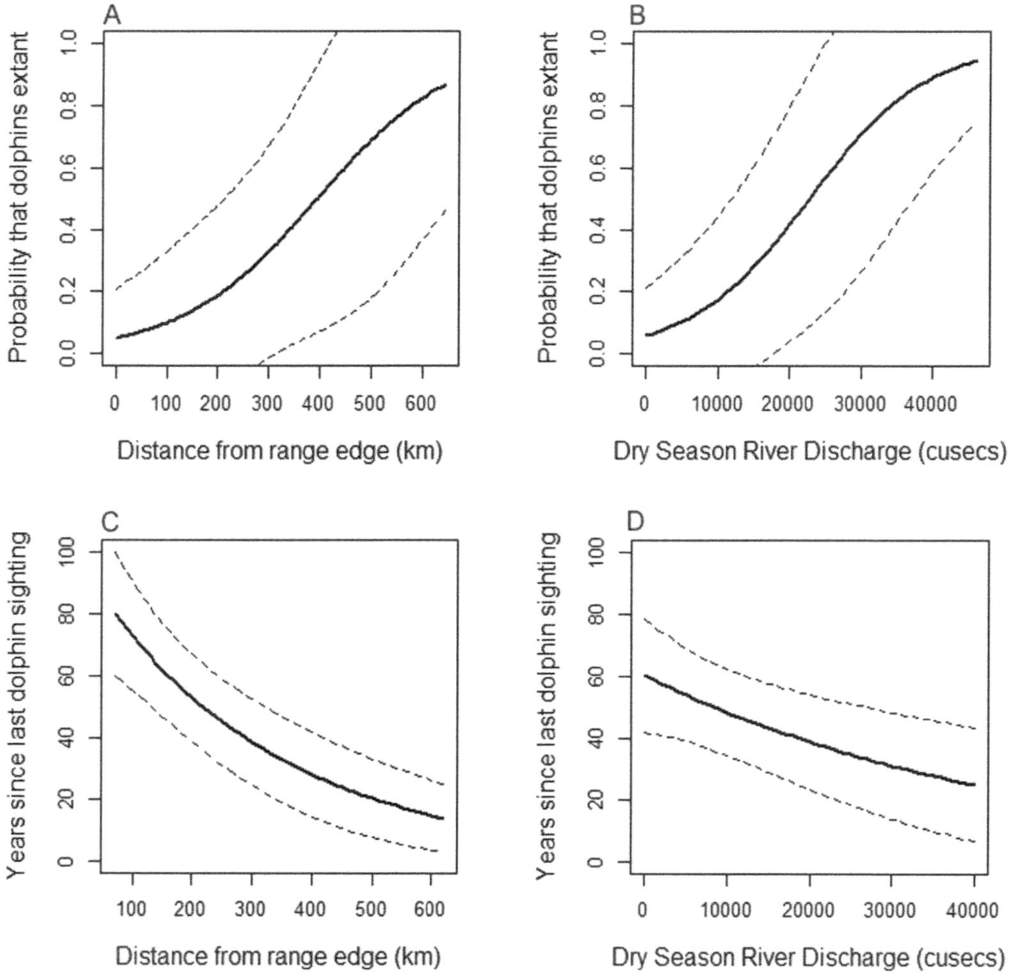

Figure 3. Significant relationships retained in GLM models of the causes of the spatial and temporal pattern of Indus dolphin decline. The figure demonstrates the probability that an Indus dolphin subpopulation is extant according to A) proximity to the edge of the former range and B) median dry season river discharge, and the relationship between the number of years since a dolphin was sighted and C) distance from the historical distributional limit, and D) median dry season river discharge.

subdivided river sections. Fifty years after Indus dolphins were isolated between barrages there is a less than a 50% chance that they will still be extant, and after 100 years this probability drops to 37%.

Discussion

Model evaluation

The river discharge data used in these models were from the last ten years but they explained well the pattern of dolphin decline that occurred decades ago. Although river discharge varies from year to year, and has generally declined, the relative discharge among barrages (e.g. the spatial relationship) has remained constant with the same locations consistently reporting high (e.g. the upper Indus) and low discharge (e.g. Indus tributaries) over time. Therefore, the assumption implicit in this analysis that the present spatial pattern of discharge reflects that present during the period leading up to dolphin extirpation is not unreasonable.

As for terrestrial habitats, such as forest fragments, we would have expected to see a relationship between species extinction and habitat patch size [5]. In fact, length of river section was one of the first variables to be excluded in candidate models. This may be because only the current configuration of 17 comparatively small habitat fragments were included in the models, and we did not consider the progression of escalating habitat fragmentation and concomitant diminishing fragment size over time. To investigate this, we constructed an additional model considering all 33 river sections that have existed since the onset of barrage construction (Table A2) in a binomial GLM, with dolphins recorded as extirpated (0) or still extent (1) in each river section at the point it was further subdivided. Explanatory variables were 1) Length of river section, 2) Isolation Date, and 3) End Date, taken as the year a new barrage was completed resulting in the sections further subdivision. When considering the entire history of habitat fragmentation, the models showed that dolphins were significantly more likely to be extirpated in smaller fragments ($p<0.05$), and that this relationship was independent of fragment creation date or duration. The link between species extirpation and habitat fragment size has been clearly established in terrestrial habitats for several species groups [5] but this is one of the first studies to show a similar relationship in riverscapes. It underlines the great importance of maintaining large sections of intact river habitat to sustain tropical aquatic biodiversity.

Table 2. Summary of temporal range decline model outputs.

Model	*AIC*	*Δ AIC*	*% explained Deviance*	*n*	*Deviance*						
					Q	**Range**	**Is. Date**	**L**	**Conf**	**Size**	**USD*Is. Date**
1	17.57	1.36	56.75	1	-	115.36	-	-	-	-	-
2	**16.21**	**-**	**76.76**	**2**	**40.67**	**115.36**	**-**	**-**	**-**	**-**	**-**
3	17.80	1.59	79.17	3	40.67	115.36	-	-	-	-	4.91
4	17.54	1.33	92.62	4	52.06	115.36	-	-	-	6.66	14.20
5	19.50	3.29	92.85	5	52.06	115.36	-	-	6.60	6.66	8.11
6	21.46	5.25	93.12	6	52.06	115.36	0.75	-	6.60	6.66	7.9
7	23.39	7.18	93.49	7	52.06	115.36	0.60	0.33	9.16	6.66	5.89

n = number of covariates, Is. Date = Isolation Date, L = Length of river section, Range = Distance from range edge, Size = River size, Conf. = confluences, Q = River discharge. Model in bold was the final selected model.

River discharge and distance from range periphery provided a good fit to the range decline data, explaining more than 76% of the deviance in the temporal model and 50% in the spatial model. However, three other aspects that may have also have played a role in the dolphins decline were not included as explanatory variables because of a lack of suitable data. These are a) water quality, b) incidental capture in fishing gear and c) hunting. The possible contributions of these to the Indus dolphin decline are discussed below.

The magnitude of surface water pollution in Pakistan has increased at a dramatic rate over the last ten years and more than 90% of industrial and municipal effluents enter water courses untreated [31]. The Indus tributaries flow through the industrial and agricultural heartland of Pakistan and are more polluted than the Indus River itself which has a greater dilution capacity and passes through remoter areas. There has been no systematic monitoring of river water quality that could provide data for this analysis. However, dolphins had already been extirpated from most areas prior to significant declines in water quality which occurred in the 1980s and 90s, and this asynchronous timing indicates that pollution was not primarily responsible for the dolphins' decline.

Mortality from accidental capture in fishing gear is considered to be the greatest threat to most cetacean populations [32]. In the past, the Indus River main channel was not intensively fished because the water was too swift for easy manoeuvrability of oar-powered boats, and instead fishing focused on side channels and adjacent pools that are rarely used by dolphins [33]. Since 2010 changing fishing practices in Sindh Province have led to an increase in dolphin mortality, however, prior to this there are very few records of incidental capture of dolphins in fishing gear and this is not likely to be a large factor in the decline of the Indus dolphin.

Indus dolphins were killed for food, oil and medicine until the late 1970s when the animal became legally protected [17,34]. Information on dolphin hunting is sparse and un-quantified and records refer only to hunting on the Indus River, where dolphins are still extant. Although it is possible that dolphins were hunted throughout the river system, there is no evidence that this was so, and the fact that they persist in the places that hunting is reported to have been intense, and have disappeared from places where hunting was not reported, suggests that this was unlikely to have been the cause of the subspecies' decline. However, the timing of reported hunting coincides with the period of decline and without more information, it is not possible to completely discount the role of hunting.

For the majority of the year the gates on all barrages are lowered to divert water into canals, and the physical opening is sufficiently small that it would be difficult or impossible for dolphins to pass through the gates and between different sections of river. It has been hypothesized that there may be consistent or frequent movements of dolphins through some barrages and between subpopulations [21,35]. It has also been theorised that due to the high water velocity and turbulence often found within the barrage gates it would be more likely for animals to move down-, rather than up-stream, and that this would lead to the downstream migratory attrition of upstream subpopulations [18,36]. The only evidence of this was obtained in 2009, when, during the annual canal maintenance period, which is one of the brief periods in the year when no water is diverted and barrage gates are fully open for several weeks, one radio-tagged Indus dolphin was recorded to move through the gates on Sukkur barrage in both an up- and down-stream direction (WWF-Pakistan unpublished). Each barrage is quite different in terms of design, location and operation and one dolphin moving through Sukkur barrage does not prove that this occurs at other barrages, or that it is a regular occurrence. Therefore, for the purposes of this study, and in the absence of evidence to the contrary, we assume no significant migration between dolphin subpopulations. It is possible that future research may demonstrate that dolphins do move across some, or all, barrages with regularity and if the movement of individuals is primarily in one direction this would be another important factor to consider in the extinction dynamics. The pattern of subpopulation persistence near the range core, and earlier extirpation of subpopulations near the range periphery could perhaps be partially explained by the consistent downstream migration of animals, but again this pattern is disrupted by the presence of dolphins in the Beas River.

Causes of Indus dolphin range decline

The clear result of this study was the relationship between low dry season river discharge and the decline of the Indus dolphin. Reduced flows directly impact dolphins by reducing the physical space available to them, reducing average water velocity and depth and increasing water temperatures. Flow regulation is also likely to indirectly impact river dolphins due to declines in fish diversity, the dominance of generalist fish species, and increased success of invasive species [37,38]. The dampened flood peaks typically associated with dams and diversions reduce the frequency, extent and duration of floodplain inundation that determines how long fish can gain access to nursery habitat and food. Water abstraction also exacerbates and concentrates existing anthropogenic threats, for example, increasing the concentration of nutrients and pollutants, and concentrating dolphins into deep pools that are also important areas for fishing, thereby increasing the chances of negative human interactions [39]. The altered hydrological regime on the Indus River has likely reduced the complexity of hydrologic and geomorphologic habitat and ultimately also diminished its carrying capacity and ability to support large numbers of aquatic megafauna. To preserve aquatic biodiversity, river management is needed that focuses on restoring both the timing and duration of flood pulses, as well as on maintaining critical minimum flows in the dry season.

The persistence of dolphins in the Beas River, India is likely to be due to the presence of constant water supplies little depleted by diversions. Dolphins in the Beas River occur in an isolated habitat fragment as the river downstream is virtually dry, and only connected with the rest of the river system for a few weeks each year during the monsoon floods. This demonstrates that in the presence of sufficient water, and an absence of other threats, river dolphins can persist for decades even in relatively small fragments of habitat near the periphery of their range. This subpopulation is of conservation importance, as all other Indus dolphins occur in a single river, the Indus, and are therefore at risk from environmentally correlated catastrophic events [40]. However, based on the historical pattern of decline, Indus dolphins are most likely to disappear in the future from locations with low river discharge located closer to the range periphery, meaning that dolphins in the Beas (close to range periphery with moderate discharge) and between Sukkur and Kotri Barrages (with low discharge located a moderate distance from the former range edge) are most at risk.

The date of habitat fragmentation was not selected by any of the models as a strong predictor of whether dolphins are still present. However depleted river discharge and habitat fragmentation by barrages are inextricably intertwined as barrages are responsible for diverting water, and they are a physical barrier that greatly impedes or prevents the dispersal of dolphins out of impacted river reaches.

Population extirpation, core habitat and conservation

Contraction of geographic range is one of the principal characteristics exhibited by declining or threatened species [41,42]. In general, at the periphery of a species geographic range, populations occupy less favourable habitat and occur at lower and more variable densities. Therefore, as a species becomes endangered it is expected that its geographic range will contract inwards, and that populations will persist in the range core until the final stages of decline [43]. However, for many endangered mammals the pattern of range decline is instead dictated by the spread of factors driving the decline, with those populations last impacted, regardless of their location, persisting longer than those that were historically large [41,43]. The range of the Indus dolphin has also contracted inwards, and dolphins persist primarily in what is assumed to be the former range core or higher density area, however this is likely to be because the greatest threat, water extraction, is concentrated in the periphery of the subspecies range. This conclusion is supported by the continued persistence of the Beas River population at the range limit. However, that animals naturally occur at lower density in upstream areas and smaller rivers and are therefore more vulnerable near the range limit is also certainly a factor. The spatial pattern of Indus River dolphin decline is very different from the gradual decline in abundance described for the Yangtze River dolphin [16].

One of the greatest challenges in conservation science involves disentangling the relative contributions of multiple factors in the decline of species, especially when causes interact or vary spatially and temporally with importance [44]. Nevertheless, the primary factor identified in these models (i.e. low dry season discharge due to water diversion at barrages) is well supported, and is also the most salient for informing current ecosystem management. In the Mekong River, numerous stressors such as fisheries bycatch, hunting and habitat destruction have reduced the resident Irrawaddy dolphin (*Orcaella brevirostris)* population to less than 100 individuals [45]. The results of our study suggests that habitat fragmentation and/or flow disruption associated with the many proposed new dams on the Mekong are likely to further drive Irrawaddy dolphin decline potentially leading to local extinction. The Indus dolphin range decline is probably most prescient for the related Ganges dolphin (*Platanista gangetica gangetica*) which occurs in the neighbouring Brahmaputra and Ganges River systems and is subject to the same threats. The Ganges River is fragmented by barrages and flow is severely reduced in many areas, and the Brahmaputra River system is the focus of massive hydropower development [46,47]. The range of the Ganges dolphin has begun to decline especially at the upper limit of the distributional range [48], and if water development continues as planned the range of the dolphin is expected to continue to shrink towards larger habitat fragments with higher discharge that are primarily in downstream locations. The results of this study suggest that other vertebrate populations in other large rivers, such as the Amazon, Orinoco and Ayeyarwady, will also respond with dramatic declines in range when dams and other water developments that reduce discharge, fragment habitat and change the hydrological regime are constructed.

The amount of habitat fragmentation and level of water withdrawals from rivers in Pakistan is extreme, negatively affecting human communities, eroding the delta, destroying fisheries and concentrating pollutants. This study indicates that if water development plans in South Asia and the wider Himalayan region proceed as currently proposed [10,13]and follow the pattern demonstrated by the Indus, the resulting habitat fragmentation and flow disruption will likely cause large declines in resident freshwater cetaceans and other freshwater dependent species. Healthy rivers are of great importance to communities, and it is critical that where developments are planned, environmental flow and impact assessments be conducted that balance human requirements for irrigation water and power with the habitat requirements of the aquatic ecosystem that are vital to humans.

Acknowledgments

We are grateful to the Office of Executive Engineers at Harike and Hussainiwala Canal Divisions and Ropar Headworks, India for allowing us river discharge data. A. Khan and R. Garstang facilitated the fieldwork. Valuable reviews were provided by P. Hammond, M. Lonergan, S. Turvey and G. Ryan.

Author Contributions

Conceived and designed the experiments: GTB. Performed the experiments: GTB MA UN. Analyzed the data: GTB SPN. Contributed to the writing of the manuscript: GTB MA UN SPN.

References

1. Sala OE, Chapin FS, Armesto JJ, Berlow R, Bloomfield J, et al. (2000) Global biodiversity scenarios for the year 2100. Science 287.
2. Dudgeon D, Arthington AH, Gessner MO, Kawabata Z-I, Knowler DJ, et al. (2006) Freshwater biodiversity: importance, threats, status and conservation challenges. Biological Reviews 81: 163–182.
3. Tilman D, May RM, Lehman CL, Nowak MA (1994) Habitat destruction and the extinction debt. Nature 371: 65–66.
4. Strayer DL, Dudgeon D (2010) Freshwater biodiversity conservation: recent progress and future challenges. Journal of the North American Benthological Society 29: 344–358.
5. Gibson L, Lynam AJ, Bradshaw CJA, He F, Bickford DP, et al. (2013) Near-complete extinction of native small mammal fauna 25 years after forest fragmentation. Science 341: 1508–1510.
6. Prugh LR, Hodges KE, Sinclair ARE, Brashares JS (2008) Effect of habitat area and isolation on fragmented animal populations. PNAS 105: 20770–20775.
7. FAO (2012) Coping with water scarcity. An action framework for agriculture and food security. Rome, Italy: FAO Water Reports No.38.
8. World Water Assessment Programme (2012) The United Nations World Water Development Report 4. Managing water under uncertainty and risk. Paris: UNESCO.
9. Dudgeon D (2000) Large-scale hydrological changes in tropical Asia: prospects for riverine biodiversity. Bioscience 50: 793–806.
10. Dutta AP (2010) Reservoir of Dams. Down to Earth May 1–15: 32–39.
11. Verma S, Kampman DA, van der Zaag P, Hoekstra AY (2009) Going against the flow: A critical analysis of inter-state virtual water trade in the context of India's National River Linking Program. Physics and Chemistry of the Earth 34: 261–269.
12. Dharmadhikary S (2008) Mountains of Concrete: Dam building in the Himalayas. Berkeley, CA: International Rivers. 48 p.
13. Ziv G, Baran E, Nam S, Rodríguez-Iturbe I, Levin SA (2012) Trading-off fish biodiversity, food security, and hydropower in the Mekong River basin. PNAS www.pnas.org/cgi/doi/10.1073/pnas.1201423109.
14. IUCN (2013) IUCN Red List of Threatened Species. version 2013.2 <http://www.iucnredlist.org> Downloaded on 6 January 2014.
15. Turvey ST, Pitman RL, Taylor BL, Barlow J, Akamatsu T, et al. (2007) First human-caused extinction of a cetacean species? Biology Letters 3: 537–540.
16. Turvey ST, Barrett LA, Hart T, Collen B, Yujiang H, et al. (2010) Spatial and temporal extinction dynamics in a freshwater cetacean. Proceedings of the Royal Society B 277: 3139–3147.
17. Anderson J (1879) Anatomical and zoological researches: comprising an account of the zoological results of the two expeditions to western Yunnan in 1868 and 1875 and a monograph of the two cetacean genera *Platanista* and *Orcella*. Bernard Quaritch, Piccadilly, London.
18. Reeves RR, Chaudhry AA, Khalid U (1991) Competing for water on the Indus Plain: Is there a future for Pakistan's river dolphins? Environmental Conservation 18: 341–349.
19. Braulik GT (2006) Status assessment of the Indus River dolphin, *Platanista gangetica minor*, March-April 2001. Biological Conservation 129: 579–590.
20. Behera SK, Nawab A, Rajkumar B (2008) Preliminary investigations confirming the occurrence of Indus River dolphin *Platanista gangetica minor* in River Beas, Punjab, India. Journal of the Bombay Natural History Society 105: 90–126.
21. Reeves RR, Leatherwood S (1994) Dams and river dolphins: can they co-exist? Ambio 23: 172–175.
22. Braulik GT, Bhatti ZI, Ehsan T, Hussain B, Khan AR, et al. (2012) Robust abundance estimate for endangered river dolphin subspecies in South Asia. Endangered Species Research 17: 201–215.
23. Collen B, Purvis A, Mace GM (2010) When is a species really extinct? Testing extinction inference from a sighting record to inform conservation assessment. Diversity and Distributions 16: 755–764.
24. Butchart SHM, Stattersfield AJ, Brooks TM (2006) Going or gone: defining 'Possibly Extinct' species to give a truer picture of recent extinctions. Bulletin-British Ornithologists Club 126: 7–24.
25. IUCN (2011) Pakistan Water Gateway: Gateway to water information about Pakistan. www.waterinfo.net.pk.
26. Braulik GT, Reichert AP, Ehsan T, Khan S, Northridge SP, et al. (2012) Habitat use by a freshwater dolphin in the low-water season. Aquatic Conservation: Marine and Freshwater Ecosystems 22: 533–546.
27. Crawley MJ (2007) The R Book. Chichester, England: John Wiley & Sons Ltd. 951 p.
28. R Development Core Team (2012) R: A language and environment for statistical computing. Vienna, Austria: R Foundation for Statistical Computing.
29. Burnham KP, Anderson DR (2002) Model selection and multi- model inference: a practical information-theoretic approach. 2nd edition. New York: Springer-Verlag. 496 p.
30. Fox J (2008) Applied regression analysis and Generalised Linear Models. 2nd edition. Los Angeles: Sage Publications, Inc. 665 p.
31. Directorate of Land Reclamation Punjab (2007) Surface water quality monitoring plan. Lahore: Irrigation and Power Department, Government of Punjab. 27 p.
32. Read AJ (2008) The looming crisis: interactions between marine mammals and fisheries. Journal of Mammalogy 89: 541–548.
33. Khan H (1947) A fishery survey of River Indus. Journal of the Bombay Natural History Society 46: 529–535.
34. Pilleri G (1972) Field observations carried out on the Indus dolphin *Platanista indi* in the Winter of 1972. Investigations on Cetacea 4: 23–29.
35. Braulik GT (2012) Conservation ecology and phylogenetics of the Indus River dolphin (*Platanista gangetica minor*). St. Andrews, UK. 267: University of St. Andrews.
36. Reeves RR (1991) Conservation of the bhulan (blind river dolphin) in the Punjab. Natura. 3–22.
37. Nilsson C, Reidy CA, Dynesius M, Revenga C (2005) Fragmentation and flow regulation of the world's large river systems. Science 308: 405–408.
38. Xenopoulos MA, Lodge DM (2006) Going with the flow: using species-discharge relationships to forecast losses in fish biodiversity. Ecology 87: 1907–1914.
39. Kelkar N, Krishnaswamy J, Choudhary S, Sutaria D (2010) Coexistence of fisheries with river dolphin conservation. Conservation Biology 24: 1130–1140.
40. Gilpin ME (1990) Extinction in finite metapopulations living in correlated environments. In: Shorrocks B, Swingland IR, editors. Living in a patchy environment. New York: Oxford University Press. 177–187.
41. Channell R, Lomolino MV (2000) Trajectories to extinction: spatial dynamics of the contraction of geographical ranges. Journal of Biogeography 27: 169–179.
42. Simberloff D (1986) The proximate causes of extinction. In: Raup DM, Jablonski D, editors. Patterns and processes in the history of life. Berlin: Springer-Verlag. 259–276.
43. Lomolino MV, Channell R (1995) Splendid isolation: patterns of geographic range collapse in endangered mammals. Journal of Mammalogy 76: 335–347.
44. Johnson PTJ, McKenzie VJ, Peterson AC, Kerby JL, Brown J, et al. (2010) Regional decline of an iconic amphibian associated with elevation, land-use change and invasive species. Conservation Biology 25: 556–566.
45. Ryan GE, Dove V, Trujillo F, Doherty Jr PF (2011) Irrawaddy dolphin demography in the Mekong River: an application of mark-resight models. Ecosphere 2: 1–15.
46. Bashir T (2010) Ganges river dolphin (Platanista gangetica) seeks help. Current Science 98: 287–288.
47. Baruah D, Hazarika LP, Bakalial B, Borah S, Dutta R, et al. (2012) A grave danger for the Ganges dolphin (Platanista gangetica, Roxburgh) in the Subansiri River due to a large hydroelectric project. Environmentalist 32: 85–90.
48. Sinha RK, Smith BD, Sharma G, Prasad K, Choudhury BC, et al. (2000) Status and distribution of the Ganges Susu (*Platanista gangetica*) in the Ganges River system of India and Nepal. In: Reeves RR, Smith BD, Kasuya T, editors. Biology and Conservation of Freshwater Cetaceans in Asia. Gland, Switzerland & Cambridge, U.K.: IUCN. 54–61.

8

Two Intense Decades of 19th Century Whaling Precipitated Rapid Decline of Right Whales around New Zealand and East Australia

Emma L. Carroll[1]*¤, Jennifer A. Jackson[2], David Paton[3], Tim D. Smith[4]

1 School of Biological Sciences, University of Auckland, Auckland, New Zealand, 2 British Antarctic Survey, Cambridge, Cambridgeshire, United Kingdom, 3 Blue Planet Marine, Canberra, Australian Capital Territory, Australia, 4 World Whaling History, Redding, California, United States of America

Abstract

Right whales (*Eubalaena* spp.) were the focus of worldwide whaling activities from the 16th to the 20th century. During the first part of the 19th century, the southern right whale (*E. australis*) was heavily exploited on whaling grounds around New Zealand (NZ) and east Australia (EA). Here we build upon previous estimates of the total catch of NZ and EA right whales by improving and combining estimates from four different fisheries. Two fisheries have previously been considered: shore-based whaling in bays and ship-based whaling offshore. These were both improved by comparison with primary sources and the American offshore whaling catch record was improved by using a sample of logbooks to produce a more accurate catch record in terms of location and species composition. Two fisheries had not been previously integrated into the NZ and EA catch series: ship-based whaling in bays and whaling in the 20th century. To investigate the previously unaddressed problem of offshore whalers operating in bays, we identified a subset of vessels likely to be operating in bays and read available extant logbooks. This allowed us to estimate the total likely catch from bay-whaling by offshore whalers from the number of vessels seasons and whales killed per season: it ranged from 2,989 to 4,652 whales. The revised total estimate of 53,000 to 58,000 southern right whales killed is a considerable increase on the previous estimate of 26,000, partly because it applies fishery-specific estimates of struck and loss rates. Over 80% of kills were taken between 1830 and 1849, indicating a brief and intensive fishery that resulted in the commercial extinction of southern right whales in NZ and EA in just two decades. This conforms to the global trend of increasingly intense and destructive southern right whale fisheries over time.

Editor: Athanassios C. Tsikliras, Aristotle University of Thessaloniki, Greece

Funding: Funding for the review and report preparation was provided by the New Zealand Ministry of Fisheries through Project ZBD200505 to the National Institute of Water and Atmospheric Research (NIWA), Oregon State University General Research Fund to JJ and CS Baker and the Lenfest Ocean Program of the Pew Charitable Trust to SR Palumbi. EC was supported by a fellowship from the Tertiary Education Commission and TS through the History of Marine Animal Populations (HMAP) project. The funders had no role in study design, data collection and analysis, decision to publish, or preparation of the manuscript.

Competing Interests: The authors have the following interests: David Paton is employed by Blue Planet Marine. This does not alter the authors' adherence to PLOS ONE policies on sharing data and materials.

* E-mail: elc6@st-andrews.ac.uk

¤ Current address: Scottish Oceans Institute, University of St Andrews, St Andrews, Fife, Scotland

Introduction

Right whales (*Eubalaena spp*) were a primary target of whalers from the mid 16th century to the late 20th century [1]. Pursued for oil, processed from their thick blubber, and for their baleen, all three species (North Pacific: *E. japonica*; North Atlantic: *E. glacialis* and southern: *E. australis*) were greatly reduced in abundance in all oceans by a series of, at times, short-term fisheries [2].

Today, the North Pacific and North Atlantic species are represented by small, remnant populations occupying a fraction of their historical range [3–5]. In contrast, the southern right whale shows spatially variable recovery: some populations are recovering at close to the biological maximum rate (e.g. South Africa [6]) whereas others are possibly functionally extinct (e.g. Chile [7]). Extant populations of southern right whales show significant differences in maternally-inherited mitochondrial DNA haplotype frequencies, consistent with female fidelity to migratory destinations [8]. This behaviourally-mediated mechanism of isolation, coupled with the spatially fragmented recovery, suggests each population underwent its own decline due to whaling and subsequent recovery (or lack thereof). Therefore, historical assessments should be attempted at a population level, to provide accurate historical records of past abundances, distribution and catch histories.

Here we focus on reconstructing the catch history for southern right whales around New Zealand and east Australia. The historical patterns of seasonal migration, distribution and abundance of southern right whales in this region are complex and not well understood. Historical records suggest there were two distinct coastal whaling grounds around New Zealand islands: New Zealand sub-Antarctic (Auckland and Campbell Islands) and mainland New Zealand (North and South Islands) [9]. Mainland New Zealand was a coastal calving ground where females would give birth in the bays and inlets during winter. It is unclear whether the New Zealand sub-Antarctic was historically a calving or feeding ground, or a mixture of both [10]. The two areas could have been linked by large-scale migration patterns, as the species moved offshore during the austral summer to feed [10]. The New Zealand sub-Antarctic is now the primary calving ground for the New Zealand population [11]. Although historically severely depleted, the New Zealand population now numbers 2,000 whales and is growing at around 7% per annum, based on a mark-recapture study [12]. The two areas appear to be inhabited by one contemporary population based on genetic studies and the movements of individuals [11,13], although it is unclear whether this was true historically.

Across the southern coast of Australia, wintering aggregations of southern right whales, particularly cows with calves, were found prior to whaling [2]. The species moved offshore during the austral summer, presumably migrating to feeding grounds. Although there is little evidence to suggest subdivision of calving grounds from the historical data [2], today there is a clear difference in recovery between southern right whales in southwest and southeast Australia. The southwest population is growing at 6.8% per annum and numbers approximately 3,000 whales [14]. In contrast, the southeast Australian population numbers approximately 500 whales, and it does not appear to be recovering at the same speed as the southwest [15].

Investigations of current population structure show there are significant differences in maternally-inherited mitochondrial DNA haplotype frequencies and bi-parentally inherited nuclear microsatellite markers between southern right whales in southwest Australia and New Zealand, indicating some degree of isolation between these two populations [13]. Preliminary findings, based on a small sample from the southeast Australia population, indicates this is a small, remnant population, distinct from New Zealand and southwest Australia. However, there is only weak genetic differentiation between the southeast Australian and New Zealand populations, suggesting the two populations could have current or historic gene flow [13]. Additionally, the two populations could have mixed during offshore migration, for example, in the Tasman Sea or during migrations to feeding grounds [16]. Therefore, there are biological reasons to consider both regions when reconstructing catch histories in the region.

In a global perspective describing shore-based and offshore whaling, seven whaling operations taking right whales around eastern Australia and New Zealand were identified [1]. Shore-based whaling began around 1805 in bays around Tasmania and the Australian mainland, and later developed in bays in New Zealand [1]. Pelagic or ship-based offshore whaling was pursued in the eastern Australian and New Zealand region by Australian and New Zealand vessels and by French, British and American registered vessels from the 1820s [1]. This was part of a much broader global whale fishery involving primarily right and sperm whales [1].

The earliest evidence of whaling in New Zealand and eastern Australian waters are reports of a 1791 port call by a whaling vessel in New Zealand and some 1805 shore whaling activity near Hobart [17]. Although little information on the magnitude of right whale catches is available for the earliest years, information on the catches of the seven whaling operations noted above is available from several primary sources beginning in 1827 [1]. One source is tabulations of returns of fisheries and the related, but less easily interpreted, records of whale oil and baleen exports, both kept by colonial or national authorities. Export records do not reflect whale oil used locally and do not necessarily relate to the year of capture.

These data were reported in barrels of oil, and usually distinguished between the more valuable oil from sperm whales (sperm oil) and the less valuable oil from right whales ("whale oil" or "black oil"). This latter term was also used for oil from pinnipeds. The reports of oil and baleen can be converted to rough numbers of whales using average numbers of barrels and pounds of baleen obtained per whale, respectively, for a subset of logbooks where those quantities were recorded for individuals or small groups of whales [17].

A second source of catch data is lists of whaling voyages, such as those tabulated for example by Starbuck [18] for American whaling vessels and by Du Pasquier [19] for French whaling vessels. These lists vary in completeness and in the information tabulated. A third source of data is daily logbooks kept by ship-based whalers, which frequently include information on numbers and species of whales captured, numbers struck and lost, and locations where whales were sought and where they were sighted or caught. Finally, there is published information about 20th century right whaling in this region, including illegal Soviet whaling [20]. Dawbin [17] estimated southern right whale catches by both shore-based and ship-based whaling using some of these sources of data. Here we review and extend his estimates using the same sources and additional sources not previously available to improve the historical catch series for New Zealand and east Australian right whales.

We consider two scenarios in order to capture the biological and historical uncertainties in the catch series: the strict New Zealand catch series, and the catch series for New Zealand plus east Australia. Vessels from Hobart and Sydney went bay whaling in New Zealand and sometimes the catch was itemised, allowing New Zealand catches to be correctly assigned to the country of origin [17,21]. This is not always the case, meaning some New Zealand catches were incorrectly assigned to the east Australian shore-based catch series. Combining the totals from colonies in southeast Australia and New Zealand should provide a complete catch series that is representative of the overall region [17,21]. In addition, the weak level of genetic differentiation between the New Zealand and east Australian stocks seen today could be due to recent divergence and the regions could have been historically linked by gene flow, with some New Zealand whales available for capture in east Australian waters and vice versa. It should be noted that the historical records available do not permit a comprehensive catch series for southwest Australia to be reconstructed [22].

We account for the distribution of catches over time for four fisheries: shore-based whaling in bays, ship-based whaling in bays, ship-based whaling offshore and whaling in the 20th century. We improve upon previous work by multiplying our estimates of catches by new, fishery-specific estimates of struck and lost whales to obtain estimates of total removals, and include explicit estimates of sampling uncertainties for some of these fisheries. These estimates of total removals of right whales are designed for use in the modelling of right whale population history in this region and represent a significant expansion on previous work, resulting in more comprehensive estimates of total catch and removals that allow insights into the development of the fishery over time.

Materials and Methods

Shore-based Whaling

Our reconstruction of the shore-based catches is based on different data for New Zealand and east Australia. In both cases, the primary reference sources were export records and returns of fisheries, as described in Dawbin [17] and in a reference bibliography [23]. For the Australian catches, we did not review the primary sources due to limitations in access and resources available, but rather used the coastal catch series presented by Dawbin [17].

Table 1. Estimated coastal catches from shore-based whaling operations, listed as number of southern right whales, for Victoria (VIC), Tasmania (TAS), New South Wales (NSW), and the catches at New South Wales that were from New Zealand (NSW-NZ) are reproduced from Dawbin [17] for convenience.

Year	VIC	TAS	NSW	NSW-NZ	NZ-Low	NZ-High
1827	0	64	0	0	0	0
1828	0	109	10	0	0	0
1829	0	131	9	0	0	24
1830	0	233	103	0	28	120
1831	0	195	201	0	30	239
1832	0	246	49	0	23	140
1833	0	346	94	62	56	295
1834	61	356	237	118	84	333
1835	170	409	279	271	98	446
1836	97	493	235	127	82	341
1837	142	815	401	198	72	226
1838	3	844	435	325	145	440
1839	60	1064	539	390	128	158
1840	0	804	17	242	86	143
1841	44	279	198	166	57	95
1842	5	167	320	249	25	61
1843	27	277	58	50	332	332
1844	35	241	114	85	276	276
1845	4	259	91	66	187	187
1846	21	85	140	54	151	151
1847	8	104	60	41	134	134
1848	3	70	77	23	83	83
1849	1	24	34	10	27	27
1850	1	46	76	15	17	17
1851	1	32	129	10	5	5
1852	0	13	24	21	17	17
1853	0	8	78	34	14	31
1854	0	0	27	1	13	13
1855	0	0	10	7	22	22
1856	0	0	23	10	34	34
1857	0	0	11	0	28	28
1858	0	0	5	0	13	13
1859	0	0	52	17	22	22
1860	0	0	42	11	2	2
1861	0	0	24	5	2	11
1862	0	0	57	5	7	9
1863	0	0	11	9	5	33
1864	0	0	22	7	3	17
1865	0	0	15	7	1	12
1866	0	0	12	0	1	7
1867	0	2	30	1	1	7
1868	0	6	45	1	4	9
1869	0	2	10	0	9	21
1870	0	1	28	0	10	23
1871–1900	0	20	136	0	226	356
1901–1930	0	0	7	0	143	143
Total	**683**	**7,745**	**4,575**	**2,638**	**2,703**	**5,104**

Two estimated coastal catch series for New Zealand (NZ-Low, NZ-High), reflecting different selections of primary sources as described in Table S1.

For New Zealand shore-based catches, the amount of whale oil and/or baleen reported was tallied for each year from 1829 for each primary source used by Dawbin [17]: Great Britain Parliamentary Records [24–25], Blue book of statistics from the Great Britain Colonial Office [26], McNab [27–30], Sherrin [21], Statistics New Zealand [31] and Wakefield [32]. To convert the oil and baleen data reported in those sources to numbers of whales, one whale was estimated to have produced 4.18 tuns of black oil or 600 pounds of baleen, based on the average yield from 413 shore-caught whales in Dawbin [17]. Rarely, the number of whales caught was available for specific years, and where available these values were used rather than values based on reported oil and baleen. The catch series reconstructed from each primary source was then compared with the catch history in Dawbin [17] and the original source reference that he used for each year was identified by this comparison. Some errors in the original catch series were identified using primary sources, and catch estimates were modified accordingly (Table 1).

For the years 1853–1930, the catch series was based on Statistics New Zealand records of baleen and right whale (black) oil exports. For each series, the amount exported each year was tallied and missing data were interpolated using a five-year moving average and the variance around this average was calculated. The export series was then converted into whales using the above conversion rates.

Ship-based Offshore Whaling

Ship-based offshore whaling and bay whaling in Australia and New Zealand was undertaken by French and American ships, and apparently to a lesser extent by British, Australian and New Zealand vessels [1]. Dawbin [17] estimated catch by French and American ships, and we first sought to reconstruct those estimates using his primary data sources. We then expanded upon that previous work on American ship-based whaling by reading a selection of logbooks that were relatively complete for location and species hunted data and by explicitly including uncertainty in our estimates.

Dawbin [17] estimated the landed catch of right whales in the vicinity of New Zealand by 19th century American and French whalers using information contained in lists of whaling voyages. For American whalers, Starbuck [18] listed summary information for individual whaling voyages including departure and subsequent arrival dates back to their home ports, customs forms entries of intended destination, and total landings of sperm and baleen products such as oil and baleen. For French whalers, Du Pasquier [19] listed similar summary information for individual whaling voyages, but also included information on the locations visited during the voyage and for many voyages the numbers of sperm and right whales taken. Dawbin [17] used these two data sources to estimate annual catches in both of the American and the French ship-based fisheries operating both offshore and in New Zealand bays.

For both US and French whalers, Dawbin identified from the two voyage lists those vessels thought to have whaled in New

Zealand or Australia [17]. For American voyages, this included those that indicated in customs forms that they were bound for the New Zealand area [18]. For French voyages, this included voyages reported to have been in New Zealand or Australian waters [19]. In both cases, all whales or whale oil reported landed from each identified voyage was assumed to be from right whales from the New Zealand area. Using the American Offshore Whaling Voyage data (see below: AOWV) [33] and our own digitization of Du Pasquier's [19] data, we were able to reproduce the estimates of both US and French right whale catches for all voyages departing in a given year [17].

We sought to improve on Dawbin's [17] estimates of catches of American whalers by using two new sets of data: the AOWV dataset [33], and the American Offshore Whaling Logbook (AOWL) dataset [34]. The AOWV dataset includes one record for each of the roughly 15,000 multi-year American whaling voyages known to have occurred from 1667 to 1927. Data recorded include voyage dates, vessel details, and amount of sperm and baleen whale oil ultimately landed. The AOWL dataset includes one record for each day at sea of a sample of roughly 10% of the American whaling voyages represented in the AOWV. The AOWL data include information on vessel tracks, whales encountered, whales killed, and at times volume of oil obtained from individual whales, all extracted from original logbooks kept by the whalers at sea.

We omitted data for some of the voyages in the AOWL sample to account for various irregularities in the original logbooks and in the completeness of the information extracted. The logbooks used to obtain these data varied in their completeness, with varying proportions of the whales identified to species, with varying proportions of voyage days being reported, and with sometimes continuous gaps in reporting. To minimize bias due to these problems, we selected logbooks for voyages where at least 75% of the whales taken were identified to species, where there was information for more than 70% of the days and where the gaps between daily entries were fewer than 10% of the total number of recorded days. Further, we only selected voyage logbooks that reported sufficient right whales to account for the whale oil reported for the voyage in the AOWV data, assuming the average barrels of whale oil per whale where oil yields for individual whales had been identified (81 barrels per right whale, SE 3.8 from AOWL data where barrels reported).

We selected AOWL data within the study area boundaries 140°E (including east Australia, which we consider to be New South Wales, Victoria and Tasmania) to 140°W and approximately 27°S to 60°S (Figure 1). Finally, we excluded voyages from the AOWV data and from the AOWL data that were of less than 16 months duration or that returned only sperm oil because in the AOWL sample such voyages did not take right whales in the study area.

These selection criteria resulted in AOWL data from 280 logbooks. We combined the AOWL and the AOWV data to estimate the total number of right whales taken by American voyages in the study areas as follows. We computed the number of voyages departing in each year (i) from the AOWV data (N_i), the number of voyages departing in each year from the AOWL data (n_i), the number of the latter that reported whaling at sometime during the voyage in New Zealand or east Australian waters (m_i), and the number of right whales caught while they were in those waters (r_i). We estimated the catches of right whale in these waters for each year of departure i as:

$$N_i\left(\frac{m_i}{n_i}\right)\left(\frac{r_i}{m_i}\right) \quad (1)$$

That is, the estimated removals are the product of the total number of voyages departing, the fraction of those that whaled in the study area, and the mean catch of right whales by those voyages from the study area. This simplifies to:

$$N_i\left(\frac{r_i}{n_i}\right), \quad (2)$$

the number of voyages departing multiplied by the mean catch of right whales in New Zealand, or east Australian, waters by all sampled voyages. The variance of these estimates is based on the variance of the takes of right whales per voyage among the sample voyages.

We then attempted to accurately reflect the year of actual take by whaling vessels. For each departure year, we estimated the number of visits to the New Zealand area by vessels in each whaling season (September to May) following a vessel's departure from home ports in the North Atlantic accounting for the time required to travel to the New Zealand area. We treated the number of visits as multinomial random variables and estimated the mean proportion in each year and the corresponding variances accordingly, ignoring the covariances. As there was no trend in the estimated proportions over departure years, we combined the data and used the aggregate proportions to allocate the estimates of total takes by vessels departing in a given year to the seasons of actual take.

British, Australian and later New Zealand whalers also operated in the New Zealand region but Dawbin had no information on them that would allow him to estimate their right whale catches. We examined information from port arrival records of whaling vessels in the Bay of Islands [35] on the type of oil obtained, but were also unable to estimate catches by these vessels.

Ship-based Bay Whaling

A substantial number of American offshore whaling vessels were known to take southern right whales while in bays in the winter where the species calved, operating alongside the shore-based whalers described above [30]. Examination of AOWL data for some of the voyages identified by McNab [30] revealed temporal gaps in the AOWL data for winter months, resulting in a lack of data on catches in the calving bays. This appears to have been a limitation of the sampling protocol used for some of the data. In the end, there were too few data on catches in New Zealand bays in the AOWL data to allow this component of the fishery to be adequately represented. Thus our estimates for catches by American ship-based whaling described above apply only when those ships were operating offshore. To estimate the catches when American ships were operating in New Zealand bays, we identified American voyages that likely engaged in bay whaling using several sources, and collected additional data from logbooks from a subset of those voyages (Sampling Protocol S1). We estimated total American ship-based catches in New Zealand calving bays by multiplying the likely number of vessels bay-whaling each winter by the average number of right whales reported in the logbooks that we read.

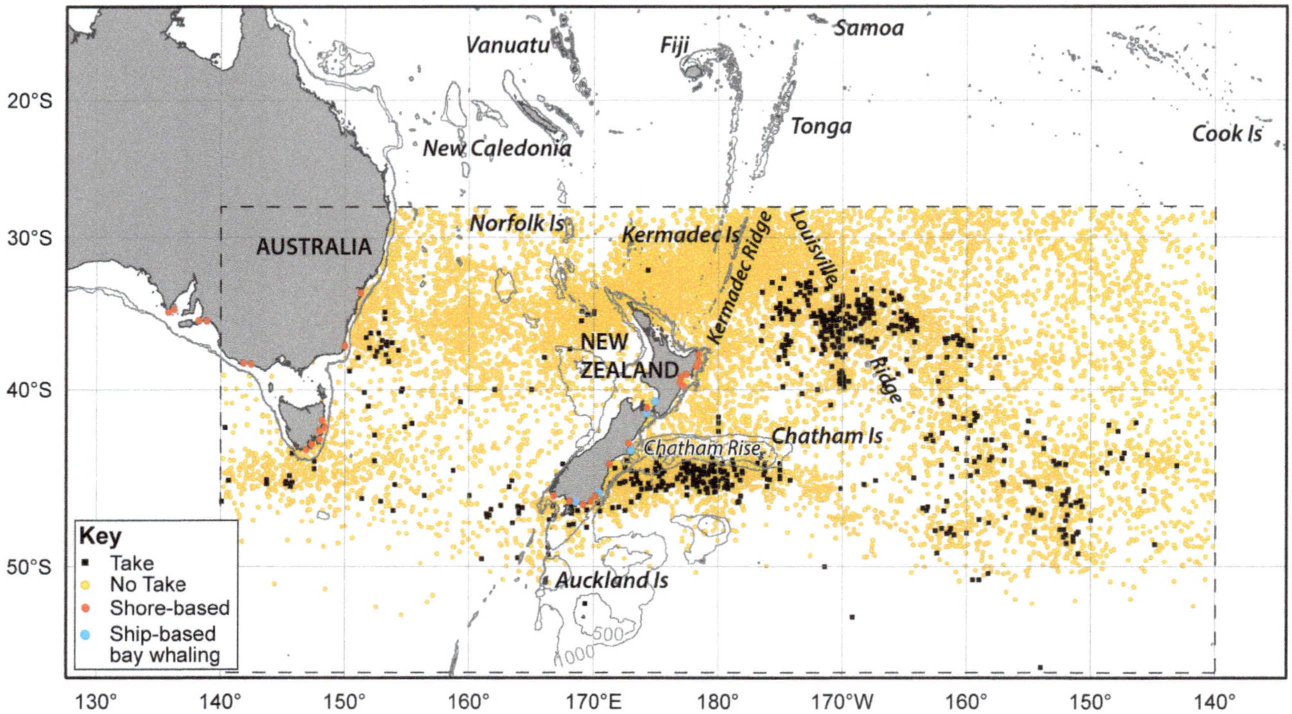

Figure 1. Observations of southern right whales in east Australian and New Zealand waters. New Zealand and east Australia study area showing coastlines, depth contours (500 and 1000 m depth), and the locations of American whaling vessels on days where right whales were taken (black dots) and were not taken (yellow dots), and the locations of winter calving areas where right whales were caught by shore-based whalers (red dots) and by ship-based bay whalers (blue dots).

20th Century Whaling

Tormosov et al. [20] described previously unreported Soviet whaling in New Zealand and east Australian waters, and these data were partitioned between the two regions. No correction for animals struck and lost was applied to these catches because 20th century factory whaling had much lower loss rates [36].

Struck and Lost Factors

In the course of whaling some animals are struck with a harpoon but not ultimately landed, as indicated in some logbooks. Although not all of these struck and lost whales died of their wounds, as shown by whales caught with evidence of having been previously harpooned, an upper bound on the total number of whales removed by whaling can be estimated using the rate at which animals are struck and lost (e.g. [37–38]). The loss rate is thought to vary depending on the conditions of whaling, with whales taken in bays less likely to be lost than those taken offshore, for example, due to a higher likelihood of recovery.

We computed the rate at which right whales were struck but lost using the numbers of whales struck and lost and the number struck and caught reported in logbooks. We treated these data as binomial random variables and estimated the overall proportion (p) of struck animals that were lost, along with its standard error. From this we estimated a loss rate factor (LRF) that can be multiplied by the estimated catches as:

$$LRF = \frac{1}{(1-p)} \tag{3}$$

The error of this estimate was approximated from the standard error of p using a Taylor Series expansion [39].

Results

We first report estimates of catches of right whales for each of the four fisheries: shore-based whaling, ship-based offshore whaling, ship-based bay whaling, and 20th century whaling. We then report estimates of struck and lost rates, and use those to estimate the total number of right whales removed.

Shore-based Whaling

All primary records used by Dawbin [17] to construct the New Zealand southern right whale coastal catch series were examined and catch series constructed from the information therein (Table 1 and Table S1). Using this information, the primary source for each year of the catch series reported in Dawbin [17] was identified in all but two years (Table S1). For the years 1871–1930, Dawbin [17] used a combination of oil and whalebone export records, in addition to the monetary value of such exports in some years.

Based on the uncertainty in the primary resources, we developed low and high catch scenarios for the New Zealand coastal fishery (Table 1). The low scenario is based on the same primary sources from Dawbin [17] for the most intensive period of the industry: years 1829–1840 [24]. This was considered the low case for these years as it records landings south of Akaroa only, excluding substantial coastal whaling operations in the Cook Strait, Cloudy Bay, and the Kapiti Coast [30]. The low scenario also uses the record with the lower export value derived from the Statistics New Zealand records (oil or baleen) for the years 1854–1930. In contrast, the high case used estimates from McNab [30] for the intensive period of whaling and the higher export values

Table 2. Summary statistics for US offshore whaling voyages departing between from 1837 to 1900, showing the numbers of voyages departing (Voy) and the number of logbooks sampled (Logs).

Dep Year	Voy	Logs	NZ M	NZ N	NZ SE	NZ RW	SE NZ RW	EA M	EAN	EA SE	EA RW	SE EA RW
1837	137	8	0.0	0	0.00	0	0	0.0	0	0.00	0	0
1838	139	6	14.7	3	8.99	2039	1249	0.5	2	0.50	70	70
1839	133	6	29.3	3	8.67	3901	1153	0.0	2	0.00	0	0
1840	149	9	8.3	3	8.33	1242	1242	0.0	1	0.00	0	0
1841	193	11	7.5	8	3.82	1448	737	3.2	5	2.96	618	571
1842	160	9	3.3	4	1.25	520	200	0.5	2	0.50	80	80
1843	193	10	1.3	6	0.61	257	119	0.4	5	0.24	77	47
1844	226	13	2.4	9	1.11	552	250	0.2	6	0.17	38	38
1845	231	17	2.3	15	0.86	539	199	0.2	5	0.20	46	46
1846	133	14	2.3	9	1.86	310	248	0.0	6	0.00	0	0
1847	145	11	0.0	6	0.00	0	0	0.3	4	0.25	36	36
1848	138	9	3.0	5	3.00	414	414	0.0	2	0.00	0	0
1849	100	7	0.0	6	0.00	0	0	0.0	2	0.00	0	0
1850	142	9	0.7	3	0.67	95	95	0.0	0	0.00	0	0
1851	231	6	0.0	4	0.00	0	0	0.0	1	0.00	0	0
1852	121	5	0.0	1	0.00	0	0	0.0	0	0.00	0	0
1853	165	4	0.0	3	0.00	0	0	0.5	2	0.50	83	83
1854	163	4	1.0	1	0.00	163	0	0.0	0	0.00	0	0
1855	134	8	1.8	4	0.75	235	101	0.0	1	0.00	0	0
1856	160	7	0.0	2	0.00	0	0	0.0	0	0.00	0	0
1857	165	5	7.5	2	0.50	1238	83	0.0	1	0.00	0	0
1858	107	5	2.7	3	0.88	285	94	0.0	1	0.00	0	0
1859	82	6	1.0	3	0.00	82	0	1.0	2	1.00	82	82
1860	86	7	0.0	1	0.00	0	0	0.0	0	0.00	0	0
1861	36	2	0.0	1	0.00	0	0	0.0	1	0.00	0	0
1862	70	8	0.0	0	0.00	0	0	0.0	0	0.00	0	0
1863	60	5	0.0	2	0.00	0	0	0.0	1	0.00	0	0
1864	68	5	0.0	0	0.00	0	0	0.0	0	0.00	0	0
1865	97	11	0.0	4	0.00	0	0	2.5	2	2.50	243	243
1866	79	5	0.0	1	0.00	0	0	1.0	1	0.00	79	0
1867	79	3	5.0	2	1.00	395	79	1.0	1	0.00	79	0
1868	75	8	0.0	0	0.00	0	0	0.0	0	0.00	0	0
1869	67	4	0.0	2	0.00	0	0	0.5	2	0.50	34	34
1870	40	1	0.0	0	0.00	0	0	0.0	0	0.00	0	0
1871	43	2	0.0	1	0.00	0	0	0.0	0	0.00	0	0
1872	34	2	0.0	1	0.00	0	0	1.0	1	0.00	34	0
1873	20	0	0.0	0	0.00	0	0	0.0	0	0.00	0	0
1874	21	0	0.0	0	0.00	0	0	0.0	0	0.00	0	0
1875	35	0	0.0	0	0.00	0	0	0.0	0	0.00	0	0
1876	37	1	0.0	0	0.00	0	0	0.0	0	0.00	0	0
1877	46	2	0.0	0	0.00	0	0	0.0	0	0.00	0	0
1878	34	1	0.0	0	0.00	0	0	0.0	0	0.00	0	0
1879	28	2	0.0	0	0.00	0	0	0.0	0	0.00	0	0
1880	36	2	0.0	0	0.00	0	0	0.0	0	0.00	0	0
1881	22	1	0.0	0	0.00	0	0	0.0	0	0.00	0	0
1882	22	2	0.0	0	0.00	0	0	0.0	0	0.00	0	0
1883	16	2	0.0	0	0.00	0	0	0.0	0	0.00	0	0
1884	18	1	0.0	0	0.00	0	0	0.0	0	0.00	0	0

Table 2. Cont.

Dep Year	Voy	Logs	NZ M	NZ N	NZ SE	NZ RW	SE NZ RW	EA M	EAN	EA SE	EA RW	SE EA RW
1885	12	0	0.0	0	0.00	0	0	0.0	0	0.00	0	0
1886	11	1	9.0	1	0.00	99	0	0.0	0	0.00	0	0
1887	13	1	0.0	0	0.00	0	0	0.0	0	0.00	0	0
1888	5	0	0.0	0	0.00	0	0	0.0	0	0.00	0	0
1889	4	0	0.0	0	0.00	0	0	0.0	0	0.00	0	0
1890	4	0	0.0	0	0.00	0	0	0.0	0	0.00	0	0
1891	11	1	0.0	0	0.00	0	0	0.0	0	0.00	0	0
1892	11	1	0.0	0	0.00	0	0	0.0	0	0.00	0	0
1893	10	2	0.0	1	0.00	0	0	0.0	0	0.00	0	0
1894	14	0	0.0	0	0.00	0	0	0.0	0	0.00	0	0
1895	9	2	0.0	0	0.00	0	0	0.0	0	0.00	0	0
1896	5	0	0.0	0	0.00	0	0	0.0	0	0.00	0	0
1897	14	3	0.0	0	0.00	0	0	0.0	0	0.00	0	0
1898	3	0	0.0	0	0.00	0	0	0.0	0	0.00	0	0
1899	10	1	0.0	0	0.00	0	0	0.0	0	0.00	0	0
1900	6	2	0.0	0	0.00	0	0	0.0	0	0.00	0	0
Total	**4,858**	**280**				**13,814**					**1,599**	

Separately for New Zealand (NZ) and east Australia (EA), the mean number of right whales taken per sampled logbook (M), and the number of voyages that took right whales in those areas (N), and the standard errors of those means (SE) are shown. Also shown are simple estimates of the total take of right whales in New Zealand (NZ RW) and East Australia (EA RW), and the standard errors of those estimated totals (SE RW).

derived from the Statistics New Zealand records for the years 1853–1930. The totals of the low and high scenarios are 2,703 and 5,104 whales, respectively. The catches of whales from New Zealand at New South Wales are also shown, totaling 2,638. Therefore the total catches for the strict New Zealand low and high scenarios are 5,341 and 7,742, respectively.

The estimated number of right whales killed in the coastal fisheries in east Australia was 13,003 whales: 683 in Victoria, 7,745 in Tasmania and 4,575 in New South Wales (Table 1). These catches were combined with the total New Zealand shore-based catch to estimate the catches for the low and high catch scenarios for east Australia plus New Zealand of 18,344 and 20,745 whales respectively.

Ship-based Offshore Whaling

For American whaling vessels, the number of voyages departing each year varied between 100 to over 200 between 1837 and the late 1850s, before beginning to decline (Table 2). The numbers of logbooks sampled increased from eight for voyages departing in 1837 and peaked at 17 in 1845. The mean number of catches of right whales per voyage varied substantially over those years. The intensity of right whaling was highest for vessels departing their home ports between 1838 and 1842 in both areas, and declined to relatively lower levels from the late 1840s onward. The sum of the estimates of total removals was 13,814 (SE 2,325) and 1,599 (SE 646), for New Zealand and east Australia, respectively (Table 2).

The proportions of American voyages whaling in the study area during each successive season since voyage departure is shown in Table 3. The proportions were relatively constant for the second through the fourth seasons and show that the estimated catches by departure year were in fact taken up to five years after the voyage departure year. We estimated the catches by calendar year from those in Table 2 by assigning the total for each departure year according to the proportions in Table 3.

Du Pasquier [19] reports both catches of individual right whales and landings of right whale products in New Zealand waters for each voyage rather than individual years. He did not have data similar to that in Table 3 but he provided data on the duration of French voyages to New Zealand. We compared that to American vessel voyage duration (Table 4). French voyages were substantially shorter than American voyages, with most lasting three years,

Table 3. The distribution of the year within a voyage (departure year = year 0) that US pelagic vessels whaled in New Zealand and east Australian waters in the 19th century, expressed as the proportion of all voyages that whaled in New Zealand in one or more seasons.

Number of years within a voyage	0	1	2	3	4	5
Proportion	0.007	0.238	0.274	0.262	0.193	0.025

Table 4. The distribution of the number of years duration of US and French pelagic whaling voyages in the 19th century, expressed as a proportion of all voyages with known length.

Country of origin	2 Years	3 Years	4 Years	5 Years	6 or more Years
USA	0.016	0.196	0.359	0.388	0.028
France	0.101	0.624	0.248	0.020	0.007

Table 5. The number of arrivals at ports in the Bay of Islands prior to 1841 for American, British, Australian and French whaling vessels and the proportions of declared cargoes of oil that were only sperm oil, only whale oil and both sperm oil and whale oil summarised by nationality.

Country of Origin	Arrivals	Arrivals with declared oil	Only Sperm Oil	Only Whale Oil	Both Sperm and Whale Oil
America	289	51	0.25	0.18	0.57
Britain	264	21	0.66	0	0.33
Australia	175	26	0.62	0.08	0.31
France	33	17	0.06	0.71	0.23

compared to five years for American voyages. We estimate French catches by departure year by allocating them equally to two, three and four years after departure, because of the relative consistency of the American proportions for voyage years three, four and five.

Although Dawbin [17] had no estimates for catches of right whales by vessels from other than France and the US, he suggested that landings by British vessels should be examined, that landings by Australian vessels were likely included in Australian landings statistics, and that other nations' ship were few. To examine this, we identified in Richards and Chisholm [35] 755 arrivals of whaling ships making port in the Bay of Islands prior to 1840. There were only seven, five and two arrivals from Germany, Canada and Portugal, respectively, confirming Dawbin's [17] conclusion that right whaling by these nations was minimal. The proportions of arriving ships with only sperm oil, only whale oil, and mixed sperm and whale oil varied among American, British, Australian and French nationalities (Table 5). For example, British and Australian vessels reported only sperm oil for 66% and 62% of arrivals, respectively, while American and French vessels reported only sperm oil much less frequently, 25% and 6% of arrivals, respectively. The numbers of port visits suggest that there were substantial numbers of vessels from Britain and Australia whaling in this area, and at least some of the time they pursued right whales. Although we have no estimates of right whales by other nations, the fact that other nationalities focused more on sperm than right whales suggests that any such catches would have been fewer than those by American and French vessels.

Ship-based Bay Whaling

We identified over 300 vessel-seasons that were in a geographical position to pursue bay whaling in New Zealand. Of these we judged that 106 and 59 were highly likely and possibly bay whaling, respectively, for a total of 165 vessels (See supplementary material: Sampling Protocol S1). We obtained the number of whales caught, and those struck and lost from some of these logbooks (Table 6). The logbook for the *Jasper* (Voyage Identification Number 7413 from [33]) reported catches for two other vessels it was working with, and as we were unable to assign whales to individual vessels, we assigned each vessel one third of the total reported. Four of the winter catches were reported in barrels of whale oil, and we divided those by 40, the average number of barrels per whale reported previously for New Zealand bay whaling [17]. This value was not statistically different from the reported barrels of oil reported for seven whales in the *Courier* logbook (mean 48 barrels, SE 9.8: Voyage Identification Number 3448 from [33]). The mean number of barrels obtained per whale during bay whaling was lower than the mean during offshore whaling (81 barrels, AOWL), which is consistent with calves being reportedly taken seven times in the bay whaling logbooks (Table 6).

The number of right whales caught per winter vessel-season for the 14 seasons averaged 23.3 (SE 3.72), but varied over the years (Table 6). For example, the average catches for the three vessels sampled in 1839 was significantly less than for the 12 vessels sampled prior to 1839 ($p<0.01$). We multiplied the number of vessel-seasons for each year that were definitely and possibly bay whaling by the average number caught per season to obtain estimates of the total right whales from 1834 to 1841. The total numbers caught over the period for the two cases were 2,404 (SE 173.5) and 3,781 (SE 274.4) for the highly likely and highly likely plus possibly bay whaling vessels, respectively (Table 7).

20th Century Whaling

Soviet whaling catches between 1963–1966 in the New Zealand sub-Antarctic islands and to the west and north totalled 294 animals. The majority were taken near the Auckland Islands [20,40]. A further 78 whales were taken south of Tasmania at 47°S and 150°E in 1969/1970 [20].

Struck and Lost Rates

Using information from the AOWL data and additional data obtained from bay whaling vessels in New Zealand calving bays (Table 8), we estimated struck and lost factors for three situations: north and south of the equator and in New Zealand whaling bays (Table 8). The proportion of right whales that were struck and lost was significantly higher ($p<0.01$) in the northern (0.50, SE 0.12) than in the southern hemisphere (0.31, SE 0.05). The proportion of right whales that were struck and lost in calving bays (0.21, SE 0.05) was significantly lower than the Southern Hemisphere offshore proportion, which is consistent with the view that the calmer and more constrained conditions in bays would facilitate catching whales [37–38].

Total 19th and 20th Century Catches and Removals

We partitioned the estimated catches in Tables 1, 2, and 7 and in Dawbin's [17] estimates of French ship-based catches among three fisheries: shore-based whaling, ship-based offshore whaling and ship-based bay whaling. We assumed that the reported Soviet removals were all of the modern 20th century catches of right whales.

Low and high scenarios for shore-based catches for New Zealand and for New Zealand plus east Australia together were summarised from Table 1. The ship-based offshore catches for New Zealand and for east Australian waters were from Table 3 (column NZ RW and column EA NZ RW) and proportions of Dawbin's French ship-based estimates. Those proportions were estimates as the fractions of American ship-based catches in New Zealand and in east Australian waters (Table 2). Within New Zealand, the estimated French ship-based catches were then

Table 6. The number of southern right whales (*Eubalaena australis*) caught, struck and lost (S&L) and calves caught during the austral winter, mentioned in a sample of logbooks of US voyage (Sampling Protocol S1), and for four vessels numbers of barrels of right whale oil obtained during those months reported in a logbook or by McNab [30], with location of logbooks shown identified by VID (Voyage Identification Number from [33]).

Year	Vessel	VID	Bay	Caught	S&L	Calves	Barrels
1836	*Columbus*	3061	Otago				1600
1836	*Erie*	4590	Cloudy Bay	20.3[A]			
1836	*Friendship*	5330	Port Cooper				1800
1836	*Gratitude*	3003	Bluff Harbour				1050
1836	*Jasper*	7413	Cloudy Bay	20.3[A]			
1836	*Martha*	9148	Otago				1700
1836	*Mary Mitchell*	9384	Otago	22	4		
1836	*South Boston*	13272	Cloudy Bay	20.3[A,B]			
1837	*Courier*	3448	Bluff Harbour	33	16	2	
1838	*Alexander Barclay*	512	Bluff Harbour	23	5	1	
1838	*Columbus*	3060	Otago	39	0	2	
1838	*Friendship*	5331	Otago	3.5	3		
1839	*Amethyst*	875	Bluff Harbour	7	4		
1839	*China*	2846	Kapiti	2	1	1	
1839	*Samuel Robertson*	12808	Cloudy Bay	6	3	1	

[A]From logbook of the *Jasper*, reporting a total of 61 whales taken by three mated vessels.
[B]Indicates the logbook reports other whalers taking cow and calf pairs.

partitioned into the offshore and bay whaling fisheries according to the fraction of American ship-based estimates in offshore waters and in bays. In making this calculation, we used the mean of the high and low estimates for American ship-based bay whaling because the differences in the proportions were minimal.

We estimated total removals by multiplying the estimates of catches (Table 9; shown by year in Table S2) by the appropriate loss rate correction factors (Table 8) for each fishery: 1.27 for bay and shore whaling and 1.45 for offshore fisheries. The total removals from New Zealand waters were between a low of 34,002 and a high of 38,800, while the total removals from New Zealand and east Australia waters combined were between a low of 53,145 and a high of 57,958 right whales. The annual estimated catches by fishery varied greatly over time (Figure 2). We have not reported the statistical uncertainties of these totals because estimates of sampling variances are not available for the estimates of shore-based whaling, but by giving low and high scenarios for these fisheries we attempted to investigate the variance.

Table 7. Numbers of American whaling vessels highly likely (H) and possibly (P) bay whaling in each winter in New Zealand, and estimates of numbers of southern right whales (*Eubalaena australis*) removed (E), with standard errors (SE).

Year	H	H+P	E(H)	SE(H)	E(H+P)	SE(H+P)
1834	1	1	23	3.7	23	3.7
1835	1	1	23	3.7	23	3.7
1836	18	19	420	67.0	444	70.7
1837	12	16	280	44.6	373	59.5
1838	22	31	514	81.9	724	115.3
1839	31	40	724	115.3	934	148.8
1840	16	47	373	59.5	1097	174.9
1841	2	7	47	7.4	163	26.0
Totals	**103**	**162**	**2404**	**173.5**	**3782**	**274.4**

Discussion

Our estimates of total removals of right whales from New Zealand and eastern Australia over the 19th and 20th centuries are substantially higher than previous estimates, increasing from the 26,000 estimated by Dawbin [17] to between 53,000 and 58,000 estimated here. Right whaling was pursued by whalers from several countries over the two centuries, however, 82% of removals were concentrated over the two decades between 1830 and 1849. Whaling was most intense over the decade from 1835 to 1844, accounting for 66% of the removals, a pattern similar to that found by Dawbin [17]. Right whaling continued in New Zealand

Table 8. Estimates of the proportion of right whales (*Eubalaena spp.*) struck but lost (P) with standard error (SE(P)) by US whalers in the northern hemisphere (North) and the southern hemisphere (South) and in New Zealand bays (New Zealand) during the 19th century.

Region	Strikes	P	SE(P)	LRF	SE(LRF)
North	302	0.50	0.029	2.00	0.115
South	328	0.31	0.026	1.45	0.054
New Zealand	171.5	0.21	0.031	1.27	0.050

Also shown are the corresponding loss rate factors (LRF, defined as 1/(1-P)) with standard errors (SE(LRF)).

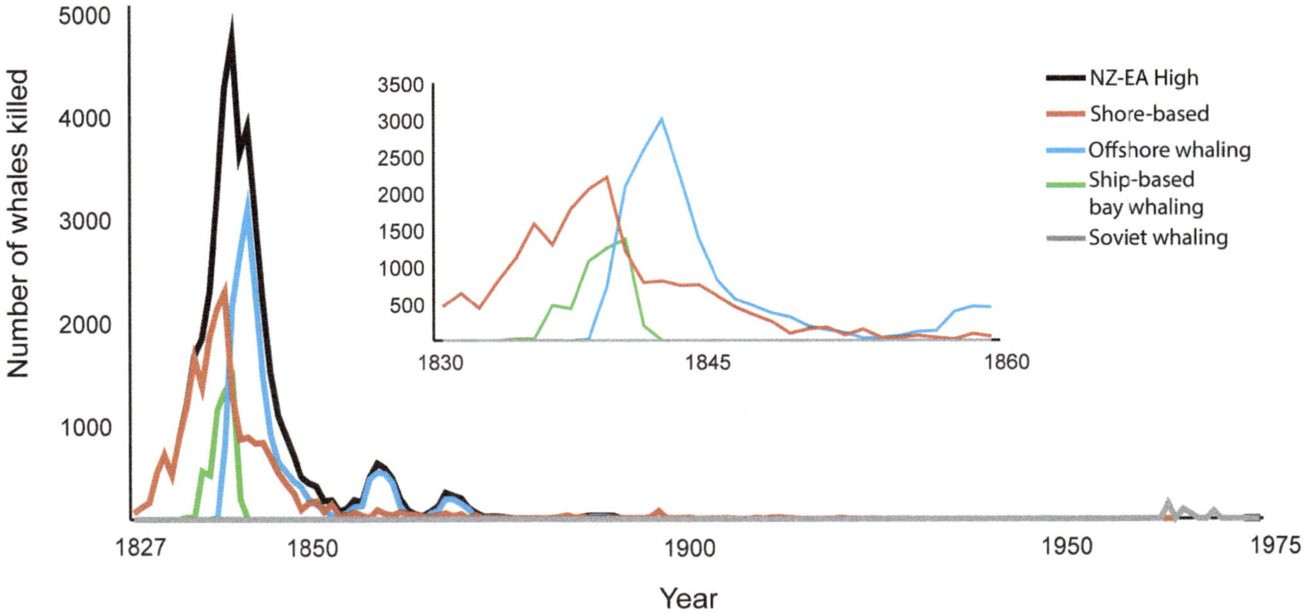

Figure 2. Catches of right whales (*Eubalaena australis*) around New Zealand and east Australia. The catches of southern right whales (number of whales killed) for the New Zealand plus east Australia high scenario are shown for the time period 1827 to 1975. No adjustment for struck and lost rate has been made to these catches.

over the remainder of the 19th century and even in the 20th century. The result was that southern rights were rare around the New Zealand mainland for most of the 19th century and were not seen at all for nearly four decades of the 20th century [41]. The effect of the initial removals was to drastically reduce the abundance of right whales. Due to the resulting low abundance, the relatively limited removals during the late 19th and 20th century were a significant contributing factor to the failure of right whales to recover in these waters for over 100 years.

The intensity of the shore, bay and offshore fisheries varied over time. Shore-based whaling declined abruptly and ship-based bay whaling by both French and American whalers ended around 1841, coincident with the British claims on New Zealand [42] and the discovery of the northwest grounds off the Alaskan coast [43]. Offshore whaling increased as bay whaling declined, and continued for some years. The ongoing low level of whaling, culminating in the Soviet catches in the late 20th century, was sufficient to keep the population at apparent low levels (Figure 2).

Table 9. Estimates of total catches and removals of southern right whales (*Eubalaena australis*) by fishery and scenario for New Zealand and east Australia.

Fishery	New Zealand - Low			New Zealand - High		
	Catches	**S&L**	**Removals**	**Catches**	**S&L**	**Removals**
Shore-based whaling	5,341	1.27	6,783	7,742	1.27	9,832
Ship-based offshore whaling	16,463	1.45	23,871	16,463	1.45	23,871
Ship-based bay whaling	2,404	1.27	3,053	3782	1.27	4,802
Modern (Soviet) whaling	294	N/A	294	294	N/A	294
Total	**24,502**		**34,002**	**29,342**		**38,800**
Fishery	**New Zealand - Low plus east Australia**			**New Zealand - High plus east Australia**		
	Catches	**S&L**	**Removals**	**Catches**	**S&L**	**Removals**
Shore-based whaling	18,344	1.27	23,297	20745	1.27	26,346
Ship-based offshore whaling	18,223	1.45	26,423	18233	1.45	26,438
Ship-based bay whaling	2404	1.27	3,053	3782	1.27	4,802
Modern (Soviet) whaling	372	N/A	372	372	N/A	372
Total	**39,343**		**53,145**	**43,122**		**57,958**

The struck and lost rate (S&L) applied to each fishery is shown and was used as a multiplier of the catches to calculate removals between 1827 and 1975. Four scenarios are considered: low and high for New Zealand and New Zealand plus east Australia.

These new estimates are a substantial improvement over previous estimates because we drew on substantial new data, particularly for the American pelagic and bay fisheries. We also accounted better for the distribution of catches over the several years of American and French voyages and we estimated total removals based on new information on whales that were struck but not landed.

Remaining uncertainties

Substantial uncertainties remain, however, and we explored the potential magnitude of two of these. One is the difference in the estimates of New Zealand and east Australian shore-based whaling by Dawbin [17] and McNab [30]. The other is the uncertainty about the numbers of American vessels participating in the winter whaling in calving bays around New Zealand. The difference in the low and high estimates based on those uncertainties is less than 2,500 whales each, <5% percent of the total removals.

Unexplored uncertainties include those that likely biased the catch series downwards, such as uncertainty in the struck and loss rate, using export rather than landing records, the killing of calves as part of the fishery, and not accounting for whaling vessels of other nationalities. There are also uncertainties that will have an unknown effect on the catch series, such as the relatively low number of extant logbooks available for the study.

Although we have now accounted for struck and loss rates, the available data are sparse: additional reading of as yet unread US voyage logbooks would help here. The proportion of the struck but lost whales that survived is unknown so there is also likely to be an upward bias in our estimates; it is not apparent how to address this uncertainty.

For shore-based whaling, the later part of the catch series is based on export records [31] as no catch data were located. These estimates were likely affected to an unknown degree by local consumption and the possibility the year of export was not the year the whale was caught. We were also unable to directly assess the east Australian shore-based catch series, which is a significant part of the catches, and instead had to rely on Dawbin's [17] records.

Futhermore, much anecdotal evidence was found that suggests coastal whalers routinely targeted cow-calf pairs, for example, Sherrin [21] states that a large proportion of the catch was calves. It was routine to kill the calf in order to secure the larger, more valuable right whale cow [21,30,32]. Our reading of American vessel-based bay whaling suggested that calves were taken frequently (7 of 15 read log books record this). This is consistent with the average barrels of oil obtained per whale indicated in one logbook (48 barrels per whale) compared to the average of 81 barrels per whale for offshore whaling indicated by the AOWL data. Similarly, this is consistent with Dawbin's [17] observation that the coastal industry had a lower yield of 4.18 tuns per whale, compared with 6 tuns per whale for the pelagic fishery. It is reasonable to suppose the lower yield is due to the smaller volume of oil extracted from the young whales. For example, Sherrin [21] states that one-year old calves produced approximately 4 tuns of oil, and given the fishery targeted the calving grounds, many calves and yearlings would have been younger and smaller whales. In addition, the size and condition of females would have decreased over the wintering season due to weight loss from lactation.

For ship-based whaling, we assumed that there were no right whale removals by ships from Britain and Australia, as well as Germany, Canada, and Portugal. While port arrival records suggest that right whales were less frequently taken by British and Australian vessels, it is likely that the Australian based landings were accounted for at least in part in the landing statistics described under shore-based whaling. The available data do not allow separate estimates to be made for either the Australian ship-based or the British ship-based whaling. Information on how many voyages such vessels made might be found in other vessel arrival and departure data in different ports. Further, additional information on the rate at which right whales were taken might be found in logbooks from such voyages if these can be located. For example, a number of logbooks from the 19th century Tasmania ship-based offshore fishery are known to exist [44], and examination of these would be useful, especially for the number, species and spatial distribution of catches, oil yield per whale, and potentially struck and lost rates. The lack of inclusion of these whaling fleets will bias the estimate low.

For American whaling, our logbook-based estimates overcame some of the uncertainties in Dawbin's [17] estimates. However, the sample of logbooks for both offshore whaling and bay whaling was sparse and the mean catches of right whales by vessels departing in the earlier years varied substantially. The low sampling intensity and rapid changes in whaling intensity mean the reliability of the annual estimates is relatively low. For example, the apparent post-bonanza pulses of catches in the mid-1800s are based on few samples, and alternative methods of pooling those data over time and alternate ways of assigning the departure year estimates to calendar year would likely result in a more protracted pattern of whaling over this period.

To determine priorities for additional historical study, it would be useful to examine the biological effects of this whaling using population modeling approaches aimed at regional rather than circumpolar assessment [15,45]. This would allow one to evaluate the potential value of improved estimates of removals for our understanding of right whales in this region.

Conclusion

Our results confirm a pattern of increasingly intense whaling and rapid depletion of populations of right whales over time. In the North Atlantic, right whaling spread out of the Bay of Biscay beginning around 1000 AD, sequentially depleting populations in the eastern and then western North Atlantic up to around 1850 (Table 2.1, [46]), nearly annihilating that species over several centuries. In the present case, beginning in the 1830s, substantial right whaling persisted for only two decades in New Zealand. Subsequently in the North Pacific, right whaling persisted for only the decade of the 1840s [43]. In the North and South Pacific in the 1830s and 1840s the AOWL data reveal that some of the same vessels pursued right whales in both hemispheres during the same voyage, seasonally shifting from one to the other, as previously described [22]. Despite this obvious pattern of intense and increasingly unsustainable right whaling, no limits were placed on hunting right whales until the 1930s. But even after that, the value of the animals prompted commercial whalers to again illegally hunt right whales in the mid-20th century, both near New Zealand and elsewhere.

Similar patterns of short duration offshore vessel whaling have been seen for other large whales. Thus gray whales in the North Pacific were depleted in 25 years, between 1845 and 1870 [37]. Humpback whales in the North Atlantic were depleted in 60 years, between 1850 and 1910 [38]. Blue whales were depleted around Iceland in a decade, between 1904 and 1914 [47–48].

Such rapid changes in abundance are not linked to the oft mentioned 'shifting baseline syndrome' [49] because the changes occurred within the working life of both whalers and their managers. Further, these changes are so rapid that their effects on whaling activities have been made evident since at least the middle of the 19th century in books and maps. The baselines have been as evident as the willingness to continue to pursue whaling to industrial extinction and whales to near extinction [50].

Supporting Information

Sampling Protocol S1 Methods for obtaining American logbook data for New Zealand calving bays.

Table S1 Estimated shore-based whaling catches, listed as number of southern right whales, with primary source listed in column to right of catch for catches at New South Wales that were from New Zealand (NSW-NZ) and two estimated coastal catch series for New Zealand (NZ-Low, NZ-High), reflecting different selections of primary sources.

Table S2 Catches by fishery by year for the New Zealand and east Australian southern right whale catch series. Shore.NZ.L: catches of the New Zealand coastal shore-based fishery under the low scenario; Shore.NZ.H: catches of the New Zealand shore-based fishery under the high scenario; Shore.EA: catches of the east Australian shore-based fishery from Dawbin [17]; Off.NZ: catches of the New Zealand ship-based offshore fishery; Off.EA: catches of the east Australian ship-based offshore fishery; Bay.L: catches of the New Zealand ship-based bay whaling fishery under the low scenario; Bay.H: catches of the New Zealand ship-based bay whaling fishery under the high scenario; Soviet.EA: modern Soviet whaling in east Australian waters from Tormosov et al. [20]; Soviet.NZ: modern Soviet whaling in New Zealand waters from Tormosov et al. [20].

Acknowledgments

The authors would like to thank Janet Dawbin and Peter Gill for their assistance with the catch series reconstruction, and Macquarie Library for access to source materials. We are also indebted to an anonymous reviewer whose comments greatly improved the manuscript and C. Scott Baker, Alison MacDiarmid and Rohan Currey for comments on earlier versions of manuscript. We thank Erika Mackay for help with the Figures. Numerous people assisted in assembling and extracting the American logbook data, including R. Reeves, J. Lund, E. Josephson, R. Richards, J. Bannister, D. Chatwin, O. Hamilton and R. Constantine who have laboriously extracted information from logbooks. We are also indebted to the institutions holding the whaling voyage logbooks, especially the Kendall Institute of the New Bedford Whaling Museum in New Bedford, Massachusetts; the Blunt White Library, the Mystic Seaport, Mystic, Connecticut, several others in the New England region, and the Pacific Manuscripts Bureau for access to copies of several logbooks examined.

Author Contributions

Conceived and designed the experiments: EC TS JJ DP. Performed the experiments: TS EC. Analyzed the data: TS JJ DP. Contributed reagents/materials/analysis tools: TS EC JJ DP. Wrote the paper: EC TS JJ DP.

References

1. Reeves RR, Smith T (2007) A Taxonomy of World Whaling: operations, eras and data sources. In: Estes JA, Brownell R, DeMaster DP, Doak DF, Williams TM, editors. Whales, whaling and ocean ecosystems. Berkley, CA: University of California Press.
2. IWC (1986) Right whales: Past and Present Status. Report of the International Whaling Commission 44: 146–152.
3. Wade P, Kennedy A, LeDuc RG, Barlow J, Carretta J, et al. (2011) The world's smallest whale population? Biology Letters 7: 83–85.
4. Fujiwara M, Caswell H (2001) Demography of the endangered North Atlantic right whale. Nature 414: 537–541.
5. Josephson E, Smith T, Reeves RR (2008) Historical distribution of right whales in the North Pacific. Fish and Fisheries 9: 155–168.
6. Brandão A, Best P, Butterworth D (2010) Estimates of demographic parameters for southern right whales off South Africa from survey data 1979 to 2006. Unpublished report (SC/62/BRG30) presented to the Scientific Committee of the International Whaling Commission.Cambridge, UK
7. Reilly SB, Bannister JL, Best P, Brown MW, Brownell R, et al. (2008) *Eubalaena australis* (Chile-Peru subpopulation). IUCN 2009 IUCN Red List of Threatened Species Version 20092. Available: http://www.iucn.redlist.org.
8. Patenaude NJ, Portway V, Schaeff C, Bannister JL, Best P, et al. (2007) Mitochondrial DNA diversity and population structure among southern right whales (*Eubalaena australis*). Journal of Heredity 98: 147–157.
9. IWC (2001) Report of the workshop on the comprehensive assessment of right whales. Journal of Cetacean Research and Management Special Issue 2: 1–60.
10. Richards R (2002) Southern right whales: a reassessment of their former distribution and migration routes in New Zealand waters, including the Kermadec grounds. Journal of the Royal Society of New Zealand 32: 355–377.
11. Carroll EL, Rayment W, Alexander AE, Baker CS, Patenaude NJ, et al. (2014) Reestablishment of former wintering grounds by the New Zealand southern right whales. Marine Mammal Science 30: 206–220.
12. Carroll EL, Childerhouse S, Fewster R, Patenaude NJ, Steel D, et al. (2013) Accounting for female reproductive cycles in a superpopulation capture recapture framework. Ecological Applications 23: 1677–1690.
13. Carroll EL, Patenaude NJ, Alexander AM, Steel D, Harcourt R, et al. (2011) Population structure and individual movement of southern right whales around New Zealand and Australia. Marine Ecology Progress Series 432: 257–268.
14. Bannister JL (2011) Population trend in right whales off southern Australia 1993–2010. Unpublished report (SC/S11/RW10) presented to the International Whaling Commission Workshop on Southern right whales, 13–16 September 2011, Buenos Aires, Argentina.
15. IWC (2012) Report of the workshop on the assessment of southern right whales, Buenos Aires, Argentinal 13–16 September 2011. Unpublished report (SC/64/Rep5) presented to the Scientific Committee of the International Whaling Commission.Cambridge, UK
16. Kato H, Matsuoka K, Nishiwaki S, Bannister JL (2007) Distributions of abundance of pygmy blue whales and southern right whales in waters off southern coast of Australia based on data from the Japan/IWC blue whale cruise 1996–96. Unpublished report (SC/59/SH10) presented to the Scientific Committee of the IWC, Cambridge, UK.
17. Dawbin W (1986) Right whales caught in waters around south eastern Australia and NZ during the nineteenth and early twentieth centuries. Report of the International Whaling Commission Special Issue 10: 261–268.
18. Starbuck A (1878) History of the American whale fishery from its earliest inception to the year 1876. Washington: United States Commission of Fish and Fisheries. Part IV. Report of the Commissioner for 1875–1876. Appendix A. Sea Fisheries.
19. Du Pasquier T (1986) Catch history of French right whaling mainly in the South Atlantic. Report of the International Whaling Commission Special Issue 10: 268–274.
20. Tormosov D, Mikhaliev Y, Best P, Zemsky V, Sekiguchi M, et al. (1998) Soviet catches of Southern right whales *Eubalaena australis* 1951–1971. Biological Conservation 86: 185–197.
21. Sherrin R (1886) Handbook of the fishes of New Zealand. Wilsons & Horton.Auckland, NZ
22. Bannister JL (1986) Notes on nineteenth century catches of southern right whales off the southern coasts of Western Australia. Report of the International Whaling Commission Special Issue 10: 255–259.
23. Dawbin W, Dawbin J (1985) Historical documents relating to right whaling in Australian and New Zealand waters during the nineteenth and twentieth centuries. Report to the Australian National Parks and Wildlife Service.Can-Canberra
24. Great Britain Parliamentary Paper (1846 (337)) Communcations between Lord Stanley and Govenor Fitzroy and Lieutenant Governor Grey, relative to New Zealand.
25. Great Britain Parliamentary Paper (1850 (1135)) Report from the Episcopal and Capitular Revenues Commissioners.
26. Great Britain Colonial Office (1853) Blue book of statistics, 1841–1853. Canberra: Australian Joint Copying Project, National Library of Australia.
27. McNab R (1907) Murihiku and the Southern Islands. Invercargill: William Smith.
28. McNab R (1908–1914) Historical records of New Zealand. Wellington: Government Printer.
29. McNab R (1909) Discovery of the Islands. The subantarctic islands of New Zealand Volume I. C.Wellington Philosophical Institute of Canterbury.
30. McNab R (1913) The old whaling days: A history of southern New Zealand from 1830 to 1840. Christchurch: Whitcombe & Tombs.
31. Statistics New Zealand (1900) Statistics of New Zealand 1854–1900. Wellington: Government Printer.
32. Wakefield EJ (1848) The handbook for New Zealand. London: John W. Parker.
33. Lund JN, Josephson E, Reeves RR, Smith T (2010) American Offshore Whaling Voyages: 1667–1927. Volume I: Voyages by Vessel; Volume II: Voyages by

Master. Old Dartmouth Historical Society – New Bedford Whaling Museum.-.New Bedford Massachusetts.
34. Smith TD, Reeves RR, Josephson E, Lund JN (2012) Spatial and seasonal distribution of American whaling and whales in the age of sail. PLoS ONE 7: e34905. doi:34910.31371/journal.pone.0034905.
35. Richards R, Chisholm J (1992) Bay of Islands shipping arrivals and departures. Canberra: Roebuck Society.
36. Clapham P, Baker CS (2002) Modern whaling. In: Perrin WF, Wursig B, Thewissen JGM, editors. Encyclopaedia of Marine Mammals. New York: Academic Press. pp. 1328–1332.
37. Reeves RR, Smith T, Lund JN, Lebo S, Josephson E (2010) Nineteenth century ship-based catches of gray whales, *Eschirichtius robustus*, in the eastern North Pacific. Marine Fisheries Review 72: 26–65.
38. Smith T, Reeves RR (2010) Historical catches of humpback whales, *Megaptera novaeangliae*, in the North Atlantic Ocean: estimates of landings and removals. Marine Fisheries Review 72: 1–43.
39. Seber G (1973) Estimation of animal abundance and related parameters. London: Griffin.
40. Patenaude NJ (2002) Demographic and genetic status of right whales at the Auckland Islands, New Zealand. PhD thesis. Auckland: University of Auckland, Auckland, New Zealand.
41. Gaskin DE (1964) Return of the southern right whale (*Eubalaena australis* Desm.) to New Zealand waters, 1963. Tuatara 12: 115–118.
42. Morton H (1982) The whale's wake. Honolulu: University of Hawaii Press.
43. Josephson E, Smith T, Reeves RR (2008) Depletion within a decade: the American 19th-century North Pacific right whale fishery. In: Starkey A, Holm P, Barnard M, editors. Oceans Past: Management insights from the History of Marine Animal Populations. London: Earthscan.
44. Richards R (2009) Sir William Crowther's whaling collection in Hobart in a global context. The Great Circle 31: 59–65.
45. Jackson JA, Patenaude NJ, Carroll EL, Baker CS (2008) How few whales were there after whaling? Inference from contemporary mtDNA diversity. Molecular Ecology 17: 236–251.
46. Reeves RR, Smith TD (2006) Near annilization of a species: Righ whales in the North Altantic. In: Kraus SD, Rolland RM, editors. The Urban Whale: Harvard University Press.
47. Hjort J (1933) Whales and whaling. Hvalradets Skrifter 7: 7–29.
48. Smith T (1994) Scaling Fisheries. Cambridge: Cambridge University Press.
49. Papworth S, Rist J, Coad L, Milner-Gulland E (2008) Evidence for shifting baseline syndrome in conservation. Conservation Letters 2: 93–100.
50. Dorsey K (2014) Whales and Nations: Environmental diplomacy on the high seas. Seattle: University of Washington Press.

9

Migratory Movements of Pygmy Blue Whales (*Balaenoptera musculus brevicauda*) between Australia and Indonesia as Revealed by Satellite Telemetry

Michael C. Double[1]*, Virginia Andrews-Goff[1], K. Curt S. Jenner[2], Micheline-Nicole Jenner[2], Sarah M. Laverick[1], Trevor A. Branch[3], Nicholas J. Gales[1]

1 Australian Marine Mammal Centre, Australian Antarctic Division, Kingston, Tasmania, Australia, **2** Centre for Whale Research (Western Australia) Inc., Fremantle, Western Australia, Australia, **3** School of Aquatic and Fishery Sciences, University of Washington, Seattle, Washington, United States of America

Abstract

In Australian waters during the austral summer, pygmy blue whales (*Balaenoptera musculus brevicauda)* occur predictably in two distinct feeding areas off western and southern Australia. As with other blue whale subspecies, outside the austral summer their distribution and movements are poorly understood. In order to describe the migratory movements of these whales, we present the satellite telemetry derived movements of eleven individuals tagged off western Australia over two years. Whales were tracked from between 8 and 308 days covering an average distance of 3,009±892 km (mean ± se; range: 832 km–14,101 km) at a rate of 21.94±0.74 km per day (0.09 km–455.80 km/day). Whales were tagged during March and April and ultimately migrated northwards post tag deployment with the exception of a single animal which remained in the vicinity of the Perth Canyon/Naturaliste Plateau for its eight day tracking period. The tagged whales travelled relatively near to the Australian coastline (100.0±1.7 km) until reaching a prominent peninsula in the north-west of the state of Western Australia (North West Cape) after which they travelled offshore (238.0±13.9 km). Whales reached the northern terminus of their migration and potential breeding grounds in Indonesian waters by June. One satellite tag relayed intermittent information to describe aspects of the southern migration from Indonesia with the animal departing around September to arrive in the subtropical frontal zone, south of western Australia in December. Throughout their migratory range, these whales are exposed to impacts associated with industry, fishing and vessel traffic. These movements therefore provide a valuable tool to industry when assessing potential interactions with pygmy blue whales and should be considered by conservation managers and regulators when mitigating impacts of development. This is particularly relevant for this species as it continues to recover from past exploitation.

Editor: Yan Ropert-Coudert, Institut Pluridisciplinaire Hubert Curien, France

Funding: This project was partly funded by the Western Australian Marine Science Institute (WAMSI). WAMSI received financial support for this project from Woodside Energy Ltd. as the operator of the Browse LNG Development. The funders had no role in study design, data collection and analysis, decision to publish, or preparation of the manuscript. No additional external funding was received for this study.

Competing Interests: The authors have declared that no competing interests exist.

* E-mail: Mike.Double@aad.gov.au

Introduction

Two blue whale subspecies are recognised in the Southern Hemisphere – the Antarctic or true blue whale (*B. m. intermedia)* and the pygmy blue whale (*B. m. brevicauda)* [1]. The pygmy subspecies was only described in the early 1960s [2] once Antarctic blue whale stocks had crashed. The vast majority of blue whale catches in the Southern Hemisphere and northern Indian Ocean were carried out prior to the identification of the pygmy blue whale subspecies and as such, the impact of commercial whaling on this subspecies is not well known [3]. However, both Japan (1959/60–1963/64) and the USSR (1962/63–1972/73) ran expeditions capitalising on their recognition [4] and it is thought that almost all historical blue whale catches north of 52°S and between 35°E and 180°E were pygmy blue whales [3,5]. It is likely therefore that whaling caused a decrease in pygmy blue whale numbers [3] with their recovery potentially slower than that of the Antarctic blue whale [5]. The current conservation status of this subspecies is highly uncertain with the IUCN Red List of Threatened Species listing it as data deficient [6].

The distribution and seasonal movements of pygmy blue whales are also poorly understood [4,7]. Prey availability clearly influences the distribution of blue whales worldwide with individuals concentrating in areas of high primary and secondary productivity and where dynamic oceanographic processes such as upwelling and frontal meandering occur [4,7]. At least four populations of pygmy blue whales are known to span the northern Indian Ocean, from Madagascar to the Subantarctic, from Indonesia to western and southern Australia and from New Zealand to the equator [4].

Branch et al. [4] hypothesised that pygmy blue whales occurring in Australian waters migrate between Australia and Indonesia via the western coast of Australia. This hypothesis is supported by acoustic recordings collected off southwestern Australia [7–9] with Australian type pygmy blue whale calls increasing between November and June and peaking between

February and May. However, Australian type pygmy blue whale calls have also been recorded in the southwest Indian Ocean subtropical frontal zone between January and June [10,11] and individuals may occupy Antarctic waters during the austral winter [12]. Off southern and western Australia, pygmy blue whales occur predictably in two distinct feeding areas throughout the austral summer [13,14]. Historically, these were areas where Soviet whalers caught large numbers of pygmy blue whales and where strandings have occurred [4]. Both of these areas are supported by complex seasonal upwelling systems producing high prey densities [13,14]. For the southern Australian feeding area, animals are only sighted between November and May [15]. Mark-recapture analyses estimate that between 569 and 1147 individuals occur in the feeding area off western Australia [16] and a line-transect survey off the southern coast of Australia estimated 671 (279–1613) individuals in a small area [17]. Genetic analyses indicate that the animals utilising these two feeding grounds are comprised of a single breeding stock [18].

For highly mobile and migratory species with dynamic and poorly understood migration routes, conservation and management remain a challenge [19]. When migratory animals routinely cross international borders, international cooperation is required to implement conservation strategies that utilise information on habitat use and movement patterns [20]. In order to describe the movement and migratory behaviours of pygmy blue whales and to investigate whether pygmy blue whales feeding in Australian waters do indeed migrate into Indonesian waters, we present here the first published satellite telemetry data for pygmy blue whales. Movement information such as this can reveal potential or unrealised impacts, especially for areas where data collection such as acoustic recordings or sightings are not possible. This information is necessary for managers to protect and conserve pygmy blue whales throughout their range and is valuable to industry when assessing the potential impacts of development, shipping and resource extraction.

Methods

Ethics Statement

This study was carried out in strict accordance with the approvals and conditions set by the Australian Antarctic Division's Antarctic Animal Ethics Committee for this project - Australian Antarctic Science project 2941. Additionally, this study was also carried out in strict accordance with the approvals and conditions set by the Western Australian Department of Environment and Conservation Animal Ethics Committee for this project - 30/2008. Fieldwork was undertaken in Commonwealth Waters with the permission of the Australian Government under EPBC permits 2007-006 and 2007-007.

Satellite Tag Deployment

The satellite tags employed are comprised of a custom-designed, anchor section joined to a stainless steel housing containing a Spot 5 transmitter manufactured by Wildlife Computers (Redmond, Washington, USA). Satellite tags were deployed using a modified version of the Air Rocket Transmitter System (ARTS, Restech) [21] at a pressure of 7.5–10 bar. Retention teeth on a purpose-designed projectile carrier grip to a metal ring fitted at the end of the tag allowing the tag to be fired from the air gun. When the tag makes contact with the whale, the rapid deceleration of the tag/projectile carrier withdraws the retention teeth releasing the projectile carrier. The metal ring then falls off in time to reduce the drag of the tag. The tag is sterilised with ethylene oxide prior to deployment and implants up to a maximum of 290 mm into the skin, blubber, interfacial layers and outer muscle mass of the whale. Retention of the tag is maintained through two actively sprung plates and a circle of passively deployed 'petals.' Once the

Table 1. Satellite tag deployment details and movement descriptors from the filtered Argos location data.

Ptt	Sex	Deployment date	First transmission	Last transmission	Tracking duration (days)	Transmitting days	Locations/day (mean ± se; range)	Track distance (km)	Distance/day (mean ± se; range)
53734	Unknown	18/03/2011	18/03/2011	19/04/2011	32	29	4.06±0.56; 0–10	1649	49.52±4.91; 8.54–125.49
53791	Female	30/03/2011	30/03/2011	28/04/2011	29	25	2.62±0.42; 0–8	2695	89.66±8.16; 10.71–250.12
98106	Unknown	26/03/2011	28/03/2011	03/05/2011	36	13	0.42±0.11; 0–2	1826	51.98±6.44; 10.26–122.78
98108	Male	30/03/2011	30/03/2011	17/05/2011	48	33	1.63±0.28; 0–8	3751	77.95±5.86; 5.00–235.96
98115	Unknown	14/03/2011	14/03/2011	11/05/2011	58	32	1.08±0.17; 0–4	1631	28.93±3.05; 3.47–107.28
98134	Unknown	17/03/2011	17/03/2011	06/04/2011	20	20	3.35±0.70; 0–10	1690	78.24±10.10; 31.08–195.15
98135	Unknown	06/04/2011	06/04/2011	08/02/2012	308	145	1.60±0.12; 0–11	10,468	57.52±3.83; 1.14–462.03
98141	Female	30/03/2011	11/06/2011	09/07/2011	28	13	0.50±0.12; 0–2	832	29.72±4.08; 3.07–93.72
88731	Male	08/04/2009	08/04/2009	15/04/2009	8	7	3.75±1.32; 0–10	382	43.46±10.56; 13.86–101.57
88740	Female	06/04/2009	06/04/2009	04/05/2009	29	29	4.58±0.33; 1–10	1,807	61.00±6.35; 21.03–161.06
88739	Male	08/04/2009	08/04/2009	18/08/2009	133	122	2.02±0.11; 0–5	6,378	47.08±2.77; 3.08–205.76

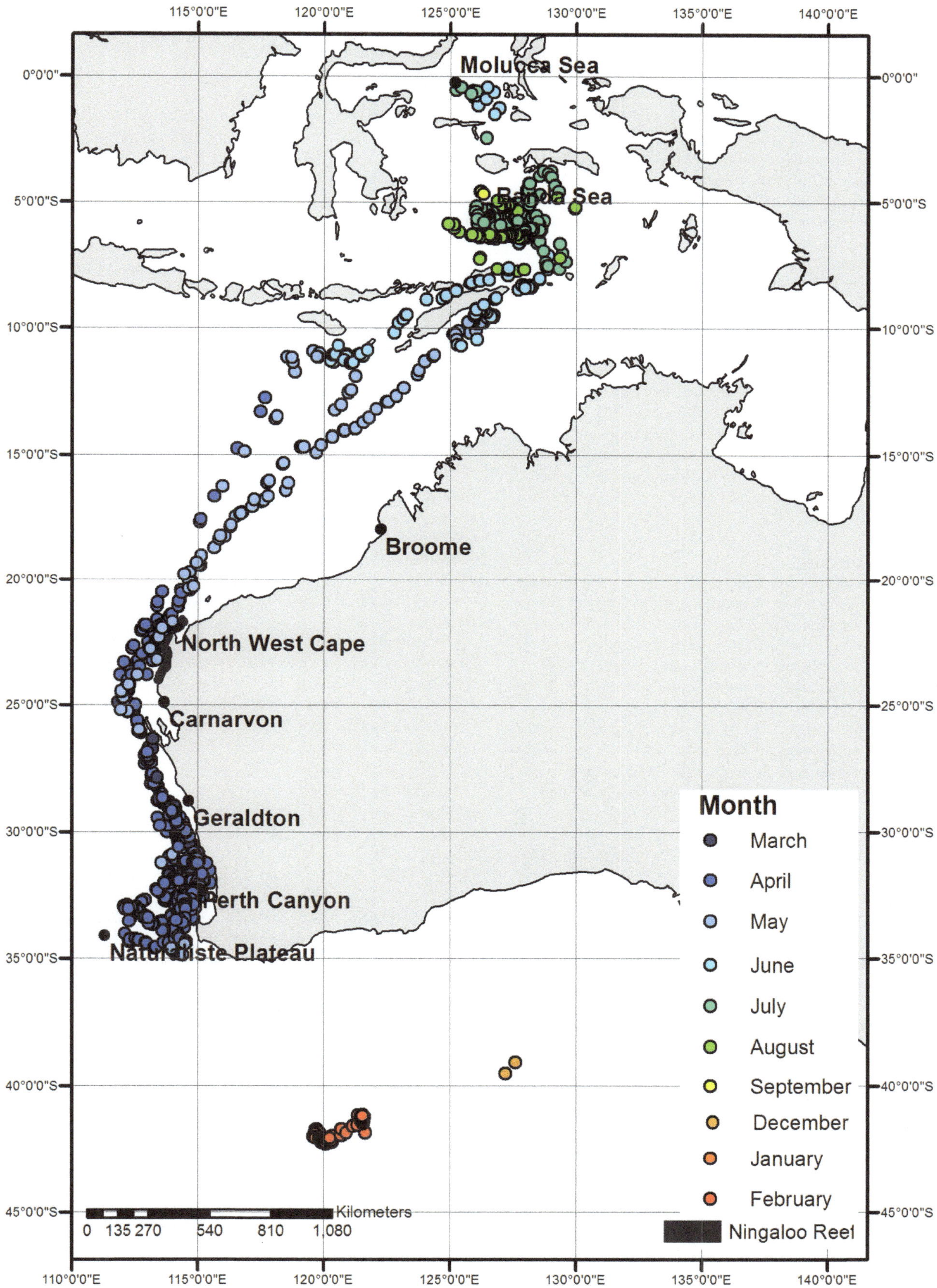

Molucca Sea
Banda Sea
Broome
North West Cape
Carnarvon
Geraldton
Perth Canyon
Naturaliste Plateau
Month
March
April
May
June
July
August
September
December
January
February
Ningaloo Reef
Kilometers
0 135 270 540 810 1,080
110°0'0"E
115°0'0"E
120°0'0"E
125°0'0"E
130°0'0"E
135°0'0"E
140°0'0"E
0°0'0"
5°0'0"S
10°0'0"S
15°0'0"S
20°0'0"S
25°0'0"S
30°0'0"S
35°0'0"S
40°0'0"S
45°0'0"S

Figure 1. Filtered satellite tag derived locations of pygmy blue whales (n = 11) by month. Individuals were tagged in March (2011: n = 7) and April (2009: n = 3; 2011: n = 1) in the Perth Canyon. The northern terminus of migration occurred in Indonesia. A single whale was tracked intermittently until February 2012 at which time it was located in the subtropical frontal zone.

tag is immersed in salt water, the salt water switch activates and the tag begins to transmit locations via the Argos satellite system when at the surface. Tags deployed in 2009 were duty cycled to a six hours on, 18 hours off transmission period at a 30 second repetition rate. In 2011, tags were duty cycled to a four hours on, eight hours off transmission period at a 30 second repetition rate. Each tag was deployed from the bow-sprit of a 5.8 m rigid-hulled inflatable boat at a range of 3–8 m and was positioned high on the body, approximately in line with the pectoral fins.

Biopsy Collection

Upon tag deployment, a small amount of skin and blubber was simultaneously collected for genetic analyses. These were collected using a biopsy dart fired from a modified.22 Paxarms system [22]. Biopsy samples were stored in 70% ethanol and DNA subsequently extracted using a Tissue DNA purification kit for the Maxwell 16 DNA extraction robot (Promega Corporation). The sex of the tagged whales was determined using a 5′ exonuclease assay of the polymorphisms in the sex-linked Zinc Finger genes as described by Morin et al. [23].

Argos Data Processing

For each Argos location, at least prior to 2013, an estimated error is calculated when at least four messages are received during a satellite pass. A location class (LC) is assigned to each location based on this estimate of error. LC 3 has an estimated error of <250 m, LC 2 has an estimated error between 250 and 500 m, and LC 1 has an estimated error between 500 and 1500 m. LC 0 has an open ended error of >1500 m whilst LC A and B have no accuracy estimation and LC Z is an invalid location. Argos locations were filtered using an algorithm based on swimming speed, distance between successive locations, and turning angles [24] to remove unlikely position estimates (default thresholds – speed of 2 ms^{-1}, spike angles of 15° and 25°, spike lengths of 2500 m and 5000 m) using R programming language (library argosfilter, function sdafilter) [25]. Although state-space based methods of filtering were attempted, due to the nature of sampling (the duty cycles employed resulted in large gaps in the data and a lack of frequent, regular locations), these methods could not successfully be applied to this data set. For further details on the impact of tag programming to the application of state-space models see [26].

Time Spent Grids

A 100 km^2 grid was applied to the tracking area. We calculated the total time spent by all individuals, and the total number of individuals present within each grid square. We generated two measures of occupancy. The first measure ("measure of occupancy 1") allocated to each individual the time spent in each grid square as a proportion of the total time spent by that individual in all grid squares. These individual values were then summed to give a sum of the proportion of time spent by all individuals in all grid squares. This first measure of occupancy was highly influenced by short tracks that occur across few grid squares therefore a second measure of occupancy was also generated. The second measure ("measure of occupancy 2") used the time spent per grid square per individual multiplied by the total time spent by that individual as a fraction of the total time spent by all individuals. This measure was then summed for all individuals across all grid squares. This second measure was influenced by long tracks placing a higher weighting on their contribution to the time spent per grid square than shorter tracks.

Environmental and Geographic Datasets

Environmental covariates were extracted for each location across the tracking period: bathymetric data (GEBCO_08 Grid, version 20100927: http://www.gebco.net; near surface chlorophyll *a* concentration (MODIS aqua satellite, 9 km and 8 day resolution): http://oceancolor.gsfc.nasa.gov/; and sea surface temperature (0.25 degree and daily resolution): http://www.ncdc.noaa.gov/sst/.

Results

Satellite tags were deployed on three blue whales in 2009 and 12 blue whales in 2011 within the Perth Canyon, approximately 32.0°S, 115.0°E, off western Australia. All tags deployed in 2009 transmitted successfully. However, of the 12 tags deployed in 2011, four performed very poorly providing either no transmissions at all (n = 2), no location data (tag was transmitting but did not generate locations; n = 1), or failed within a day (n = 1). Poor tag performance may possibly result from electronic failure or structural failure resulting in rapid migration of the tag out of the whale.

The remaining 11 whales were tracked from between 8 and 308 days (mean = 66.3 days, standard error = 26.1, Table 1). A total of 2793 locations (including LC Z) were received of which 1378 locations were retained post filtering. Throughout the tracking period, each satellite tagged whale covered an average distance of 3,009±892 km (mean ± se; range: 382–10,467 km) at a rate of 21.9±0.7 km per day (0.09–455.8 km).

All whales were tagged during March (2011: n = 7) and April (2009: n = 3; 2011: n = 1) and ultimately migrated northwards post tag deployment (Fig. 1). Five whales (2009: n = 2; 2011: n = 3) initially travelled in a loop southwards towards the Naturaliste Plateau (34.1°S, 111.3°E) prior to a direction change north (2009: n = 3; tag ceased transmitting: n = 2) whilst another five whales (2009: n = 1; 2011: n = 4) travelled northwards immediately post tagging. One satellite tag deployed in 2011 did not initiate transmission of locations until June at which time the whale had reached the northern terminus of its migration in the Banda Sea. All 11 tracks are detailed individually within the supplementary material (Fig. S1).

The tagged whales travelled relatively near to the western Australian coastline (100±1.7 km; Fig. 2a) throughout March (Fig. 1; 2011: n = 6) and April (Fig. 1; 2009: n = 3; 2011: n = 7) until reaching North West Cape (22.23°S, 113.96°E)–a prominent peninsula in the northwest of the Australian state of Western Australia. Animals continued to travel northwards and offshore (238.0±13.9 km; Fig. 2a) during May (Fig. 1; 2009: n = 2; 2011: n = 4) towards Indonesia. By June, whales were travelling through the Savu and Timor Seas (Fig. 1; 2009: n = 1; 2011: n = 2) and were located within 137.3±3.1 km of the nearest Indonesian coastline. The northern terminus of migration for the remaining whales was within the Banda (2009: n = 1; 2011: n = 1; 6°S, 127°E) and Molucca Seas (2011: n = 1; 0.42°S, 125.42°E), just south of the equator. Only one satellite tag (deployed in 2011) continued to relay location information after August 1st, locating the whale

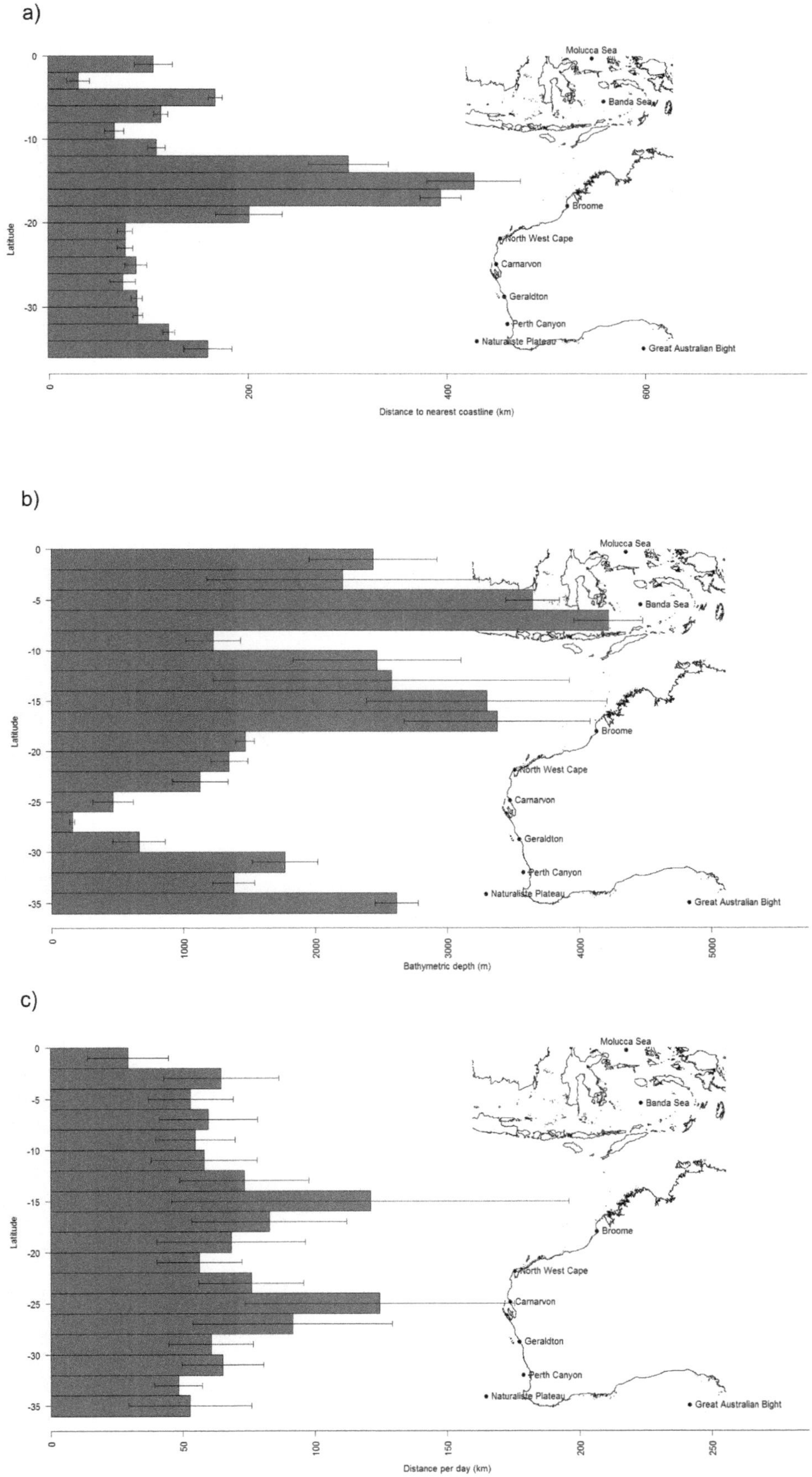
a)
Latitude
Distance to nearest coastline (km)
Molucca Sea
Banda Sea
Broome
North West Cape
Carnarvon
Geraldton
Perth Canyon
Naturaliste Plateau
Great Australian Bight
b)
Latitude
Bathymetric depth (m)
Molucca Sea
Banda Sea
Broome
North West Cape
Carnarvon
Geraldton
Perth Canyon
Naturaliste Plateau
Great Australian Bight
c)
Latitude
Distance per day (km)
Molucca Sea
Banda Sea
Broome
North West Cape
Carnarvon
Geraldton
Perth Canyon
Naturaliste Plateau
Great Australian Bight

Figure 2. Pygmy blue whale migration (n = 11) towards Indonesia in relation to 2° bins of latitude. a) Distance to coastline (km), b) bathymetric depth (m) and c) distance travelled per day (km). All bars represent mean values with 95% confidence intervals. Map of Australia is included to indicate the position of each longitudinal bin in relation to the Australian and Indonesian coastline.

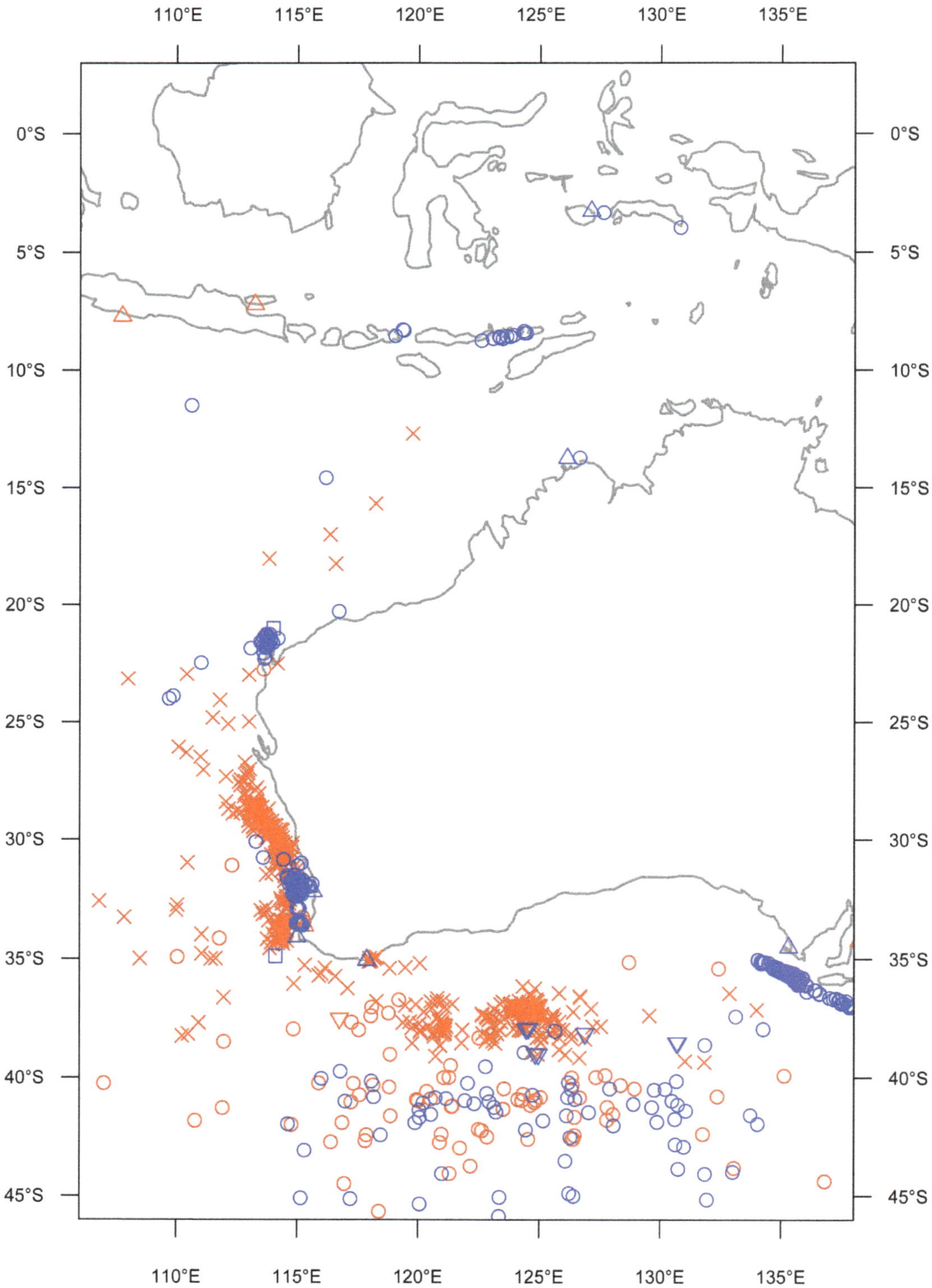

Figure 3. Positional information for pygmy blue whales. Catches (x), sightings (○), strandings (Δ), acoustic recordings (□) and discovery marks (∇) collected up to and including 1973 when whaling stopped (red) and after 1973 (blue). Modified from [4].

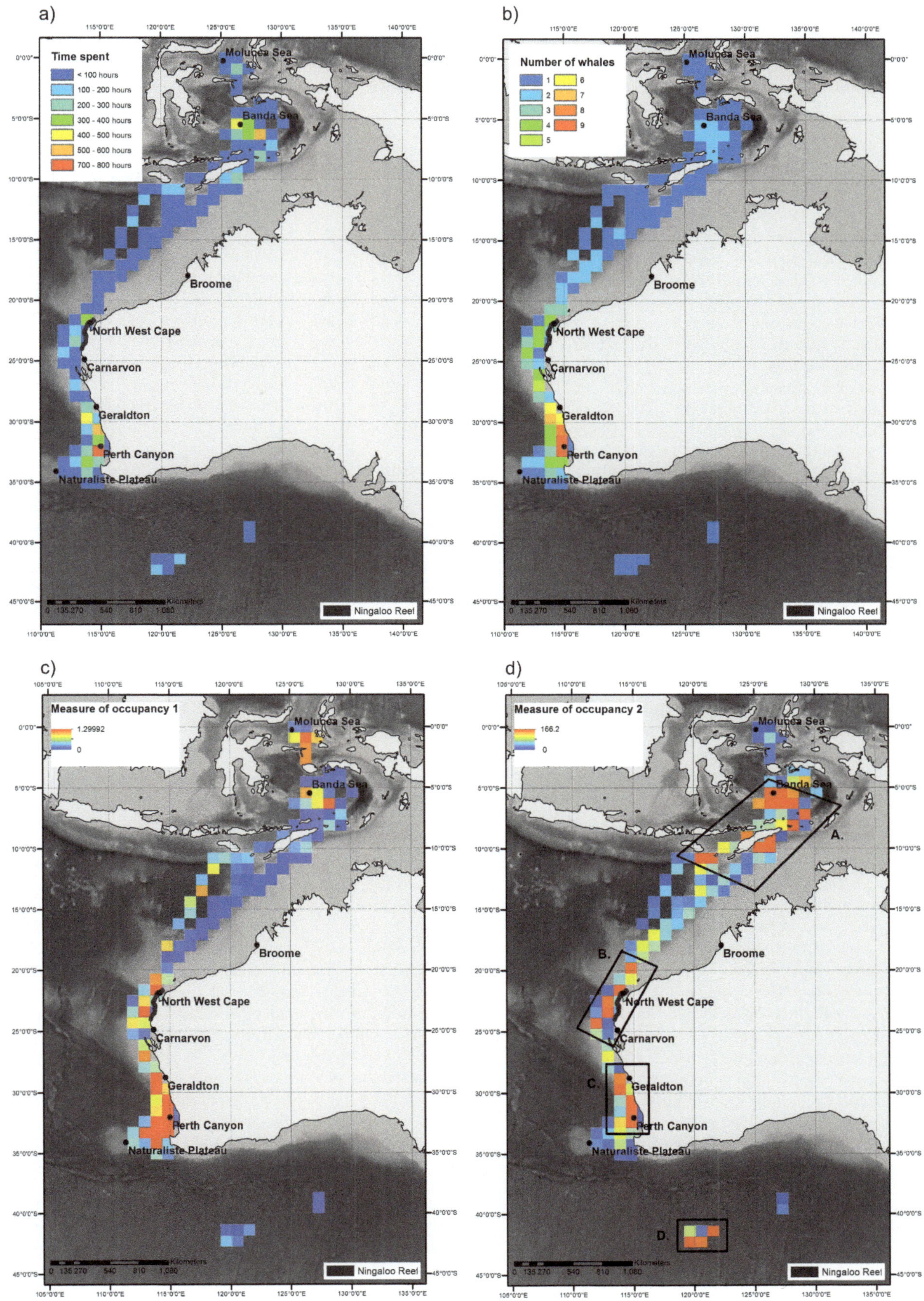
a)
Time spent
< 100 hours
100 - 200 hours
200 - 300 hours
300 - 400 hours
400 - 500 hours
500 - 600 hours
700 - 800 hours
Molucca Sea
Banda Sea
Broome
North West Cape
Carnarvon
Geraldton
Perth Canyon
Naturaliste Plateau
Ningaloo Reef
b)
Number of whales
1
2
3
4
5
6
7
8
9
Molucca Sea
Banda Sea
Broome
North West Cape
Carnarvon
Geraldton
Perth Canyon
Naturaliste Plateau
Ningaloo Reef
c)
Measure of occupancy 1
1.29992
0
Molucca Sea
Banda Sea
Broome
North West Cape
Carnarvon
Geraldton
Perth Canyon
Naturaliste Plateau
Ningaloo Reef
d)
Measure of occupancy 2
166.2
0
Molucca Sea
Banda Sea
A.
B.
C.
D.
Broome
North West Cape
Carnarvon
Geraldton
Perth Canyon
Naturaliste Plateau
Ningaloo Reef

Figure 4. Gridded measures of time spent and occupancy for satellite tagged pygmy blue whales (n = 11). a) Total time spent, b) number of whales, c) measure of occupancy 1: proportion of time spent per grid square per individual summed across all individuals and d) measure of occupancy 2: sum of individual time spent per grid square adjusted by the contribution to total time spent by all individuals for all pygmy blue whales throughout the tracking period. Four regions of potentially higher occupancy are indentified in Fig. 4d): A. Indonesia, B. Ningaloo Reef, C. Perth Canyon/Naturaliste Plateau and D. Subtropical frontal zone. The grid presented is 100 km×100 km. GEBCO bathymetry is also shown.

within the Banda Sea until September after which a temporary pause in location transmission occurred until December. This whale migrated southwards throughout the three months in which location information was not transmitted. Once the tag resumed transmitting, the whale was located at the approximate position of the subtropical front [27], south of western Australia for at least three months until tag failure in February (Fig. 1). These 11 satellite tracks line up closely with blue whale positional information collected both pre and post whaling including sightings, strandings, acoustic recordings and Discovery mark retrievals (Fig. 3; modified from [4]).

Pygmy blue whales occupied shallowest (1369.5±47.4 m) waters when closest to the coast throughout March and April (Fig. 2a and 2b), migrating into progressively deeper waters offshore (2617.0±143.5 m) in May and then often deep coastal Indonesian waters (3788.5±66.4 m) from June to September. Rate of travel increased slightly from 66.7±3.6 km/day in March and April to 77.0±7.4 km/day in May. Once the animals reached the northern end of the migration, rate of travel slowed to 58.3±7.4 km/day (June–September).

When examining total time spent by all whales across the duration of the tracking period, there were areas with distinctly higher occupancy times (Fig. 4a). Even accounting for the larger number of active tags earlier in the tracking period, areas such as the Perth Canyon (single grid cell time spent: 773.8 hours – Fig. 4a; number of whales: 9– Fig. 4b), the Naturaliste Plateau region (time spent across grid cells occurring in the Naturaliste Plateau area: at least 1100 hours – Fig. 4a; number of whales: at least 4– Fig. 4b) and North West Cape (single grid cell time spent: 330.3 hours – Fig. 4a; number of whales: 4– Fig. 4b) appeared to be areas in which occupancy was relatively prolonged. Areas of high occupancy time were also found within the Banda (single grid cell time spent: 599.8 hours – Fig. 4a; number of whales: 3– Fig. 4b) and Molucca Seas (single grid cell time spent: 232.4 hours – Fig. 4a; number of whales: 1– Fig. 4b), Indonesia. Upon examination of the number of whales that occurred within each grid square (Fig. 4b), it was apparent that tracks converged around the North West Cape region whilst milling was evident in the Perth Canyon/Naturaliste Plateau region.

The first measure of occupancy highlights the importance of the tagging location - the Perth Canyon/Naturaliste Plateau region (Fig. 4c). Even though this measure was highly influenced by short tracks, the Perth Canyon/Naturaliste Plateau region was an area in which at least five of the tagged individuals milled (Fig. 1; Fig. 4b) before migrating north. By accounting for the contribution of each individual to the entire dataset, the second measure of occupancy (Fig. 4d) highlights the importance of areas through which the longer tracks traversed such as Indonesian waters and the subtropical frontal zone. To represent potential areas of higher occupancy throughout the entire tracking range therefore, we examined all areas in which the grid square value for the second measure of occupancy was greater than the mean value (mean = 48.2, equating to all dark yellow, orange and red grid squares in Fig. 4d). This resulted in the demarcation of four regions: A. Indonesia, B. Ningaloo Reef, C. Perth Canyon/Naturalist Plateau and D. Subtropical frontal zone.

The environmental and oceanographic properties of these four regions of potentially higher occupancy differed (Fig. 5). Individuals occupied shallow water in Ningaloo Reef (1088±47 m) and Perth Canyon/Naturaliste Plateau (1176±58 m) as compared to Indonesia (3192±80 m) and the subtropical frontal zone (4749±16 m; Fig. 5a) and occupied warmer waters off Ningaloo Reef (28.2±0.1°C) than when in Indonesia (27.3±0.03°C; Fig. 5b). Pygmy blue whales in the Indonesia region occupied waters with comparable productivity to the subtropical frontal zone (Indonesia: 0.24±0.005 mg m^{-3}, subtropical frontal zone: 0.28±0.005 mg m^{-3}; Fig. 5d). However, the highest values of productivity occurred in the Perth Canyon/Naturaliste Plateau region (max = 1.69 mg m^3; Fig. 5d) with Ningaloo Reef producing the lowest productivity values (min: 0.04 mg m^{-3}; Fig. 5d).

Discussion

These satellite tag derived movements of pygmy blue whales provide important new information potentially indicating the general migration pattern and breeding grounds of pygmy blue whales that feed off the western coast of Australia. Assuming these movements are representative of the animals that feed off the western Australian area as a whole, pygmy blue whales migrate north from the Perth Canyon/Naturaliste Plateau region in March/April reaching Indonesia by June where they remain until at least September. Southern migration from Indonesia may occur from September and finish by December in the subtropical frontal zone after which the animals may make their way slowly northwards towards the Perth Canyon by March/April. This temporal pattern of migration as revealed by satellite telemetry is supported by acoustic recordings of Australian type pygmy blue whale calls off south western Australia [7–9,28,29]. However these calls have also been recorded well to the west of this proposed migratory path within Subantarctic waters [10,11] and potentially Antarctic waters [11,12] indicating multiple migration routes, elasticity in migratory behaviour or the potential for calls to be detected hundreds of kilometres from the source. Elasticity in migratory behaviour has been demonstrated by eastern North Pacific blue whales and is hypothesised to be related to changes in prey availability mediated by oceanographic processes [30].

The single satellite tagged whale that was tracked to the subtropical frontal zone south of Australia was about 900 km west of the other well described pygmy blue whale feeding area, the Bonney Upwelling (between 139.75°E and 143°E) [14]. This subtropical frontal zone region is also an area in which blue whales were caught and sightings recorded historically. Genetic evidence indicates mixing between the animals in the feeding areas of the Perth Canyon (off western Australia) and Bonney Upwelling (off southern Australia) [18], and animals photographed in the Bonney Upwelling have also been resighted in the Perth Canyon [15]. This indicates the potential for individuals from the Bonney Upwelling to follow a similar migration route to those animals feeding in the Perth Canyon.

When whales were migrating north, lower rates of travel and relatively higher occupancy were recorded in the Perth Canyon/Naturaliste Plateau region, North West Cape/Ningaloo Reef region, Indonesian waters and the subtropical frontal zone. It is known that feeding occurs off the Perth Canyon [13] and that the

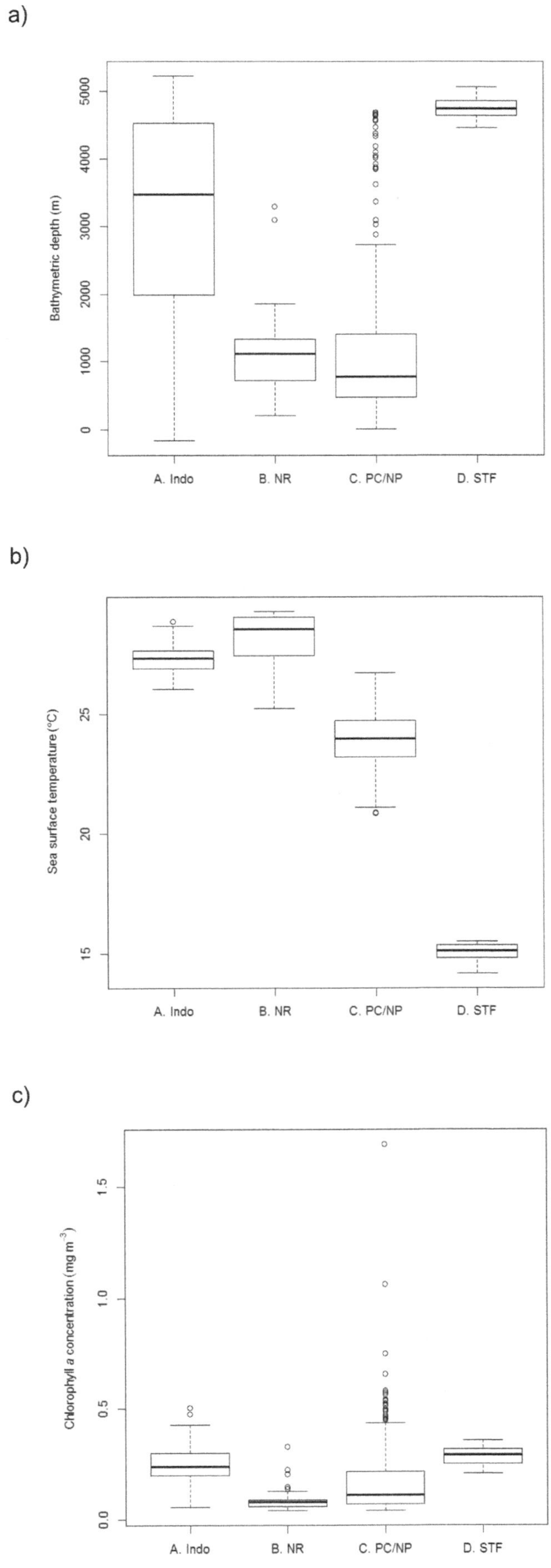

Figure 5. Boxplots describing the oceanographic characteristics of the four regions of potentially higher occupancy as identified in Fig. 4d. A. Indonesia (n = 324), B. Ningaloo Reef (n = 106), C. Perth Canyon/Naturaliste Plateau (n = 396) and D. Subtropical frontal zone (n = 78): a) bathymetric depth (m), b) sea surface temperature (°C) and c) chlorophyll *a* concentration (mg m^{-3}).

Indian Ocean sector of the subtropical frontal zone is a productive region in which phytoplankton blooms occur [31]. The North West Cape/Ningaloo Reef region has the capacity to offer feeding opportunities as primary production rates are equal to those recorded in current upwelling systems [32] and the reef possesses unique biophysical conditions able to support large biomasses of marine species [33] including other migratory krill feeding specialists [34,35]. Indeed, surface lunge feeding of pygmy blue whales has been observed at North West Cape and Ningaloo Reef in June (C. Jenner and M-N Jenner, unpublished data, 2001). However, the low values of productivity experienced by the whales in this study indicate that the higher measure of occupancy attributed to the North West Cape/Ningaloo Reef region may be a result of tracks narrowly converging as individuals round the peninsula on their way north.

As with most baleen whales species, blue whales are thought to migrate between high and low latitudes: at high latitudes they feed and at low latitudes they calve [36,37]. The breeding regime of pygmy blue whales caught in the waters around Kerguelen Island may offer insight into the breeding regime of pygmy blue whales occupying Australian waters considering Australian pygmy blue whale type calls have been detected as far west as the Crozet Islands [10] and around Kerguelen Island [11]. For these animals, foetal lengths and pregnancy ratios indicate that mating occurs between November and January whilst calving occurs in the austral winter with another minor calving event in the austral summer [2]. The timing of movement as suggested by the satellite tagging data presented here indicates that pygmy blue whales may calve in the Banda and Molucca Seas, highlighting the need for ongoing conservation and management efforts for blue whales in Indonesian waters.

Behavioural plasticity as a reflection of changing prey fields is common to a wide variety of marine vertebrates [38] therefore presumably pygmy blue whales are adapted to exploit widely dispersed and ephemeral temperate and tropical food sources. Archival dive data from a tag deployed on a blue whale in Indonesia between May and July provided evidence of feeding within Indonesian waters (B. Kahn, unpublished data). Our results also indicate that pygmy blue whales occupy relatively productive waters when in the Indonesia region. Frontal formation occurs within the Banda and Molucca Seas [39] and upwelling is enhanced during the south-east monsoon that occurs between May and August in the eastern Banda Sea [40] leading to high productivity throughout the time in which pygmy blue whales are located in the area. The Banda and Molucca Seas therefore may also act as a feeding ground for pygmy blue whales and have certainly been indicated as important foraging locations for another migrating marine vertebrate [41].

Throughout their migratory range, satellite telemetry demonstrates that pygmy blue whales move through areas posing multiple risks. Along the western Australian coastline, whales migrate through an area that contains the majority of Australia's gas resources and in which production and development is ongoing [42]. As such, pygmy blue whales occupying or travelling through these areas are exposed to anthropogenic noise associated with industry, shipping and seismic exploration. In Indonesian waters, pygmy blue whales encounter fisheries related risks such as entanglement from offshore gillnets and acoustic habitat degradation from reef bombing (B. Kahn, unpublished data). Baleen whales use sound for communication and to gain information about the environment they occupy [43]. Blue whales are more likely to emit calls when in the presence of noise generated by ship traffic [44] and to increase calling rates associated with social encounters and feeding when exposed to noise generated by seismic surveys [45]. Elevated ambient noise potentially reduces an individual's ability to detect socially relevant sounds therefore an increase in call rate is likely a compensatory behaviour to maintain communication with conspecifics in the presence of increased noise levels [44,45].

Whilst anthropogenic noise may alter blue whale behaviour, it is unlikely to pose a conservation risk unless it causes population level consequences such as changes in growth, reproduction and survival of individuals [46]. Elevated ambient noise has been responsible for abandonment or avoidance of critical habitat by a number of cetacean species including gray whales [47], bowhead whales [48] and killer whales [49]. Critical habitat includes habitat used to meet essential life cycle requirements such as foraging and breeding (http://www.environment.gov.au/cgi-bin/sprat/public/publicregisterofcriticalhabitat.pl), both of which are activities likely to be impacted by elevated ambient noise for these pygmy blue whales. As such, displacement from critical habitat would likely impact growth, reproduction and survival. In addition, exposure to ship noise has been linked to chronic stress in right whales with physiological implications that may impact the recovery of this endangered population [50]. Whilst population level impacts such as displacement and chronic stress do pose a conservation risk, current studies of large cetaceans are unlikely to provide evidence of noise and stress directly impacting on growth, reproduction and survival due to difficulties associated with establishing long term studies of known individuals that can be predictably tracked throughout their lifetime. With additional tracking data, the potential exists to generate habitat models to describe where pygmy blue whales are found and therefore assess and mitigate anthropogenic effects. In the meantime, the description of movements as presented here can be used by managers within a precautionary principle framework to identify risks within the pygmy blue whale migratory range such as increased ambient noise from development, shipping and fishing and therefore assist in mitigating these risks. For industry, these movements are invaluable when examining potential interactions with pygmy blue whales and will enable informed decisions regarding aspects of development such as vessel traffic, seismic exploration and field locations. A combined approach by industry and managers when accounting for the movements of the pygmy blue whales utilising Australian waters will ensure the ongoing recovery of this previously exploited species.

Supporting Information

Figure S1 Filtered satellite tag derived locations of pygmy blue whales by month. a) 53734, b) 53791, c) 88731, d) 88739, e) 88740, f) 98106, g) 98108, h) 98115, i) 98134, j) 98135, k) 98141, l) all tracks.

Acknowledgments

This paper is dedicated to the memory of Eric King. Eric, an engineer at the Australian Antarctic Division, continually displayed remarkable ingenuity, skill and dedication in the development of satellite tags for large baleen whales – a contribution that is now sadly missed. We also thank workshop engineers S. Whiteside, C. Broinowski and all others that invested a great deal of effort and energy into the successful design and application of this satellite tag. Tag attachment developments were enhanced through photographic contributions by W. and P. Osborn, while V. Boladeras efficiently maintained the field database during often hectic moments. Many thanks go to D. Smith of the Australian Antarctic Data Centre for assistance with data curation and GIS respectively. We thank B. Miller and S. Bestley for their valuable comments and input to various stages of this manuscript. We thank the anonymous reviewer for their constructive input and the consequent improvements to this manuscript.

Author Contributions

Conceived and designed the experiments: MCD KCSJ MNJ SML NJG. Performed the experiments: MCD KCSJ MNJ SML NJG. Analyzed the data: VAG. Wrote the paper: VAG MCD TAB.

References

1. Rice DW (1998) Marine mammals of the world: systematics and distribution. The Society for Marine Mammalogy Special Publication 4. Kansas: Allen Press, Inc. 234p.
2. Ichihara T (1966) The pygmy blue whale, *Balaenoptera musculus brevicauda*, a new subspecies from the Antarctic. In: Norris KS, editor. Whales, dolphins and porpoises: 79–113.
3. Branch TA, Abubaker E, Mkango S, Butterworth D (2007) Separating southern blue whale subspecies based on length frequencies of sexually mature females. Mar Mamm Sci 23: 803–833.
4. Branch TA, Stafford KM, Palacios DM, Allison C, Bannister JL, et al. (2007) Past and present distribution, densities and movements of blue whales Balaenoptera musculus in the Southern Hemisphere and northern Indian Ocean. Mammal Rev 37: 116–175.
5. Branch TA, Mikhalev YA, Kato H (2009) Separating pygmy and Antarctic blue whales using long-forgotten ovarian data. Mar Mamm Sci 25: 833–854.
6. Cetacean Specialist Group (1996) Balaenoptera musculus ssp. brevicauda. IUCN 2013 IUCN Red List of Threatened Species Version 20131.
7. Stafford KM, Chapp E, Bohnenstiel DR, Tolstoy M (2011) Seasonal detection of three types of "pygmy" blue whale calls in the Indian Ocean. Mar Mamm Sci 27: 828–840.
8. McCauley R, Bannister J, Burton C, Jenner C, Rennie S, et al. (2004) Western Australian exercise area blue whale project. Final summary report. Milestone 6 for Australian Defence.
9. Gavrilov AN, McCauley RD (2013) Acoustic detection and long-term monitoring of pygmy blue whales over the continental slope in southwest Australia J Acoust Soc Am 134: 2505–2513.
10. Samaran F, Adam O, Guinet C (2010) Discovery of a mid-latitude sympatric area for two Southern Hemisphere blue whale subspecies. Endanger Spec Res 12: 157–165.
11. Samaran F, Stafford KM, Branch TA, Gedamke J, Royer J-Y, et al. (2013) Seasonal and Geographic Variation of Southern Blue Whale Subspecies in the Indian Ocean. PLoS ONE 8: e71561.
12. Attard CR, Beheregaray LB, Jenner KCS, Gill PC, Jenner MN, et al. (2012) Hybridization of Southern Hemisphere blue whale subspecies and a sympatric area off Antarctica: impacts of whaling or climate change? Mol Ecol 21: 5715–5727.
13. Rennie S, Hanson CE, McCauley RD, Pattiaratchi C, Burton C, et al. (2009) Physical properties and processes in the Perth Canyon, Western Australia: Links to water column production and seasonal pygmy blue whale abundance. J Mar Syst 77: 21–44.
14. Gill PC (2002) A blue whale (Balaenoptera musculus) feeding ground in a southern Australian coastal upwelling zone. J Cetacean Res Manag 4: 179–184.
15. Gill PC, Morrice MG, Page B, Pirzl R, Levings AH, et al. (2011) Blue whale habitat selection and within-season distribution in a regional upwelling system off southern Australia. Mar Ecol Prog Ser 421: 243–263.
16. Jenner C, Jenner M, Burton C, Sturrock V, Salgado Kent C, et al. (2008) Mark recapture analysis of pygmy blue whales from the Perth Canyon, Western Australia 2000–2005. Paper presented to the Scientific Committee of the International Whaling Commission: SC/60/SH16.
17. Kato H, Matsuoka K, Nishiwaki S, Bannister J (2007) Distribution and abundances of pygmy blue whales and southern right whales in waters off the southern coast of Australia, based on data from the Japan/IWC bue whale cruise 1995–96. Paper presented to the Scientific Committee of the International Whaling Commission: SC/59/SH10.

18. Attard CR, Beheregaray LB, Jenner C, Gill P, Jenner M, et al. (2010) Genetic diversity and structure of blue whales (Balaenoptera musculus) in Australian feeding aggregations. Conserv Genet 11: 2437–2441.
19. Hyrenbach KD, Forney KA, Dayton PK (2000) Marine protected areas and ocean basin management. Aquat Conserv: Mar Freshwat Ecosyst 10: 437–458.
20. Shillinger GL, Palacios DM, Bailey H, Bograd SJ, Swithenbank AM, et al. (2008) Persistent leatherback turtle migrations present opportunities for conservation. PLoS Biol 6: e171.
21. Heide-Jørgensen MP, Kleivane L, ØIen N, Laidre KL, Jensen MV (2001) A new technique for deploying satellite transmitters on baleen whales: Tracking a blue whale (Balaenoptera musculus) in the North Atlantic. Mar Mamm Sci 17: 949–954.
22. Krutzen M, Barre LM, Moller LM, Heithaus MR, Simms C, et al. (2002) A biopsy system for small cetaceans: darting success and wound healing in Tursiops spp. Mar Mamm Sci 18: 863–878.
23. Morin PA, Nestler A, Rubio-Cisneros NT, Robertson KM, Mesnick SL (2005) Interfamilial characterization of a region of the ZFX and ZFY genes facilitates sex determination in cetaceans and other mammals. Mol Ecol 14: 3275–3286.
24. Freitas C, Kovacs KM, Ims RA, Fedak MA, Lydersen C (2008) Ringed seal post-moulting movement tactics and habitat selection. Oecologia 155: 193–204.
25. R Core Team (2012) R: A language and environment for statistical computing. Vienna, Austria: R Foundation for Statistical Computing. ISBN 3-900051-07-0, URL http://www.R-project.org/.
26. Breed GA, Costa DP, Goebel ME, Robinson PW (2011) Electronic tracking tag programming is critical to data collection for behavioral time-series analysis. Ecosphere 2: art10.
27. De Deckker P, Moros M, Perner K, Jansen E (2012) Influence of the tropics and southern westerlies on glacial interhemispheric asymmetry. Nat Geosci 5: 266–269.
28. Gavrilov AN, McCauley RD, Salgado-Kent C, Tripovich J, Burton C (2011) Vocal characteristics of pygmy blue whales and their change over time. J Acoust Soc Am 130: 3651–3660.
29. Salgado Kent CP, Gavrilov AN, Recalde-Salas A, Burton CLK, McCauley RD, et al. (2012) Passive acoustic monitoring of baleen whales in Geographe Bay, Western Australia. Acoust Aust: 1–8.
30. Calambokidis J, Barlow J, Ford JK, Chandler TE, Douglas AB (2009) Insights into the population structure of blue whales in the Eastern North Pacific from recent sightings and photographic identification. Mar Mamm Sci 25: 816–832.
31. Moore JK, Abbott MR (2000) Phytoplankton chlorophyll distributions and primary production in the Southern Ocean. JGR 105: 28709–28728,28722.
32. Furnas M (2007) Intra-seasonal and inter-annual variations in phytoplankton biomass, primary production and bacterial production at North West Cape, Western Australia: Links to the 1997–1998 El Niño event. Cont Shelf Res 27: 958–980.
33. Sleeman JC, Meekan MG, Wilson SG, Jenner CK, Jenner MN, et al. (2007) Biophysical correlates of relative abundances of marine megafauna at Ningaloo Reef, Western Australia. Mar Freshw Res 58: 608–623.
34. Taylor J (1996) Seasonal occurrence, distribution and movements of the whale shark, Rhincodon typus, at Ningaloo Reef, Western Australia. Mar Freshw Res 47: 637–642.
35. Wilson SG, Polovina JJ, Stewart BS, Meekan MG (2006) Movements of whale sharks (Rhincodon typus) tagged at Ningaloo Reef, Western Australia. Mar Biol 148: 1157–1166.
36. Mackintosh NA (1966) The distribution of southern blue and fin whales. In: Norris KS, editor. Whales, dolphins and porpoises: 125–145.
37. Brown SG (1954) Dispersal in blue and fin whales. Discovery Reports: 355–384.
38. Hays GC, Hobson VJ, Metcalfe JD, Righton D, Sims DW (2006) Flexible foraging movements of leatherback turtles across the North Atlantic Ocean. Ecology 87: 2647–2656.
39. Belkin IM, Cornillon PC (2007) Fronts in the world ocean's large marine ecosystems. ICES CM 500: 21.
40. Longhurst AR (2007) Ecological geography of the sea. Amsterdam, Netherlands: Academic Press. 560 p.
41. Benson SR, Eguchi T, Foley DG, Forney KA, Bailey H, et al. (2011) Large-scale movements and high-use areas of western Pacific leatherback turtles, Dermochelys coriacea. Ecosphere 2: art84.
42. Geoscience Australia and BREE (2012) Australian Gas Resource Assessment 2012. Canberra.
43. Tyack PL, Miller EH (2002) Vocal anatomy, acoustic communication and echolocation. In: Hoelzel AR, editor. Marine mamml biology: an evolutionary approach. Oxford, England: Blackwell Science. Pp. 142–184.
44. Melcón ML, Cummins AJ, Kerosky SM, Roche LK, Wiggins SM, et al. (2012) Blue whales respond to anthropogenic noise. PLoS ONE 7: e32681.
45. Di Iorio L, Clark CW (2010) Exposure to seismic survey alters blue whale acoustic communication. Biol Lett 6: 51–54.
46. Tyack PL (2008) Implications for marine mammals of large-scale changes in the marine acoustic environment. J Mammal 89: 549–558.
47. Bryant PJ, Lafferty CM, Lafferty SK (1984) Reoccupation of Laguna Guerrero Negro, Baja California, Mexico, by gray whales. In: Jones ML, Swartz SL, Leatherwood S, editors. The gray whale *Eschrichtius robustus*. Orlando, Florida: Academic Press. 375–387.
48. Richardson WJ, Würsig B, Greene Jr CR (1986) Reactions of bowhead whales, Balaena mysticetus, to seismic exploration in the Canadian Beaufort Sea. J Acoust Soc Am 79: 1117.
49. Morton AB, Symonds HK (2002) Displacement of Orcinus orca (L.) by high amplitude sound in British Columbia, Canada. ICES J. Mar. Sci. 59: 71–80.
50. Rolland RM, Parks SE, Hunt KE, Castellote M, Corkeron PJ, et al. (2012) Evidence that ship noise increases stress in right whales. Proc. R. Soc. B 279: 2363–2368.

10

Remote Acoustic Monitoring of North Atlantic Right Whales (*Eubalaena glacialis*) Reveals Seasonal and Diel Variations in Acoustic Behavior

Leanna P. Matthews*, Jessica A. McCordic, Susan E. Parks

Department of Biology, Syracuse University, Syracuse, New York, United States of America

Abstract

Remote acoustic monitoring is a non-invasive tool that can be used to study the distribution, behavior, and habitat use of sound-producing species. The North Atlantic right whale (*Eubalaena glacialis*) is an endangered baleen whale species that produces a variety of stereotyped acoustic signals. One of these signals, the "gunshot" sound, has only been recorded from adult male North Atlantic right whales and is thought to function for reproduction, either as reproductive advertisement for females or as an agonistic signal toward other males. This study uses remote acoustic monitoring to analyze the presence of gunshots over a two-year period at two sites on the Scotian Shelf to determine if there is evidence that North Atlantic right whales may use these locations for breeding activities. Seasonal analyses at both locations indicate that gunshot sound production is highly seasonal, with an increase in the autumn. One site, Roseway West, had significantly more gunshot sounds overall and exhibited a clear diel trend in production of these signals at night. The other site, Emerald South, also showed a seasonal increase in gunshot production during the autumn, but did not show any significant diel trend. This difference in gunshot signal production at the two sites indicates variation either in the number or the behavior of whales at each location. The timing of the observed seasonal increase in gunshot sound production is consistent with the current understanding of the right whale breeding season, and our results demonstrate that detection of gunshots with remote acoustic monitoring can be a reliable way to track shifts in distribution and changes in acoustic behavior including possible mating activities.

Editor: Michael L. Fine, Virginia Commonwealth Univ, United States of America

Funding: Right Whale Grants Program of the (U.S.) National Oceanographic and Atmospheric Administration and by Office of Naval Research grant #N00014-03-1-0099. The funders had no role in study design, data collection and analysis, decision to publish, or preparation of the manuscript.

Competing Interests: The authors have declared that no competing interests exist.

* E-mail: lematthe@syr.edu

Introduction

Understanding patterns of animal communication can reveal critical information about life history [1]. Many species rely on particular signals to locate and select mates, a process essential to individual fitness and survival of the species. In marine systems, sound is the most efficient modality for long-range communication, and acoustic signaling is the predominant form of communication for most marine species [2]. The production of loud acoustic signals for long-range communication provides an opportunity for using remote acoustic sensing to capture annual trends and seasonal changes in signal production of a variety of species [3–8].

Remote acoustic sensing, commonly referred to as passive acoustic monitoring (PAM), uses autonomous recorders to collect data for long periods of time with minimal disturbance to the environment. Such recording units allow researchers to collect acoustic data in remote locations or during weather conditions or seasons that would otherwise prohibit direct observation [9,10].

In both terrestrial and marine systems, PAM has been used to investigate various aspects of habitat use and behavior by sound-producing species. For instance, passive acoustics has been used to investigate spatial and temporal dynamics of Mexican antthrush (*Formicarius moniliger*) [11] and assess the density and composition of orthopteran species assemblages [12]. Acoustic monitoring can also be used to detect species-specific signals, thereby using these signals as reliable indicators of the presence of a species within acoustic range of the receiver. For example, passive acoustics has been used to investigate the presence, behavior, and population structure of African elephant (*Loxodonta africana*) populations [13]. It has also been used in aquatic environments to track annual distribution of North Atlantic right whales (*Eubalaena glacialis*) [14], document seasonal and geographic variation in fin whale (*Balaenoptera physalus*) song [15], and monitor reproductive behavior of teleost fishes [16]. Along with presence-absence information, context-specific vocalizations enable observers to make inferences about behavior and use of a particular habitat. If the production of vocalizations associated with mating behavior shows clear seasonal trends, this can be used to identify the timing of the breeding season and the breeding locations for different species, including baleen whales [3,4,17]. Detection of signals with known behavioral functions can be particularly informative for identifying critical habitat areas or seasons for the protection of endangered or threatened species.

For over a decade, the critically endangered North Atlantic right whale (*Eubalaena glacialis*) has been monitored using autonomous recorders (for reviews, see [18,19]). The right whale

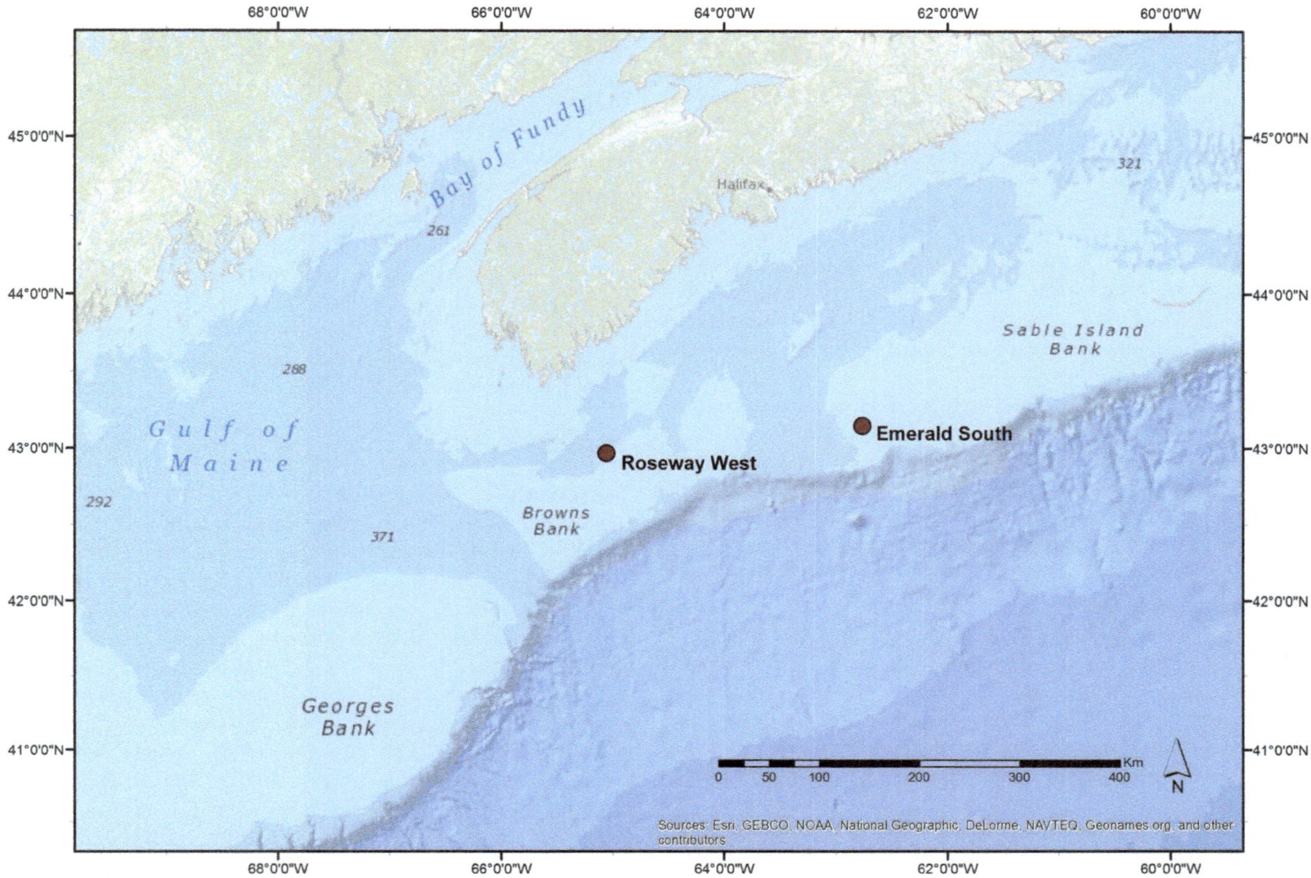

Figure 1. Locations of autonomous recording units on the Scotian Shelf. Roseway West: 42°58′00″ N 65°03′24″W; Emerald South: 43°8′54″N 62°45′42″W. Numbers correspond to water depth in meters.

produces a species-specific stereotyped contact call, known as an "upcall" [20,21]. This call has been recorded from both sexes and from all age classes, and it is therefore a reliable signal to detect the presence of right whales in an area. This is the primary call used for PAM detection studies of right whales [20,22,23]. Using detections of upcalls, previous studies [5] confirmed the presence of right whales in two sites on the Scotian Shelf during the late summer and early autumn. This study and others using upcalls have revealed the presence of right whales in locations or seasons otherwise overlooked from traditional visual surveys. However, the broad usage of the upcall by all individuals in the species limits the behavioral information that can be obtained from detection of this signal alone. Previous preliminary studies have indicated significant seasonal changes in right whale usage of different call types [18].

Along with tonal upcalls, right whales also produce several other stereotyped acoustic signals [20,24,25]. Among these is a high-intensity (>185 dB re 1μPa p-p), broadband sound known as a "gunshot" [22,25]. The gunshot sound is a short-duration signal (<0.02 s) with a peak frequency near 1.19 kHz and a known frequency range of 20 Hz to 20 kHz, with the upper range limited by the sampling rate of the recording equipment [22,25]. In the North Atlantic right whale, extended bouts of gunshot sound production have only been recorded from adult males on the feeding grounds, suggesting that it may function as a reproductive advertisement for females or an agonistic signal toward other males [25,26]. Previous studies have reported recordings of individual gunshot sounds produced by nursing females in Southern right whales (*Eubalaeana australis*) [20]. It is possible that female North Atlantic right whales are also capable of gunshot sound production. However, the long stereotyped sequences of gunshot sounds observed from males have not been described in any female for any right whale species.

Male right whales also produce gunshots during surface active groups (SAGs), right whale aggregations that typically involve males competing for access to a focal female [22,27]. Although SAGs can form in the absence of reproductively mature animals [28], the production of gunshots by adult males in SAGs with potentially reproductive compositions strengthens the association of the gunshot call with mating behavior [22].

Previous studies have detected gunshots using passive acoustic recorders during the spring, summer, and autumn months, with a marked increase in the rate of signal detection later in the season [18,26]. Estimates of the gestation length, seasonal shifts in vocal behavior, and hormone analysis all indicate a probable right whale breeding season beginning in the autumn and lasting through the winter [18,29–31]. These prior studies suggest that long-term recordings may clarify patterns of seasonality in the gunshot call and its possible association with the breeding season in right whales. In this study, we analyzed the presence of gunshots over a two-year period at two sites on the Scotian Shelf to determine whether there is any evidence that North Atlantic right whales use these habitat areas for mating-related activities.

Materials and Methods

Data Collection

Autonomous acoustic recording systems (NOAA/PMEL autonomous hydrophones, [32]) were deployed in two locations off the coast of Nova Scotia in June and July of 2004 (Figure 1) [5]. The recording systems recorded continuously for two years (July 2004–June 2006) at a sampling rate of 2 kHz, with a flat frequency response of ±3 dB from 30–840 Hz. This low frequency range includes the majority of right whale sound production, including the peak energy of right whale gunshot sounds [22,33,34]. Because water depth in the Scotian Shelf is less that 300 m, hydrophones were fixed 7 m above the sea floor to allow for optimum acoustic reception [5].

Recorders were positioned in two historical whaling areas in the North Atlantic [35]. One recording site, "Roseway West," is a protected marine conservation area [36] and a known location for North Atlantic right whale surface-active groups [37]. The second recorder, "Emerald South," was located approximately 200 km northeast of Roseway West on Emerald Bank.

Seasonality Analysis

For each site, 24-hour periods were randomly selected from each week of data for the duration of the acoustic recordings. Spectrograms for each of these periods were visually inspected using RavenPro 1.5 [38] to detect gunshot presence in each hour (2 kHz sampling, Hann window, discrete Fourier transform (DFT) size 128, analysis resolution = 15.6 Hz and 0.032 s, 50% overlap). The first 60 gunshots of a given hour were selected for subsequent diel analysis. Based on previous observations of gunshot bouts [26], a maximum of 60 gunshots was determined to be a reliable indicator of high gunshot activity. To test for seasonality, we determined the number of hours per day in which gunshots were detected. This did not take into consideration the total number of individual gunshots but rather assessed only the presence or absence of gunshot sounds in each hour (e.g. [6,39]). The two sites, Roseway West and Emerald South, were analyzed separately.

Diel Analysis

Diel patterns of right whale gunshot signals were also analyzed for both locations. For months of peak gunshot activity (August, September, October, November), each vocalization was coded based on light regime, and hourly averages (gunshots per hour) were calculated with a maximum of 60 gunshots per hour. The light code for each gunshot corresponded to one of four non-overlapping three-hour time blocks. These time blocks refer to sunrise/dawn, midday, sunset/dusk, and night, and were determined using data from the United States Naval Observatory [40].

Statistics

For seasonal analysis, data from the two years of recordings were pooled and monthly averages of gunshot presence were calculated. These averages were compared via ANOVA to test for significant differences among months. To control for Type I error rate across multiple comparisons, Tukey's HSD was used as a one-step pairwise comparison procedure to test for differences between each of the months. For the diel analysis, the data were blocked by day and compared via ANOVA. Tukey's HSD was used to test for pairwise significance among the light regimes. All statistical tests were done using R v. 2.15.2 [41].

Results

Seasonality

Gunshot sounds (Figure 2) were detected seasonally in both sites in both years. The average and standard deviation of hours per day of gunshot activity were calculated for each site (Table 1). Gunshot activity was detected most of the year at Emerald South. Gunshots occurred in all months of the year at the Emerald location, while only occurring June through December at the Roseway West site. However, gunshot activity levels were higher in the Roseway West site, with more hours per day with gunshot presence detection. The peak month of gunshot activity for Roseway West was August ($\bar{\bar{x}} \pm SD$: 10.78±7.61 hours/day), while the peak month of activity in Emerald South was October ($\bar{\bar{x}} \pm SD$: 2.20±1.69 hours/day). The maximum number of hours observed in a single day for Roseway West was 21 hours/day, while the maximum for Emerald South was 10 hours/day. ANOVA results (Table 2) for both sites indicated significance at $\alpha = 0.05$. Further analysis via Tukey's HSD revealed that for Roseway West, the months of August, September, October, and November had a significantly higher number of hours of gunshot presence than other months. This indicates a strong seasonal trend in the Roseway site. The Emerald South site also showed a seasonal trend. Because this trend was weaker than Roseway, months were grouped into seasons (Jan-Mar, Apr-Jun, Jul-Sept, Oct-Dec) for pairwise analysis. Tukey's HSD indicated that the second half of the year (July-December) had a significantly higher number of hours of gunshots. Seasonal trends for both sites can be seen in Figure 3.

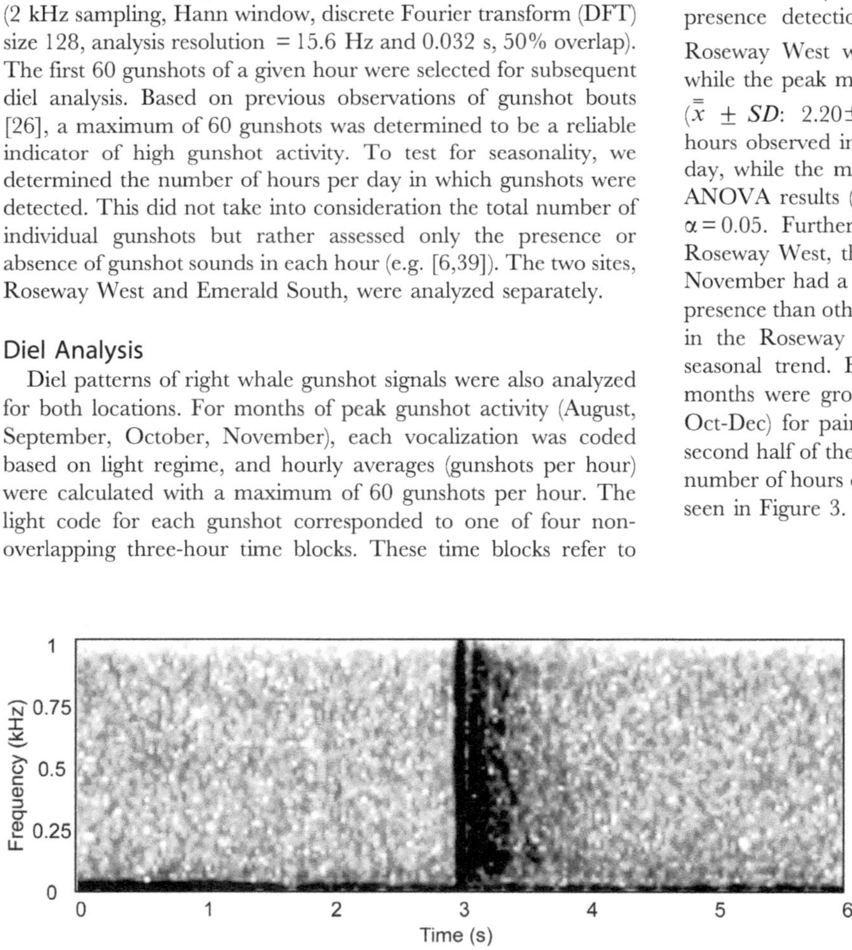

Figure 2. Spectrogram of a gunshot sound recorded in Roseway West. Spectrogram parameters: 2 kHz sampling, Hann window, discrete Fourier transform (DFT) size 128, analysis resolution = 15.6 Hz and 0.032 s, 50% overlap.

Table 1. Average and standard deviation of hours per day of gunshot activity for each site.

	Hours per Day in that Gunshots are Present ($\bar{\bar{x}} \pm SD$)											
	Jan	Feb	Mar	Apr	May	June	July	Aug	Sept	Oct	Nov	Dec
Roseway West	0.00±0.00	0.00±0.00	0.00±0.00	0.00±0.00	0.00±0.00	0.75±1.04	1.10±1.20	10.78±7.61	9.11±7.39	9.62±6.21	8.11±8.54	1.13±3.18
Emerald South	0.11±0.33	0.25±0.71	0.11±0.33	0.11±0.33	0.29±0.49	0.38±1.06	1.67±1.94	1.00±0.93	3.11±4.11	2.20±1.69	1.88±2.80	0.33±0.71

Diel Trends

Diel trends were analyzed for both sites. The average gunshot rates for each light regime (dawn, midday, dusk, night) for both sites can be seen in Table 3. ANOVA results (Table 4) revealed no significant diel trend for Emerald South ($p = 0.551$). There was, however, a trend for Roseway West ($p = 1.83e\text{-}05$). In Roseway West, gunshot activity peaked during the night hours ($\bar{\bar{x}} \pm SD$: 24.30±21.67 GS/hr), and was lowest during midday ($\bar{\bar{x}} \pm SD$: 6.78±12.62). Figure 4 demonstrates that gunshot rate increases from midday to night before tapering off during the dawn hours. Pairwise comparisons (Tukey's HSD) indicated significant differences in gunshot activity.

Discussion

Gunshot sound production is highly seasonal, with a significant increase in gunshot sound detection in the autumn. The timing of this increase is consistent with the current understanding that the right whale breeding season peaks in October-December [18,29–31]. Detection of gunshot sounds can therefore be used as a reliable way to track the seasonal mating activities of right whales. Previously, the location of the majority of the right whale population during the breeding season was unknown [27,42]. Data presented here indicate that the sites along the Scotian Shelf, in particular Roseway West, are potentially important locations for right whale breeding activities. Both seasonal and diel trends indicate variation in the usage of right whale gunshot sounds between the two sites. Roseway West has significantly more gunshot sounds overall, with a clear diel trend in production of these signals at night. Emerald South, while also showing seasonal trends in signal production, shows lower levels of gunshot sound production suggesting that either fewer whales utilize this habitat, or that the behavior of whales in this location varies from those on Roseway West.

Seasonality

There was a similar seasonal trend at both recording sites, with the majority of gunshots occurring in August through November (Figure 3). This trend is consistent with the movement patterns of North Atlantic right whales, which move from the Bay of Fundy into areas off the Scotian Shelf during the middle to late summer [43,44]. The gunshot seasonality observed here is also similar to the seasonal trend observed for right whale contact calls – i.e. upcalls – in these same locations in Roseway Basin [5].

Although right whale gunshot production in Emerald South exhibited seasonality, it was not as strong of a seasonal occurrence as observed at the Roseway West site. Previous studies [45] have used satellite tag data to indicate that right whale movements around the Scotian Shelf vary from year to year. This could be one possible explanation for the differences between the strength of the seasonal trend in Roseway West and Emerald South. It is also possible that differences in habitat use and prey abundance could account for the inter-site variation, with right whales preferring Roseway West to Emerald South because of denser prey aggregations [5].

There are three potential explanations for the seasonal pattern of gunshots. First, the increase in gunshot production could be directly tied to an increase in the number of whales at that location. The two recording sites analyzed here are located in areas historically associated with right whales [35], and the timing of high gunshot presence matches the annual movement patterns of right whales [44]. The movement of whales into and out of the Scotian Shelf could account for seasonal gunshot variation.

Table 2. ANOVA results for the seasonality analysis of both sites.

	Df	Sum Sq	Mean Sq	F Value	P
Roseway West	11	1952	177.48	8.949	1.14e-10
Residuals	93	1844	19.83		
Emerald South	11	101.1	9.193	3.089	0.000145
Residuals	91	270.8	2.976		

Second, the seasonal increase in gunshots could be due to an increase in signal production by individual whales. This would indicate a shift in the behavior of right whales, possibly for reproductive purposes. Gunshots have only been documented in adult male North Atlantic right whales and this signal is thought to serve a reproductive function, whether that be as advertisement to females or as male-male agonistic displays [22]. An increase in gunshot production by individual males in the Scotian Shelf could indicate that right whales are using this area as a breeding ground, and the strong seasonal trend associated with Roseway West points to this location as a likely area for reproductive activity. A third explanation for gunshot seasonality is a combination of both increased whale density and increased acoustic signaling.

To better understand the reason behind the observed seasonality of gunshot sound production in these sites, it would be of interest to use passive acoustic monitoring to estimate North Atlantic right whale population density. These types of population estimations from passive acoustics have been done with other large cetaceans [46–48]. A more comprehensive knowledge of North Atlantic right whale population flux in a specific habitat would provide insight into the basis of observed seasonal acoustic trends.

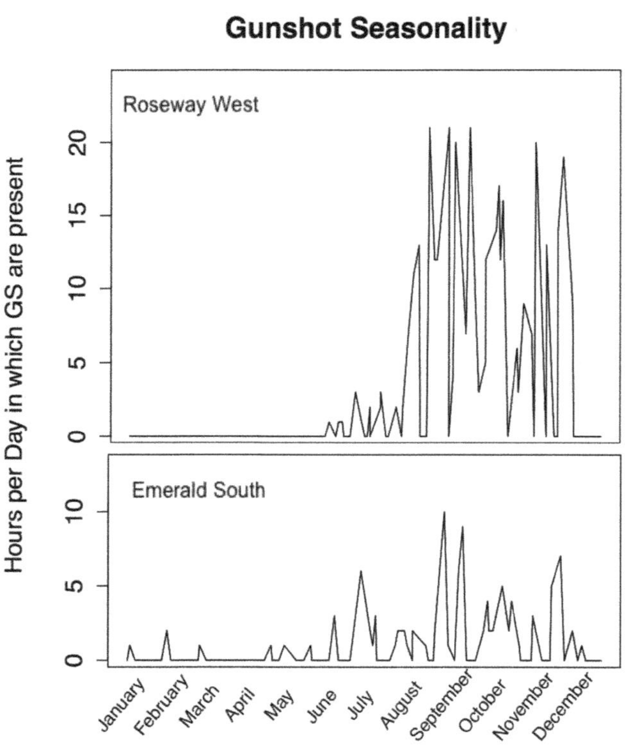

Figure 3. Seasonal trends of gunshot signal production. The hours per day with gunshots (GS) showed a seasonal pattern in both Roseway West (top) and Emerald South (bottom). In Roseway West, August-November had significantly more gunshot presence than other months. Emerald South, despite having fewer hours per day with gunshots overall, still exhibits a seasonal trend, with significantly more gunshot hours during the second half of the year.

Diel Trends

Previous studies have indicated diel patterns in sound production in multiple cetacean species, including North Atlantic right whales [7,8,24,49–51]. The data on North Atlantic right whales show results similar to those presented here for Roseway West, with increased calling activity at night and reduced calling activity during daytime hours. This periodicity may be due, in part, to prey abundance and distribution [7,8]. North Atlantic right whales forage primarily on late-stage copepods, *Calanus finmarchicus* [52]. In Roseway Basin, these late-stage copepods aggregate in discrete layers [53]. It has been shown that in the summer feeding grounds (Bay of Fundy, Roseway Basin), North Atlantic right whales forage during the day when the prey is concentrated in deeper water [52]. The diel pattern of gunshot signaling in Roseway West supports the idea of an inverse relationship between calling and foraging. Similar diel trends in gunshot production have also been observed in the Bay of Fundy, another summer foraging location for North Atlantic right whales [26]. In both of these sites, aggregations of late-stage *C. finmarchicus* are likely composed of individuals in diapause [53]. It is possible that right whale individuals use visual cues to detect prey layers during foraging dives [52], thereby limiting efficient foraging to daytime hours. Increased levels of gunshot production during the nighttime hours allow individuals to signal without a great loss in foraging time.

Our results show that there is no significant diel trend for gunshots at Emerald South. It is possible that the differences between sites are due to food availability or differences in the

Table 3. Average and standard deviation of gunshot rates for the light regimes in both sites.

Site	Light Regime	Gunshot Rate (GS/hr; $\bar{\bar{x}} \pm SD$)
Roseway West	Midday	6.78±12.62
	Dusk	11.85±15.51
	Night	24.00±21.09
	Dawn	19.98±15.86
Emerald South	Midday	0.69±1.71
	Dusk	0.21±0.45
	Night	2.74±9.30
	Dawn	7.02±15.21

Table 4. ANOVA results for the diel analysis of both sites.

Site		Df	Sum Sq	Mean Sq	F Value	P
Roseway West	Day	29	14535	501.2	3.167	1.63e-05
	Light Regime	3	4597	1532.5	9.683	1.40e-05
	Residuals	87	13769	158.3		
Emerald South	Day	13	972	74.76	0.910	0.551
	Light Regime	3	405	134.91	1.642	0.195
	Residuals	39	3204	82.16		

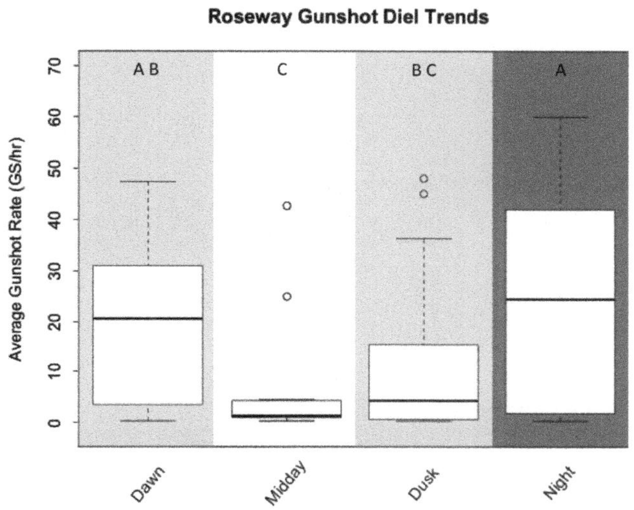

Figure 4. Gunshot diel trends in Roseway West. Gunshot rate (GS/hr) was significantly higher during night hours when compared to midday and dusk. There is a significant increase of gunshots from dusk to night and a clear decreasing trend of gunshot rate between night and dawn. Shading corresponds to light regime, and different letters indicate significant differences between them as determined by Tukey's Test.

demographics of the whales present at this site. There is an abundance of late-stage *C. finmarchicus* around Roseway West, while the Emerald South site is not in an area known for intensive right whale foraging [5]. While there was no significant diel trend for gunshots at this site, previous studies have shown diel trends in upcall production at Emerald South [5]. These results, however, show an opposite trend, with significantly more upcalls in the daytime hours. It has been hypothesized that, because of the low level of food availability in Emerald South, right whales spend more time socializing at this location [5].

Conclusions

Remote acoustic monitoring is an important tool for understanding patterns in animal communication, and studies on the seasonality of context-specific acoustic signals allow inferences to be made about the behavior and habitat use of certain species. Our results further support the hypothesis that the North Atlantic right whale breeding season occurs in the later part of the year (August – November) and point to Roseway West as a potentially important breeding location for the North Atlantic right whale population. Future studies should focus on the behavioral context of gunshots on the Scotian Shelf and on the movements of right whales into and out of the recording locations. The hydrophone array used to collect this data allows for potential estimates of population size and individual calling rates [16]. Pairing population density measurements with the data presented here will allow a better understanding of right whale behavior and will be valuable information for future conservation efforts.

Acknowledgments

Special thanks to Dr. David Mellinger and Dr. Sharon Nieukirk for collecting and providing access to the recordings used in this study. Thanks to the captains of the R/V Wanderbird, Rick and Karen Miles, and her crew for assistance with this instrument deployment and recovery. Thanks to Alex Hay of Dalhousie and Glen Herbert of DFO-MPO for logistical and permitting support.

Author Contributions

Conceived and designed the experiments: LPM JAM SEP. Performed the experiments: LPM JAM. Analyzed the data: LPM. Wrote the paper: LPM JAM SEP.

References

1. Bradbury JW, Vehrencamp SL (1998) Principles of Animal Communication. Sunderland, MA: Sinauer Associates, Inc. 697 p.
2. Au WWL, Hastings MC (2008) Principles of Marine Bioacoustics. New York, NY: Springer. 677 p.
3. Payne RS, Webb D (1971) Orientation by means of long range acoustic signaling in baleen whales. Ann NY Acad Sci 188: 110–141.
4. Tremain SB, Swiston KA, Mennill DJ (2008) Seasonal Variation in Acoustic Signals of Pileated Woodpeckers. Wilson J Ornithol 120: 499–504.
5. Mellinger DK, Nieukirk SL, Matsumoto H, Heimlich SL, Dziak RP, et al. (2007) Seasonal Occurrence of North Atlantic right whale (*Eubalaena glacialis*) vocalizations at two sites on the scotian shelf. Mar Mam Sci 23: 856–867.
6. Stafford KM, Nieukirk SL, Fox CG (2001) Geographic and seasonal variation of blue whale calls in the North Pacific. J Cetacean Res Manag 3: 65–76.
7. Munger LM, Wiggins SM, Moore SE, Hildebrand JA (2008) North Pacific right whale (*Eubalaena japonica*) seasonal and diel calling patterns from long-term acoustic recordings in the southeastern Bering Sea, 2000–2006. Mar Mam Sci 24: 795–814. doi:10.1111/j.1748-7692.2008.00219.x
8. Mussoline SE, Risch D, Clark CW, Hatch LT, Weinrich MT, et al. (2012) Seasonal and diel variation in North Atlantic right whale up-calls: implications for management and conservation in the northwestern Atlantic Ocean. Endanger Species Res 17: 17–26. doi:10.3354/esr00411
9. Blumstein DT, Mennill DJ, Clemins P, Girod L, Yao K, et al. (2011) Acoustic monitoring in terrestrial environments using microphone arrays: applications, technological considerations and prospectus. J App Ecol 48: 758–767. doi:10.1111/j.1365-2664.2011.01993.x
10. Mellinger DK, Stafford KM, Moore SE, Dziak RP, Matsumoto H (2007) An overview of fixed passive acoustic observation methods for cetaceans. Oceanography 20: 36–45.

11. Kirschel ANG, Cody ML, Harlow ZT, Promponas V, Vallejo EE, et al. (2011) Territorial dynamics of Mexican Antthrushes revealed by individual recognition of thier songs. Ibis 153: 255–268.
12. Fisher FP, Schulz U, Schubert H, Knapp P, Schmoger M (1997) Quantitative assessment of grassland quality: acoustic determination of population sizes of Orthopteran indicator species. Ecol Appl 7: 909–920.
13. Payne KB, Thompson M, Kramer L (2003) Elephant calling patterns as indicators of group size and composition: the basis for an acoustic monitoring system. Afr J Ecol 41: 99–107.
14. Morano JL, Rice AN, Tielens JT, Estabrook BJ, Murray A, et al. (2012) Acoustically Detected Year-Round Presence of Right Whales in an Urbanized Migration Corridor. Conserv Biol 26: 698–707. doi:10.1111/j.1523-1739.2012.01866.x
15. Morano JL, Salisbury DP, Rice AN, Conklin KL, Clark CW (2012) Seasonal and geographical patterns of fin whale song in the western North Atlantic Ocean. J Acoust Soc Am 132: 1207–1212.
16. Rountree RA, Gilmore RG, Goudey CA, Hawkins AD, Luczkovich JJ, et al. (2006) Listening to Fish: Applications of Passive Acoustics to Fisheries Science. Fisheries 31: 433–446.
17. Bridges AS, Dorcas ME (2000) Temporal variation in Anuran Calling Behavior: Implications for Surveys and Monitoring Programs. Copeia 2: 587–592.
18. Van Parijs SM, Clark CW, Sousa-Lima RS, Parks SE, Rankin S, et al. (2009) Management and research applications of real time and archival passive acoustic sensors over varying temporal and spatial scales. Mar Ecol Prog Ser 395: 21–36. doi:10.3354/meps08123
19. Clark CW, Brown MW, Corkeron PJ (2010) Visual and acoustic surveys for North Atlantic right whales, Eubalaena glacialis, in Cape Cod Bay, Massachusetts 2001–2005: Management implications. Mar Mam Sci 26: 837–854. doi:10.1111/j.1748-7692.2010.00376.x
20. Clark CW (1983) Acoustic Communication and Behavior of the Southern Right Whale (*Eubalaena australis*). In: Payne R, editor.Communication and Behavior of Whales.Washington, D.C.: Westview Press, Inc. pp. 163–198.
21. Tellechea JS, Norbis W (2012) A note on recordings of Southern right whales (*Eubalaena australis*) off the coast of Uruguay. J Cetacean Res Manag (Spec Iss) 12: 361–364.
22. Parks SE, Tyack PL (2005) Sound production by North Atlantic right whales (*Eubalaena glacialis*) in surface active groups. J Acoust Soc Am 117: 3297. doi:10.1121/1.1882946
23. Parks SE, Searby A, Celerier A, Johnson MP, Nowacek DP, et al. (2011) Sound production behavior of individual North Atlantic right whales: implications for passive acoustic monitoring. Endanger Species Res 15: 63–76. doi:10.3354/esr00368
24. Matthews JN, Brown S, Gillespie D, Johnson M, McLanaghan R, et al. (2001) Vocalisation rates of the North Atlantic right whale. J Cetacean Res Manag 3: 271–282.
25. Parks SE, Hamilton PK, Kraus SD, Tyack PL (2005) The gunshot sound produced by male North Atlantic right whales (*Eubalaena glacialis*) and its potential function in reproductive advertisement. Mar Mam Sci 21: 458–475.
26. Parks SE, Hotchkin CF, Cortopassi KA, Clark CW (2012) Characteristics of gunshot sound displays by North Atlantic right whales in the Bay of Fundy. J Acoust Soc Am 131: 3173. doi:10.1121/1.3688507
27. Kraus SD, Hatch JJ (2001) Mating strategies in the North Atlantic right whale (*Eubalaena glacialis*). J Cetacean Res Manage (Spec Iss 2): 237–244.
28. Parks SE (2003) Response of North Atlantic right whales (*Eubalaena glacialis*) to playback of calls recorded from surface active groups in both the North and South Atlantic. Mar Mam Sci 19: 563–580.
29. Hunt KE, Rolland RM, Kraus SD, Wasser SK (2006) Analysis of fecal glucocorticoids in the North Atlantic right whale (*Eubalaena glacialis*). Gen Comp Endocriol 148: 260–272.
30. Best PB (1994) Seasonality of reproduction and the length of gestation in southern right whales *Eubalaena australis*. J Zool 232: 175–189.
31. Hamilton PK, Knowlton AR, Marx MK, Kraus SD (1998) Age structure and longevity in North Atlantic right whales (*Eubalaena glacialis*) and their relation to reproduction. Mar Ecol Prog Ser 171: 285–292.
32. Fox CG, Matsumoto H, Lau T-KA (2001) Monitoring Pacific Ocean seismicity from an autonomous hydrophone array. J Geophys Res-Sol Ea 106: 4183–4206.
33. Clark CW (1982) The acoustic repertoire of the southern right whale, a quantitative analysis. Anim Behav 30: 1060–1071.
34. McDonald MA, Moore SE (2002) Calls recorded from North Pacific right whales in the eastern Bering Sea. J Cetacean Res Manag 43: 261–266.
35. Mitchell E, Kozicki VM, Reeves RR (1986) Sightings of right whales, *Eubalaena glacialis*, on the Scotian Shelf, 1966–1972. Rep Int Whal Comm (Spec Iss) 10: 83–107.
36. Brown MW, Allen JM, Kraus SD (1995) The designation of seasonal right whale conservation zones in the waters of Atlantic Canada. In: Shackell NL, Willison MJH, editors. Marine Protected Areas and Sustainable Fisheries. Proceedings of a symposium on marine protected areas and sustainable fisheries conducted at the Second International Conference on Science and the Management of Protected Areas. Wolfville, Nova Scotia, Canada: Science and Management of Protected Areas Association.pp. 90–98
37. Brown MW, Brault S, Hamilton PK, Kenney RD, Knowlton AR, et al. (2001) Sighting heterogeneity of right whales in the western North Atlantic: 1980–1992. J Cetacean Res Manag (Spec Iss) 2: 251–260.
38. Bioacoustics Research Program (2012) Raven Pro: Interactive Sound Analysis Software (Version 1.5). Ithaca, NY: The Cornell Lab of Ornithology. Available: http://www.birds.cornell.edu/raven.
39. Nieukirk SL, Stafford KM, Mellinger DK, Dziak RP, Fox CG (2004) Low-frequency whale and seismic airgun sounds recorded in the mid-Atlantic Ocean. J Acoust Soc Am 115: 1832–1843.
40. United States Naval Observatory. Available: http://aa.usno.navy.mil/ data/docs/RS_OneYear.php. Accessed 2013 June.
41. R Development Core Team (2011) R: A Language and Environment for Statistical Computing (Version 2.15.2). Vienna, Austria: R Foundation for Statistical Computing. Available: http://www.R-project.org/.
42. IWC (International Whaling Commisson) (2001) Report of the workshop on status and trends of western North Atlantic right whales. J Cetacean Res Manag (Spec Iss) 2:1–87.
43. Cetacean and Turtle Assessment Program (CETAP) (1982) A characterization of marine mammals and turltes in the mid-Atlantic areas of the U.S. outer continental shelf. Final Report of the Cetacean and Turtle Assessment Program to the U.S. Department of the Interior. Contract AA5551-CT8-48. 570 pp.
44. Winn HE, Price CA, Sorensen PW (1986) The distributional biology of the right whale (*Eubalaena glacialis*) in the western North Atlantic. Rep Int Whal Comm (Spec Iss): 129–138.
45. Mate BR, Nieukirk SL, Kraus SD (1997) Satellite-monitored movements of the Northern right whale. J Wildl Manage 61: 1393–1405.
46. McDonald MA, Fox CG (1999) Passive acoustic methods applied to fin whale population density estimation. J Acoust Soc Am 105: 2643–2651.
47. Marques TA, Munger L, Thomas L, Wiggins SM, Hildebrand JA (2011) Estimating North Pacific right whale *Eubalaena japonica* density using passive acoustic cue counting. Endanger Species Res 13: 163–172.
48. Ackleh AS, Ioup GE, Ioup JW, Ma B, Newcomb JJ, et al. (2012) Assessing the Deepwater Horizon oil spill impact on marine mammal population through acoustics: Endangered sperm whales. J Acoust Soc Am 131: 2306. doi:10.1121/1.3682042
49. Baumgartner MF, Fratantoni DM (2008) Diel periodicity in both sei whale vocalization rates and the vertical migration of their copepod prey observed from ocean gliders. Limnol Oceanogr 53: 2197–2209.
50. Stafford KM, Moore SE, Fox CG (2005) Diel variation in blue whale calls recorded in the eastern tropical Pacific. Anim Behav 69: 951–958.
51. Wiggins SM, Oleson EM, McDonald MA, Hildebrand JA (2005) Blue Whale (*Balaenoptera musculus*) Diel Call Patterns Offshore of Southern California. Aquat Mam 31: 161–168. doi:10.1578/AM.31.2.2005.161
52. Baumgartner MF, Mate BR (2003) Summertime foraging ecology of North Atlantic right whales. Mar Ecol Prog Ser 264: 123–135. doi:10.3354/meps264123
53. Baumgartner MF, Cole TVN, Campbell RG, Teegarden GJ, Durbin EG (2003) Associations between North Atlantic right whales and their prey, *Calanus finmarchicus*, over diel and tidal time scales. Mar Ecol Prog Ser 264: 155–166.

11

Humpback Whale Populations Share a Core Skin Bacterial Community: Towards a Health Index for Marine Mammals?

Amy Apprill[1]*, Jooke Robbins[2], A. Murat Eren[3], Adam A. Pack[4,5], Julie Reveillaud[3], David Mattila[6], Michael Moore[1], Misty Niemeyer[7], Kathleen M. T. Moore[7], Tracy J. Mincer[1]*

1 Woods Hole Oceanographic Institution, Woods Hole, Massachusetts, United States of America, **2** Center for Coastal Studies, Provincetown, Massachusetts, United States of America, **3** Josephine Bay Paul Center, Marine Biological Laboratory, Woods Hole, Massachusetts, United States of America, **4** University of Hawaii at Hilo, Hilo, Hawaii, United States of America, **5** The Dolphin Institute, Honolulu, Hawaii, United States of America, **6** Hawaiian Islands Humpback Whale National Marine Sanctuary, Kihei, Hawaii, United States of America, **7** International Fund for Animal Welfare, Yarmouth Port, Massachusetts, United States of America

Abstract

Microbes are now well regarded for their important role in mammalian health. The microbiology of skin – a unique interface between the host and environment - is a major research focus in human health and skin disorders, but is less explored in other mammals. Here, we report on a cross-population study of the skin-associated bacterial community of humpback whales (*Megaptera novaeangliae*), and examine the potential for a core bacterial community and its variability with host (endogenous) or geographic/environmental (exogenous) specific factors. Skin biopsies or freshly sloughed skin from 56 individuals were sampled from populations in the North Atlantic, North Pacific and South Pacific oceans and bacteria were characterized using 454 pyrosequencing of SSU rRNA genes. Phylogenetic and statistical analyses revealed the ubiquity and abundance of bacteria belonging to the *Flavobacteria* genus *Tenacibaculum* and the *Gammaproteobacteria* genus *Psychrobacter* across the whale populations. Scanning electron microscopy of skin indicated that microbial cells colonize the skin surface. Despite the ubiquity of *Tenacibaculum* and *Psychrobater* spp., the relative composition of the skin-bacterial community differed significantly by geographic area as well as metabolic state of the animals (feeding versus starving during migration and breeding), suggesting that both exogenous and endogenous factors may play a role in influencing the skin-bacteria. Further, characteristics of the skin bacterial community from these free-swimming individuals were assembled and compared to two entangled and three dead individuals, revealing a decrease in the central or core bacterial community members (*Tenacibaculum* and *Psychrobater* spp.), as well as the emergence of potential pathogens in the latter cases. This is the first discovery of a cross-population, shared skin bacterial community. This research suggests that the skin bacteria may be connected to humpback health and immunity and could possibly serve as a useful index for health and skin disorder monitoring of threatened and endangered marine mammals.

Editor: Christina A. Kellogg, U.S. Geological Survey, United States of America

Funding: A.A. was funded by a WHOI Ocean Life Institute post-doctoral scholar award, and this research was supported by a grant to A.A. and T.J.M. from Woods Hole Oceanographic Institution's (WHOI) Marine Mammal Center. The funders had no role in study design, data collection and analysis, decision to publish, or preparation of the manuscript.

Competing Interests: The authors have declared that no competing interests exist.

* E-mail: apprill@whoi.edu (AA); tmincer@whoi.edu (TJM)

Introduction

It is now recognized that microbes play an important role in the health of mammals, especially humans [1,2]. This theme has led to the emergence of comparative studies targeting other disparate mammalian species, aiming to understand the connections between microbiota and exogenous (environmental) and endogenous (host-associated) influences [3]. The largest mammalian organ in direct contact with the environment is the skin, and the skin of marine mammals is a particularly interesting surface because it is in constant contact with seawater microorganisms, which are generally orders of magnitude more populous than airborne microorganisms surrounding terrestrial mammals [4,5]. Additionally, seawater microorganisms can exhibit population shifts related to altered environmental conditions [6,7] that have the potential to influence the composition of microbes residing on marine mammal skin.

Recently, the skin of humpback whales (*Megaptera novaeangliae*) were shown to associate with diverse communities of bacteria that differ from the communities present in the seawater [8]. Further, specific groups of bacteria were associated with most individuals studied, regardless of whale age or sex [8]. These findings suggest that humpback whales may harbor specific skin-bacterial associates. However, documenting if and how skin bacterial communities vary between humpbacks from different populations and geographic areas is necessary to understand the extent of this host-bacterial specificity. Importantly, documenting the presence of a core, or central, microbial community across all members of a population of marine mammals may provide an indication of a relationship that is fundamental to the host [9,10].

The skin microbiome of humpback whales is particularly attractive for cross-population studies. This species is found in all oceans, and the structure, dynamics and ecology of many populations have been well-studied [11,12]. Humpback whales migrate annually from mid- to high latitude feeding grounds to low latitude breeding and calving grounds [13–17], and some individuals hold the most extensive mammalian migration on record [18]. The ubiquity and migratory nature of this species provides an opportunity to study the impact of various oceanic conditions on the composition of the skin microbiota. Furthermore, humpback whales fast while on migration and throughout the breeding season, and so there is also an opportunity to evaluate how routine metabolic stress may affect microbial communities. Lastly, this species is vulnerable to human impacts in many parts of its range, and the effects of those impacts can be cryptic. Understanding the relation of the skin microbiome to health and geography may enhance monitoring and conservation efforts of populations.

The idea that humpback whale skin harbors specific bacterial associates may have larger implications for health or health monitoring of these cryptic animals. At first mention, it might seem nonsensical that the surface epidermis, a layer of dead skin, has any connection to health. However, in humans there is evidence suggesting links between the composition of the skin-bacteria and inflammatory, allergic conditions [19]. Also, shifts in skin-bacteria together with an altered host immune response may lead to the development of skin diseases or disorders [20]. There has been much less attention towards the study of marine mammal skin, microbes and health. Historically, microbes have been observed on the skin of odontocetes, and antibiotic production by these cells was thought to prevent colonization of the skin by pathogens [21]. Additionally, captive cetaceans exposed to water purified with oxidizing agents (chlorine or ozone - aimed to reduce microbial abundances in the seawater) resulted in the animals developing skin infections that may have been linked to a lack of beneficial microbes on their skin [22]. It is clear that an understanding of skin-bacteria on normal, healthy marine mammals is necessary before any connections about health can be explored. Importantly, this framework will be useful to recognize if and how skin-bacteria differ on unhealthy animals, including the growing number of marine mammals displaying skin lesions and disorders [23].

The goal of this study was to characterize the skin-associated bacteria from humpback whales, and explore whether animals from diverse geographic areas and populations share a core, or central, bacterial community. Additionally, this investigation examined the extent that geographic area and host-specific factors (population, age, sex and metabolic state) may contribute to variability within the skin-bacteria. Lastly, the diversity and composition of skin-bacteria from the apparently healthy animals was compared to entangled and recently deceased individuals (with potentially reduced or extinct immune responses), and used to examine the potential for skin-bacteria to provide any clues about a compromised health state.

Materials and Methods

Ethics Statement

Collection of skin samples was conducted under NOAA permits #633-1778, 633-1778, 932-1489, 774-1714, 1071-1770-00, 1000-1617, and the approval of the Government of American Samoa. All samples were collected according to permit guidelines.

Samples

Skin samples were collected from two geographic areas in the North Pacific: one feeding ground (Southeast Alaska, summer 2009) and the breeding ground (Hawaiian Islands, winter 2007 and 2008) that those individuals share with others in the North Pacific. Samples were also collected from a South Pacific breeding area (American Samoa, winter 2009) and a North Atlantic feeding area (the Gulf of Maine, summer 2009) (Table S1). Skin was obtained by biopsy sampling techniques on the upper flank near the animal's dorsal fin [24]. Additionally, humpback whales shed pieces of skin from during high-energy surface behaviors [25] and these can originate from any skin surface on the body. Freshly dislodged samples were collected using nets and sieves. Marine mammals have been shown to shed the surface epidermis regularly [22], and therefore sloughed skin is likely only slightly older in age than the biopsied skin. Additionally, some samples were obtained from the suction cup used to attach acoustical tags to whales (Table S1). Skin samples were handled using sterile tools, with the exception of sloughed skin nets and sieves which were only rinsed with seawater between uses. However, skin was in contact with these nets and sieves for a very short amount of time (seconds). Skin was kept on ice for no more than 10 hours before freezing. Samples of seawater were also obtained with the Hawaii samples, and collection methods outlined previously [8]. We also requested samples from stranding and entanglement response networks to compare free-swimming whales to individuals that were potentially health-impaired. As detailed in Table S2, the samples available to this study included two live whales entangled in fishing gear and three whales sampled after death.

Data on whale age, sex, metabolic state and health state were made at the time of sampling or obtained from long-term population monitoring programs. Sex was known from molecular genetic analysis [26], visual observation of the genital slit [27], the presence of a dependent calf (for females) or sex-linked stereotypical behaviors on the breeding grounds. Exact age was known for individuals first documented as dependent calves. Individuals were considered independent juveniles from ages 1–4 and adults from 5 years onward. When exact age was not known, age class was inferred based on its minimum age or visually assessed from size and behavior. Metabolic state was determined by geographic area (for non-nursing calves), with animals in their breeding and feeding grounds displaying catabolic and anabolic metabolisms, respectively.

Pyrosequencing of V1–V3 regions of bacterial SSU rRNA genes

Nucleic acids were extracted from 2–70 mg (generally 25 mg) of skin or 220–240 ml of surface seawater (<0.5 m) using previously described methods [8]. DNA was quantified using the PicoGreen fluorescent assay (Invitrogen Corp., Carlsbad, CA, USA) on a SpectraMax M2 plate reader (Molecular Devices Corp., Sunnyvale, CA, USA). Barcoded primers targeting the V1–V3 regions of the SSU rRNA genes, 27FB and 519R, were utilized for pyrosequencing analysis (primers and barcodes described in [28]). Triplicate 20 µl PCR reactions contained 2.5 U of Pwo SuperYield DNA polymerase (Roche Applied Science, Indianapolis, IN USA), 1 × Pwo SuperYield buffer, 200 µM of each dNTPs, 200 nM of each barcoded fusion primer, and 4–12 ng of genomic template. After an initial denaturation step at 95°C for 3 min, the reaction conditions were: 30–35 cycles of 95°C for 15 s, 55°C for 30 s, and 72°C for 2 min, concluding with an extension at 72°C for 7 min. The reactions were carried out in an Eppendorf Mastercycler (Eppendorf, San Diego, CA, USA). A subset of each reaction (14 µl) was run onto a 1% agarose/TBE gel, and the

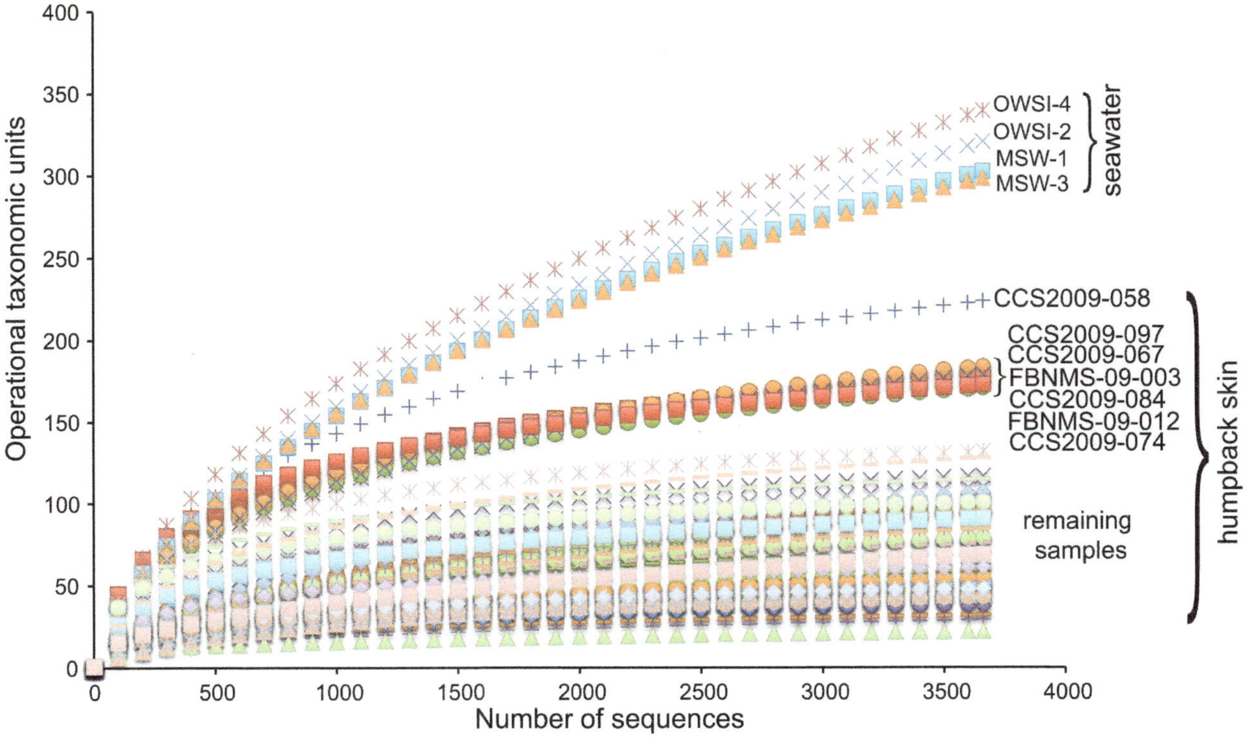

Figure 1. Rarefaction analysis of bacterial SSU rRNA gene sequences. This analysis represents data subsampled at 3,662 sequences per sample. Samples with OTU richness greater than 130 are listed by name.

replicate reactions were combined, purified using the Qiagen MinElute Gel Extraction kit (Qiagen Inc.), and quantified using the PicoGreen fluorescent assay and a Bio-Rad CXF96 Real-Time system (Bio-Rad Laboratories, Hercules, CA, USA). Barcoded amplicons were combined in two separate equimolar ratio libraries and shipped to the University of Illinois W.M. Keck Center for Comparative and Functional Genomics for sequencing. Each library was sequenced on one-half of a picotiter plate using a 454 Life Sciences Genome Sequencer FLX instrument with Titanium series chemistry (Roche, Branford, CT, USA). Sequencing was conducted from primer B using an amplicon protocol.

Sequence processing

Processing of SFF files was conducted using Mothur [29]. After barcode and primer removal, 754,593 sequences were resolved from the combined libraries with an average length of 498 bp. Sequences were then aligned to the 16S rRNA molecule using the Silva database alignment template [30], trimmed at similar molecule locations and pre-clustered to eliminate outliers. After removal of sequencing and PCR-related noise and quality trimming [31], the total number of sequences was reduced to 536,055 with an average read length of 261 bp. An additional 10,768 sequences were found to be chimeric and removed. In order to identify non-bacterial sequences, the data were classified using first the RDP [32] and secondly using the Silva taxonomy training sets with a Bayesian classifier. Sequences identified as chloroplasts (1,885) were subsequently removed from the dataset. A distance matrix was constructed of the sequences, and sequences were clustered into operational taxonomic units (OTUs) using the average neighbor algorithm at a 97% similarity level. In order to verify the taxonomy of sequences, representative sequences from unclassified OTUs (~2000 sequences) were aligned using the SINA web aligner and imported into a Silva 106 non-redundant database using the ARB software [33]. After aligning the

Table 1. Observed OTUs (97% similarity grouped) and diversity indices from 3,662 sequences for the different sample categories (data for normal animals only).

Samples (samples, individuals)	OTUs (St. Dev.)	OTU Range	Shannon Diversity (H′) (St. Dev.)	Simpson Diversity (1−λ′) (St. Dev.)	Pielou's evenness (J′) (St. Dev.)
Seawater (4 s.)	316 (19)	299–340	3.021 (0.27)	0.805 (0.06)	0.550 (0.004)
Gulf of Maine (21 s., 19 ind.)	99 (56)	32–224	2.638 (0.64)	0.850 (0.10)	0.592 (0.09)
Hawaii (18 s., 14 ind.)	71 (33)	22–132	2.192 (0.95)	0.717 (0.24)	0.521 (0.193)
American Samoa (13 s., 12 ind.)	90 (46)	19–179	2.384 (0.74)	0.805 (0.11)	0.544 (0.11)
Alaska (5 s., 3 ind.)	72 (19)	51–86	2.058 (0.67)	0.717 (0.24)	0.471 (0.14)

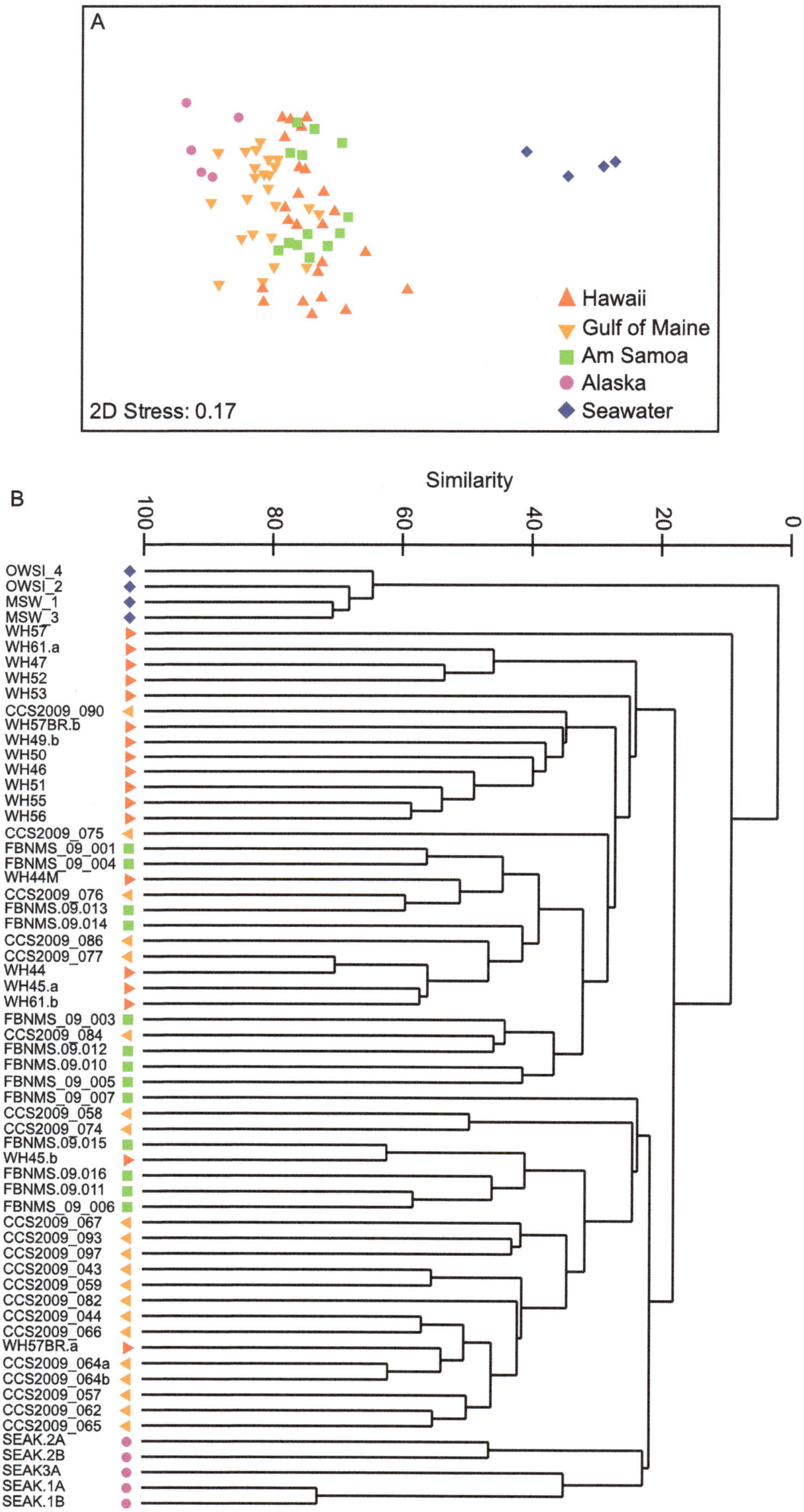
A
Hawaii
Gulf of Maine
Am Samoa
Alaska
Seawater
2D Stress: 0.17
B
Similarity
100
80
60
40
20
0
OWSI_4
OWSI_2
MSW_1
MSW_3
WH57
WH61.a
WH47
WH52
WH53
CCS2009_090
WH57BR.b
WH49.b
WH50
WH46
WH51
WH55
WH56
CCS2009_075
FBNMS_09_001
FBNMS_09_004
WH44M
CCS2009_076
FBNMS.09.013
FBNMS.09.014
CCS2009_086
CCS2009_077
WH44
WH45.a
WH61.b
FBNMS_09_003
CCS2009_084
FBNMS.09.012
FBNMS.09.010
FBNMS_09_005
FBNMS_09_007
CCS2009_058
CCS2009_074
FBNMS.09.015
WH45.b
FBNMS.09.016
FBNMS.09.011
FBNMS_09_006
CCS2009_067
CCS2009_093
CCS2009_097
CCS2009_043
CCS2009_059
CCS2009_082
CCS2009_044
CCS2009_066
WH57BR.a
CCS2009_064a
CCS2009_064b
CCS2009_057
CCS2009_062
CCS2009_065
SEAK.2A
SEAK.2B
SEAK3A
SEAK.1A
SEAK.1B

Figure 2. Similarity analysis of whale skin-bacteria. (A) Clustering dendogram, and (B) nonmetric multidimensional scaling analysis of 97% similarity grouped OTUs of bacterial communities associated with whale skin sampled from different populations and locations, and seawater.

sequences in the database, sequences (152) further identified as chloroplasts or mitochondria were removed, and sequences belonging to a number of candidate groups that were not included in the training set were re-classified. The finalized dataset included 522,954 sequences with an average length of 261 bp, grouped into 9,088 OTUs. In order to compare the alpha and beta diversity of samples of equal sample size, the samples were randomly sub-sampled to a depth of 3,663 sequences, which corresponded to the lowest number of sequences recovered in the dataset for an individual sample. All analyses were conducted on the sub-sampled sequence set. Raw sequence data were archived under NCBI Sequence Read Archive BioSample accessions SAMN02566578–SAMN02566647.

Phylogenetic trees were constructed separately for sequences classifying to the *Tenacibaculum* and *Psychrobacter* genera, and included the most dominant sequences from this study as well as reference sequences of recognized species from each genus and previously published environmental gene clones of high similarity. Because of the short nature of the sequences from this study, the trees were first constructed using the RAxML maximum likelihood method [34] with nearly full-length reference sequences, with rapid bootstrap analysis on 1,000 runs. The amplicon sequences obtained from this study, as well as short sequences from the bottlenose dolphin study [35] were then added to the tree without changing the topology using the ARB parsimony interactive method. All analyses were done using the ARB software [33].

The diversity of closely related organisms within the genera *Tenacibaculum* and *Psychrobacter* was further investigated using oligotyping, [36] a supervised computational method that allows the identification of closely related bacterial organisms that are grouped into one taxon [37,38]. Oligotyping analyses were performed on 148,974 *Tenacibaculum* and 140,729 *Psychrobacter* reads using the oligotyping pipeline version 0.9 (available from http://oligotyping.org). To reduce the impact of noise, minimum substantive abundance (*M*) of an oligotype was set to 50 for both runs, as a result of which, any oligotype with most abundant unique sequence with less than 50 reads was removed from the final results. Oligotyping analysis of *Tenacibaculum* and *Psychrobacter* spp. reads resulted in 98 and 53 oligotypes, respectively. Noise removal with *M* discarded 2.00% and 0.73% of *Tenacibaculum* and *Psychrobacter* spp. reads, respectively.

Statistical analyses

For testing differences in the bacterial community richness between samples, non-parametric comparison tests were conducted using the software Minitab version 13 (Cleverbridge, Inc, Chicago, IL). For assessing beta diversity of the bacterial community among sample groupings, a relative abundance matrix of the 97% similarity grouped OTUs were square root transformed and further analyzed using the PRIMER 6 version 6.1.13 software (PRIMER-E Ltd., Plymouth, UK) [39] using statistical tests and parameters designed for comparison of ecological datasets [40]. A distance matrix was constructed using Bray-Curtis similarity, and both nonmetric multidimensional scaling (NMS) ordination (2-dimensional representation of bacterial community structure) and hierarchical clustering analysis (CLUSTER) (similarity dendrogram) were used to explore groupings of the samples without pre-defined sample classes. PERMANOVA tests were used to test for significant differences in bacterial community composition between pre-defined sample

Table 2. PERMANOVA analysis of the effect of geographic area, metabolic state, age class, sex and sample type on the 97% OTU grouped bacterial SSU rDNA data from humpback skin (GOM = Gulf of Maine).

Data	Variation	d.f.	SS	Pseudo-F^f or t-value^t	p-value
Normal whales	Location (Hawaii, GOM, Alaska, Am Samoa)	3	20279	2.9121^f	0.001***
Hawaii vs. GOM				1.5052^t	0.006**
Hawaii vs. Am Samoa				1.2228^t	0.073
Hawaii vs. Alaska				1.4257^t	0.004*
GOM vs. Am Samoa				1.7878^t	0.001***
Am Samoa vs. Alaska				2.1007^t	0.001***
Normal whales	Sample type (biopsy, sloughed, tag)	2	15231	3.2809^f	0.001***
Biopsy vs. sloughed				2.2608^t	0.001***
Biopsy vs. tag				1.1763^t	0.123
Sloughed vs. tag				1.1683^t	0.150
Normal whales	Age (calf, juvenile, adult)	2	7217.8	1.4042^f	0.065
Calves vs. juveniles				1.0371^t	0.342
Calves vs. adults				1.1616^t	0.141
Juveniles vs. adults				1.3665^t	0.028*
Normal whales	Metabolic state (anabolism, catabolism)	1	12670	3.6307^f	0.001***
Normal whales	Sex (male, female)	1	2444.1	0.90575^f	0.513

***$p \leq 0.001$,
**$p \leq 0.01$,
*$p \leq 0.05$.

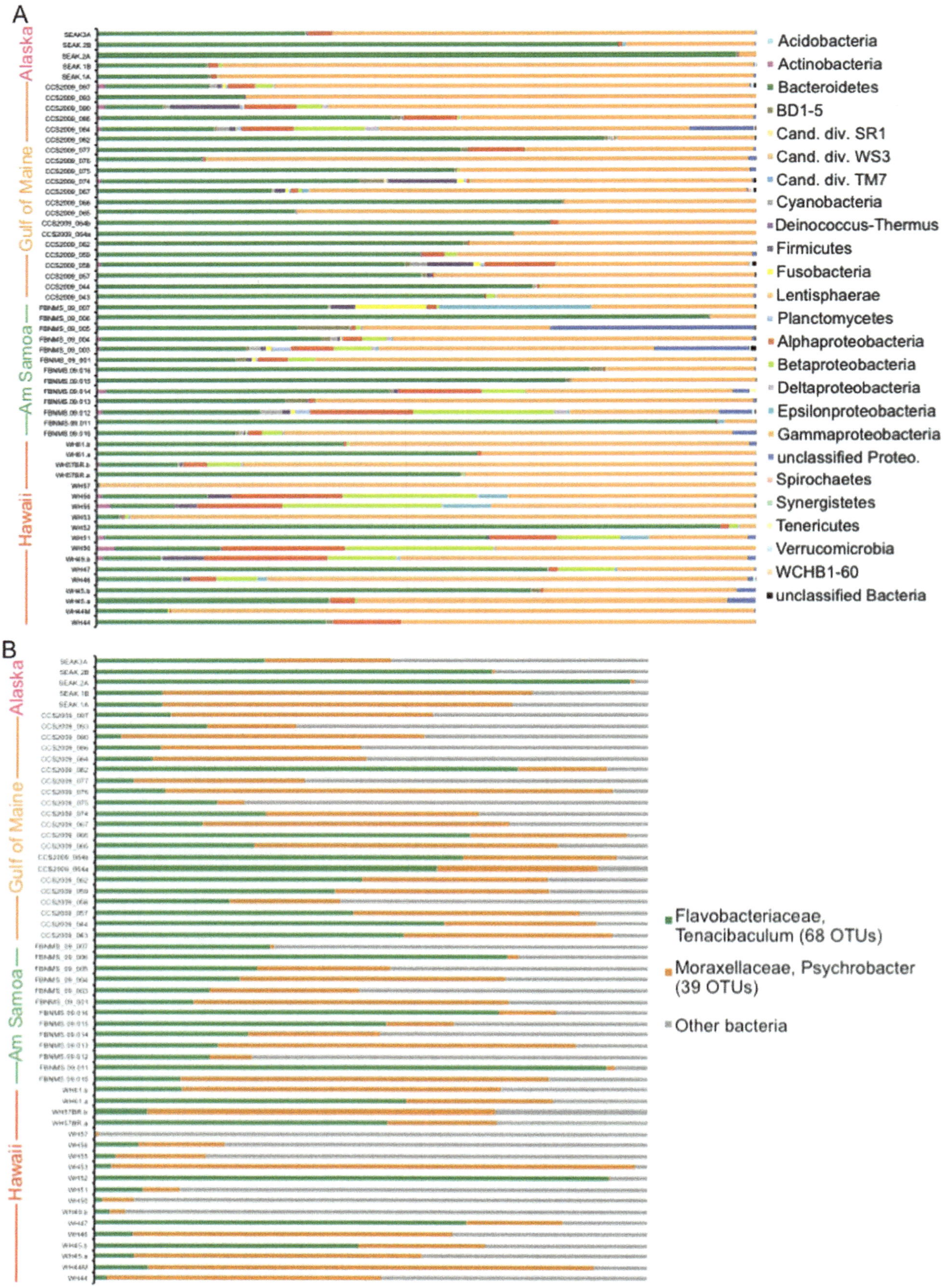
A
Alaska
Gulf of Maine
Am Samoa
Hawaii
Acidobacteria
Actinobacteria
Bacteroidetes
BD1-5
Cand. div. SR1
Cand. div. WS3
Cand. div. TM7
Cyanobacteria
Deinococcus-Thermus
Firmicutes
Fusobacteria
Lentisphaerae
Planctomycetes
Alphaproteobacteria
Betaproteobacteria
Deltaproteobacteria
Epsilonproteobacteria
Gammaproteobacteria
unclassified Proteo.
Spirochaetes
Synergistetes
Tenericutes
Verrucomicrobia
WCHB1-60
unclassified Bacteria
B
Alaska
Gulf of Maine
Am Samoa
Hawaii
Flavobacteriaceae, Tenacibaculum (68 OTUs)
Moraxellaceae, Psychrobacter (39 OTUs)
Other bacteria

Figure 3. Taxonomic composition whale skin-bacteria. The 97% similarity grouped OTU bacterial communities were associated with humpback whale skin sampled from different populations and locations. Graphs represent total composition (A), and abundances of skin-bacterial sequences specifically belonging to the *Tenacibaculum* and *Psychrobacter* lineages (B). Each bar graph represents the relative abundance of each major taxonomic group, and sums to 100%.

groups, e.g., whales sampled in different geographic areas, and were conducted using a Bray-Curtis similarity matrix of relative abundance data with 9999 permutations under Type II sum of squares and a reduced model. Significance levels were confirmed using Monte Carlo simulations. When significant differences were identified between sample categories, similarity percentages test (SIMPER) was preformed to assess each OTU's contribution to a pre-defined sample group. Multidimensional scaling analysis on samples using *Tenacibaculum* and *Psychrobacter* spp. oligotypes were performed using Morisita-Horn distance measure.

Scanning electron microscopy

Skin biopsy samples collected from seven individuals in the Gulf of Maine during 2010 and 2011 (Table S1) were preserved in a 2% glutaraldehyde and 2% formalin solution in phosphate buffered saline (PBS) and stored at 4°C. In preparation for microscopic analysis, ~30 mg samples containing the skin surface and several mm below were rinsed at room temperature 3× in PBS solution for 20 min, followed by 10 min in a 50% ethanol and PBS solution, and stored overnight in a 70% ethanol and PBS solution. The following morning, samples were rinsed for 10 min each in 85% ethanol and PBS and 95% ethanol and PBS. Samples were subsequently rinsed 3× for 15 min each in ethanol. Samples were then dried to the critical point using a Samdri 780A (Tousimis Research Corporation, Rockville MD, USA) and sputter-coated with a 10 nm coat using a Leica EM QSG100 Modular High Vacuum Coating System (Leica Microsystems, Buffalo Grove, IL, USA). Samples were viewed on a JEOL 840 scanning electron microscope (JEOL, Peabody, MA, USA) with representative images captured.

Table 3. Abundances of whale skin-bacterial sequences corresponding to the *Tenacibaculum* and *Psychrobacter* genera.

Genus	Range	Mean (St. dev)
Alaska		
Tenacibaculum	12–97%	44% (38)
Psychrobacter	0.4–67%	31% (32)
Tenacibaculum+Psychrobacter	53–98%	75% (16)
Gulf of Maine		
Tenacibaculum	5–76%	35% (23)
Psychrobacter	5–81%	36% (17)
Tenacibaculum+Psychrobacter	27–96%	71% (23)
American Samoa		
Tenacibaculum	15–92%	39% (26)
Psychrobacter	0.7–67%	27% (24)
Tenacibaculum+Psychrobacter	28–94%	65% (21)
Hawaii		
Tenacibaculum	0.1–93%	22% (28)
Psychrobacter	0.3–95%	33% (29)
Tenacibaculum+Psychrobacter	0.8–98%	55% (33)

Results

Richness of whale skin-associated bacteria is similar across geographic areas

SSU rRNA gene pyrosequencing data were used to examine the skin-bacteria from the normal (apparently healthy) whales. Rarefaction analysis of the 3,662 SSU rRNA gene sequences per sample revealed that the majority of the whale-skin bacterial communities reached a plateau at <100 OTUs (Figure 1). The average observed number of bacterial OTUs associated with whale skin varied among 71–99 with some variation among samples (Table 1). There was no significant difference in the richness of the skin-bacterial communities (measured by OTU abundance) among geographic areas (Mood's Median Test, chi-square = 2.92, df = 3, p = 0.404), and other measurements of diversity and evenness were also comparable (Table 1). OTU abundance from all of the humpback samples was significantly less than seawater (Mann Whitney Test, W = 2346, p<0.01, Table 1, Figure 1), suggesting that factors shaping the skin and seawater microbial communities are distinct.

Composition of skin-bacterial community in healthy individuals is primarily related to geographic area, metabolic state and sample type

Nonmetric multidimensional scaling analysis comparisons of the 97% similarity grouped OTUs revealed that skin-bacteria generally grouped by geographic area, and were highly similar in composition compared to the surface seawater samples (Figure 2a). Clustering analysis of the OTUs further revealed that the similarity between the whale skin-bacterial samples was broad, ranging from 10–65% (Figure 2b). Skin taken from whales in Hawaii, Gulf of Maine and Alaska each generally comprised a distinct cluster (Figure 2b). PERMANOVA with pair-wise tests revealed that skin-bacteria between all sampling locations were significantly distinct, except for American Samoa and Hawaii (Table 2).

Phylogenetic identifications of the bacterial SSU rDNA OTUs demonstrated that 23 major taxonomic groups associated with the skin of healthy humpback whales (Figure 3a). The majority of whale skin samples were associated with bacteria belonging to the *Bacteroidetes* and *Gammaproteobacteria*, and many skin samples also contained bacteria belonging to the *Firmicutes* and *Alphaproteobacteria*. Two genera, *Tenacibaculum* (*Bacteroidetes*) and *Psychrobacter* (*Gammaproteobacteria*) were found to be abundant within the majority of whale skin samples (Figure 3b), and collectively made up a large portion, 55–75%, of the skin-bacterial community (Table 3). SIMPER analysis showed that dissimilarities in the abundance of the different *Tenacibaculum* and *Psychrobacter* spp. OTUs, as well as several other bacterial groups, contributed to the differences in the skin-bacteria of whales from different geographic areas (Table S2). SIMPER analysis also demonstrated that the locations with more similar skin-bacteria (American Samoa and Hawaii) harbored sequences affiliated with *Cardiobacteria* and *Aquabacterium* spp., which were not present in the results of samples from the other locations (Table S2). Animals sampled in American

Table 4. Results of SIMPER analysis describing 97% similarity grouped OTUs contributing to the differences between the whales with anabolic and catabolic metabolisms (>1% contribution only).

OTU	Av. abundance anabolic whales	Av. abundance catabolic whales	Average Diss.	Diss. S.D.	Contribution to differences (%)	Taxonomy
2	9.14	14.22	8.11	0.86	9.39	*Gammaproteobacteria, Psychrobacter*
3	10.61	4.31	5.78	0.90	6.69	*Bacteroidetes, Tenacibaculum*
1	4.39	7.44	5.37	0.49	6.21	*Bacteroidetes, Tenacibaculum*
4	4.23	5.83	4.10	0.63	4.75	*Bacteroidetes, Flavobacteriaceae*, uncultured humpback group
8	6.24	0.02	3.12	0.37	3.61	*Bacteroidetes, Tenacibaculum*
12	2.88	5.13	3.08	0.55	3.57	*Bacteroidetes, Tenacibaculum*
10	4.05	2.34	2.70	0.56	3.13	*Bacteroidetes, Tenacibaculum*
27	1.49	5.06	2.65	0.75	3.06	*Bacteroidetes, Tenacibaculum*
9	3.10	3.56	2.59	0.73	2.99	*Gammaproteobacteria, Moraxellaceae*
7	1.65	4.19	2.35	0.64	2.72	*Gammaproteobacteria, Psychrobacter*
19	4.26	0.16	2.10	0.57	2.43	*Gammaproteobacteria, Psychrobacter*
11	1.70	3.24	2.07	0.47	2.40	*Gammaproteobacteria, Psychrobacter*
21	3.61	0.01	1.81	0.40	2.09	*Gammaproteobacteria, Psychrobacter*
14	0.96	2.85	1.81	0.34	2.09	*Gammaproteobacteria, Psychrobacter*
15	0.70	3.11	1.61	0.71	1.87	*Betaproteobacteria, Aquabacterium*
28	3.12	0.34	1.60	0.74	1.85	*Gammaproteobacteria, Psychrobacter*
5	2.26	1.60	1.53	0.41	1.78	*Gammaproteobacteria, Cardiobacteria*
24	0.00	2.87	1.43	0.19	1.66	*Gammaproteobacteria, Acinetobacter*
29	2.81	0.30	1.38	0.61	1.60	*Bacteroidetes, Tenacibaculum*
41	1.44	1.47	1.29	0.50	1.50	*Bacteroidetes, Flavobacteriaceae*
22	0.51	2.00	1.17	0.39	1.35	*Proteobacteria*, Incertae Sedis (humpback-specific)
37	0.39	1.84	1.02	0.43	1.18	*Bacteroidetes, Flavobacteriaceae*
20	1.30	1.10	0.97	0.67	1.12	*Alphaproteobacteria*, Roseobacter clade AS-21
33	1.88	0.08	0.93	0.60	1.08	*Gammaproteobacteria, Colwellia*

Samoa and Hawaii were not actively feeding (except for one nursing calf which were not included in the analysis), and therefore their metabolism is catabolic compared to the anabolic (feeding) metabolism of whales in Alaska and the Gulf of Maine. This metabolic difference was found to correspond significantly to the composition of the skin-bacteria, regardless of geographic area (Table 2). The differences in skin-bacteria related to metabolic state were largely attributed to different sequence variants of *Tenacibaculum* and *Psychrobacter* spp. (in addition to some other taxa, Table 4).

Significant differences were also evident between the bacterial communities on sloughed skin versus biopsy samples (Table 2), which may be related to the anatomical locations from which the samples originated. Biopsies were generally focused on the flank whereas sloughed skin was from unknown locations on the animal. With the exception of dependent calves, samples from juveniles harbored significantly different bacteria from adults (Table 2). No relationship was found between whale sex and skin-bacteria.

Tenacibaculum and *Psychrobacter* spp. are core members of the humpback skin bacterial community, but exhibit regional-specific sequence variations

Sequences affiliated with the genus *Tenacibaculum* accounted for average abundances of 44, 35, 39 and 22% of the whale-skin communities from Alaska, Gulf of Maine, American Samoa and Hawaii, respectively (Table 3). In total, 44 OTUs (at 97% similarity grouping; represented in two or more samples) comprised this group, with the most cosmopolitan OTU present in 53 of the 57 samples from different regions (OTU 12, Figure 4A). Five of the OTUs (27, 41, 436, 257 and 403) were present in all samples except those from Alaska. A phylogenetic analysis demonstrated that the sequences from these OTUs clustered into several distinct lineages, and are related to sequences recovered from the blowholes of bottlenose dolphins [35] (Figure 5).

Oligotyping analysis was used to examine more subtle variations with the SSU rRNA gene of the *Tenacibaculum* spp. sequences. This analysis identified 98 *Tenacibaculum* spp. oligotypes in the whales, which differed significantly by region, but with the most overlap between samples from Hawaii and American Samoa (Figure 6A). Similar to the trends for the entire community, PERMANOVA analysis revealed that the *Tenacibaculum* spp. oligotypes differed in composition with the distinct metabolic states of the whales (Table 5). Additionally, the *Tenacibaculum* spp. oligotypes differed by sample type (biopsy or sloughed skin), suggesting that skin age and/or anatomical location of the skin plays a role in shaping these communities (Table 5).

Sequences within the *Psychrobacter* genus were present at average abundances of 31, 36, 27 and 33% of the skin bacterial

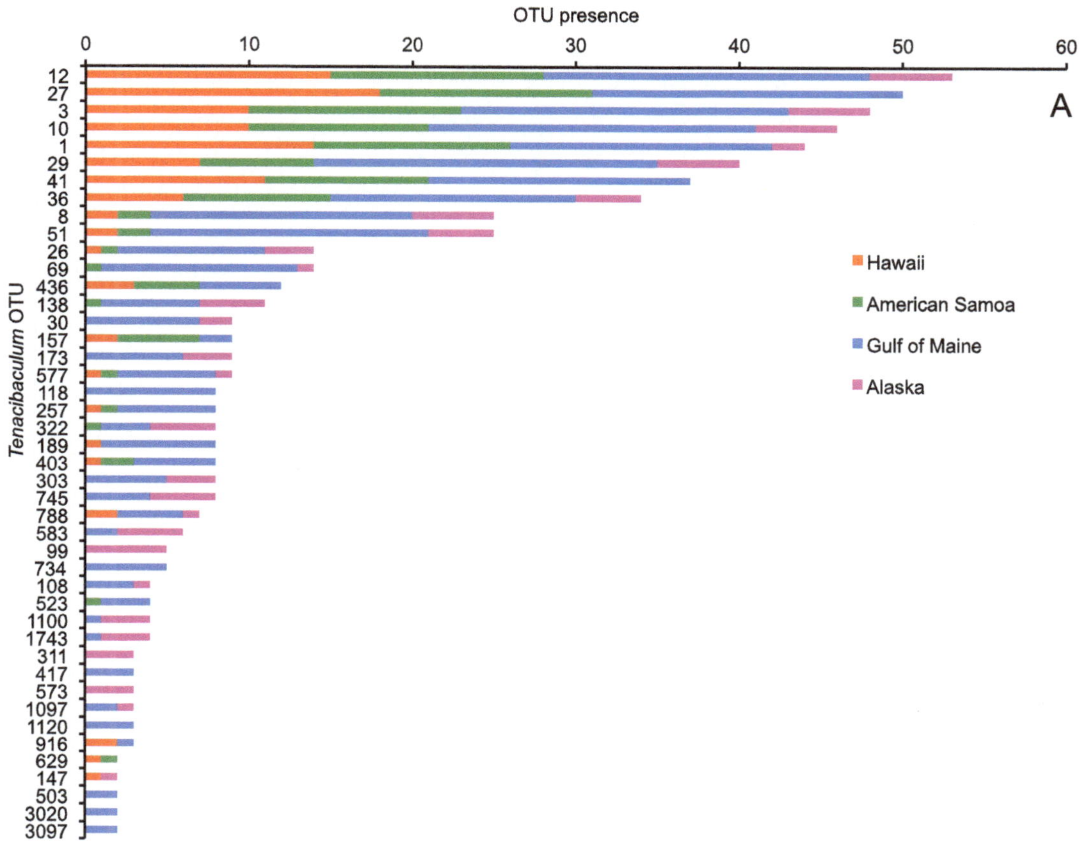
OTU presence
0
10
20
30
40
50
60
A
Tenacibaculum OTU
12
27
3
10
1
29
41
36
8
51
26
69
436
138
30
157
173
577
118
257
322
189
403
303
745
788
583
99
734
108
523
1100
1743
311
417
573
1097
1120
916
629
147
503
3020
3097
Hawaii
American Samoa
Gulf of Maine
Alaska

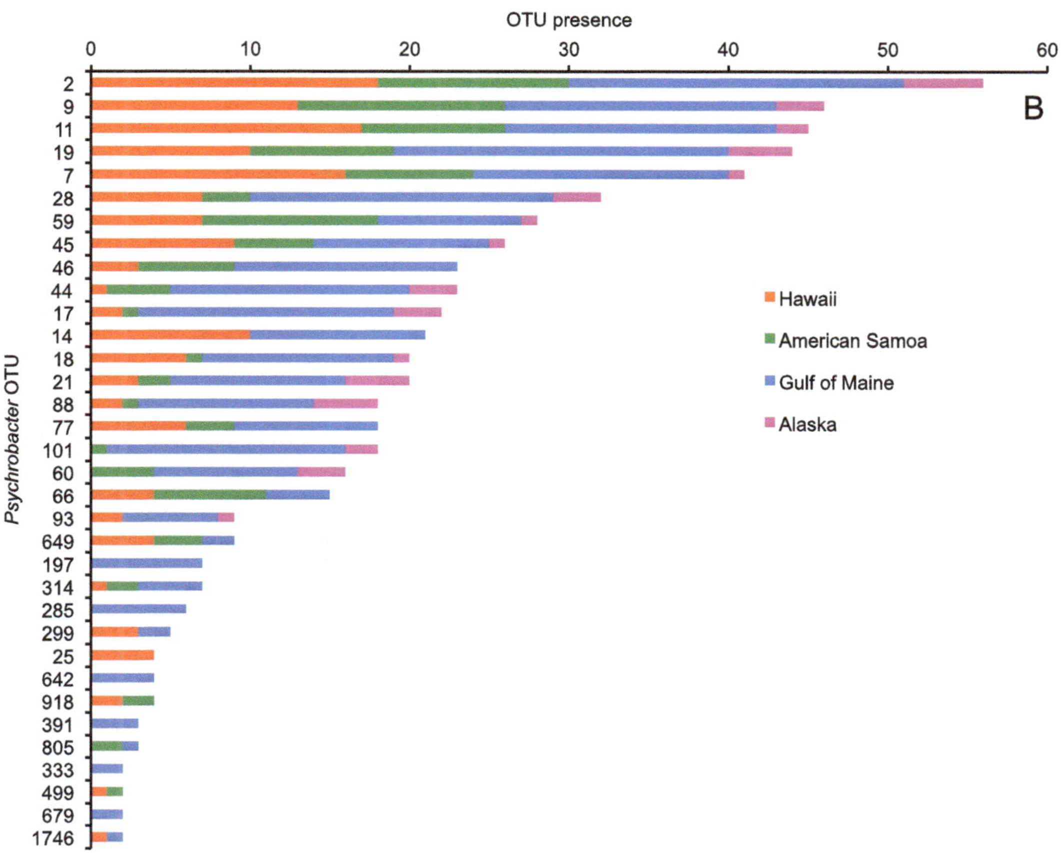
OTU presence
0
10
20
30
40
50
60
B
Psychrobacter OTU
2
9
11
19
7
28
59
45
46
44
17
14
18
21
88
77
101
60
66
93
649
197
314
285
299
25
642
918
391
805
333
499
679
1746
Hawaii
American Samoa
Gulf of Maine
Alaska

Figure 4. Abundances of *Tenacibaculum* and *Psychrobacter*-affiliated sequences. Presence of skin-bacterial OTUs (97% similarity grouped) classified as *Tenacibaculum* (A) and *Psychrobacter* (B), in relation to the geographic region of the whales.

communities from Alaska, Gulf of Maine, American Samoa and Hawaii, respectively (Table 3). There were 34 OTUs within the *Psychrobacter* (at 97% similarity grouping; represented in two or more samples), and six OTUs (46, 14, 77, 66, 649 and 314) were present in all samples except those obtained from Alaska (Figure 4B). The *Psychrobacter* spp. sequences grouped into several distinct phylogenetic lineages, and were also related to sequences recovered from dolphin blow (Figure 7). Oligotyping analysis of the sequence variations within the *Psychrobacter* spp. OTUs revealed 53 distinct sequence types, and these were found to differ significantly by geographic location as well as metabolic state (Table 5). As seen for *Tenacibaculum* spp., the *Psychrobacter* spp. oligotypes were most similar for whales sampled in American Samoa and Hawaii (Figure 6B).

Microscopic visualization of skin-bacteria

In a subset of skin samples (n = 7), microbial sized cells were identified on the skin surface using scanning electron microscopy. Variations within the density and morphologies of cells did vary between samples (Figure 8). Additionally, differences in the 'smoothness' of the skin, including presence of organic-matter type flocking material as well as brittleness of the skin was also evident between samples. In some samples, phytoplankton-sized cells including diatoms were visible and appeared to harbor their own smaller microbial associates (Figure 8). These results suggest that these microbes were viable and actively adhered to the skin surface, and were at least able to stay attached during the extensive sample handling and processing.

Figure 5. Phylogenetic relationship of *Tenacibaculum*-affiliated sequences. Represented SSU rRNA gene sequences include *Tenacibaculum*-affiliated sequences recovered from whale skin, cultivated *Tenacibaculum* isolates, and relatives within the *Tenacibaculum* lineage. Red sequences are short-reads from this study and blue are full-length sequences from a previous study on humpback whale skin [8]. The scale bar corresponds to 0.10 substitutions per nucleotide position, and only bootstrap values >70 are listed. Sequences from *Caldilinea aerophila* (AB067647), *Roseiflexus castenholzii* (CP000804) and *Actinomyces oris* (GQ421308), were used to form the outgroup. Full accession numbers for the dolphin respiratory-bacteria sequences are available in [35].

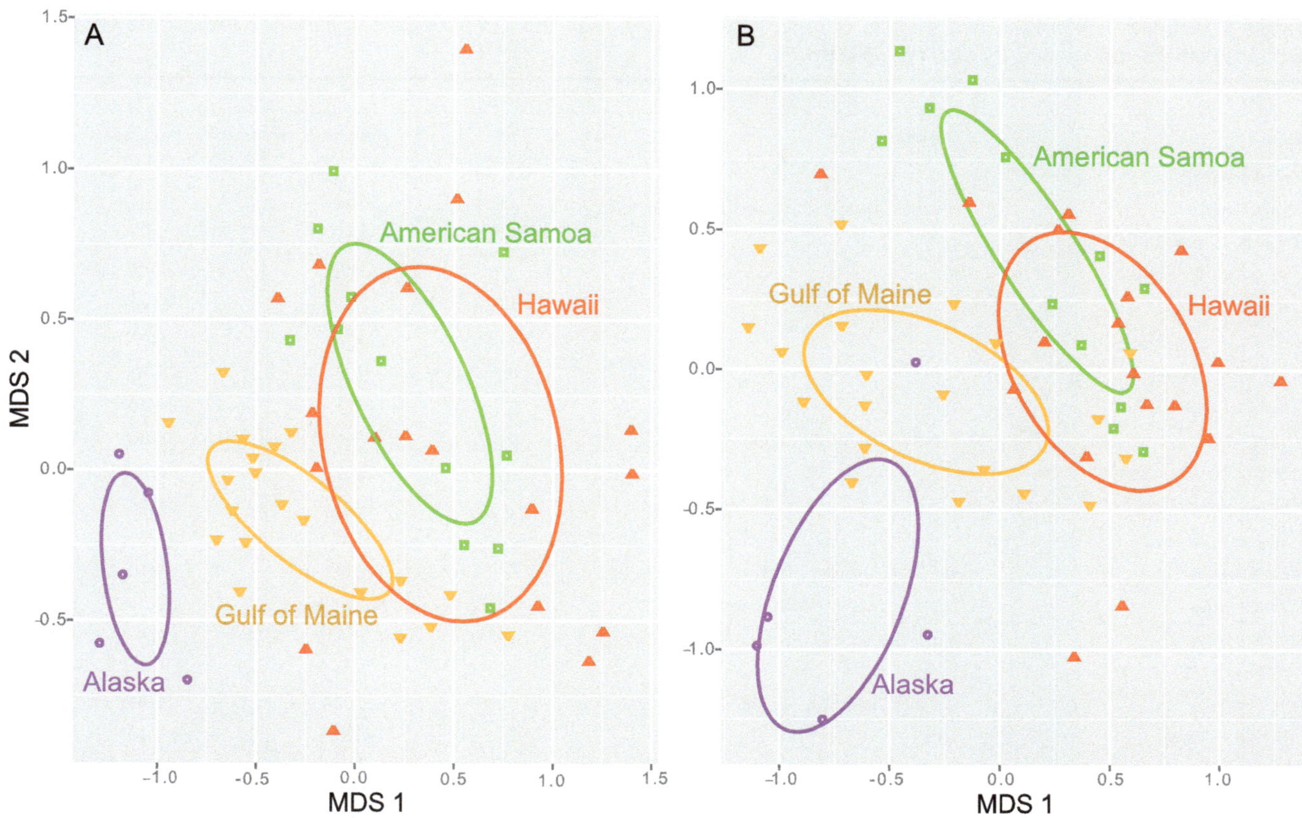

Figure 6. Oligotyping analysis of major humpback skin-bacterial groups. MDS analysis of the healthy animals from four regional groups based on *Tenacibaculum* (A) and *Psychrobacter* (B) oligotypes. Centroid and shape of each ellipse is defined by the distribution and standard deviation of points in the defined regional groups.

Using the core bacterial community concept to develop a health-index for whale skin

The bacterial communities associated with the normal individuals were compared to determine indicator criteria for normal, 'healthy' skin, including: OTU richness of the entire community and abundances of the main core bacterial community members (*Tenacibaculum* spp.+*Psychrobacter* spp.) (Table 6). These criteria were compared to samples obtained from a small set of entangled and/or dead individuals (Table S2). In the latter animals, OTU richness and the percentage of the core bacterial community sequences fell outside of the range of the normal animals, but there were no consistent trends. Several entangled and dead individuals also harbored pathogenic-type bacteria, including members of the *Clostridium* spp., *Lachnospiraceae* spp., and *Staphylococcus* spp. and the predatory bacteria, *Bdellovibrio* spp., that were not detected in the skin samples from free-swimming animals (Table 6).

Discussion

Tenacibaculum and *Psychrobacter* spp. are cosmopolitan associates of whales

Humpback whales from three oceanic populations and four distinct geographic areas were found to exhibit *Tenacibaculum* and *Psychrobacter* spp. on their skin. These sequences appear to be specific to humpbacks, and are secondly related to sequences recovered from dolphin blow. The dolphin blow interacts with the blowhole epidermis and thus it is possible that dolphins also harbor specific strains of *Tenacibaculum* and *Psychrobacter* on their skin. The role or interactions of these microorganisms with humpback skin, and possibly with other marine mammals, is of considerable interest as their interactions may be linked to host health.

Bacteria within the genus *Tenacibaculum* have been previously isolated from marine habitats [41–44] and a variety of marine animals [45,46] including diseased fish [47–49]. It is possible that humpback-specific *Tenacibaculum* spp. may be particularly adapted to a host-associated lifestyle, but their prominence (in both diversity and abundance) on healthy skin indicates an unlikely role in disease. One study did demonstrate a predatory function for *Tenacibaculum* spp. [50], and this feature might be aiding the whales in removal of fouling microbes.

Members of the genus *Psychrobacter* are generally a widespread and evolutionarily successful group of organisms. Sequences belonging to this genus have most commonly been recovered from cold environments including Antarctic soil, sea ice and deep-sea environments, and the genus is generally referred to as psychrophilic and osmotolerant [51,52]. Additionally, sequences have been found associated with krill and ascidians [53,54]. One possible reason that *Psychrobacter* spp. are so prevalent on humpback skin may be related to their tolerance of the 'extreme' type conditions of the skin surface. In the course of a single year, these animals are exposed to a large temperature range, on the order of ~6–25°C between their seasonal feeding and breeding areas. As noted previously, some migrations are the longest of any known mammal [18], and likely create a stressful type environment for survival of many bacteria. The genome of *P. arcticus* strain 273-4 does provide genetic evidence for survival under cold and stress conditions, including changes in membrane composition and

Table 5. PERMANOVA analysis of the effect of geographic area, age, sex and sample type on *Tenacibaculum* and *Psychrobacter* spp. populations determined by oligotyping.

Data	Variation	d.f.	SS	Pseudo-F[f] or t-value[t]	p-value
Tenacibaculum					
Normal whales	Location (Hawaii, GOM, Alaska, Am Samoa)	3	24929	3.6248[f]	0.001***
Hawaii vs. GOM				1.7933[t]	0.001***
Hawaii vs. Am Samoa				1.4839[t]	0.012*
Hawaii vs. Alaska				1.5359[t]	0.008**
GOM vs. Am Samoa				2.4447[t]	0.001***
Am Samoa vs. Alaska				2.4634[t]	0.001***
GOM vs. Alaska				1.8131[t]	0.001***
Normal whales	Sample type (biopsy, sloughed, tag)	2	13077	3.8521[f]	0.001***
Biopsy vs. sloughed				1.8692[t]	0.002**
Biopsy vs. tag				1.045[t]	0.361
Sloughed vs. tag				1.5504[t]	0.012*
Normal whales	Age (calf, juvenile, adult)	2	4727.2	1.0917[f]	0.360
Normal whales	Metabolic state (catabolic, anabolic)	1	15172	5.706[f]	0.001***
Normal whales	Sex (male, female)	1	3027.6	1.3518[f]	0.165
Psychrobacter					
Normal whales	Location (Hawaii, GOM, Alaska, Am Samoa)	3	13889	3.5707[f]	0.001***
Hawaii vs. GOM				1.4225[t]	0.012*
Hawaii vs. Am Samoa				1.7009[t]	0.002**
Hawaii vs. Alaska				1.5197[t]	0.01**
GOM vs. Am Samoa				2.8567[t]	0.001***
Am Samoa vs. Alaska				2.1002[t]	0.001***
GOM vs. Alaska				2.095[t]	0.001***
Normal whales	Sample type (biopsy, sloughed, tag)	2	4854	1.1335[f]	0.274
Normal whales	Age (calf, juvenile, adult)	2	4778.1	1.2284[f]	0.207
Normal whales	Metabolic state (catabolic, anabolic)	1	15867	6.349[f]	0.001***
Normal whales	Sex (male, female)	1	2761.8	1.2785[f]	0.198

***$p \leq 0.001$,
**$p \leq 0.01$,
*$p \leq 0.05$.

synthesis of cold shock proteins [55], and it is possible that humpback-associated *Psychrobacter* spp. also possess these qualities.

A recent metagenomic study focused on the functional-potential of the human skin-associated bacteria revealed that the resident bacteria exploited compounds produced by the skin (sugars and lipids) [56]. The same study also suggested that these bacteria might be contributing to host health by aiding in the regulation of skin acidity and epidermal permeability. Investigations into the genomic potential of *Tenacibaculum* and *Psychrobacter* spp. associated with whales could aid in understanding if human and whale skin bacteria exhibit similar functions, as well as help in revealing the potential for whale-specific interactions.

Geographic area and/or metabolic state are related to major bacterial community differences, and more subtle variations within Tenacibaculum and Psychrobacter spp. populations

Geographic area affected the structure of the skin-bacterial community, and also drove more subtle differences in the structure of *Tenacibaculum* and *Psychrobacter* spp. populations. Water temperature as well as seawater nutrients typically vary between the subtropical (Hawaii, American Samoa) and mid- to high-latitudes (Alaska, Gulf of Maine) examined in this study, and these environmental differences may impact skin-bacteria. The structure of the bacterial community was most similar between animals sampled in Hawaii and American Samoa, and it appears that a feature of the oligotrophic tropical water, and/or a change in animal physiology, may be related to this trend. The animals from Hawaii and American Samoa exhibited catabolic metabolisms at the time of sampling. Similar shifts in microbes have been associated with changes in metabolic state and nutrient flux in other animals, but have previously only been observed in the gut microbiome [57]. For the whales, these catabolic periods may be associated with physiological-based factors previously observed in humans, which may directly affect the skin. These include decreased repair and replacement of the skin cells (and other cells with typically high turnover), decreased production of immune-related cells as well as decreased adiposity, and increased blood pressure [58]. These catabolic states may also be related to decreased wound healing [59]. The connection between metabolism, skin-bacteria and wound healing is particularly important

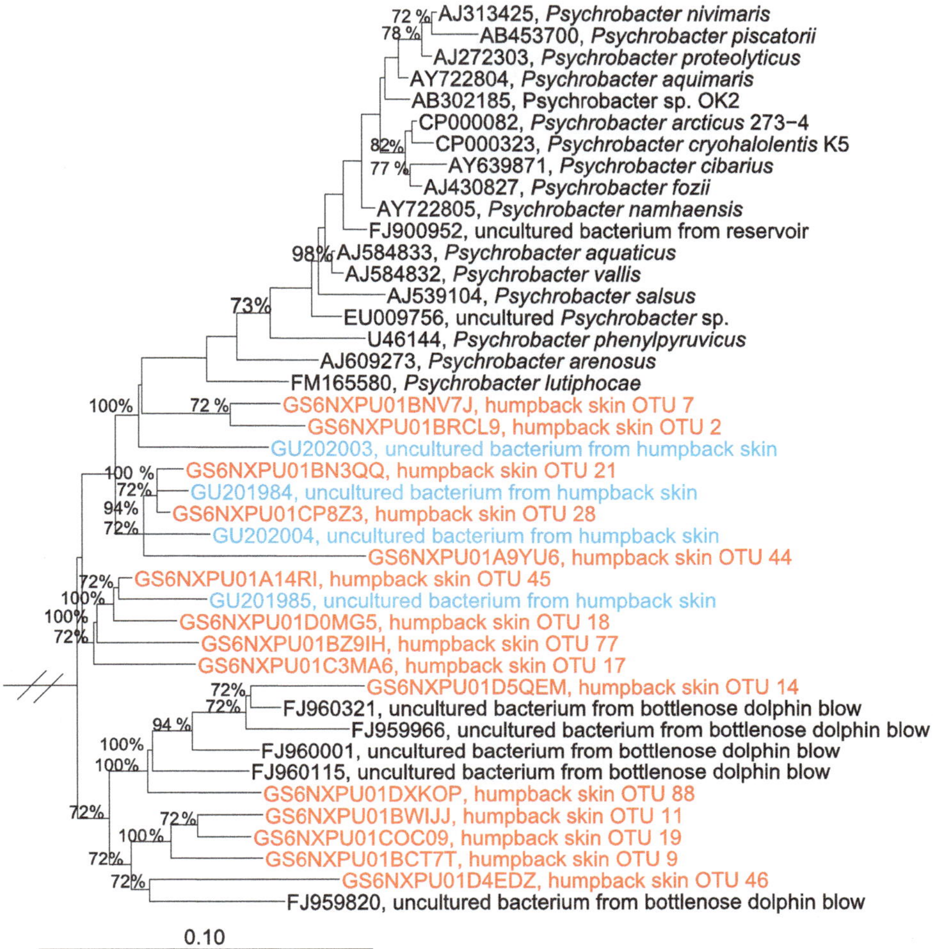

Figure 7. Phylogenetic relationship of Psychrobacter-affiliated sequences. Represented SSU rRNA gene sequences include *Psychrobacter*-affiliated sequences recovered from whale-skin, cultivated *Psychrobacter* isolates, and relatives within the *Gammaproteobacteria*. Red colored sequences are short-reads from this study, and blue full-length sequences are from a previous study [8]. The scale bar corresponds to 0.10 substitutions per nucleotide position, and only bootstrap values >70 are listed. Sequences from *Hymenobacter soli* (AB251884), *Salegentibacter flavus* (AY682200) and *Ureaplasma urealyticum* were used to form the outgroup.

for marine mammals due to the high incidences of lesions reported in many populations [23]. For example, it is possible that wound healing is prolonged when animals are in their breeding grounds. In this scenario, skin bacteria producing antibiotics or other defenses may be critical for protection of the whales.

This study also revealed a relationship between sample type (biopsy and sloughed skin) and bacterial community composition. This finding could be linked to the fact that the biopsied skin contained epidermis and dermis down to just before the blubber, whereas the sloughed skin was more representative of surface dead skin (stratum corneum). However, the stratum corneum does appear to be a major location in which the bacteria colonize. Anatomical location may also play a role in structuring the bacterial community. The biopsy samples were all taken from the same location on the animal (near dorsal fin), whereas the sloughed skin was removed during breaching activities from unknown locations on the animal. Repeated skin samplings of different anatomical locations from captive animals may be necessary to fully resolve the more subtle influences of anatomy on skin-bacteria. It should also be noted that the nets used to collect the sloughed skin were not sterilized prior to sampling and therefore some bacterial cells may have been transferred among samples or from the environment. Future studies using sloughed skin should consider net sterilization or single-use samplers in order to minimize the potential for cross-contamination.

Organization of microbial cells on whale skin

In contrast to the SSU rDNA data that revealed a bacterial community dominated by two major genera, scanning electron micrographs of the humpback whale skin surface revealed large differences in the densities and morphology of surface-associated microbes. Although all samples were treated equally, it is possible that some cells as well as organic matter were dislodged during the extensive washing and processing of these tissues. Due to the rapid turnover rate of marine mammal skin, it is conceivable that recently dead versus older skin harbor different properties, that may have also affected cell adhesiveness during processing.

It is not known if the humpback-associated bacterial cells are only superficially associated, or if they are found deeper into their thick epidermis. Histological analysis of sectioned tissues may further reveal cells embedded within the epidermis. Human skin primarily supports microbial cells on the epidermis, but microbes are also found within the dermis, including hair shafts, sweat glands and sebaceous glands [60]. The specific structure of

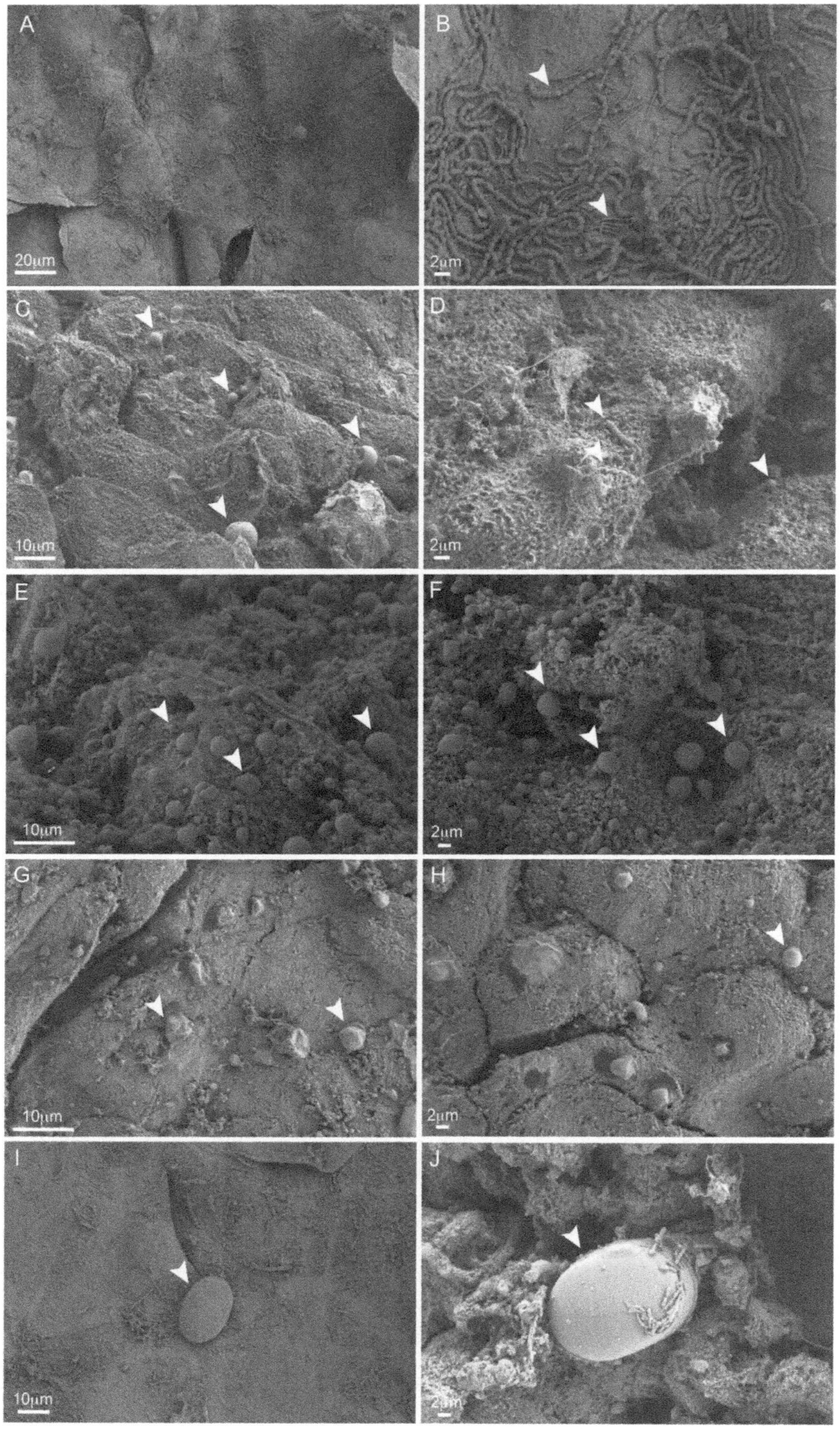

Figure 8. Scanning electron micrographs of whale skin-bacteria. Overview (A, C, E, G) and detailed (B, D, F, H) scanning electron micrographs of the surface of humpback whale skin from individuals CCS2010-100 (A, B), CCS2010-99 (C,D), CCS2010-96 (E,F) and CCS2010-98 (G,H). Diatoms, and microbes residing on these cells, were present on the surfaces of CCS2010-100 (I) and CCS2010-97 (J). Arrows indicate possible unique microbial cell morphotypes.

humpback skin has not been previously studied, but the epidermis of other species of whales are 15–20 times thicker than humans and excessively smooth, thus providing barrier properties to the environment, as well as limiting the attachment by microbes and penetration by pathogens [22,23,61,62]. Additionally, marine mammal skin is thought to provide an innate barrier as well as programmed immune protection [63]. These features collectively suggest that bacteria are most likely restricted to the epidermal surface.

Table 6. Comparison of skin-bacteria among normal, health-compromised and dead humpback whales, (n/a) indicates not applicable.

Sample name (# samples or replicates)	Condition	Number of bacterial OTUs	Abundance of core bacteria (*Tenacibaculum+Psychrobacter* spp.)	Bacteria not detectable in healthy individuals
Normal, all populations (57)	Normal	71–99	55–75%	n/a
IFAW10-188Mn (2)	Dead	35, 51	2%, 17%	*Clostridium* spp.
Ent1 (2)	Entangled,	159, 172	44%, 32%	*Lachnospiraceae* spp., undescribed *Clostridia, Bdellovibrio* spp.
SxMn1023 (1)	Dead	28	90%	none
Ent2 (1)	Entangled	93	36%	*Staphlococcus* spp., *Deferribacterale*, SAR406
Calf (3)	Dead	55, 111, 130	50%, 60%, 70%	none

Trends from the core bacterial community aid development of an index for 'normal' health

The identification of *Tenacibaculum* and *Psychrobacter* spp. as central and cosmopolitan (or 'core') members of the humpback skin bacterial community may aid in the development of a gauge for 'normal' skin health for these animals. Our results represent a wide demographic sampling of humpback whales with no obvious health impairment. It indicates some notable differences from live entangled and dead humpback whales, although available sample sizes are small. Entanglement can cause acute death, as well as chronic health effects from injury, infection and impaired mobility [64]. The live entangled whales in this study were thought to have been entangled for an extended period and exhibited mild external signs of impairment in large whales, such as pale skin, a proliferation of cyamids (whale lice) and diminished body condition [65,66]. The differences in skin bacteria from free-swimming whales also suggest an altered health state, although we cannot exclude the possibility that bacteria were introduced and exacerbated by the foreign materials entangling the whales. Skin-bacteria from dead animals also differed from the trends reported for the free-swimming animals. Most striking was the appearance of pathogenic and predatory bacteria such as *Clostridia* and *Bdellovibrio* spp. in the deceased and entangled individuals. The emergence of these bacteria is probably related to the decay of the skin as well as a likely decrease in host immune functions and possibly also the alteration in body temperature. Interestingly, these animals tended to harbor fewer of the core bacterial community members (compared to normal animals), which further suggests that host-controlled features are probably related to the presence of these possible mutualistic or commensal populations. In laboratory studies, an alteration in the abundance of the major skin-bacterial associates has been demonstrated in mice experiencing skin disorders [67]. The microbe-host dysbiosis model introduced by Scharschmidt and Fischbach [20] explains how an imbalance in the composition of the skin microbes (perhaps caused by exogenous factors) can lead to an alteration of the host immune response. Alternatively, there may be host-susceptibility factors that cause an alteration of the host immune response (e.g., excess inflammation), which in turn triggers an endogenous shift in the skin-microbial community. Changes in both microbiota and immune response could cause a negative feedback cycle, in which both parameters become even more significantly altered.

Sightings of skin lesions on many species of marine mammals appear to be increasing, with possible correlations to changes in the marine ecosystem including pathogens [23]. This skin surface barrier and its associated core bacterial community may be an important component to maintaining health and preventing lesion formation. However, many features of this microbial association remain uncharacterized. Future studies examining the metabolism and antibiotic production of widespread *Tenacibaculum* and *Psychrobacter* spp. associates on humpbacks are needed, as well as determining whether these bacteria are more broadly found on other species of marine mammals. Additionally, examining how these cells are able to maintain residence on humpback skin is important for a broader understanding of the features that pathogen or lesion-associated bacteria may use to preserve their invasion. Finally, future studies of animals with skin lesions or disorders and other characterized health and inflammation disorders will greatly advance our understanding of the connection between the skin microbiota and health. Some of these future studies may have broader implications for mammalian skin health, including humans, and offer an opportunity to study skin-microbial adaptations under exposure to remarkably diverse environmental conditions. Overall, understanding the connections between the skin surface bacterial communities, lesion formation and animal health may aid our conservation efforts of these important animals in a time of unprecedented change to the ocean environment.

Acknowledgments

We are grateful to M. Lammers, S. Landry, E. Lyman, T.A. Mooney, D. Rogers, A. Stimpert, J. Tackaberry and A. Tagarino for facilitating sample collections. Where available, molecular genetic sexes were provided by D. Steel (South Pacific samples) and P. Palsbøll/M. Bérubé (North Atlantic samples).

Author Contributions

Conceived and designed the experiments: AA TJM J. Robbins AAP MN KMTM MM. Performed the experiments: AA. Analyzed the data: AA AME J. Reveillaud. Contributed reagents/materials/analysis tools: TJM J. Robbins AME AAP J. Reveillaud DM MM MN KMTM. Wrote the paper: AA TJM J. Robbins AME AAP J. Reveillaud DM MM MN KMTM.

References

1. Willing BP, Dicksved J, Halfvarson J, Andersson AF, Lucio M, et al. (2010) A pyrosequencing study in twins shows that gastrointestinal microbial profiles vary with inflammatory bowel disease phenotypes. Gastroenterology 139: 1844–1854.e1841.
2. Wikoff WR, Anfora AT, Liu J, Schultz PG, Lesley SA, et al. (2009) Metabolomics analysis reveals large effects of gut microflora on mammalian blood metabolites. Proceedings of the National Academy of Sciences of the United States of America 10: 3698–3703.
3. Ley RE, Hamady M, Lozupone C, Turnbaugh PJ, Ramey RR, et al. (2008) Evolution of mammals and their gut microbes. Science 320: 1647–1651.
4. Bowers RM, McCubbin IB, Hallar AG, Fierer N (2012) Seasonal variability in airborne bacterial communities at a high-elevation site. Atmospheric Environment 50: 41–49.
5. DeLong EF, Taylor LT, Marsh T, Preston CM (1999) Visualization and enumeration of marine planktonic Archaea and Bacteria by using polyribonucleotide probes and fluorescent in situ hybridization. Appl Environ Microb 65: 5554–5563.
6. Bernhardt AE, Colbert D, McManus J, Field KG (2005) Microbial community dynamics based on 16S rRNA gene profiles in a Pacific Northwest estuary and its tributaries. FEMS Microbiology Ecology 51: 115–128.
7. Yeo SK, Huggett MJ, Eiler A, Rappé MS (2013) Coastal bacterioplankton community dynamics in response to a natural disturbance. PLoS ONE 8: e56207.
8. Apprill A, Mooney TA, Lyman E, Stimpert AK, Rappé MS (2011) Humpback whales harbour a combination of specific and variable skin bacteria. Environmental Microbiology Reports 3: 223–232.
9. Douglas AE (1998) Nutritional Interactions in Insect-Microbial Symbioses: Aphids and Their Symbiotic Bacteria Buchnera. Annual Review of Entomology 43: 17–37.
10. Muscatine L, Cernichiari E (1969) Assimilation of photosynthetic products of zooxanthellae by a reef coral. Biol Bull 137: 506–523.
11. Fleming A, Jackson J (2011) Global review of humpback whales (*Megaptera novaeangliae*). In: U.S. Department of Commerce NOAA, National Marine Fisheries Service, Southwest Fisheries Science Center, editor. NOAA Technical Memorandum NMFS.
12. Clapham PJ (2000) The humpback whale: seasonal feeding and breeding in a baleen whale. In: Mann J, Connor RC, Tyack PL, Whitehead H, editors. Cetacean socities: field studies of dolphins and whales. Chicago: University of Chicago Press. pp. 173–196.
13. Baker CS, Herman LM, Perry A, Lawton WS, Straley JM, et al. (1986) Migratory movement and population structure of humpback whales (*Megaptera novaeangliae*) in the Central and Eastern North Pacific. Marine Ecology Progress Series 31: 105–119.
14. Chittleborough RG (1965) Dynamics of two populations of the humpback whale *Megaptera novaeangliae* (Borowski). Australian Journal of Marine and Freshwater Research 16: 33–128.
15. Craig AS, Herman LM, Gabriele CM, Pack AA (2003) Migratory timing of humpback whales (*Megaptera novaeangliae*) in the central North Pacific varies with age, sex and reproductive status. Behaviour 140: 981–1001.
16. Dawbin WH (1966) The seasonal migratory cycle of humpback whales. In: Norris KS, editor. Whales, dolphins and porpoises. Berkeley: University of California Press. pp. 145–170.
17. Katona S, Beard JA (1990) Population size, migrations, and feeding aggregations of the humpback whale (*Megaptera novaeangliae*) in the western Atlantic. Report of the International Whaling Committee Special Issue 12: 295–305.
18. Robbins J, Dalla Rosa L, Allen JM, Mattila DK, Secchi ER, et al. (2011) Return movement of a humpback whale between the Antarctic Peninsula and American Samoa: a seasonal migration record. Endangered Species Research 13: 117–121.
19. Hanski I, von Hertzen L, Fyhrquist N, Koskinen K, Torppa K, et al. (2012) Environmental biodiversity, human microbiota, and allergy are interrelated. Proceedings of the National Academy of Sciences.
20. Scharschmidt TC, Fischbach MA (2013) What lives on our skin: ecology, genomics and therapeutic opportunities of the skin microbiome. Drug Discovery Today: Disease Mechanisms 10: e83–e89.
21. Harrison RJ, Thurley KW (1974) Structure of the epidermis in *Tursiops*, *Delphinus* and *Phocoena*. In: Harrison RJ, editor. Functional anatomy of marine mammals. London and New York: Academic Press. pp. 45–71.
22. Geraci JR, St. Aubin DJ, Hicks BD (1986) The epidermis of odontocetes: a vew from within. In: Bryden MM, Harrison R, editors. Research on Dolphins. Oxford: Clarendon Press. pp. 3–21.
23. Mouton M, Botha A (2012) Cutaneous Lesions in Cetaceans: An Indicator of Ecosystem Status?
24. Palsbøll P, Larsen F, Hansen E (1991) Sampling of skin biopsies from free-ranging large cetaceans in West Greenland: Development of new biopsy tips and bolt designs. Report of the International Whaling Committee Special Issue 13: 71–79.
25. Clapham PJ, Palsbøll PJ, Mattila DK (1993) High energy behaviors in humpback whales as a source of sloughed skin for molecular analysis. Marine Mammal Science 9: 214–220.
26. Bérubé M, Palsbøll PJ (1996) Identification of sex in cetaceans by multiplexing with three ZFX and ZFY specific primers: erratum. Molecular Ecology 5: 602.
27. Glockner DA, Venus S. Determining the sex of humpback whales (*Megaptera novaeangliae*) in their natural environment; 1983. pp. 447–464.
28. Apprill A, Hughen K, Mincer TJ (2013) Major similarities in the bacterial communities associated with lesioned and healthy Fungiidae corals. Environmental Microbiology 15: 2063–2072.
29. Schloss PD, Westcott SL, Ryabin T, Hall JR, Hartmann M, et al. (2009) Introducing mothur: open-source, platform-independent, community-supported software for describing and comparing microbial communities. Appl Environ Microbiol 75: 7537–7541.
30. Pruesse E, Quast C, Knittel K, Fuchs B, Ludwig W, et al. (2007) SILVA: a comprehensive online resource for quality checked and aligned ribosomal RNA sequence data compatible with ARB. Nucleic Acids Res 35: 7188–7196.
31. Schloss PD (2010) The effects of alignment quality, distance calculation method, sequence filtering, and region on the analysis of 16S rRNA gene-based studies. PLoS Comput Biol 6: e1000844.
32. Cole JR, Wang Q, Cardenas E, Fish J, Chai B, et al. (2009) The Ribosomal Database Project: improved alignments and new tools for rRNA analysis. Nucleic Acids Research 37: D141–D145.
33. Ludwig W, Strunk O, Westram R, Richter L, Meier H, et al. (2004) ARB: a software environment for sequence data. Nucleic Acids Res 32: 1363–1371.
34. Stamatakis A (2006) RAxML-VI-HPC: maximum likelihood-based phylogenetic analyses with thousands of taxa and mixed models. Bioinformatics 22: 2688–2690.
35. Johnson WR, Torralba M, Fair PA, Bossart GD, Nelson KE, et al. (2009) Novel diversity of bacterial communities associated with bottlenose dolphin upper respiratory tracts. Environmental Microbiology Reports 1: 555–562.
36. Eren AM, Maignien L, Sul WJ, Murphy LG, Grim SL, et al. (2013) Oligotyping: Differentiating between closely related microbial taxa using 16S rRNA gene data. Methods in Ecology and Evolution 4: 1111–1119.
37. Eren AM, Zozaya M, Taylor CM, Dowd SE, Martin DH, et al. (2011) Exploring the Diversity of *Gardnerella vaginalis* in the Genitourinary Tract Microbiota of Monogamous Couples Through Subtle Nucleotide Variation. PLoS ONE 6: e26732.
38. McLellan SL, Newton RJ, Vandewalle JL, Shanks OC, Huse SM, et al. (2013) Sewage reflects the distribution of human faecal *Lachnospiraceae*. Environmental Microbiology 15: 2213–2227.
39. Clarke KR, Warwick RM (2001) Change in Marine Communities: An Approach to Statistical Analysis and Interpretation. Plymouth: PRIMER-E.
40. McCune B, Grace J (2002) Analysis of ecological communities. Gleneden Beach, Oregon MJM Software Design. 304 p.
41. Frette L, Jorgensen NOG, Irming H, Kroer N (2004) *Tenacibaculum skagerrakense* sp. nov., a marine bacterium isolated from the pelagic zone in Skagerrak, Denmark. International Journal of Systematic and Evolutionary Microbiology 54: 519–524.
42. Yoon J-H, Kang S-J, Oh T-K (2005) *Tenacibaculum lutimaris* sp., nov., isolated from a tidal flat in the Yellow Sea, Korea. Int J Syst Evol Microbiol 55: 793–798.
43. Choi DH, Kim Y-G, Hwang CY, Yi H, Chun J, et al. (2006) *Tenacibaculum litoreum* sp. nov., isolated from tidal flat sediment. International Journal of Systematic and Evolultionary Microbiology 56: 635–640.
44. Jung S-Y, Oh T-K, Yoon J-H (2006) *Tenacibaculum aestuarii* sp. nov., isolated from a tidal flat sediment in Korea. International Journal of Systematic and Evolutionary Microbiology 56: 1577–1581.
45. Heindl H, Wiese J, Imhoff JF (2008) *Tenacibaculum adriaticum* sp. nov., from a bryozoan in the Adriatic Sea. International Journal of Systematic and Evolutionary Microbiology 58: 542–547.
46. Wang J-T, Chou Y-J, Chou J-H, Chen CA, Chen W-M (2008) *Tenacibaculum aiptasiae* sp. nov., isolated from a sea anemone *Aiptasia pulchella*. International Journal of Systematic and Evolutionary Microbiology 58: 761–766.
47. Piñeiro-Vidal M, Carballas CG, Gomez-Barreiro O, Riaza A, Santos Y (2008) *Tenacibaculum soleae* sp. nov., isolated from diseased sole (*Solea senegalensis* Kaup). International Journal of Systematic and Evolutionary Microbiology 58: 881–885.
48. Piñeiro-Vidal M, Gijón D, Zarza C, Santos Y (2012) *Tenacibaculum dicentrarchi* sp. nov., a marine bacterium of the family Flavobacteriaceae isolated from European sea bass. International Journal of Systematic and Evolutionary Microbiology 62: 425–429.
49. Piñeiro-Vidal M, Riaza A, Santos Y (2008) *Tenacibaculum discolor* sp. nov. and *Tenacibaculum gallaicum* sp. nov., isolated from sole (*Solea senegalensis*) and turbot (*Psetta maxima*) culture systems. International Journal of Systematic and Evolutionary Microbiology 58: 21–25.
50. Banning E, Casciotti K, Kujawinski E (2010) Novel strains isolated from a coastal aquifer suggest a predatory role for Flavobacteria. FEMS Microbiology Ecology 73: 254–270.
51. Bowman JP, Cavanagh J, Austin JJ, Sanderson K (1996) Novel *Psychrobacter* Species from Antarctic Ornithogenic Soils. International Journal of Systematic Bacteriology 46: 841–848.
52. Bowman JP, Nichols DS, McMeekin TA (1997) *Psychrobacter glacincola* sp. nov., a Halotolerant, Psychrophilic Bacterium Isolated from Antarctic Sea Ice. Systematic and Applied Microbiology 20: 209–215.

53. Denner EBM, Mark B, Mark B, Busse H-J, Busse H-J, et al. (2001) *Psychrobacter proteolyticus* sp. nov., a Psychrotrophic, Halotolerant Bacterium Isolated from the Antarctic Krill Euphausia superba Dana, Excreting a Cold-Adapted Metalloprotease. Systematic and Applied Microbiology 24: 44–53.
54. Romanenko LA, Schumann P, Rohde M, Lysenko AM, Mikhailov VV, et al. (2002) *Psychrobacter submarinus* sp. nov. and *Psychrobacter marincola* sp. nov., psychrophilic halophiles from marine environments. International Journal of Systematic and Evolutionary Microbiology 52: 1291–1297.
55. Ayala-del-Río HL, Chain PS, Grzymski JJ, Ponder MA, Ivanova N, et al. (2010) The Genome Sequence of *Psychrobacter arcticus* 273-4, a Psychroactive Siberian Permafrost Bacterium, Reveals Mechanisms for Adaptation to Low-Temperature Growth. Applied and Environmental Microbiology 76: 2304–2312.
56. Mathieu A, Delmont TO, Vogel TM, Robe P, Nalin R, et al. (2013) Life on Human Surfaces: Skin Metagenomics. PLoS ONE 8: e65288.
57. Costello EK, Gordon JI, Secor SM, Knight R (2010) Postprandial remodeling of the gut microbiota in Burmese pythons. ISME J 4: 1375–1385.
58. Nixon PGF (1989) Human functions and the Heart. In: Seedhouse D, Cribb A, editors. Changing Ideas in Health Care. New York: John Wiley and Sons.
59. Demling RH (2009) Nutrition, Anabolism, and the Wound Healing Process: An Overview. Eplasty 9: e9.
60. Grice EA, Segre JA (2011) The skin microbiome. Nature Reviews Microbiology 9: 244–253.
61. Brown WR, Geraci JR, Hicks BD, St.-Aubin DJ, Schroeder JP (1983) Epidermal cell proliferation in the bottlenose dolphin (*Tursiops truncatus*). Canadian Journal of Zoology 61: 1587–1590.
62. Hicks BD, St. Aubin DJ, Geraci JR, Brown WR (1985) Epidermal growth in the bottlenose dolphin, *Tursiops truncatus*. Journal of Investigative Dermatology 85: 60–65.
63. Zabka TS, Romano TA (2003) Distribution of MHC II (+) cells in skin of the Atlantic bottlenose dolphin (*Tursiops truncatus*): An initial investigation of dolphin dendritic cells. The Anatomical Record Part A: Discoveries in Molecular, Cellular, and Evolutionary Biology 273A: 636–647.
64. Cassoff RM, Moore KM, McLellan WA, Barco SG, Rotstein DS, et al. (2011) Lethal entanglement in baleen whales. Diseases of Aquatic Organisms 96: 175–185.
65. Gulland FMD, Nutter FB, Dixon K, Calambokidis J, Schorr G, et al. (2008) Health Assessment, Antibiotic Treatment, and Behavioral Responses to Herding Efforts of a Cow-Calf Pair of Humpback Whales (*Megaptera novaeangliae*) in the Sacramento River Delta, California. Aquatic Mammals 32: 182–192.
66. Pettis HM, Rolland RM, Hamilton PK, Brault S, Knowlton AR, et al. (2004) Visual health assessment of North Atlantic right whales (*Eubalaena glacialis*) using photographs. Canadian Journal of Zoology 82: 8–19.
67. Scharschmidt TC, List K, Grice EA, Szabo R, Renaud G, et al. (2009) Matriptase-Deficient Mice Exhibit Ichthyotic Skin with a Selective Shift in Skin Microbiota. J Invest Dermatol 129: 2435–2442.

12

Hybrid Speciation in a Marine Mammal: The Clymene Dolphin (*Stenella clymene*)

Ana R. Amaral[1,2]*, Gretchen Lovewell[3], Maria M. Coelho[1], George Amato[2], Howard C. Rosenbaum[2,4]

1 Centro de Biologia Ambiental, Faculdade de Ciências, Universidade de Lisboa, Lisbon, Portugal, **2** Sackler Institute for Comparative Genomics, American Museum of Natural History, New York, New York, United States of America, **3** Mote Marine Laboratory, Sarasota, Florida, United States of America, **4** Ocean Giants Program, Wildlife Conservation Society, New York, United States of America

Abstract

Natural hybridization may result in the exchange of genetic material between divergent lineages and even the formation of new taxa. Many of the Neo-Darwinian architects argued that, particularly for animal clades, natural hybridization was maladaptive. Recent evidence, however, has falsified this hypothesis, instead indicating that this process may lead to increased biodiversity through the formation of new species. Although such cases of hybrid speciation have been described in plants, fish and insects, they are considered exceptionally rare in mammals. Here we present evidence for a marine mammal, *Stenella clymene*, arising through natural hybridization. We found phylogenetic discordance between mitochondrial and nuclear markers, which, coupled with a pattern of transgressive segregation seen in the morphometric variation of some characters, support a case of hybrid speciation. *S. clymene* is currently genetically differentiated from its putative parental species, *Stenella coeruleoalba* and *Stenella longisrostris*, although low levels of introgressive hybridization may be occurring. Although non-reticulate forms of evolution, such as incomplete lineage sorting, could explain our genetic results, we consider that the genetic and morphological evidence taken together argue more convincingly towards a case of hybrid speciation. We anticipate that our study will bring attention to this important aspect of reticulate evolution in non-model mammal species. The study of speciation through hybridization is an excellent opportunity to understand the mechanisms leading to speciation in the context of gene flow.

Editor: Norman Johnson, University of Massachusetts, United States of America

Funding: AR Amaral was funded by a grant from the Portuguese Foundation for Science and Technology (SFRH/BPD/79002/2011). The authors thank NOAA Fisheries for funding to initiate this project. The funders had no role in study design, data collection and analysis, decision to publish, or preparation of the manuscript.

Competing Interests: The authors have declared that no competing interests exist.

* E-mail: aramaral@fc.ul.pt

Introduction

The consequences of the exchange of genetic material between individuals belonging to different species have long been a matter of debate among biologists [1,2]. Natural hybridization among plant species has been seen as a potential mechanism for the spread of beneficial mutations and the formation of new taxa. In contrast, hybridization in animals has traditionally been viewed as a rare and evolutionarily destructive process leading to the prevention or reversal of divergence between incipient species (e.g. [3]). In recent years, however, hybridization has been well documented across several animal groups, such as insects, birds and fishes, suggesting that this phenomenon may be more important for the evolutionary history and speciation of animals than previously thought [4]. In some cases, hybridization has been shown to increase the genetic diversity of populations [5], providing them with a higher adaptability to environmental change. In other cases, it can lead to the formation of new species, able to explore new niches [6]. Hybridization leading to speciation is a phenomenon that has attracted the attention of scientists for some time now, although it is still poorly understood [7,8]. It has been reported in plants, fishes, insects and birds, but very rarely in mammals and never in marine mammals [9,10]. Some argue that the unique genetic (e.g. high rates of gene rearrangements that can lead to a rapid change in gene expression and post-zygotic isolation mechanisms) and morphological mechanisms mammals have, can reduce the potential for the production of viable hybrids [11]. Others argue that hybridization in mammals can be as frequent as in other animals, although its detection can be hindered by cryptic morphological characteristics and by a general lack of extensive studies [4]. Because morphological variation may not always have a genetic basis, hybrid individuals can have the exact same morphotype as one of the parental species [4], thus leading to instances in which hybrids are cryptic. Their identification is thus sometimes only possible with the use of molecular tools.

Cetaceans are a group of marine mammals that have diverged from their terrestrial ancestors around 53 Mya [12]. Their evolution has been characterized by some rapid radiation events, which in some groups has led to a confusing taxonomy and a difficulty in clarifying phylogenetic relationships due to the confounding effects of incomplete lineage sorting and possibly hybridization [13]. Hybridization in cetaceans has been reported to occur both in captivity and in the wild [14], but the extent to which it has contributed to the evolutionary history of these species remains unexplored. Because cetaceans are known to exhibit prominent karyological uniformity [15], they may have the potential to produce viable hybrid offspring more easily than other mammals. This has in fact been confirmed in captivity for at least a cross between *Tursiops truncatus* and *Delphinus capensis*, which produced a fertile hybrid [16]. In the wild there have been several documented cases of hybridization, across whale and porpoise species, but also among dolphin species, including within the genus *Stenella* [17]. While hybridization in whales and porpoises has been confirmed using molecular tools [18,19], such confirmation is lacking for hybridization among dolphin species.

The clymene dolphin, *Stenella clymene*, Gray (1846) is endemic to the tropical and subtropical Atlantic Ocean. Its cranial features closely resemble those of *Stenella coeruleoalba*, but its external appearance and behaviour are more similar to those of *Stenella longirostris* [20], which has led to some confusion regarding its recognition as a full species. Early molecular studies showed an additional uncertainty in the phylogenetic position of *S. clymene*, with mitochondrial DNA phylogenetic trees placing it as sister taxa of *S. coeruleoalba* [21,22], and a nuclear DNA tree based on Amplified Fragment Length Polymorphism (AFLPs) placed it as sister taxa of *S. longirostris* [21]. This conflicting evidence between different molecular and morphological characters has led to the present investigation. With the aim to understand the evolutionary mechanisms that may be behind the origin of *S. clymene*, namely a possible origin through natural hybridization, we sequenced one mitochondrial and six nuclear loci and followed a phylogenetics and population genetics approach. If *S. clymene* is of recent hybrid origin, we expect to find genetic intermediacy between the two parental forms, nuclear admixture, mitochondrial capture and unique mtDNA variation.

Materials and Methods

Ethics statement

Procedures for ensuring animal welfare during biopsy sampling were approved as part of the Scientific Research permits issued by the National Marine Fisheries Service under the authority of the Marine Mammal Protection Act of 1972 (16 U.S.C. 1361 et seq), the regulations governing the taking and importing of marine mammals (MMPA) (50 CPR part 216), the Endangered Species Act of 1973 (ESA) (16 U.S.C. 1531 et seq.), and the regulations governing endangered fish and wildlife permits (50 CFR parts 222–226). Biopsies were taken under NMFS permit numbers 14097, 774–1437, 774–1714, 1026/689424, and 873 issued to the National Marine Fisheries Southwest Fisheries Science Center. The samples originating from outside US jurisdiction were imported under CITES Import permit numbers US774223 and US689420, and under CITES Certificate of Scientific Exchange #690343. CITES permits are issued by the U.S. Fish & Wildlife Service. The Southwest Fisheries Science Center is a Registered Scientific Institution under CITES (US052).

DNA extraction and sequencing

Tissue samples (skin or muscle) from *Stenella longirostris*, *Stenella coeruleoalba* and *Stenella clymene* were obtained from free-range dolphins using a dart biopsy system or from dead, stranded individuals. Fifty-eight samples were received as DNA extraction from the Southwest Fisheries Science Center, Marine Mammal and Turtle Research Sample Collection (SWFSC-NOAA, La Jolla, CA). DNA from the remaining samples was extracted following a standard proteinase K and two phenol-chloroform-isoamyl extractions [23]. Samples from *Stenella coeruleoalba* and *S. longirostris* used in this study were obtained from different locations in the Atlantic, Pacific and Indian oceans in order to provide an overall estimate of genetic diversity for each species. Samples from *Stenella clymene* were mostly obtained from the Gulf of Mexico and adjacent Atlantic Ocean waters and included a mass stranding that occurred near Tarpon Springs, in Florida (USA), in 1995.

For the mitochondrial DNA (mtDNA), a total of 72 individuals (*Stenella clymene*, n = 15; *S. longirostris*, n = 21; *S. coeruleoalba*, n = 36) were amplified and sequenced for the cytochrome *b* gene (1106 bp) using the L-strand primer on tRNA glutamine (L14724, 5′-TGACTTGAARAACCAYCGTTG 3′) and the H-strand primer on tRNA threonine (5′CCTTTTCCGGTTTA-CAAGAC 3′) [22] (GenBank Accession Numbers KF691950 - KF692018). Polymerase Chain Reactions (PCR) and sequencing conditions used are described in [24]. In addition, four anonymous nuclear loci [Del_05, Del_10, Del_12 and Del_16 [25]] and two introns [BTN [26] and PLP [27]] were PCR amplified and sequenced as described in [28] (GenBank Accession Numbers KF691817 – KF691949). Some samples failed to sequence for some of the loci. All PCR products were cleaned by adding 0.5 U of Shrimp Alkaline Phosphatase and 5 U of Exonuclease I and incubated at 37°C for 30 min and 80°C for 15 min. Both strands were directly sequenced (BigDye Terminator CycleSequencing; Applied Biosystems) on an ABI 3730 automated sequencer. All sequences obtained were aligned using the software Sequencher, version 5.0 (Gene Codes Corporation). An extended dataset, including additional samples from *Delphinus delphis, D. capensis, D. capensis tropicalis, Stenella attenuata, S. frontalis, S. longirostris, S. coeruleoalba, S. clymene, Tursiops truncatus, T. aduncus, Lagenodelphis hosei, Sousa chinensis, Sotalia fluviatilis, Globicephala melas* and *Phocoena phocoena*, sequenced for nine nuclear loci (the ones mentioned above plus CHRNA1, Del_04 and Del_11, see [13]) was used to estimate a species tree of the subfamily Delphininae using the method implemented in the software package *BEAST [29]. 1 000 million MCMC generations sampling every 100 000 generations were run in the program BEAST v. 1.7.4. [30], choosing the Yule process as the species tree prior, the Piecewise constant and linear model for population size estimates and a strict molecular clock with an uncorrelated lognormal distribution. The program Tracer v.1.5 was run to ensure mixing and that there was no lack of convergence of the posterior distribution and parameters by examining effective population size (ESS) values. TreeAnnotator v. 1.6.1. was subsequently used to summarize the obtained trees in a single tree that best represents the posterior distribution.

The same dataset sequenced for the cytochrome *b* gene was used to estimate a Bayesian phylogenetic tree using the program MrBayes v. 3.2.1 [31]. 5 million MCMC generation sampling every 1 000 generations were run. The dataset was partitioned by codon positions and the Tamura-Nei model of nucletodide substitution was chosen, according to the results given by Modeltest v. 3.8. [32]. Sequences from *Globicephala melas* and *Phocoena phocoena* were used as outgroups.

Heterozygous individuals found in the nuclear loci were resolved using the program PHASE v.2.1. [33,34]. Alleles were inferred setting the phase-certainty threshold to 90%.

Nucleotide and haplotype diversities for each mitochondrial and nuclear locus, and for each species, were estimated in DNAsp v. 5.10.01 [35]. In order to assess the degree of genetic differentiation between the three *Stenella* species, pairwise F_{ST} was estimated for each locus with 10 000 random generations using Arlequin v. 3.5 [36]. Genealogical relationships at the haplotype level for each locus were inferred using the median-joining network as implemented in Network v. 4.6.0.0 [37].

In order to understand the phylogenetic relationships of the three species within the subfamily Delphininae, a species tree based on nine nuclear loci was estimated using a multi-locus species tree approach. A Bayesian phylogenetic tree was additionally estimated for the cytochrome *b* data.

Morphological data were collected from 12 *S. clymene* individuals that were part of the mass stranding in Florida. Photographic data were available, which allowed the external appearance of the individuals to be compared and confirm the identity of two individuals that contain mitochondrial DNA of *S. longisrostris*. In addition, these dataset allowed us to verify if skull measurements correspond to the ones described in the species description of *Stenella clymene* [20]. Skull measurements were taken as described in [20].

Results

In total, 72 sequences were obtained for the mitochondrial cytochrome *b* gene for the three *Stenella* species, which grouped into 61 haplotypes (Table 1). No haplotypes were shared among the three species. For the six nuclear loci, the number of sequences obtained for each species for each locus varied from 3 *S. clymene* sequences obtained for Del_05 to 17 *S. coeruleoalba* sequences obtained for BTN (Table 1). Some samples were not successfully phased into alleles and were therefore removed from subsequent analyses (Table 1). We shared haplotypes among the three species in two of the six nuclear genes, as also seen in the haplotype networks (see below).

Levels of genetic diversity found in the mtDNA were high for all species, with *S. clymene* showing the highest nucleotide diversity, but the lowest haplotype diversity (Table 2). For the nuclear DNA, levels of genetic diversity obtained varied among loci, as expected given the stochasticity of the nuclear genome (Table 2). Nevertheless, levels found are within the range reported for nuclear loci obtained in other dolphin species (e.g. [28]).

The three species showed high levels of differentiation among them (Table 3). In the mtDNA, *S. clymene* and *S. coeruleoalba* were the least differentiated. In the nuclear DNA, levels of differentiation varied among loci, with *S. clymene* and *S. longirostris* being the least differentiated in 3 of the 6 loci (Table 3).

The mtDNA haplotype network showed a clear separation between *S. coeruleoalba* and *S. longirostris*, with most *S. clymene* haplotypes being clustered among *S. coeruleoalba* (Figure 1). Two *S. clymene* haplotypes clustered with *S. longirostris*. The same pattern is seen in the mtDNA phylogenetic tree (Figure 2). In order to rule out misidentification or errors in amplification and sequencing of these individuals, DNA was re-extracted and amplification and sequencing was conducted in two different laboratories, and the same results were obtained.

Table 1. Number of *Stenella clymene, S. coeruleoalba* and *S. longirostris* specimens sequenced for this study.

	mtDNA	nuDNA					
	cytb	**BTN**	**PLP**	**Del_05**	**Del_10**	**Del_12**	**Del_16**
S. clymene	15	6/5	4/4	3/3	4/4	5/4	5/5
S. coeruleoalba	36	17/16	7/7	8/8	7/7	10/10	11/8
S. longirostris	21	11/10	10/9	8/7	6/6	9/9	11/11
Total samples	72	34/31	21/20	19/18	17/17	24/23	27/24
Total sites	1107	495	646	653	401	735	718
Variable sites	121	8	12	12	4	17	17

Numbers indicated after the slash correspond to the total number of samples used for analyses after removing samples that failed to be phased.

Table 2. Levels of genetic diversity obtained for *Stenella clymene, S. coeruleoalba* and *S. longirostris* for the mitochondrial DNA (mtDNA) and for six nuclear loci.

a) mtDNA	π	**Hd**
Scly	0.01319+−0.00475	0.886+−0.069
Scoer	0.00872+−0.00134	0.967+−0.019
Slong	0.00884+−0.00104	0.995+−0.016
b) BTN		
Scly	0.00214+−0.00142	0.756+−0.130
Scoer	0.00247+−0.00030	0.806+−0.048
Slong	0.00250+−0.00028	0.805+−0.054
c) Del_05		
Scly	0.00155+−0.00047	0.800+−0.172
Scoer	0.00222+−0.00058	0.750+−0.107
Slong	0.00085+−0.0003	0.495+−0.151
d) Del_10		
Scly	0.00196+−0.00063	0.607+−0.164
Scoer	0.00101+−0.00043	0.385+−0.149
Slong	0.00138+−0.00016	0.545+−0.062
e) Del_12		
Scly	0.00569+−0.00095	0.964+−0.077
Scoer	0.00145+−0.00035	0.705+−0.111
Slong	0.00101+−0.00024	0.601+−0.113
f) Del_16		
Scly	0.00350+−0.00056	0.978+−0.054
Scoer	0.00186+−0.00030	0.850+−0.060
Slong	0.00239+−0.00034	0.874+−0.038
g) PLP		
Scly	0.00370+−0.00056	0.929+−0.084
Scoer	0.00167+−0.00032	0.758+−0.084
Slong	0.00197+−0.00042	0.732+−0.096

π nucleotide diversity; Hd – haplotype diversity.

Haplotype networks obtained for the six nuclear loci showed different patterns from the one obtained for the mtDNA (Figure 3). Two of the loci showed very little differentiation among the three species (BTN and Del_10), with shared haplotypes among all three. In contrast, the other four loci (PLP, Del_05, Del_12 and Del_16) showed a more clear separation, with fewer shared haplotypes. In PLP and Del_12, only *S. coeruleoalba* and *S. longirostris* share haplotypes, with *S. clymene* haplotypes occupying more external positions. In Del_05 and Del_16, *S. clymene* shares haplotypes with *S. longirostris*, clearly being more related to this species than to *S. coeruleoalba*. This pattern strongly contrasts from the one obtained in the mtDNA, where *S. clymeme* showed to be more related with *S. coeruleoalba* (Figure 1).

The two *S. clymene* individuals that clustered with *S. longirostris* in the mtDNA network share haplotypes with both *S. coeruleoalba* and *S. longirostris* in the least variable nuclear loci, but in the more informative loci are located in the tips of the networks, with the exception of Del_05 where they share a haplotype with *S. longirostris*.

These two individuals, along with others that were sequenced in this study, were part of a mass stranding near Tarpon Springs, Florida (USA) in June 1995. In order to confirm their

Table 3. Pairwise fixation index values (F_{ST}) obtained between *Stenella clymene, S. coeruleoalba* and *S. longirostris* for the mitochondrial DNA (mtDNA) and for six nuclear loci.

a) mtDNA		
	Scly	Scoer
Scly		
Scoer	0.43892***	
Slong	0.71155***	0.80717***
b) BTN		
	Scly	Scoer
Scly		
Scoer	0.10393*	
Slong	0.01041	0.11836***
c) Del_05		
	Scly	Scoer
Scly		
Scoer	0.66941***	
Slong	−0.01483	0.56731***
d) Del_10		
	Scly	Scoer
Scly		
Scoer	0.38882**	
Slong	0.41950**	0.44734**
e) Del_12		
	Scly	Scoer
Scly		
Scoer	0.45211***	
Slong	0.488556***	0.44955***
f) Del_16		
	Scly	Scoer
Scly		
Scoer	0.53299***	
Slong	0.15699**	0.648***
g) PLP		
	Scly	Scoer
Scly		
Scoer	0.36231***	
Slong	0.33817***	0.15106**

$^{**}P<0.01$; $^{***}P<0.001$.

morphological identification as *S. clymene*, we examined the available skulls, photographs of external appearance and biometric data collected by the Mote Marine Laboratory at the time of the stranding. Twelve skulls were measured in order to obtain the same characteristics described in [20] (Table S1), and a graph of the preorbital width against the length of the upper toothrow was plotted (Figure 4). This relationship summarizes differences in shape known to differentiate the three species [20]. The scatterplot shows that all individuals fit within the measurements described for *S. clymene*. Preorbital width varied between 152 and 170.5 mm (described range is 156–171 mm) and length of the upper toothrow varied between 174.5 and 208.75 mm (described range is 183–210 mm). Information on these twelve individuals is included in Table S1. The available photographs of the external appearance do not shed much light into the coloration patterns (Figure S1). Apart from the two putative hybrids, the other individuals that were part from the mass stranding had mitochondrial DNA of *Stenella clymene*.

The species tree obtained with the nuclear loci resulted in a topology similar to the one recently published by [13], except for the branches including the genera *Delphinus, Lagenodelphis* and *Tursiops* and the three *Stenella* species analysed in this study (Figure 5). These branches had a low posterior probability value, reflecting an uncertainty of the method positioning the taxa. The study mentioned above did not include *Stenella clymene*, which may explain the differences observed in the tree here presented. Nevertheless, the results confirm that *S. clymene* is phylogenetically closer to *S. longirostris* than to *S. coeruleoalba*, although a relatively low posterior probability value (0.86) separates these two clades.

Discussion

Our results suggest that the evolution of *Stenella clymene* does not follow a simple bifurcating tree, but more likely the result of reticulation through the admixture between two other closely related species, *S. coeuruleolba* and *S. longisrostris*. This finding has important implications not only for the clarification of the evolutionary relationships among species of the genus *Stenella* but also for our understanding of hybrid speciation in mammal species.

We confirmed that the mitochondrial genome of *Stenella clymene* is more closely related to that of *S. coeruleoalba* whereas the nuclear genome seems to be more closely related to *S. longirostris*. This result had already been described in previous phylogenetic studies, albeit with a smaller sample size and different molecular markers [21,22]. This discrepancy of results between mitochondrial and nuclear markers suggests a hybrid origin of *S. clymene*, as a result of an ancient hybridization between a female *S. coeruleoalba* and a male *S. longirostris*. Our results further indicate that *S. clymene* is currently genetically distinct from its parental species, although backcrosses may still occur. We found two *S. clymene* haplotypes clustered with *S. longirostris* in the mitochondrial trees. These two individuals were part of a mass stranding of clymene dolphins in Florida. Their identification was confirmed by skull anatomy, external biometric measurements and coloration patterns, as explained in the Results section. The majority of the individuals have a mitochondrial DNA of *S. clymene*, except for the two individuals that appear to be the result of introgression between *S. clymene* and *S. longirostris*.

The morphological characteristics of *S. clymene* further support its origin through hybrid speciation. The morphometric variation seen in this species is outside that of both its putative parental species [20], suggesting a pattern of transgressive segregation [38,39]. This pattern arises in later generation hybrids where parental alleles have recombined to form new genetic associations. These new combinations lead to novel phenotypes, which can sometimes explore new habitats and resources, contributing to niche divergence [40]. Although transgressive phenotypes can also be the result of selection following speciation, the most widely accepted phenomenon to originate evolutionary novelty is hybridization [41].

Ecological divergence has been suggested as one of the main drivers of hybrid speciation, leading to reproductive isolation between the hybrid and the parental species [5,42]. This mechanism could explain the origin of *S. clymene*, although our knowledge on the ecology of this species is scarce and the few data

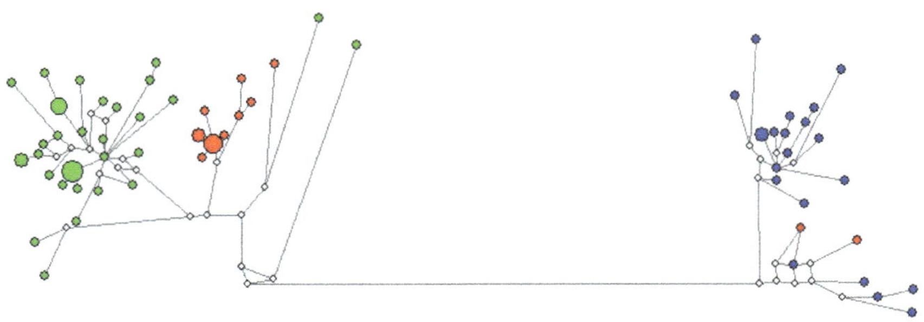

Figure 1. Median-joining network of cytochrome *b* haplotypes of *Stenella clymene* (red circles), *S. coeruleoalba* (green circles) and *S. longirostris* (blue circles). Circle size is proportional to the number of individuals exhibiting the correspondent haplotype. Length of lines separating the haplotypes is proportional to the number of mutational steps. White circles indicate missing, intermediate haplotypes.

available suggests that at least in some areas, the clymene dolphin may be exploring the same resources as *S. longirostris*. Mixed schools of both species have been observed in Florida and in the Caribbean [43]. Although ecological divergence has been suggested as critical for the persistence of a species of hybrid origin [5,42], we suggest that behavioral mechanisms instead may be responsible for the maintenance of *S. clymene* as an independent lineage. Sexual selection, for example, can lead to assortative mating, resulting in a significant decrease in female mating with individuals from other species, and generating reproductive isolation [44–46]. This mechanism has been well demonstrated in primate, rodent and seal hybrid zones [11]. It is possible that mechanisms of female choice may have evolved in *S. clymene* as well. In this instance, *S. clymene* females would more often choose conspecific (hybrid) individuals rather than individuals from the parental species, even if encounters were frequent. Little is known about female choice in cetacean species, but this mechanism occurs among many other mammals. Baboon females, for example, have a preference for males that are phenotypically similar to themselves, regardless of ancestry [47]. Alternative

Figure 2. Bayesian phylogenetic tree generated in MrBayes for the cytochrome *b* gene. Posterior probability values are above nodes. Branch lengths are in substitutions/site.

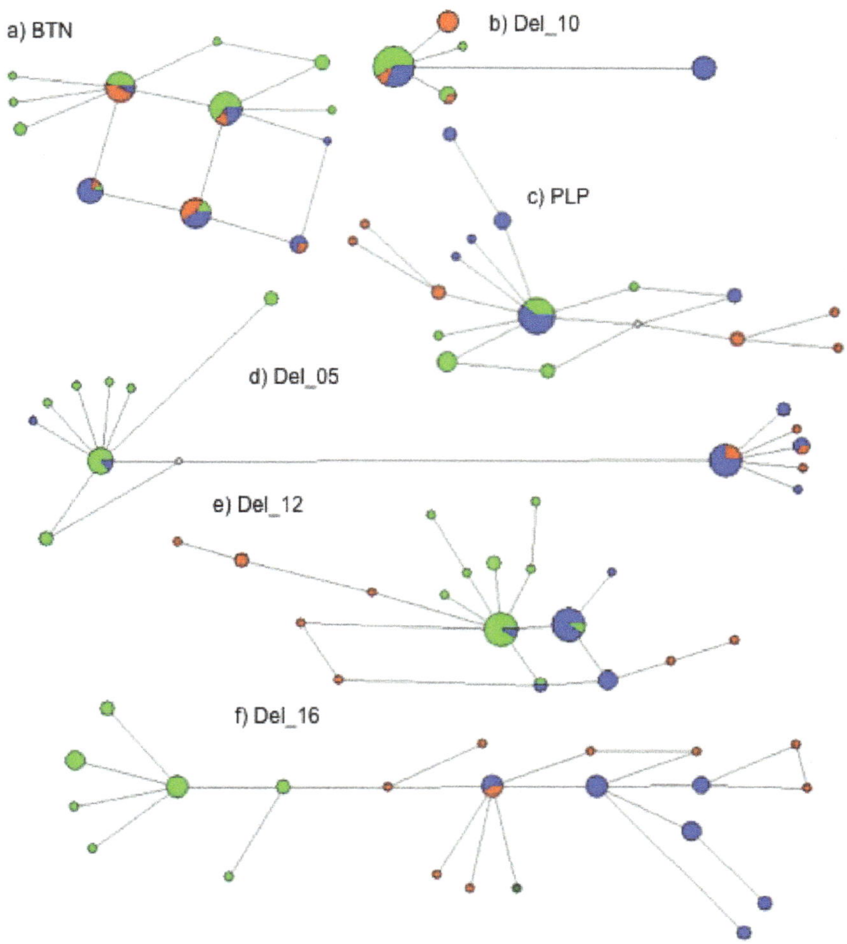

Figure 3. Median-joining networks of nuclear gene haplotypes of *Stenella clymene* (red circles), *S. coeruleoalba* (green circles) and *S. longirostris* (blue circles). a) BTN, b) Del_10, c) PLP, d) Del_05, e) Del_12, f) Del_16. Circle size is proportional to the number of individuals exhibiting the correspondent haplotype. Each species within each haplotype is coloured according to the legend. Length of lines separating the haplotypes is proportional to the number of mutational steps. White circles indicate missing, intermediate haplotypes.

mechanisms that have been proposed to explain the establishment of reproductive isolation between a hybrid and its parental species include chromosomal rearrangements and spatial isolation from the parental species. We do not consider these to be likely for *Stenella clymene*. Cetaceans, and dolphin species in particular, present strikingly similar karyotypes[15], which suggests a lack of major chromosomal rearrangement differences among the various species. The three *Stenella* species occur in sympatry, which also excludes the possibility of spatial isolation leading to reproductive isolation.

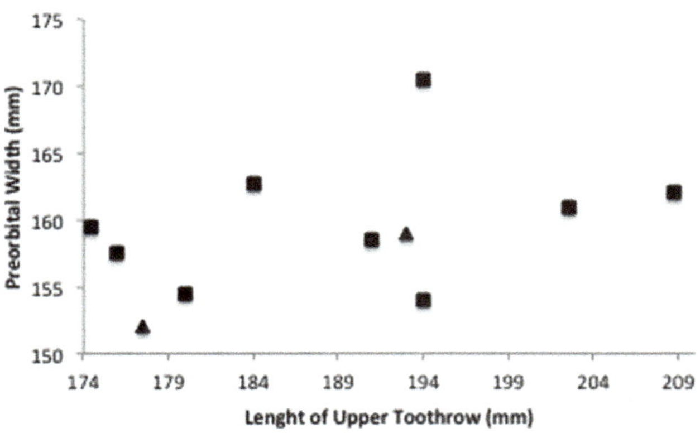

Figure 4. Scatterplot of preorbital width of the skull against length of the upper toothrow for 12 specimens of *Stenella clymene* stranded in Florida in 1995. Triangles represent the two hybrid individuals identified in this study.

Several cases of hybrid speciation have been documented in animals [9], suggesting that this phenomenon may be more important than previously thought. Some of these cases have been found due to discrepancies in molecular markers, as the ones we found here (e.g. [45]). In mammals, only one case has been fully confirmed [38] but a handful of others suggested [48,49], which shows that reticulate evolution is a viable mechanism for speciation in this group.

There are two alternative scenarios that could explain our results, although we consider them less likely: introgressive hybridization between *S. clymene* and *S. longirostris* such as the hybrid individuals that were part of the mass stranding in Florida, and non-reticulating forms of speciation including incomplete lineage sorting. If we had a case of introgressive hybridization, we would expect the hybrids to show intermediate morphological characters between the two species, which they do not. *S. clymene*

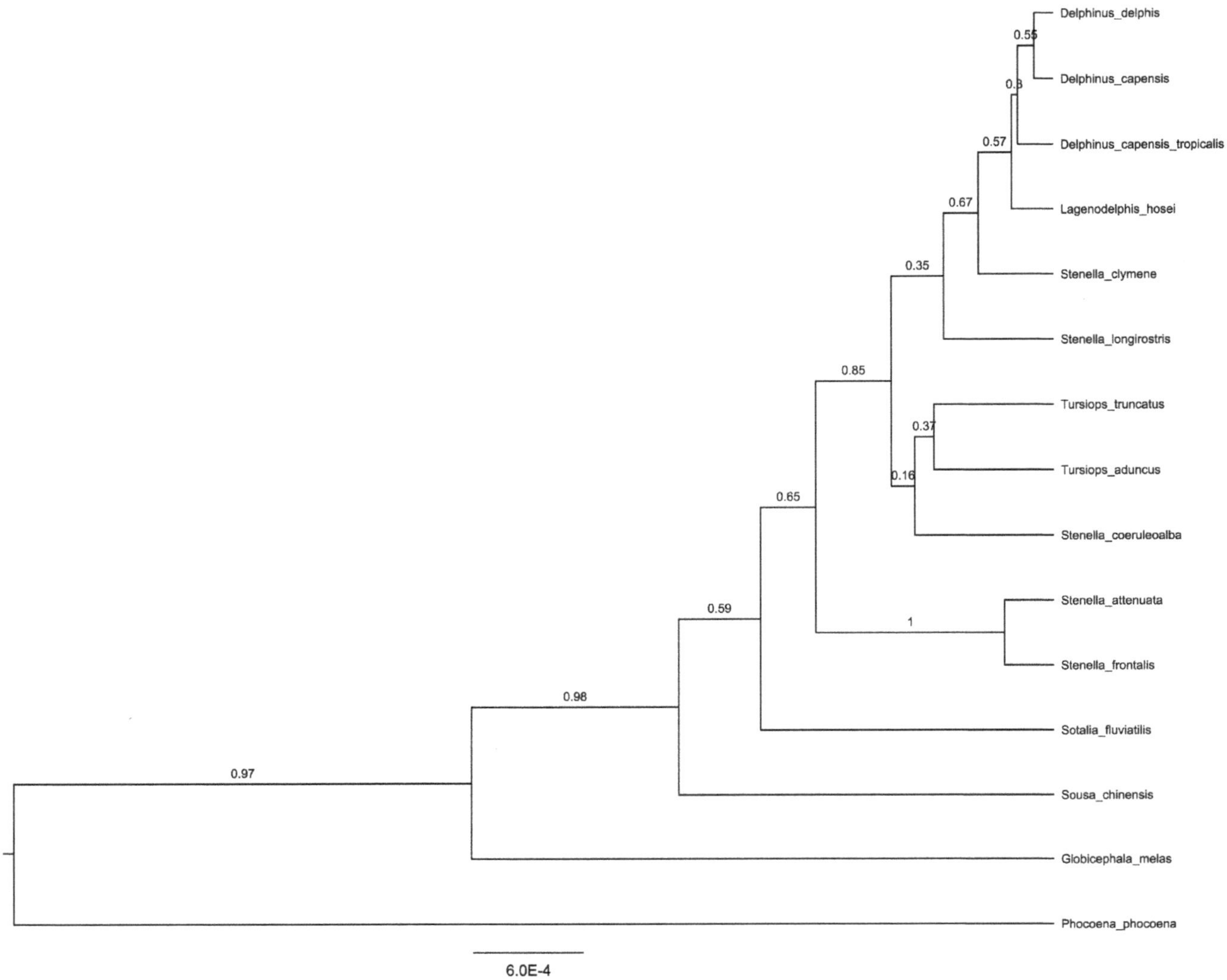

Figure 5. Species tree estimated with the *BEAST method. Posterior probability values are above nodes.

has morphological characteristics of both *S. coeruleoalba* and *S. longirostris,* which lead to a novel phenotype. In addition, sporadic hybridization does not explain the strong phylogenetic discordance seen between mtDNA and nuDNA markers. Incomplete lineage sorting could explain the different patterns seen in the nuclear haplotype networks, and it is likely still affecting the nuclear genome of these species, given their recent divergence. However, it does not entirely explain the strong discordance between mtDNA and nuDNA markers or the morphological trait variation seen in *S. clymene.* Incomplete lineage sorting is usually found at any stage during a process of divergence or speciation, until populations or species are completely differentiated. However, even when this differentiation seems complete, we may still find portions of the genome where alleles are shared among populations or species, specially if there is ongoing gene flow (hybridization). Dolphin species within the subfamily Delphininae are known to have diverged quite recently, possibly during a rapid radiation event [13]. This coupled with a slow nuclear genome evolution which seems to be characteristic of cetacean species, leads to a general lack of polymorphism in nuclear loci and subsequent low levels of nuclear genetic differentiation among species [50].

The evolutionary relationships among species of the genus *Stenella* have been contentious to date, with morphological and nuDNA characters disagreeing with mtDNA [13,22]. We suggest that this is likely due to the occurrence of natural hybridization among these species. This phenomenon is not easily incorporated into phylogenetic analyses, although some recent methods have been developed (e.g. [51,52]), and until it is taken into account, a clear understanding of the relationships among these will not be possible.

Our study provides the first evidence of a marine mammal species that originated through hybridization between two other species. We suggest that sexual selection through assortative mating is likely the mechanism responsible for maintaining *Stenella clymene* as an independent lineage, despite ongoing backcrosses with the parental species. The permeability of the genome and the karyological uniformity of these species, coupled with a complex social structure and behavior that likely contribute to the establishment of reproductive isolation, suggest that hybrid speciation may be a more common evolutionary process in these species than previously thought. As more molecular studies become available, we expect to see additional cases of hybrid speciation reported in the near future. Such reports will provide evidence that, as with other animal complexes, reticulate evolution has been an important evolutionary process contributing to the diversity of marine mammals as well.

Supporting Information

Figure S1 Photographs of external appearance of a *Stenella clymene* x *Stenella longirostris* hybrid (a, b) and of *Stenella clymene* individuals (c).

Table S1 Skull measurements for 12 *Stenella clymene* specimens stranded in Florida in 1995.

Acknowledgments

We would like to thank K. M. Robertson and the Southwest Fisheries Science Center for providing samples for this study; M. Arnold for providing comments to the manuscript; W. F. Perrin and R. G. LeDuc for first putting the hypothesis of hybrid speciation in the clymene dolphin forward. We are grateful to the staff of the AMNH New York, in particular to Rebecca Hersch. We also thank Matthew Leslie and Catherine Munsch for previous contributions in the laboratory.

Author Contributions

Conceived and designed the experiments: ARA HCR. Performed the experiments: ARA. Analyzed the data: ARA GL. Contributed reagents/materials/analysis tools: GL MMC GA HCR. Wrote the paper: ARA.

References

1. Arnold ML (1992) Natural hybridization as an evolutionary process. Annu Rev Ecol Syst 23: 237–261.
2. Barton NH (2001) The role of hybridization in evolution. Mol Ecol 10: 551–568.
3. Mayr E (1963) Animal Species and Evolution. Cambridge, Massachusetts: Harvard University Press.
4. Mallet J (2005) Hybridization as an invasion of the genome. Trends Ecol Evol 20: 229–237.
5. Buerkle CA, Morris RJ, Asmussen MA, Rieseberg LH (2000) The likelihood of homoploid hybrid speciation. Heredity 84: 441–451.
6. Baskett ML, Gomulkiewicz R (2011) Introgressive hybridization as a mechanism for species rescue. Theor Ecol 4: 223–239.
7. Mallet J (2007) Hybrid speciation. Nature 446: 279–283.
8. Rieseberg LH (1997) Hybrid origins of plant species. Annu Rev Ecol Syst 28: 359–389.
9. Mavarez J, Linares M (2008) Homoploid hybrid speciation in animals. Mol Ecol 17: 4181–4185.
10. Baker RJ, Bradley RD (2006) Speciation in mammals and the genetic species concept. J Mammal 87: 643–662.
11. Shurtliff QR (2013) Mammalian hybrid zones: a review. Mamm Rev 43: 1–21.
12. Arnason U, Gullberg A, Janke A (2004) Mitogenomic analyses provide new insights into cetacean origin and evolution. Gene 333: 27–34.
13. Amaral AR, Jackson JA, Moeller LM, Beheregaray LB, Manuela Coelho M (2012) Species tree of a recent radiation: The subfamily Delphininae (Cetacea, Mammalia). Mol Phylogenet Evol 64: 243–253.
14. Bérubé M (2002) Hybridism. In: Perrin WF, Wursig B, Theweissen JGM, editors. Encyclopedia of marine mammals. San Diego, Calif.: Academic Press Inc. pp. 596–600.
15. Arnason U (1980) C-banded and G-banded karyotypes of 3 delphinids - *Stenella clymene*, *Lagenorhynchus albirostris* and *Phocoena phocoena*. Hereditas 92: 179–187.
16. Zornetzer HR, Duffield DA (2003) Captive-born bottlenose dolphin x common dolphin (*Tursiops truncatus* x *Delphinus capensis)* intergeneric hybrids. Can J Zool 81: 1755–1762.
17. Silva-Jr JM, Silva FJL, Sazima I (2005) Presumed Interspecific hybrids in the genus *Stenella* (Delphinidae) in the Tropical West Atlantic. Aquatic Mammals 31.
18. Berube M, Aguilar A (1998) A new hybrid between a blue whale, *Balaenoptera musculus*, and a fin whale, *B-physalus*: Frequency and implications of hybridization. Mar Mamm Sci 14: 82–98.
19. Willis PM, Crespi BJ, Dill LM, Baird RW, Hanson MB (2004) Natural hybridization between Dall's porpoises (*Phocoenoides dalli*) and harbour porpoises (*Phocoena phocoena*). Can J Zool-Rev Can Zool 82: 828–834.
20. Perrin WF, Mitchell ED, Mead JG, Caldwell DK, Vanbree PJH (1981) *Stenella clymene*, a rediscovered tropical dolphin of the Atlantic J Mammal 62: 583–598.
21. Kingston SE, Adams LD, Rosel PE (2009) Testing mitochondrial sequences and anonymous nuclear markers for phylogeny reconstruction in a rapidly radiating group: molecular systematics of the Delphininae (Cetacea: Odontoceti: Delphinidae). BMC Evol Biol 9: 245.
22. LeDuc RG, Perrin WF, Dizon AE (1999) Phylogenetic relationships among the delphinid cetaceans based on full cytochrome b sequences. Mar Mamm Sci 15: 619–648.
23. Rosel PE, Block BA (1996) Mitochondrial control region variability and global population structure in the swordfish, *Xiphias gladius*. Mar Biol 125: 11–22.
24. Amaral AR, Sequeira M, Cedeira-Martínez J, Coelho MM (2007) New insights on population genetic structure of *Delphinus delphis* from the northeast Atlantic and phylogenetic relationships within the genus inferred from two mitochondrial markers. Mar Biol 151: 1967–1976.
25. Amaral AR, Silva MC, Moller LM, Beheregaray LB, Coelho MM (2010) Anonymous nuclear markers for cetacean species. Conservat Gen 11: 1143–1146.
26. Roca AL, Georgiadis N, Pecon-Slattery J, O'Brien SJ (2001) Genetic evidence for two species of elephant in Africa. Science 293: 1473–1477.
27. Lyons LA, Kehler JS, O'Brien SJ (1999) Development of comparative anchor tagged sequences (CATS) for canine genome mapping. J Hered 90: 15–26.
28. Amaral AR, Beheregaray LB, Bilgmann K, Freitas L, Robertson KM, et al. (2012) Influences of past climatic changes on historical population structure and demography of a cosmopolitan marine predator, the common dolphin (genus *Delphinus*). Mol Ecol 21: 4854–4871.
29. Heled J, Drummond AJ (2010) Bayesian Inference of Species Trees from Multilocus Data. Mol Biol Evol 27: 570–580.
30. Drummond AJ, Suchard MA, Xie D, Rambaut A (2012) Bayesian Phylogenetics with BEAUti and the BEAST 1.7. Mol Biol Evol 29: 1969–1973.
31. Ronquist F, Teslenko M, van der Mark P, Ayres DL, Darling A, et al. (2012) MrBayes 3.2: Efficient Bayesian Phylogenetic Inference and Model Choice Across a Large Model Space. Syst Biol 61: 539–542.
32. Posada D, Crandall KA (1998) MODELTEST: testing the model of DNA substitution. Bioinformatics 14: 817–818.
33. Stephens M, Donnelly P (2003) A comparison of Bayesian methods for haplotype reconstruction from population genotype data. Am J Hum Genet 73: 1162–1169.
34. Stephens M, Smith NJ, Donnelly P (2001) A new statistical method for haplotype reconstruction from population data. Am J Hum Genet 68: 978–989.
35. Librado P, Rozas J (2009) DnaSP v5: a software for comprehensive analysis of DNA polymorphism data. Bioinformatics 25: 1451–1452.
36. Excoffier L, Lischer HEL (2010) Arlequin suite ver 3.5: a new series of programs to perform population genetics analyses under Linux and Windows. Mol Ecol Resour 10: 564–567.
37. Bandelt HJ, Forster P, Rohl A (1999) Median-joining networks for inferring intraspecific phylogenies. Mol Biol Evol 16: 37–48.
38. Larsen PA, Marchan-Rivadeneira MR, Baker RJ (2010) Natural hybridization generates mammalian lineage with species characteristics. Proc Natl Acad Sci U S A 107: 11447–11452.
39. Rieseberg LH, Archer MA, Wayne RK (1999) Transgressive segregation, adaptation and speciation. Heredity 83: 363–372.
40. Tobler M, Carson EW (2010) Environmental variation, hybridization, and phenotypic diversification in Cuatro Cienegas pupfishes. J Evol Biol 23: 1475–1489.
41. Dittrich-Reed DR, Fitzpatrick BM (2013) Transgressive Hybrids as Hopeful Monsters. Evolutionary Biology 40: 310–315.
42. Gross BL, Rieseberg LH (2005) The ecological genetics of homoploid hybrid speciation. J Hered 96: 241–252.
43. Jefferson TA, Curry BE (2003) *Stenella clymene*. Mammalian Species 726: 1–5.
44. Abbott R, Albach D, Ansell S, Arntzen JW, Baird SJE, et al. (2013) Hybridization and speciation. J Evol Biol 26: 229–246.
45. Meyer A, Salzburger W, Schartl M (2006) Hybrid origin of a swordtail species (Teleostei: *Xiphophorus clemenciae*) driven by sexual selection. Mol Ecol 15: 721–730.
46. Ritchie MG (2007) Sexual selection and speciation. Annual Review of Ecology Evolution and Systematics. pp. 79–102.
47. Bergman TJ, Beehner JC (2003) Hybrid zones and sexual selection: insights from the Awash baboon hybrid zone. In: Jones C, editor. Sexual Selection and Primates: New Insights and Directions. Norman, Oklahoma, USA: American Society of Primatologists. pp. 500–537.
48. Tosi AJ, Morales JC, Melnick DJ (2003) Paternal, maternal, and biparental molecular markers provide unique windows onto the evolutionary history of macaque monkeys. Evolution 57: 1419–1435.
49. Yonekawa H, Moriwaki K, Gotoh O, Miyashita N, Matsushima Y, et al. (1988) Hybrid origin of Japanese mice *Mus musculus molossinus* - evidence from restriction analysis of mitochondrial DNA. Mol Biol Evol 5: 63–78.
50. Jackson JA, Baker CS, Vant M, Steel DJ, Medrano-Gonzalez L, et al. (2009) Big and Slow: Phylogenetic Estimates of Molecular Evolution in Baleen Whales (Suborder Mysticeti). Mol Biol Evol 26: 2427–2440.
51. Kubatko LS (2009) Identifying Hybridization Events in the Presence of Coalescence via Model Selection. Syst Biol 58: 478–488.
52. Meng C, Kubatko LS (2009) Detecting hybrid speciation in the presence of incomplete lineage sorting using gene tree incongruence: A model. Theor Pop Biol 75: 35–45.

13

Satellites, the All-Seeing Eyes in the Sky: Counting Elephant Seals from Space

Clive R. McMahon[1]*, Hamish Howe[1], John van den Hoff[2], Rachael Alderman[3], Henk Brolsma[2], Mark A. Hindell[1]

1 Institute for Marine and Antarctic Studies, University of Tasmania, Hobart, Tasmania, Australia, **2** Australian Antarctic Division, Channel Highway, Kingston, Tasmania, Australia, **3** Threatened Species & Marine Section, Department of Primary Industries, Water and the Environment, Hobart, Tasmania, Australia

Abstract

Regular censuses are fundamental for the management of animal populations but, are logistically challenging for species living in remote regions. The advent of readily accessible, high resolution satellite images of earth mean that it is possible to resolve relatively small (0.6 m) objects, sufficient to discern large animals. To illustrate how these advances can be used to count animals in remote regions, individual elephant seals (*Mirounga leonina*) were counted using satellite imagery. We used an image taken on 10/10/2011 to count elephant seals (n = 1790±306 (95%CL)) on the isthmus of Macquarie Island, an estimate which overlapped with concurrent ground counts (n = 1991). The number of individuals *per* harem estimated using the two approaches were highly correlated, with a slope close to one and the estimated intercept also encompassing zero. This proof of concept opens the way for satellites to be used as a standard censusing technique for inaccessible and cryptically coloured species. Quantifying the population trends of higher order predators provides an especially informative and tractable indicator of ecosystem health.

Editor: Yan Ropert-Coudert, Institut Pluridisciplinaire Hubert Curien, France

Funding: The authors have no support or funding to report.

Competing Interests: The authors have declared that no competing interests exist.

* E-mail: clive.mcmahon@utas.edu.au

Introduction

The population trends of apex predators provide invaluable information on the state of the environment in which animals live because the status of a population is an integrated signal of the state of the lower trophic levels and the environments that sustain them. Determining the bio-physical drivers of such change, and identifying which elements are changing, can be difficult, because long times series are needed to accurately describe population trends and to determine the relationships between population fluctuations and broader environmental changes [1]. These time series rely on regular and accurate censuses which are difficult to maintain, especially for animals that occur on remote oceanic islands or in inaccessible locations such as Antarctica [2], because of logistical constraints in accessing these locations.

Recently launched satellites such as Geo-Eye-1 (panchromatic, 0.5 m resolution and multispectral imagery 1.65 m resolution), WorldView-1 (panchromatic, 0.6 m resolution), WorldView2 (panchromatic, 0.46 m resolution and 8-band multispectral imagery 1.8 m resolution) and QuickBird-2 (2.4 m multispectral, 0.6 m panchromatic) provide easily accessible high resolution images of the Earth's surface. This imagery increasingly provides a vital tool for the collection of information that addresses local to global-scale research and policy priorities [3]. High resolution satellite images have allowed ecologists and wildlife demographers to undertake desktop censuses of the distribution and abundances of animal populations, especially those in remote regions of the globe or those that are sensitive to disturbance [4,5,6]. Satellite images have been particularly useful in identifying new seabird nesting habitat [4,7–9], determining rookery size and species composition [10] and quantifying temporal changes in colony size [11]. Most seabird studies are presently restricted to identifying entire colonies [12], not individuals, and therefore need to make several assumptions about animal density and intra-colony distribution to make population estimates. Species that are larger than the current satellite imagery resolution avoid such requirements.

Several studies have used satellite images to count larger species at the individual level [5,13], including marine mammals, although pinnipeds have proven to be difficult [13]. The sole exception is the Weddell seal (*Leptonochotyes weddellii*) whose dark bodies were highly contrasted against the snow covered Antarctic sea-ice, making these an ideal test for the efficacy of satellite imagery to inform about seal abundance and change over time [5]. Southern elephant seals (*Mirounga leonina*) are the largest pinniped species whose terrestrial breeding sites occur on remote sub-Antarctic islands scattered throughout the Southern Ocean. Elephant seals are a relatively well-studied species and are especially tractable for detecting and recording significant changes in the Antarctic marine ecosystem because they are wide-ranging and therefore integrate environmental signals across ocean basins [14]. While it is possible to quantify their population status through regular breeding season censuses surprisingly little is known on the global population status and trends of this species especially the larger populations that breed on South Georgia and islands on the Kerguelen Plateau (>60% of the global population) [15]. Those islands are logistically extremely difficult to census regularly given their size, remoteness and the treacherous terrain that separates the breeding beaches on these islands. Consequently population information at those locations is sparse, collected infrequently and total island abundance estimates and population trends are often based on smaller sub-areas.

Typically, the number of number of breeding female elephant seals present in mid-October is taken as a reliable index of population size and used to monitor trends [16]. This is conventionally done by through annual ground counts which require considerable logistical investment that is difficult to guarantee from year-to-year. Unlike Weddell seals that are highly contrasted with their breeding substrate, female elephant seals are grey/brown coloured and aggregate in harems established on dark volcanic-sand beaches that offer little contrast, which may make them difficult to distinguish in satellite images.

One of the best-studied southern elephant seal populations breeds on Macquarie Island. About 15% of all females breeding at Macquarie Island congregate in spatially stable harems on the Isthmus study area. Elephant seals found within this study area have been censused by foot on a near-daily basis throughout each breeding season from 1988 onward [17]. Such a committed survey effort within an accessible and well defined area makes the Isthmus an ideal site against which to test the idea that southern elephant seals breeding at very remote, difficult to access locations could be accurately censused from satellite imagery.

This study aimed to use satellite imagery of the Macquarie Island isthmus study area captured during the October breeding season to count the number of female harems, the number of seals *per* harem and the total number of seals in the area. These estimates are then ground-truthed against counts of seals made on the same day. From this study we assess the applicability of the method to census seals observed at terrestrial breeding locations where their detection rates might be poor compared to ice-breeding species.

Methods

This study was carried out at Macquarie Island under ethics approval from the Australian Antarctic Animal Ethics Committee (ASAC 2265) and the Tasmanian Parks and Wildlife Service.

To count the number of elephant seals from space we acquired an image of the Macquarie Island isthmus study area (−54.504531°, 158.934330°, Fig. 1) taken on a relatively cloud free day (10/10/2011) by the Geo-Eye satellite. The image is a GeoEye-1 satellite image (DigitalGlobe Catalog ID for the image inumber: 10504100009C3600) captured 10 October 2011 and the off nadir angle is 25.8° (Fig. 2). Image processing was courtesy of the Australian Antarctic Division Data Centre by Angela Bender to produce a pan-sharpened orthorectified image (GE_10Oct2011_ps_orc). The resolution (pixel size) of the pan-sharpened orthorectified image is 0.5 metres. For our study we used the panchromatic Geoeye1 image given this cloud free image coincided with concurrent ground counts on the island. While this was sufficient for our purposes, multispectral sensors such as WorldView2 may be useful for other applications.

A niave observer (HH) searched this image for harems within the study area. We deliberatly used a niave observer to avoid biases introduced by prior knowledge of the island or the distribution and number of harems. Each harem was delineated by a polygon encompassing all animals and assigned a random number. Three counts in random order were made of each harem from a constant virtual height of 300 m and all seals in the harems were counted, using simple image viewing software (Google Earth). The mean of these three counts was taken to represent the number of seals in that harem.

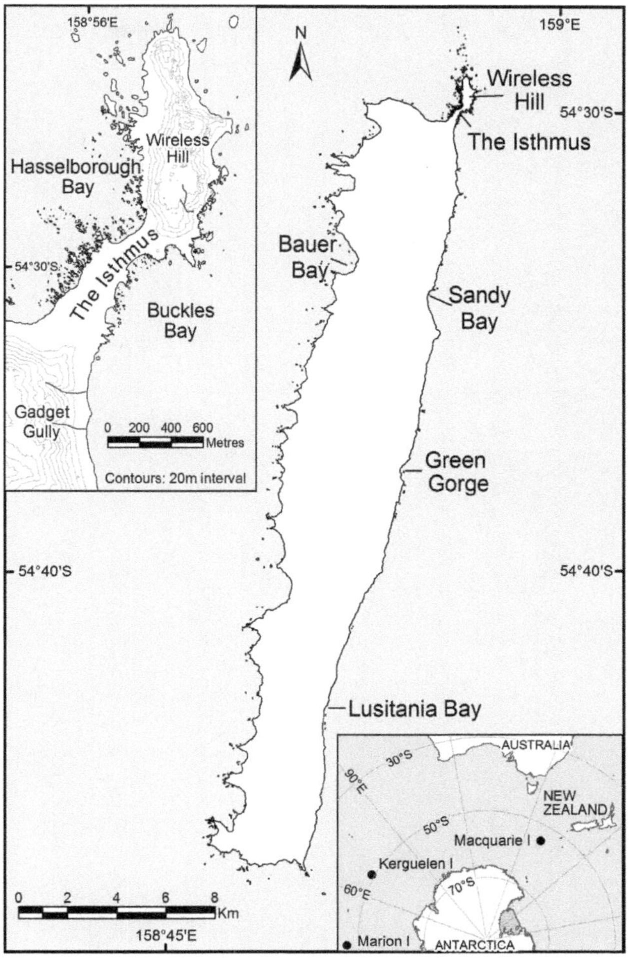

Figure 1. Macquarie Island and the main isthmus study area (upper left panel) and the location of Macquarie Island in the southern Pacific Ocean (lower right panel).

Isolated males and juveniles were excluded from the satellite counts because female counts are the standard way of quantifying population size in elephant seals. New born and nursing pups are approximately 1 m long at this time and coloured black, therefore it is unlikely they would have been counted along with other larger harem seals. No effort was made to distinguish between adult male and female seals within the harem polygons. Including males would not greatly affect counts within the harems because typically there is approximately one male per 50 female seals [18] *i.e.* the total count for the isthmus would be biased by only 40 seals given a total isthmus population of approximately 2000 adult female seals. It is also important to note that there is no diel pattern to elephant seal haulout as the seals remain ashore with their pups for 3 weeks, so time of day that the image was taken relative to the ground count is not a consideration in this study.

Only once the satellite counts had been completed did we compare them to the ground counts. Consequently, there was no prior knowledge of the ground counts at the time of the analyses of the satellite images. Ground counts of individual harems of female seals on the Macquarie Island isthmus were made by two observers. When individual ground counts differed, subsequent counts were done until the final tallies were within 5% [17]. These counts formed part of the regular annual monitoring programme whereby female elephant seals are counted throughout the

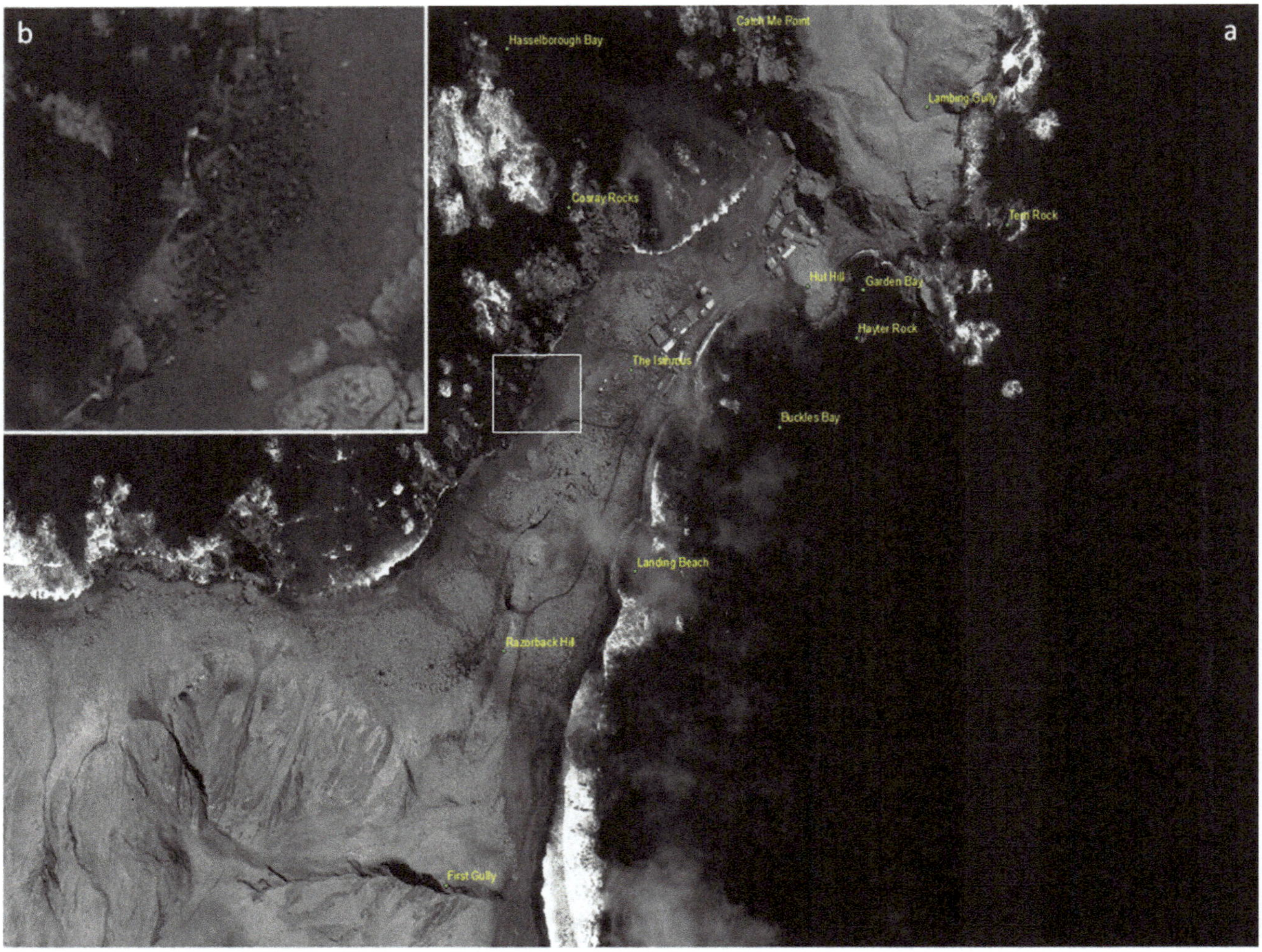

Figure 2. Southern elephant seals - *Mirounga leonina*, (a) congregate to breed annually in large groups known as harems on remote sub-Antarctic islands. These harems can be seen from space and individual seals are large enough (2.5 m long) to be counted (b).

breeding season at daily to weekly intervals. Fortuitously the ground count made on the 10th October 2011 coincided with acquisition of a cloud-free satellite image.

The satellite counts for each harem were compared to the corresponding ground counts for each harem using linear regression. The satellite counts and the ground counts of seal numbers were taken to be equivalent when the 95% CL of the mean satellite estimate overlapped the ground count.

Results

Here we demonstrate that despite their cryptic colouration, vital demographic information in the form of population censuses for elephant seals can be collected remotely by satellites, and that the abundance of seals estimated from satellite imagery were an accurate representation of numbers of seals counted on the beach on the same day. The Geo-Eye satellite image contained sufficient resolution to firstly locate all 12 harems on the isthmus study area (Fig. 2). Further, the mean satellite count of elephant seals within the study area was 1790±306 (n = 3 counts; 95% confidence limit) seals, compares well with the ground count of the study area made on the same day: 1991 adult females within 12 harems, which lies within the confidence limits of the satellite estimate.

A linear regression relating the number of seals per harem from satellite imagery and ground counts had a highly significant relationship ($F_{1,10} = 107.3$, $r^2 = 0.9062$, *Ground Counts* = 22.03+0.946**Satellite counts*). The slope of this relationship was 0.964±0.21(95% confidence limit) which encompasses 1.0 and the intercept of the relationship was 22.03±37.79 (95% confidence limit) thereby also encompassing zero. The line of parity (where ground counts equal satellite counts) is also contained within the 95% confidence intervals of the relationship (the dotted line in Fig 3).

Discussion

We show that individual elephant seals can be reliably and accurately counted from simple, single-spectrum satellite images using manual counting, thereby demonstrating the quick simple and relatively inexpensive utility of the technique. While it was undoubtedly fortuitous that an image was available that was taken on a cloud free day corresponding to a concurrent ground count, this limitation can be resolved given that many of the satellites can be directed to take specific images at specific times, although at some cost to the user. We found no difference between satellite and ground counts for: the total number of harems, the total number of seals on the isthmus, or the number of seals *per* harem,

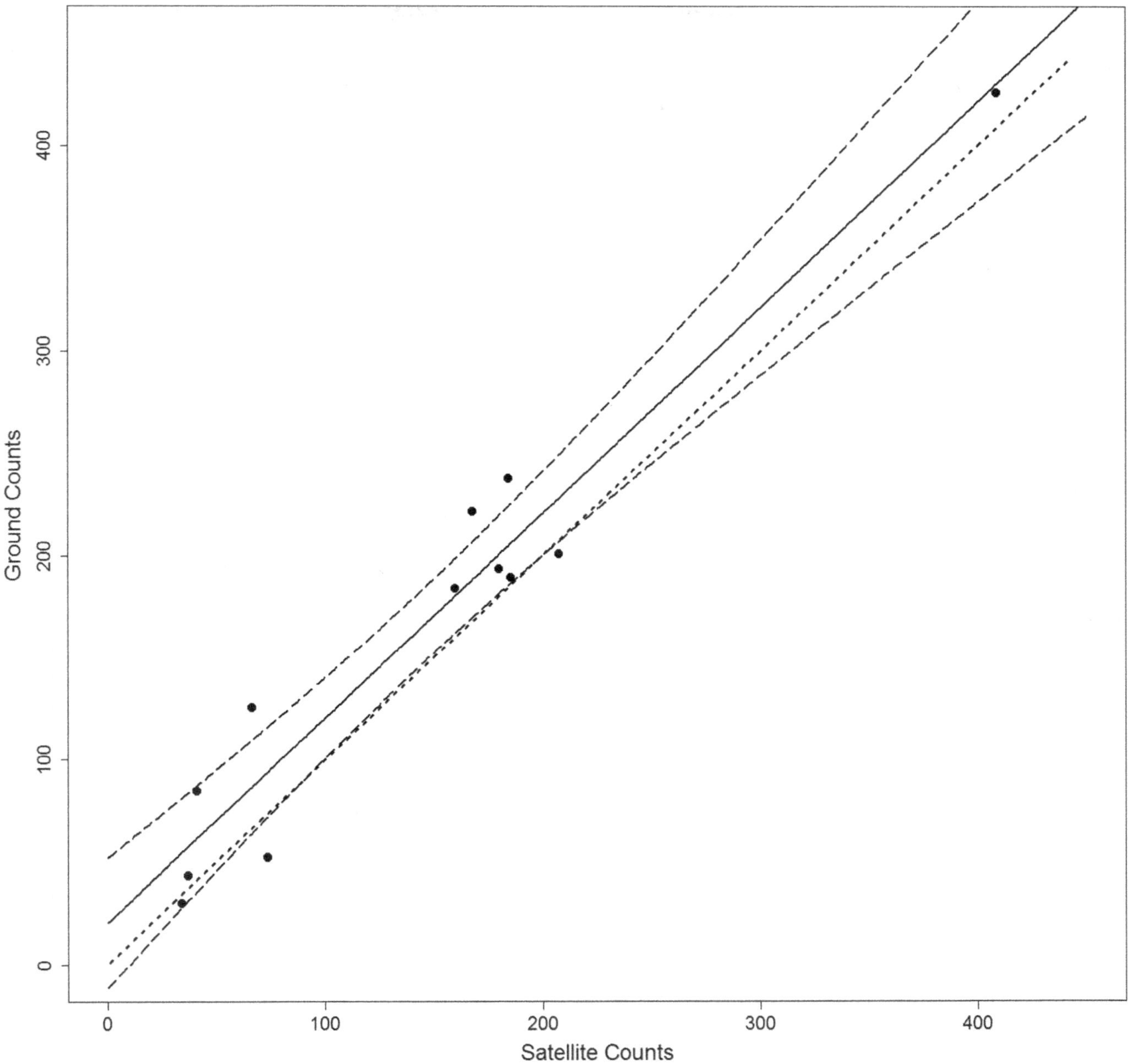

Figure 3. The number of seals counted from space was highly correlated to the actual numbers on beach determined by simultaneous ground counts (*Ground Counts* = 20.11+1.0003Satellite counts*).** The solid line represents the line of best fit, the dashed lines indicated the 95% Confidence limits of that line, and the dotted line is the line of parity (*i.e. Ground Counts = Satellite counts*).

demonstrating that remotely sensed images can reliably be used to census robustly elephant seals on remote sub-Antarctic islands. Earlier work by La Rue [5] demonstrated that individual Weddell seals could be accurately counted from space, but our study is the first to successfully count animals with poor contrast against their background *i.e.* dark bodies on a dark background. This finding greatly broadens the utility of the method, which can potentially be used in many terrestrial and coastal situations.

Despite the congruence of the mean satellite estimates with the ground counts there was nonetheless some disparity between the satellite derived estimates and ground counts. The degree of this disparity *i.e.* the difference between the two types of counts, is important because the power to detect inter-annual variability and abundance trends relies on the precision of annual census data.

There are several sources of variance within the estimates. The first is discerning the individual seals. Given that the pixel size of the images is 0.6 m, and that the average length of an adult female seal is 2.4 m and the width is 1.4 m, the seals will be represented in 4–6 pixels, it is unlikely that individual seals were missed. Consequently count variance is most likely due to our in ability to distinguish seals from similar sized rocks and from shading due to sun angle and from animals casting shadows onto adjacent seals in the tightly packed breeding harems [18]. While the harems are tightly packed, elephant seals do not lie on top of one another during the breeding season, and missing seals because they are stacked on top of each other is unlikely.

Resolving the inherent errors of detecting seals in natural settings from satellite images can be relatively easily overcome by

(i) increasing the number of replicate counts of the images, thereby reducing the variance around the counts, (ii) by obtaining images on other days during the breeding season when detection parameters may have improved and (iii) by adjusting the image to improve contrast and sharpen edges. This would improve our capacity to distinguish males within the harems and therefore give more accurate counts of females. Alternative methods such as aerial surveys could also improve the accuracy of counts but while aerial surveys give better resolution and hence higher accuracy, such surveys are expensive and often sub-Antarctic islands are not within range of survey aircraft.

Our findings illustrate the general utility of using satellite acquired census information to accurately enumerate the numbers of individual large-bodied animals in the wild. While our findings are a manifestly useful tool for counting elephant seals at remote and rarely visited islands such as Heard Island and the remote beaches of South Georgia which contain about 62% of the World's elephant seals population for which there are no contemporary counts. This techniques of counting seals using satellites can be extended to other Sothern Ocean pinnipeds such as pack-ice seals, other marine mammals such as whales or large terrestrial animals such as zebra, camels, elephants, bison and savannah ungulates that occur in open terrain [19,20]. This has the potential to revolutionize how animals are censused at a time when robust demographic information including longitudinal count series to quantify population trends and growth rates are sorely needed, especially in the light of the current biodiversity crisis [21–22].

Acknowledgments

We thank the 2011 ANARE expeditioners to Macquarie Island who conducted the ground counts. The satellite imagery is provided courtesy of DigitalGlobe 2013. DigitalGlobe confirm that publication of the satellite image under the agreement with the Australian Antarctic Division is permitted. We thank the Australian Antarctic Division Data centre (David Smith) for providing Figure 1 and Angela Bender for the pan-sharpened orthorectified image.

Author Contributions

Conceived and designed the experiments: CRM JvdH MAH. Performed the experiments: CRM HH JvdH HB MAH. Analyzed the data: CRM HH JvdH HB MAH. Contributed reagents/materials/analysis tools: JvdH RA CRM HB MAH. Wrote the paper: CRM HH JvdH RA HB MAH. Naive observer: HH. Gave informed consent to participate in the study verbally, this was not documented in writing: HH. Supervised HH's study: JvdH MAH.

References

1. McMahon CR, Bester MN, Hindell MA, Brook BW, Bradshaw CJA (2009) Shifting trends: detecting environmentally mediated regulation in long-lived marine vertebrates using time-series data Oecologia 159: 69–82.
2. Schofield O, Ducklow HW, Martinson DG, Meredith MP, Moline MA, et al. (2010) How do polar marine ecosystems respond to rapid climate change? Science 328: 1520–1523.
3. Horning N, Robinson J, Sterling E, Turner W, Spector S (2010) Remote sensing for ecology and conservation: Oxford University Press. 448 p.
4. Hughes BJ, Martin GR, Reynolds SJ (2011) The use of Google Earth (TM) satellite imagery to detect the nests of masked boobies Sula dactylatra. Wildl Biol 17: 210–216.
5. LaRue MA, Rotella JJ, Garrott RA, Siniff DB, Ainley DG, et al. (2011) Satellite imagery can be used to detect variation in abundance of Weddell seals (Leptonychotes weddellii) in Erebus Bay, Antarctica. Polar Biol 34: 1727–1737.
6. Platonov NG, Mordvintsev IN, Rozhnov VV (2013) The possibility of using high resolution satellite images for detection of marine mammals. Biol Bull 40: 197–205.
7. Fretwell PT, Trathan PN (2009) Penguins from space: faecal stains reveal the location of emperor penguin colonies. Glob Ecol Biogeogr 18: 543–552.
8. Fretwell PT, LaRue MA, Morin P, Kooyman GL, Wienecke B, et al. (2012) An Emperor Penguin Population Estimate: The First Global, Synoptic Survey of a Species from Space. Plos One 7.
9. Trathan PN, Fretwell PT, Stonehouse B (2011) First Recorded Loss of an Emperor Penguin Colony in the Recent Period of Antarctic Regional Warming: Implications for Other Colonies. Plos One 6.
10. Lynch HJ, White R, Black AD, Naveen R (2012) Detection, differentiation, and abundance estimation of penguin species by high-resolution satellite imagery. Polar Biol 35: 963–968.
11. Naveen R, Lynch HJ, Forrest S, Mueller T, Polito M (2012) First direct, site-wide penguin survey at Deception Island, Antarctica, suggests significant declines in breeding chinstrap penguins. Polar Biol 35: 1879–1888.
12. LaRue MA, Ainley DG, Swanson M, Dugger KM, Lyver POB, et al. (2013) Climate Change Winners: Receding Ice Fields Facilitate Colony Expansion and Altered Dynamics in an Adélie Penguin Metapopulation. Plos One 8: e60568.
13. Laliberte AS, Ripple WJ (2003) Automated wildlife counts from remotely sensed imagery. Wildl Soc Bull 31: 362–371.
14. Hindell MA, Bradshaw CJA, Guinet C, Harcourt RG (2003) Ecosystem monitoring and modelling: can marine mammals signal or predict change? In: Gales N, Hindell MA, Kirkwood R, editors. Marine mammals and humans: towards a sustainable balance. Melbourne: CSIRO Publishing.
15. McMahon CR, Bester MN, Burton HR, Hindell MA, Bradshaw CJA (2005) Population status, trends and a re-examination of the hypotheses explaining the recent declines of the southern elephant seal *Mirounga leonina*. Mamm Rev 35: 82–100.
16. Hindell MA, Burton HR (1987) Past and present status of the southern elephant seal (*Mirounga leonina*) at Macquarie Island. J Zool (Lond) 231: 365–380.
17. Van den Hoff J, Burton HR, Raymond B (2007) The population trend of southern elephant seals (Mirounga leonina L.) at Macquarie Island (1952–2004). Polar Biol 30: 1275–1283.
18. McMahon CR, Bradshaw CJA (2004) Harem choice and breeding experience of female southern elephant seals influence offspring survival. Behav Ecol Sociobiol 55: 349–362.
19. Gutro R (2005) Satellite Data to Track Wildlife: Elephants in Space. Available: http://www.nasa.gov/vision/earth/lookingatearth/elephants_space.html. Accessed 2013 October 1.
20. Yang Z (2012) Evaluating high resolution GeoEye-1 satellite imagery for mapping wildlife in open savannahs. Enschede, The Netherlands University of Twente 1–61 p.
21. Reich PB, Tilman D, Isbell F, Mueller K, Hobbie SE, et al. (2012) Impacts of biodiversity loss escalate through time as redundancy fades. Science 336: 589–592.
22. Hooper DU, Adair EC, Cardinale BJ, Byrnes JEK, Hungate BA, et al. (2012) A global synthesis reveals biodiversity loss as a major driver of ecosystem change. Nature 486: 105–108.

14

First Long-Term Behavioral Records from Cuvier's Beaked Whales (*Ziphius cavirostris*) Reveal Record-Breaking Dives

Gregory S. Schorr[1]*, Erin A. Falcone[1], David J. Moretti[2], Russel D. Andrews[3,4]

1 Cascadia Research Collective, Olympia, Washington, United States of America, **2** Naval Undersea Warfare Center, Newport, Rhode Island, United States of America, **3** School of Fisheries and Ocean Sciences, University of Alaska Fairbanks, Fairbanks, Alaska, United States of America, **4** Alaska SeaLife Center, Seward, Alaska, United States of America

Abstract

Cuvier's beaked whales (*Ziphius cavirostris*) are known as extreme divers, though behavioral data from this difficult-to-study species have been limited. They are also the species most often stranded in association with Mid-Frequency Active (MFA) sonar use, a relationship that remains poorly understood. We used satellite-linked tags to record the diving behavior and locations of eight *Ziphius* off the Southern California coast for periods up to three months. The effort resulted in 3732 hr of dive data with associated regional movements – the first dataset of its kind for any beaked whale – and included dives to 2992 m depth and lasting 137.5 min, both new mammalian dive records. Deep dives had a group mean depth of 1401 m (s.d. = 137.8, n = 1142) and duration of 67.4 min (s.d. = 6.9). The group mean time between deep dives was 102.3 min (s.d. = 30.8, n = 783). While the previously described stereotypic pattern of deep and shallow dives was apparent, there was considerable inter- and intra-individual variability in most parameters. There was significant diel behavioral variation, including increased time near the surface and decreased shallow diving at night. However, maximum depth and the proportion of time spent on deep dives (presumed foraging), varied little from day to night. Surprisingly, tagged whales were present within an MFA sonar training range for 38% of days locations were received, and though comprehensive records of sonar use during tag deployments were not available, we discuss the effects frequent acoustic disturbance may have had on the observed behaviors. These data better characterize the true behavioral range of this species, and suggest caution should be exercised when drawing conclusions about behavior using short-term datasets.

Editor: Andreas Fahlman, Texas A&M University-Corpus Christi, United States of America

Funding: Field work was funded by a grant to Cascadia Research Collective from the United States Navy (N45) via the Naval Postgraduate School (Grant Number N00244-10-1-0050) (http://www.nps.edu/). Development of the Mk10-A LIMPET tag was funded by Wildlife Computers and the Office of Naval Research (Grant Number N000140811203) with additional support from the Alaska SeaLife Center and National Oceanic and Atmospheric Administration/Northwest Fisheries Science Center. Marine Mammal Monitoring on Navy Ranges (M3R) is funded by the United States Navy (N45) via the Living Marine Resources program (http://www.lmr.navy.mil/). The funders had no role in study design, data collection and analysis, decision to publish, or preparation of the manuscript.

Competing Interests: DJM is employed by the United States Navy, though in a different division than the funders of this work. No part of the analysis or write up was influenced by the funders or sponsors of the research. This does not alter the authors' adherence to PLOS ONE policies on sharing data and materials.

* E-mail: gschorr@cascadiaresearch.org

Introduction

The beaked whales (family *Ziphiidae*), Cuvier's beaked whale (*Ziphius cavirostris*, hereafter *Ziphius)* in particular, have exceptional diving capabilities. Though it is the most broadly distributed beaked whale species, they remain poorly understood, as their behavior, especially their preference for deep water habitat typically far from shore, makes them notoriously difficult to study. To date, published dive data from *Ziphius* have come primarily from short-term data logging tags deployed in three different regions: the Ligurian Sea, Hawaii, and most recently Southern California. These deployments have averaged only 12 hr in length and in total represent less than 215 hrs and 327 dives [1–3]. Despite their limitations, these data suggest *Ziphius* routinely conduct some of the deepest and longest dives of any mammal [1,2], many of which exceed the estimated aerobic dive limit for the species by a factor of two or more [2]. *Ziphius* also appear unusually sensitive to acoustic disturbance; it is the species most frequently stranded coincident with Mid-Frequency Active (MFA) sonar exposure during military exercises [4]. Though it has generated a great deal of scientific interest, the causal relationship between these strandings and sonar remains elusive [3,5].

The U.S. Navy utilizes the Southern California Range Complex (SOCAL Range Complex) to conduct various training activities [6], including Anti-Submarine Warfare (ASW) training which frequently includes the use of MFA sonar (Figure 1). Between 2009 and 2014, the Navy is authorized to use up to 3408 hours of MFA sonar annually [7]. While MFA use can occur anywhere within the SOCAL Range Complex, the Southern California Anti-submarine Warfare Range (SOAR) (Figure 1), is specifically instrumented to facilitate ASW training, with 172 bottom-mounted hydrophones covering nearly 1800 km^2.

A recent study has sought to directly measure the behavioral responses of *Ziphius* to MFA by exposing suction-cup tagged whales to a simulated naval sonar signal. The first two successful playback experiments of this type occurred within the SOCAL Range Complex, and in both experiments, the tagged whale

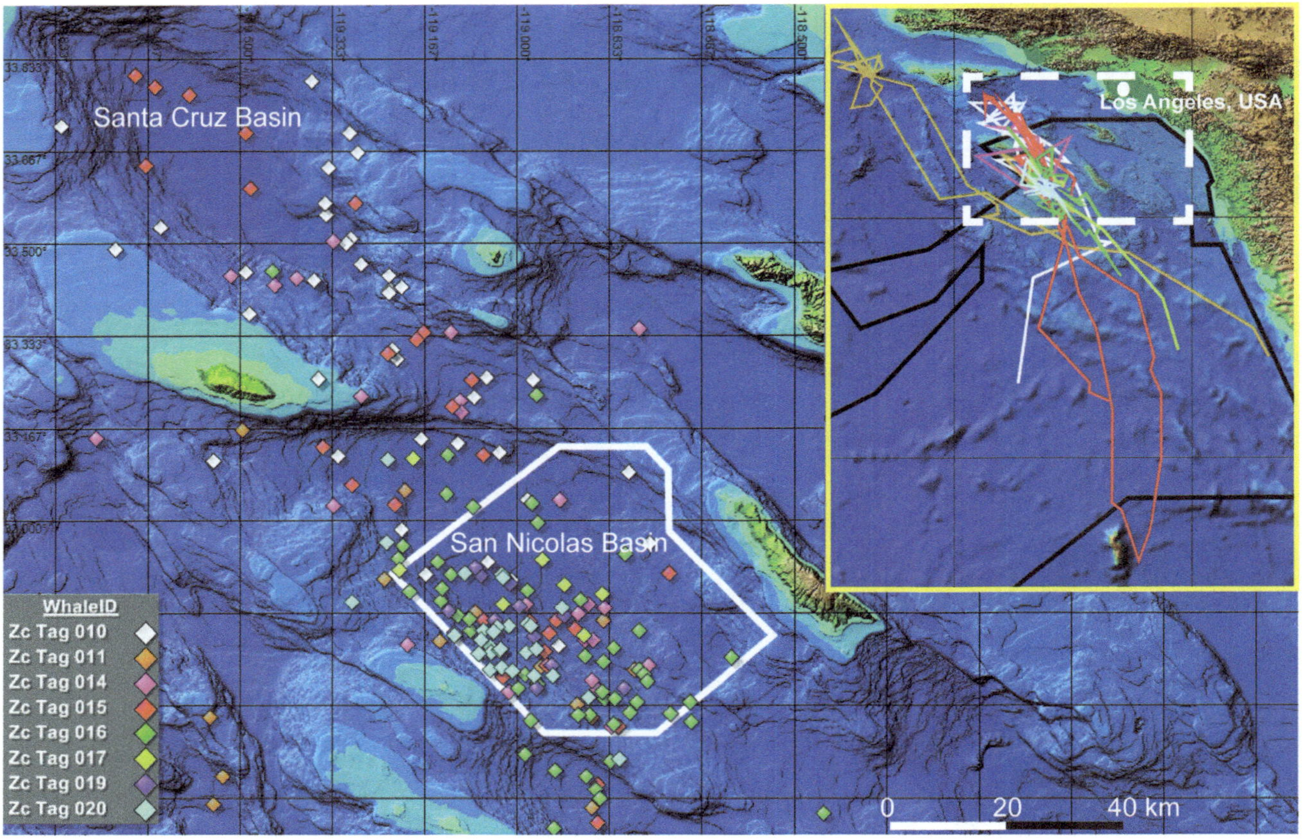

Figure 1. Map of the study area displaying one daily position estimate of each tagged whale. All eight whales were tagged within the Southern California Anti-submarine Warfare Range (outlined in solid white). Tagged whales were within the San Nicolas Basin for 51% of all days tags transmitted, and within the SOAR boundaries for 71% of the days when in the basin, suggesting site fidelity to the MFA sonar training range. Inset map shows the entire track-line of each tagged whale with the primary map region indicated by a dashed box. The SOCAL Range Complex, which encompasses all areas of authorized MFA sonar use [6], is lightly shaded and outlined in black in the inset map. Map created using Mysticetus Observation System, v1.8.0.124 (Entiat River Technologies, Preston, WA, USA).

responded strongly to simulated sonar at received levels as low as 89 dB re 1 μPa [3]. However, one whale was also incidentally exposed to real naval sonar from a distant ship at a received level of 106 dB re 1 μPa and showed no response, suggesting that the behavior of this species in the presence of sonar is likely influenced by more than simply received level [3]. Ultimately, the implications of the behavioral responses captured in these short-term tagging records are limited by a lack of long-term context in which they occur, since the behavior of individual *Ziphius* beyond a matter of hours post-tagging has not been documented.

In 2006, we began a cooperative visual-acoustic study combining the Marine Mammal Monitoring on Navy Ranges (M3R) program [8] and small boat visual surveys in order to identify which cetacean species utilize the SOAR range. Surprisingly, 37 *Ziphius* groups were encountered on the SOAR range during five surveys between 2006–2008, and re-sightings of photo-identified individuals suggested some degree of site-fidelity to this MFA training range [9]. In an effort to better characterize the distribution and diving behavior of *Ziphius* on the SOAR range, satellite tagging was added to the study as the technology became available. Here we describe the first long-term records of diving behavior from eight *Ziphius* tagged on SOAR (Figure 1), which provide insight into the true behavioral range of this species in a region of regular acoustic disturbance. We calculate basic parameters of dives and intervening surfacing bouts, describe patterns in the dive/surfacing cycle, and assess variability within and between individuals. We compare these values to published data from this and other regions, and discuss the relevance of these data to studies of the behavioral impacts of MFA sonar.

Materials and Methods

Ethics Statement

All field work was conducted under U.S. National Marine Fisheries Service research permits No. 540–1811 and 16111. Tagging was approved by the Cascadia Research Collective Institutional Animal Care and Use Committee.

Fieldwork

Small boat surveys were conducted as described in Falcone *et al.* [9]. We used a modified air rifle to deploy Mk10-A Argos-linked dive recorders (Wildlife Computers, Redmond, WA, USA) in the Low Impact Minimally Percutaneous External-electronics Transmitter (LIMPET) configuration [10,11] on or near the dorsal fin. The tags were affixed to the dorsal fin by two barbed darts constructed from medical-grade titanium which were gas-sterilized prior to implantation. Tagged whales were assigned to putative sex and age classes based on body size, pigmentation, and markings from photographs taken at the time of deployment, and compared to an existing regional catalog to assess sighting history, age and sex [9].

Tag Data

Dive data were collected using the Wildlife Computers Behavior Log (BL) function [12], which compiled records of user-defined "dives" and "surfacing bouts" and transmitted them via the Argos satellite system. Gaps in the BL occurred when one or more Argos message failed to be received due to factors such as no satellite coverage, surfacing behavior, weather, or duty cycling. Dives were defined as any submergence that exceeded 50 m depth and lasted longer than 30 s, and each dive record provided the start time, end time, maximum depth reached, and dive shape. Dive shapes were automatically classified by Wildlife Computers' Dive Analysis Program (DAP) version 3.0 build 141 [13] as "Square", "V", or "U" based on the proportion of time spent in the bottom phase of the dive, and were used to create a visual representation of the BL as an approximated dive trace. Maximum dive depths were reported in resolution steps of up to $+/-1.5\%$ of the actual maximum depth measured during the dive. The depth accuracy for the type of pressure transducers used in these tags was independently verified in a pressure chamber to 3000 m, which demonstrated a maximum error of$<+/-2.5\%$ of the recorded value.

Surfacing bouts represented the time between qualifying dives (i.e., when the whale did not descend below 50 m for more than 30 s), and included the start/end times only. Start/end times were determined by the transition from wet to dry at the surface for five tags, and crossing the 5 m depth threshold upon descent or ascent of a qualifying dive for the remaining three tags (Table 1). Because surfacing bouts were typically brief, the average value of surfacing bout durations differed significantly between tags using the wet/dry sensor and tags using the depth sensor to log start/end times (median 1.83 min wet/dry $n = 3380$, versus median = 2.39 min depth $n = 3429$, Mann-Whitney U Test, $p<0.001$). Thus statistical analyses of surfacing bout duration used only wet/dry sensor tags, as these more accurately characterize short respiratory bouts. The durations of dives, which were much longer than surfacing bouts, did not vary significantly by sensor type.

Tags were programmed to transmit daily for 28 days, and then to transmit on alternating days thereafter (duty cycle) to extend battery life (at the expense of data continuity later in the transmission period). Argos position estimates were filtered for plausibility with the Douglas Argos Filter [14] using the same parameters outlined by Schorr *et al.* [11]. The single best daily position estimates per whale (highest quality Argos location), as determined by the Douglas Argos Filter, were used to generate Figures 1 and 5; all other spatial analyses used the complete set of Argos positions that passed the filter.

BL records were assigned to day (solar elevation >0) or night classes (solar elevation <0) by calculating the solar elevation at the start time and the location of the nearest filtered Argos position estimate in time. Solar elevation was calculated using tools available from http://www.esrl.noaa.gov/gmd/grad/solcalc/calcdetails.html (accessed 21 August 2011). Geographic analyses were conducted using the Mysticetus Observation System, v1.8.0.124 (Entiat River Technologies, Preston, WA, USA).

Dive and Surfacing Bout Classification

Previous authors have partitioned beaked whale dives beyond 50 m into two classes based on their depth, duration, and the presence of foraging behavior as indicated by acoustic and/or fine-scale movement data [1,2]: long, deep dives associated with foraging by the presence of echolocation clicks and/or evidence of

Table 1. Summary of satellite tag deployments by individual.

Whale ID	Age Class	Sex	Event Start/End Time By	Deployment Date	Transmission Duration (days)	Total BL Data (hr)	Filtered Argos Positions (n)	Distance to Deployment Location (km) median (range)
Zc010	Adult	Female	Wet/Dry	29Jun2010	53.6	499.2	171	68.1 (5.3–265.5)
Zc011	Adult	Male	Wet/Dry	29Jun2010	89.8	544.7	174	198.7 (3.5–289.5)
Zc014	Sub-adult	Unknown	5 m Depth	06Jan2011	22.5	91.2	80	17.5 (2.4–94.4)
Zc015	Adult	Female	5 m Depth	06Jan2011	70.6	1015.9	291	83.8 (3.5–452.3)
Zc016	Adult	Male	5 m Depth	06Jan2011	88.7	857.5	194	20.0 (0.8–103.2)
Zc017	Adult	Female	Wet/Dry	23Jul2011	9.7	110.1	42	29.2 (6.3–235.7)
Zc019	Adult	Female	Wet/Dry	15Jan2012	12.0	172.5	49	12.2 (0.9–33.1)
Zc020	Adult	Male	Wet/Dry	15Jan2012	26.4	441.2	122	16.8 (0.8–49.7)

prey chases; and intervening bouts of shorter, shallower, silent dives with no indication of prey searching or chase (the purpose of which remains unclear [15–17]). Based on these descriptions, we performed a K-means cluster analysis on the BL dives for each individual, using dive depth and duration to partition all dives into "deep" and "shallow" classes, though the presence or absence of foraging behavior could not be determined for these dives with the sensors available. While these parameters provided clear separation into deep and shallow clusters for most dives (one-way ANOVA by individual, all p-values <0.0001), we visually verified the classification accuracy for all long dives which fell below the 5th percentile of depth or duration for that class and all shallow dives that fell above the 5th percentile for depth or duration for that class by individual dataset based on the pattern of surrounding dives in the record (Figure 2). Where there were gaps in the record surrounding a questionable dive or ambiguity in the pattern, the original K-means classification was retained. Eight total dives (0.1% of all dives) from only two individuals were reclassified from their original K-means cluster.

Following Tyack *et al.* [2], the Inter-Deep Dive Interval (IDDI) was calculated as the time between the end of one deep dive and the start time of the next deep dive when there were no data gaps in the intervening period. The overall deep dive rate was calculated by dividing the total number of deep dives by the total hours of BL data for each whale. The depths and durations of deep and shallow dives, surfacing bout durations and IDDIs were not normally distributed, so individual medians with ranges were calculated and the group mean of these values used to characterize events across all individuals.

Surfacing bouts were classified based on their position within the sequence of deep and shallow dives. Surfacing bouts immediately following a deep dive were called first surfacing bouts, those that separated sequential shallow dives were called intermediate surfacing bouts, those that immediately preceded a deep dive were called terminal surfacing bouts, and those that separated back-to-back deep dives were called single surfacing bouts.

Results

Eight tags were deployed on *Ziphius* at SOAR from 2010 to 2012, providing data for periods of up to three months (Table 1). The resulting dataset consisted of 3732 hrs of dive and surfacing bout records with 1123 concurrent animal location estimates (Table 1, Figure 1). In total, tagged whales performed 1142 deep dives to a group mean depth of 1401 m and duration of 67.4 min, and 5685 shallow dives averaging 275 m and 21.0 min (Table 2). The deepest dive reached 2992 m, and the longest dive lasted 137.5 min.

The stereotypical dive pattern previously described for *Ziphius* – single deep, foraging dives separated by a series of shallow dives [1,2] (Figure 3A) – was prevalent throughout the dive records of all eight whales, though there was substantial behavioral variation both within and among individuals (Figures 3B,3C). Occasional back-to-back deep dives were performed by all but one whale. In the combined data, deep dives ranged from 656–2992 m in depth and from 24.0–137.5 min in duration (Table 2). The group mean duration of the IDDI was 102 min and consisted of 4.13 shallow dives (Table 3), though this value ranged widely (0–21 shallow dives per IDDI). The group mean rate of deep dives was 0.30 per hour (s.d. = 0.05).

The durations of all surfacing bouts captured by the five tags using the wet/dry sensor were pooled and compared by type (Figure 4). Terminal surfacing bouts were significantly longer (median = 3.13 min, range = 0.33–79.75 $n = 499$) than both first surfacing bouts (median = 1.93 min, range = 0.02–79.73, $n = 509$) and IS (median = 1.66 min, range = 0.02–76.53, $n = 2188$) (Kruskal-Wallis multiple comparison, Z-value >14.6, $\alpha = 0.05$). Terminal surfacing bouts were not significantly longer than single surfacing bouts (Kruskal-Wallis multiple comparison, Z-val-

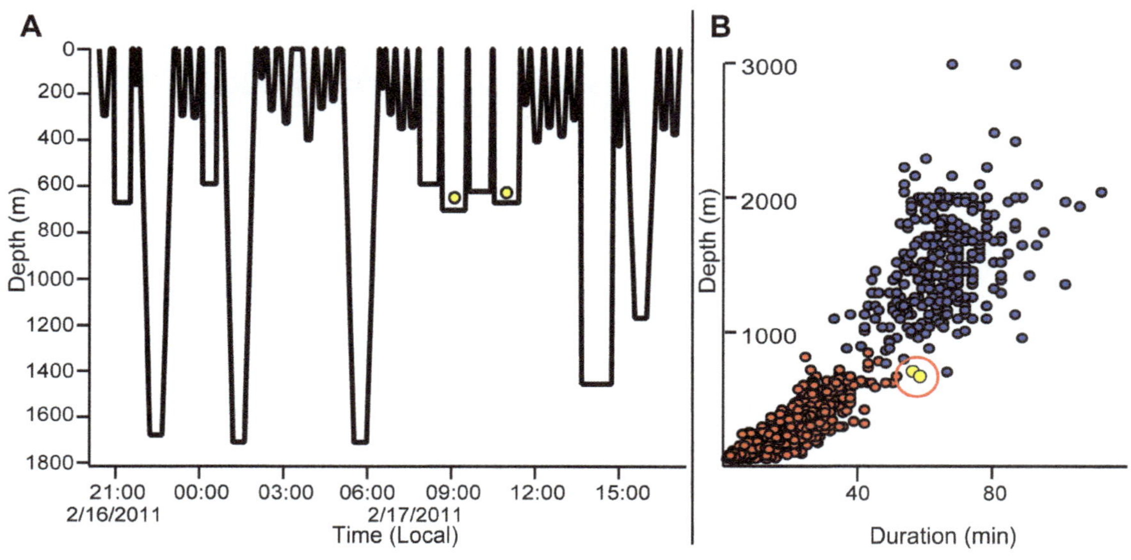

Figure 2. Classification of dive type. (**A**)Sample dive trace from Zc015 containing two reclassified dives. The four consecutive dives in the series beginning at 07:49 were all unusually long for shallow dives, but unusually shallow for deep dives, and thus difficult to classify using depth and duration alone. The K-means cluster analysis for Zc015 (**B**) classified the two dives indicated by yellow circles in each panel as deep dives and the other two as shallow dives. All four of these dives in figure A were similar and presumably of the same class. Other deep dives in the adjacent period were up to 1000 m deeper than these two dives, and four deep dives in series would be exceedingly rare based on this dataset and prior studies. Thus, we believe all four of these dives were much more likely to be atypically long, shallow dives, and the two indicated with yellow circles were reclassified as such.

Table 2. Summary of dive and surfacing bout parameters by individual.

	Deep Dives			Shallow Dives			Surfacing bouts	
Whale ID	*n*	*Depth (m)*	*Duration (min)*	*n*	*Depth (m)*	*Duration (min)*	*n*	*Duration (min)*
Zc010	187	1328 (880–1924)	59.2 (24.0–137.5)	908	280 (54–784)	17.6 (1.9–40.0)	1089	1.87 (0.02–38.1)
Zc011	129	1464 (768–2295)	78.4 (51.7–114.7)	846	280 (52–880)	24.0 (2.0–57.1)	963	1.59 (0.02–71.1)
Zc014	32	1520 (656–1872)	58.7 (36.8–82.7)	148	272 (58–592)	19.5 (2.0–41.1)	180	2.13 (0.50–38.1)*
Zc015	315	1488 (704–2992)	64.5 (33.1–112.5)	1617	288 (50–848)	21.3 (1.1–58.1)	1931	2.39 (0.30–205.5)*
Zc016	263	1360 (784–1840)	72.0 (35.7–110.4)	1054	312 (50–720)	25.6 (2.1–66.7)	1318	2.39 (0.08–123.1)*
Zc017	25	1616 (1072–1840)	76.3 (56.0–99.7)	184	256 (54–576)	19.7 (4.4–37.9)	209	1.99 (0.02–79.7)
Zc019	53	1264 (912–1808)	64.5 (38.9–91.2)	282	256 (52–1168)	18.7 (0.9–43.2)	334	1.99 (0.02–70.4)
Zc020	138	1168 (768–1872)	65.6 (36.8–114.7)	646	256 (58–784)	21.9 (1.0–57.1)	784	1.93 (0.25–84.8)
Total	**1142**	**1401 (137.8)**	**67.4 (6.9)**	**5685**	**275 (18.3)**	**21.0 (2.6)**	**6808**	**1.87 (0.17)/2.30 (0.2)***

Individual table values are medians, with ranges in parentheses. * indicates tags that logged the start and end of dives and surfacing bouts when the whale passed through 5 m depth on descent/ascent of qualifying dives; others used the wet/dry sensor on the tag for time stamps. The latter likely represent a more accurate measure of actual respiratory periods in surfacing bout metrics. Datasets are summed and group means (s.d.) are calculated across individuals at the bottom of the table.

ue = 1.14), but this is likely influenced by small sample size ($n = 8$) and high variability of single surfacing bout durations, which ranged from 2.4–84.8 min in length.

Diel difference in deep dive depth and duration were compared using a paired T-test of the individual median values. Foraging dive depths were significantly deeper at night (1487 m) than during the day (1345 m) (p one-tailed = 0.02, $df = 7$, T-value = −2.48), while deep dive durations were longer during the day (69.8 min) than at night (64.9 min) (p one-tailed = 0.001, $df = 7$, T-value = 4.53). A series of paired T-tests was run comparing the proportion of time spent in each behavior mode (deep dive, shallow dive and surfacing bout) for each whale during the day and night (Table 4). All tests were significant at $p = 0.05$, some strongly so.

Tagged whales utilized the San Nicolas Basin, which encompasses SOAR, on over 51% of days location estimates were received (Figure 1) and were documented in the basin during eight different months of the year. Thirty-six percent of all 1123 locations received were within SOAR boundaries. Group mean distance to tag deployment location was 56 km (range = 12–199, $n = 8$) (Table 1, Figure 5).

Discussion

As with past studies, the animals tagged here exhibited profound diving capabilities; however, the dive depths and durations reported here far exceed the prior records for this species. The deepest dive of 2992 m exceeded the previous maximum depth from this species by 1104 m [2], and the longest dive lasted 137.5 min, a 45% increase over the previous maximum dive duration [18]. These values also represent new mammalian dive depth and duration records, previously 2388 m [19] and 120 min [20] from southern elephant seals.

Surfacing bouts averaged less than 2 min (Table 2), reflecting this species' exceptionally short gas exchange interval, especially relative to their typically long dives. Surfacing bouts preceding deep dives were significantly, but not dramatically, longer than average (group mean = 3.29 min, $n = 5$), suggesting *Ziphius* may prepare for these longer submergences. Surfacing bouts that separated back-to-back deep dives were among the longest observed as a class (group mean = 35.7 min, range = 2.39–115.7, $n = 5$), though some of these were no longer than average (Figure 4). Extended surfacing bouts, some lasting several hours (max 205.5 min from a depth sensor tag), were uncommon, and occurred at various points in the dive cycle.

Diel patterns in dive behavior were strongly evident in this dataset, confirming previous observations from short overnight deployments [18] that *Ziphius* spend significantly more time in waters above 50 m at night than they do during the day. In fact, all but one of 22 surfacing bouts longer than 60 min occurred at night. Tagged whales spent significantly more time engaged in shallow diving during the day, but significantly less time above 50 m. Though deep dives were significantly deeper at night, the difference was only 142 m. The percentage of time spent deep diving was not significantly different from day to night (Table 4), suggesting whales foraged around the clock, targeting prey that did not vertically migrate.

The overall rate of deep dives (0.30 per hr, s.d. = 0.05), indicated our whales foraged approximately 7 times per day. *Ziphius* in Hawai'i conducted approximately 10 deep dives daily [1,18], and whales in the Ligurian Sea conducted 11–12 foraging dives per day [2]. Assuming the deep dive rate approximates foraging effort, whales in this study appear to forage less often – potentially by as much as 40% – than whales in the other study

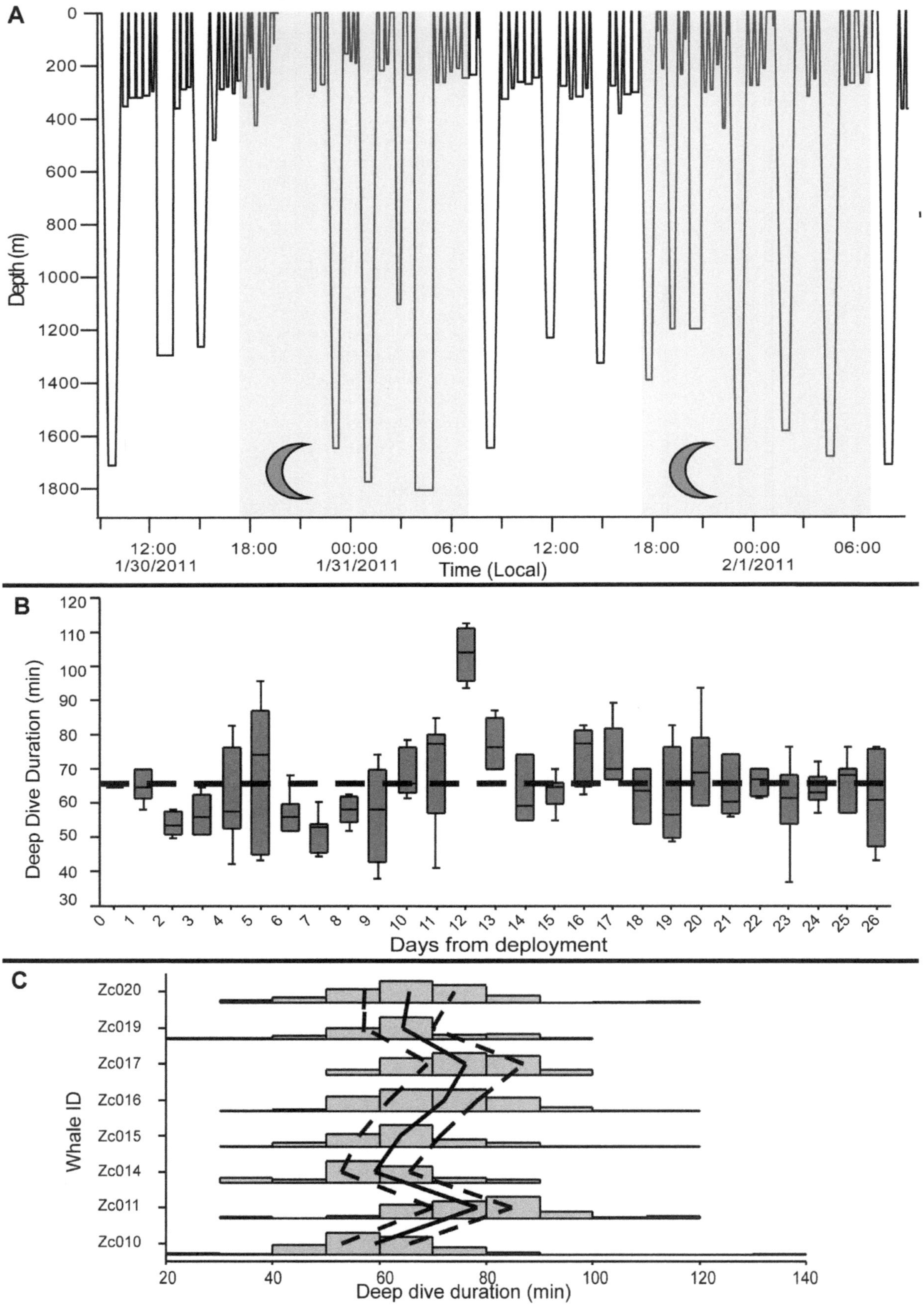

Figure 3. Variation in diving behavior. (**A**) A 48-hr BL dive trace from Zc015, illustrating the pattern of "deep" and "shallow" dives. Darkened areas represent night. The gap in the dive record during the first night period reflects data that were not received via Argos. (**B**) Box plot of deep dive

duration by day for the first 26 days of Zc015. The dashed line represents the overall median dive duration for Zc015 of 64.0 min. (**C**) Histogram plot of "deep" dive duration by individual, demonstrating the variability both within and among individuals. The solid black line connects the median value for each individual; the dashed lines connect the upper and lower 25th percentiles.

areas. This could be driven by several factors. The deeper and longer dives documented in this study may require longer recovery periods between them, which would in turn reduce the number of deep dives whales can conduct each day. However, on average, deep dives in this study were not dramatically deeper and were of similar duration to those observed elsewhere [18], so this seems unlikely to result in such a comparatively large decrease in foraging effort. Another possibility is that whales in our study area are more successful at foraging than whales elsewhere, and don't need to forage as often. This could be due to alternate foraging tactics, higher prey densities, or better prey quality. Unfortunately, direct measures of foraging activity are not available from these tags, and even general measures of prey type and density are difficult to collect with existing technology at the depths these whales are diving, so the role these factors play in foraging rate remains unknown, though undoubtedly important.

It is also possible that these apparent differences in regional foraging rates are an artifact of small sample sizes inherent in suction cup tag deployments. This view into longer-term behavioral patterns of *Ziphius* suggests caution should be exercised when comparing short-term behavior records from one tagged individual against others. Most parameters displayed greater variability than limited datasets from this species have captured. Figure 3A is a sample dive trace demonstrating variation in the depth, duration, and frequency of dives in one whale over a relatively short period of 48 hours – which is still a longer continuous period of data than any previous tag deployment on this species has captured [1–3]; the reduction in shallow diving and increased surfacing bout durations that occur more commonly at night are also evident. Figure 3B displays variation in deep dive duration within and between days for a single individual over four weeks. The total distributions of deep dive duration also varied considerably among individuals (figure 3C). All of this variation suggests that the behavioral range of this species is broader than their stereotypical patterning might suggest, both within and among individuals, and that a longer-term behavioral record from a given individual is the most appropriate context in which to evaluate a given behavior. Lacking this, an extensive behavioral sample from the population of interest might be the next most suitable alternative to evaluating the magnitude of a behavioral change.

The two *Ziphius* off Southern California that were exposed to simulated MFA sonar in an experimental context both appeared to exhibit strong responses at low received levels when comparing exposure dives to a small number of baseline dives [3]. Responses included increased dive duration, prolonged cessation of foraging, and rapid swimming away from the stimulus, though one whale did not respond similarly to very distant real MFA sonar at similar received levels (due to the disparity in transmission levels between the simulated source and actual Navy sonar). DeRuiter *et al.* [3] suggest that proximity to the source may have accounted for the lack of response to real MFA sonar. If similar behaviors occur in response to real MFA sonar but at closer proximity and/or higher received levels, it is possible that sonar exposure may have contributed to the longer average dive durations and apparently, reduced foraging rates observed in this study. While tags utilized in DeRuiter *et al.* [3] collect data on a very different scale than ours, a comparison of the basic parameters of the reactive behaviors against our more extensive regional data set is warranted, as is an evaluation of how such reactions might influence trends in our data, given the prevalence of sonar in the region.

Seven of the whales tagged in this study conducted a total of 75 dives longer than the longest reactive dive in the experimental exposures [3]; in fact, the longest reactive dive was only about 10 min longer than the median deep dive duration for two of our tagged whales. Thus, the reactive dive durations observed in the experimental exposures do not appear far outside the normal behavioral range of some *Ziphius* in this region. The longest average deep dive duration in our study was associated with the whale (Zc011) that remained farthest from SOAR and spent the most time well outside the area where the U.S. Navy is authorized to use MFA (Figures 1 and 5, Tables 1 and 2) [6]. These observations suggest that MFA exposure is unlikely to be a primary factor in the long average dive durations from this dataset, though it may influence some dives.

Table 3. Inter-deep dive intervals (IDDI) and deep dive rates.

	Inter-Deep Dive-Intervals			
Whale ID	*n*	*Duration (min)*	*# Shallow Dives*	**Deep Dives per hr**
Zc010	127	87.3 (2.7–276.3)	4 (0–10)	0.37
Zc011	68	132.2 (2.4–267.1)	5 (0–9)	0.24
Zc014	11	33.1 (3.7–116.9)	2 (0–4)	0.35
Zc015	249	116.7 (2.8–431.1)	5 (0–15)	0.31
Zc016	176	100.0 (26.3–399.2)	4 (0–10)	0.31
Zc017	13	139.6 (57.6–227.6)	5 (2–10)	0.23
Zc019	34	99.6 (28.6–369.3)	4 (0–21)	0.31
Zc020	105	110.4 (25.4–287.5)	4 (0–19)	0.31
Total	**783**	**102.3 (30.8)**	**4.13 (0.99)**	**0.3 (0.05)**

Individual table values are medians, with ranges in parentheses. The IDDI represents the period from the end of one deep dive to the beginning of the next deep dive, where there was no gap in dive data. Datasets are summed and group means (s.d.) are calculated across individuals at the bottom of the table.

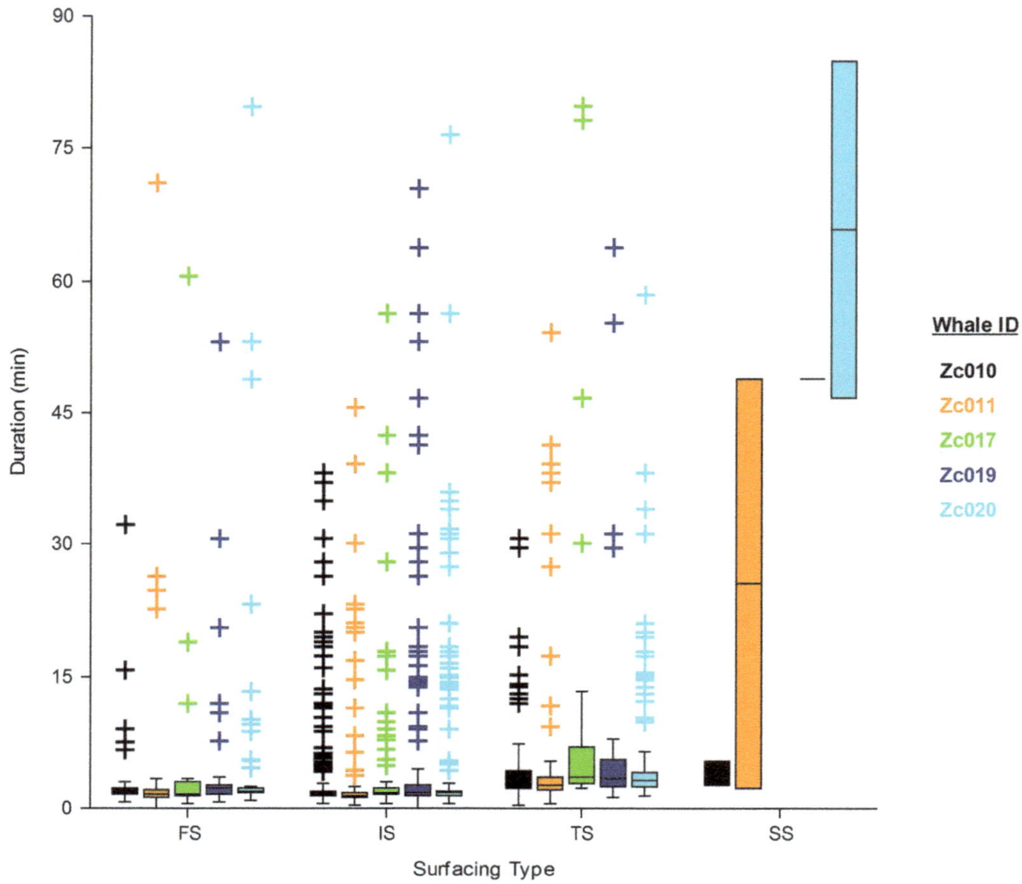

Figure 4. Box plot of surfacing bout duration by surfacing type, by individual. First Surfacing bouts (FS), Intermediate Surfacing bouts (IS), Terminal surfacing bouts (TS), and Single Surfacing bouts between deep dives (SS). Only the five tags using the wet/dry sensor are included (Table 2). Zc019 had only one SS, and Zc017 did not conduct any consecutive deep dives. Boxes represent the median and inter-quartile range, with whiskers at 1.5 times the inter-quartile range. Severe outliers are indicated by plus symbols. Surfacing bouts were typically very short, however occasional severe outliers (representing unusually long surfacing bouts) occurred in all surfacing bout types.

The post-exposure IDDIs in the experimentally-exposed whales (6.6 and 7.6 hrs) [3] were, however, on par with the longest IDDI captured in this dataset (7.2 hrs) (Table 3), and thus did likely represent a major deviation from normal behavioral patterning for whales in this region. If sonar exposure is responsible for the longer average IDDI in our data, it would negatively influence foraging rates. As whales in this region may be conducting upwards of 40% fewer foraging dives than *Ziphius* elsewhere [2,18], the impact of sonar could be significant if foraging success and caloric needs are otherwise similar among these populations. However, if we again look to Zc011 as an example, its foraging rate was among the lowest of all tagged whales (0.24 deep dives per hr), and it was arguably the least likely to have been exposed to MFA sonar during most of the deployment (Figure 1), suggesting that factors other than sonar also influence IDDI.

All whales tagged in this study were tagged within the boundaries of SOAR, an area where training exercises including the use of MFA sonar and a variety of other loud anthropogenic sound sources (e.g. explosions, acoustic counter-measures) occur regularly. The U.S. Navy Letter of Authorization suggests this could be as high as 2471 hours of ship-based MFA per year (an average of 6.8 hours/day), not including other sources such as active sonobuoys, dipping helicopter sonar, or submarine-based sonar [6]. Previous studies have suggested *Ziphius* may even alter their diving behavior in the presence of loud ships [21], which regularly pass through this region on their way to Los Angeles/ Long Beach Harbor.

Most of our tagged whales predominately used SOAR, and the waters of the San Nicolas Basin which encompass it, during the time of tag transmission. In total, they were present within the San Nicolas Basin on 53% of days tags transmitted, and spent 71% of their time within the boundaries of the SOAR range when in the basin (Figure 1). While four whales left the Southern California Bight, two returned to SOAR during the transmission period. Zc016 was inside the San Nicolas Basin on 74% of days over the three months her tag was active. Previous and ongoing photo-identification related to this tagging effort have revealed the same whales on SOAR repeatedly across years, and at times *Ziphius* occur in higher densities on SOAR than have been reported anywhere else along the US West Coast – the region across which this population is managed [9,22]. These findings suggest the San Nicolas Basin represents important habitat for these whales, despite its high level of acoustic disturbance. The reasons for this remain unclear, as the Southern California Bight is an area of complex oceanographic processes, and most productivity studies have occurred well above the depths at which these whales forage [23].

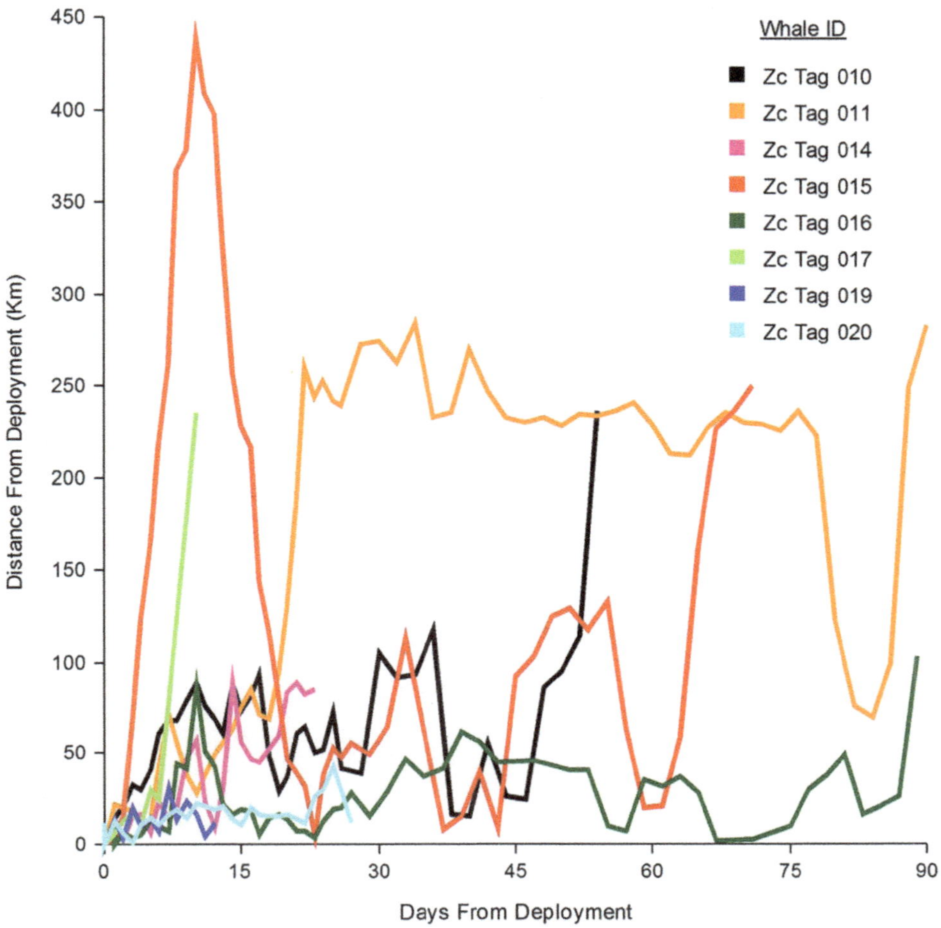

Figure 5. Distance to deployment location from day of deployment, by individual. Distance from deployment was calculated using the best location per day, as determined by the Douglas Argos Filter [14]. All individuals were tagged on the SOAR MFA sonar training range, and most remained in close proximity to the area in which they were tagged. While Zc015 made an extra-regional excursion over 450 km away, she returned to within 5 km of her tagging location.

Given the acoustic sensitivity of beaked whales and other odontocetes, it is hard to imagine that sonar use does not displace whales at least occasionally, an effect that was suggested by the experimental exposures [3] and also by a satellite-tagged Blainville's beaked whale that was exposed to MFA sonar in the Bahamas [24]. Though studies on longer-term, or larger scale displacement of disturbed cetaceans are limited, Morton and Symonds [25] demonstrated a significant decline in use of an area by killer whales while acoustic harassment devices were present, and harbor porpoises have been shown to abandon an area during loud pile driving operations [26]. Whatever displacement does occur in this population appears to be temporary. Unfortunately, with no data on beaked whale prey assemblages within the San Nicolas Basin, it is presently very difficult to quantify the effect this displacement might have on population health. All whales

Table 4. Diel differences in diving behavior.

Variable	Day	Night	*T*	*p*
Deep Dives per Hour	0.27	0.33	−3.28	<0.01
% Time Surface bout	0.070	0.175	−6.68	<0.01
% Time Shallow Diving	0.610	0.465	8.03	<0.01
% Time Deep Diving	0.320	0.360	−2.28	0.06

Behavior modes were partitioned into day and night classes for each tagged whale, and the total time spent within each class was calculated for each whale by day and night. A series of paired T-tests was then run ($df = 7$ for all tests) to assess diel differences. In general, whales spent more time at the surface at night and more time on shallow dives during the day. Though they conducted significantly more deep dives at night, the total time spent deep diving was not significantly different, as night time deep dives were often shorter in duration.

continued to forage even outside the San Nicolas Basin, so there is clearly suitable prey elsewhere.

Given the regular use of MFA sonar in the region [6,7], tagged whales were almost certainly exposed at some point during data collection. Without a doubt, a valuable application of this data set would be a detailed assessment of whale behavior in the presence and absence of acoustic disturbance. Unfortunately, detailed, accurate records of sonar use across the entire spatial and temporal scales these tags were active are much less readily available than one might expect, particularly when the animals were outside of SOAR. Previous authors correlating beaked whale strandings with sonar records have encountered these limitations both in this region and elsewhere [4,27]. Where available, records may include only the start time and location of MFA sonar use, despite activity that may last multiple hours, and during which ships may move large distances over variable courses. Single-ship events are often not captured. In light of this, an independent, comprehensive analysis of the entire dataset presented here in the context of sonar use is unlikely to ever be possible, especially given the apparent complexities of *Ziphius* response [3], which may be mediated by both source level and proximity, and would require detailed knowledge of vessel movements and sonar transmission schedules. Efforts to identify known sonar exposures in this dataset are underway, but they will involve only a subset of this dataset where major sources of acoustic disturbance - or just as importantly, lack thereof –can be accurately documented and independently verified.

Ultimately, the prolonged and recurrent use of the San Nicolas Basin by whales in this population suggests that *Ziphius* in this region have likely adapted to life with a certain amount of acoustic disturbance, and that local advantages may outweigh the costs it imposes. In light of this, the behavior of whales tagged in this region may not be representative of whales that are naive to such intense sounds. Given that whales tagged in this study far exceeded diving behavior previously described as extreme [2], the role humans might play in shaping this behavior can't be discounted. Regardless, this long-term data set provides a much more complete picture of the true diving capabilities of *Ziphius* in a region of regular acoustic disturbance.

Acknowledgments

We appreciate the support of H. Nevitt, R. Tahimic, D.J. Pascua, R. Svenson, C. Harr, D. Yamashita, G. Rice, D. Greenhoe, F. Stone, R. Gisner, Wildlife Computers, and the crew from CSC/SCORE. Many individuals assisted, but in particular we wish to thank M.B. Hanson, J. Foster, E. McCarthy, R. Morrissey, S. Jarvis, N. Di Marzio, A. Zerbini, R.W. Baird, J. Calambokidis, A. Dilley, F. Falcone, J. Falcone, S. Mahaffy, D. Holzer, D.L. Webster, and the Whale Acoustics Lab at Scripps Institution of Oceanography.

Author Contributions

Conceived and designed the experiments: GSS EAF DJM. Performed the experiments: GSS EAF DJM. Analyzed the data: GSS EAF RDA DJM. Contributed reagents/materials/analysis tools: GSS EAF RDA DJM. Wrote the paper: GSS EAF RDA DJM.

References

1. Baird RW, Webster DL, McSweeney DJ, Ligon AD, Schorr GS, et al. (2006) Diving behaviour of Cuvier's (Ziphius cavirostris) and Blainville's (Mesoplodon densirostris) beaked whales in Hawai'i. Can J Zool 84: 1120–1128. doi:10.1139/z06-095.
2. Tyack PL, Johnson M, Soto NA, Sturlese A, Madsen PT (2006) Extreme diving of beaked whales. J Exp Biol 209: 4238–4253. doi:10.1242/jeb.02505.
3. DeRuiter SL, Southall BL, Calambokidis J, Zimmer WM, Sadykova D, et al. (2013) First direct measurements of behavioural responses by Cuvier's beaked whales to mid-frequency active sonar. Biol Lett 9: 20130223. doi: 10.1098/rsbl.2013.0223.
4. D'Amico A, Gisner RC, Ketten DR, Hammock JA, Johnson C, et al. (2009) Beaked whalel strandings and naval exercises. Aquat Mamm 34: 452–472. doi:10.1578/AM.35.4.2009.452.
5. Cox TM, Ragen T, Read A, Vos E, Baird R, et al. (2006) Understanding the impacts of anthropogenic sound on beaked whales. J Cetacean Res Manag 7: 177–187.
6. Anonymous (2008) Southern California Range Complex Environmental Impact Statement. Available: http://hstteis.com/. Accessed 15 October 2013.
7. Anonymous (2009) Taking and importing marine mammals U.S. Navy Training in the Southern California Range Complex: final rule. Available: http://www.nmfs.noaa.gov/pr/permits/incidental.htm#loa. Accessed 2013 October 15.
8. Moretti D, Morrissey R, DiMarzio N, Ward J (2006) Verified passive acoustic detection of beaked whales (Mesoplodon densirostris) using distributed bottom-mounted hydrophones in the tongue of the ocean, Bahamas. J Acoust Soc Am 119: 3374.
9. Falcone E, Schorr G, Douglas A, Calambokidis J, Henderson E, et al. (2009) Sighting characteristics and photo-identification of Cuvier's beaked whales (*Ziphius cavirostris*) near San Clemente Island, California: a key area for beaked whales and the military? Mar Biol 156: 2631–2640. doi:10.1007/s00227-009-1289-8.
10. Andrews R, Pitman R, Ballance L (2008) Satellite tracking reveals distinct movement patterns for Type B and Type C killer whales in the southern Ross Sea, Antarctica. Polar Biol 31: 1461–1468. doi:10.1007/s00300-008-0487-z.
11. Schorr GS, Baird RW, Hanson MB, Webster DL, McSweeney DJ, et al. (2009) Movements of satellite-tagged Blainville's beaked whales off the island of Hawai'i. Endang Species Res 10: 203–213.
12. Wildlife Computers (2012) Wildlife Computers Mk10 Tag and Host User Guide. Available: http://wildlifecomputers.com/downloads.aspx.
13. Wildlife Computers (2013) Wildlife Computers Data Analysis Programs. Available: http://wildlifecomputers.com/downloads.aspx.
14. Douglas DC, Weinzierl R, Davidson SC, Kays R, Wikelski M, et al. (2012) Moderating Argos location errors in animal tracking data. Methods Ecol Evol 3: 999–1007.
15. Jepson PD, Arbelo M, Deaville R, Patterson IAP, Castro P, et al. (2003) Gas-bubble lesions in stranded cetaceans. Nature 425: 575–576. doi:10.1038/425575a.
16. Zimmer WMX, Tyack PL (2007) Repetetive shallow dives pose decompression risk in deep-diving beaked whales. Mar Mammal Sci 23: 888–925. doi:10.1111/j.1748-7692.2007.00152.x.
17. Hooker SK, Baird RW, Fahlman A (2009) Could beaked whales get the bends?: Effect of diving behaviour and physiology on modelled gas exchange for three species: Ziphius cavirostris, Mesoplodon densirostris and Hyperoodon ampullatus. Respir Physiol Neurobiol 167: 235–246. doi:10.1016/j.resp.2009.04.023.
18. Baird RW, Webster DL, Schorr GS, McSweeney DJ, Barlow J (2008) Diel variation in beaked whale diving behavior. Mar Mammal Sci 24: 630–642. doi:10.1111/j.1748-7692.2008.00211.x.
19. Costa DP, Huckstadt LA, Crocker DE, McDonald BI, Goebel ME, et al. (2010) Approaches to Studying Climatic Change and its Role on the Habitat Selection of Antarctic Pinnipeds. Integr Comp Biol 50: 1018–1030. doi:10.1093/icb/icq054.
20. Hindell MA, Slip DJ, Burton HR, Bryden MM (1992) Physiological implications of continuous, prolonged, and deep dives of the southern elephant seal (Mirounga leonina). Can J Zool 70: 370–379. doi:10.1139/z92-055.
21. Aguilar Soto N, Johnson M, Madsen PT, Tyack PL, Bocconcelli A, et al. (2006) Does intense ship noise disrupt foraging in deep-diving Cuvier's beaked whales (Ziphius cavirostris)? Mar Mammal Sci 22: 690–699. doi:10.1111/j.1748-7692.2006.00044.x.
22. Carretta J, Oleson E, Weller D, Lang A, Forney K, et al. (2012) US Pacific Marine Mammal Stock Assessments: 2012. NOAA.
23. McClatchie S (2013) Regional Fisheries Oceanography of the California Current System. Springer. 253 p.
24. Tyack PL, Zimmer WMX, Moretti D, Southall BL, Claridge DE, et al. (2011) Beaked Whales Respond to Simulated and Actual Navy Sonar. PLoS ONE 6: e17009. doi:10.1371/journal.pone.0017009.
25. Morton AB, Symonds HK (2002) Displacement of Orcinus orca (L.) by high amplitude sound in British Columbia, Canada. ICES J Mar Sci J Cons 59: 71–80.
26. Brandt MJ, Diederichs A, Betke K, Nehls G (2011) Responses of harbour porpoises to pile driving at the Horns Rev II offshore wind farm in the Danish North Sea. Mar Ecol Prog Ser 421: 205–216.
27. Filadelfo R, Pinelis YK, Davis S, Chase R, Mintz J, et al. (2009) Correlating whale strandings with Navy exercises off Southern California. Aquat Mamm 35: 445–451. doi:DOI 10.1578/AM.35.4.2009.445.

15

Fine-Scale Variability in Harbor Seal Foraging Behavior

Kenady Wilson[1*¤], Monique Lance[2], Steven Jeffries[2], Alejandro Acevedo-Gutiérrez[1]

1 Western Washington University, Department of Biology, Bellingham, Washington, United States of America, **2** Washington Department of Fish and Wildlife, Wildlife Science Program, Lakewood, Washington, United States of America

Abstract

Understanding the variability of foraging behavior within a population of predators is important for determining their role in the ecosystem and how they may respond to future ecosystem changes. However, such variability has seldom been studied in harbor seals on a fine spatial scale (<30 km). We used a combination of standard and Bayesian generalized linear mixed models to explore how environmental variables influenced the dive behavior of harbor seals. Time-depth recorders were deployed on harbor seals from two haul-out sites in the Salish Sea in 2007 (n = 18) and 2008 (n = 11). Three behavioral bout types were classified from six dive types within each bout; however, one of these bout types was related to haul-out activity and was excluded from analyses. Deep foraging bouts (Type I) were the predominant type used throughout the study; however, variation in the use of bout types was observed relative to haul-out site, season, sex, and light (day/night). The proportional use of Type I and Type II (shallow foraging/traveling) bouts differed dramatically between haul-out sites, seasons, sexes, and whether it was day or night; individual variability between seals also contributed to the observed differences. We hypothesize that this variation in dive behavior was related to habitat or prey specialization by seals from different haul-out sites, or individual variability between seals in the study area. The results highlight the potential influence of habitat and specialization on the foraging behavior of harbor seals, and may help explain the variability in diet that is observed between different haul-out site groups in this population.

Editor: Andreas Fahlman, Texas A&M University-Corpus Christi, United States of America

Funding: This project was supported in part by Award #0550443 from the National Science Foundation (http://www.nsf.gov/) (A.A.) and by funds from the Western Washington University Department of Biology (http://www.biol.wwu.edu/biology/ (K.W.), the Office of Research for Sponsored Programs (http://www.wwu.edu/rsp/) (K.W.) and the Washington Department of Fish and Wildlife (http://wdfw.wa.gov) (S.J. and M.L.). The funders had no role in study design, data collection and analysis, decision to publish, or preparation of the manuscript.

Competing Interests: The authors have declared that no competing interests exist.

* E-mail: kenady.wilson@duke.edu

¤ Current address: Duke University Marine Lab, Beaufort, North Carolina, United States of America

Introduction

Harbor seals are abundant marine predators throughout the northern hemisphere, yet we still have a limited understanding of their fine-scale behavior and ecological impacts in many regions. Although harbor seals dive for reasons unrelated to foraging (mating or resting at the bottom) diving is typically used as a proxy for understanding the foraging behavior of these mammals [1–3]. Modeling the fine-scale changes in dive behavior enables the prediction of future behaviors under varying environmental conditions. Further, understanding the variability of foraging behaviors within a population is important to fully understand the role of predators in the environment and identify the size and scale of specialization that may occur within a population or group of animals.

Harbor seals are ideal candidates for analyzing behavior on a fine spatial scale because the species has already been studied extensively around the world. Previous studies have revealed differences in diving behavior among different regional populations [4], and among age and sex classes [5–7]. They have also shown that harbor seals are opportunistic predators that feed on locally abundant prey and commonly switch foraging behaviors as prey abundances change seasonally and annually [8,9]. Studies have also revealed species-wide similarities in optimal foraging depths [10,11] and the importance of available habitat in determining foraging behavior [10,12]. Most research has identified one foraging pattern throughout a respective study area. For example, Bjørge et al. [13] and Tollit et al. [12] found that harbor seals fed mostly on benthic prey with little diurnal variation in dive depths or types. In contrast, studies conducted in Alaska and Canada indicated that foraging occurred most often at dusk and that seals demonstrated a distinct diurnal foraging pattern, using square-shaped dives as they followed the diurnal vertical migration of prey [4,14,15]. These seasonal, spatial, and diel variations in behavior [7,15,16] show that foraging behavior differs both among and between populations and geographic regions (e.g. Sable Island, a single large haul-out site, and SE Alaska, multiple haul-out sites in the same region), but these findings do not address small-scale behavioral variation that may occur between haul-out sites within a population.

The inland waters of Washington and British Columbia, a region known as the Salish Sea (Figure 1), are composed of diverse oceanographic and biological features. The Salish Sea includes three major basins: the Strait of Georgia, Puget Sound, and the Strait of Juan de Fuca, each with its own response to forcing mechanisms within the oceanographic system [17]. There are convoluted networks of islands, shallow tidal passes, estuaries, and a relatively large tidal range all within a restricted geographic area. The San Juan Islands are located near the convergence of all 3 of the major basins listed above, creating a dynamic and variable environment. Harbor seals are the only year-round resident pinnipeds in the Salish Sea and have an estimated population of 12 000 in the inland waters of Washington [18], and 39 000 in the Strait of Georgia in British Columbia [19]. There are numerous harbor seal haul-out sites throughout the San Juan Islands, where diet and population monitoring has been conducted since the late 1970s [18,20,21]. These sites tend to fall into two broad categories: estuarine (soft-bottomed bays) and non-estuarine (rocky-reef islands) [22,23]. The presence and abundance of prey species in certain areas is typically correlated with the habitat available and the sediment type [24]. Therefore, the variation in prey associated with different haul-out sites (rocky vs. sandy) in the region, suggests that the foraging behavior and diet of harbor seals may vary between site types. Given the diverse and dynamic ecosystem and the ubiquity of harbor seals in the area, the San Juan Islands is an excellent system in which to study the variability of foraging behaviors within a population of predators.

Harbor seals in the San Juan Islands can move long distances (> 100 km) and haul-out in multiple locations; however, they tend to travel to and from a single site and appear to be faithful to these locations [25–27]. Identifying differences between haul-out site groups, may be indicative of individual variability; however, it may also indicate prey or habitat specialization within the study area [28]. In the second case, seals may have adopted distinct foraging strategies depending on where they haul-out and which prey is readily available, even though seals hauling out less than 5 km away, in an area they could easily exploit, have adopted different strategies for use in a different habitat. The primary goal of this study was to identify differences in diving behavior between relatively close haul-out sites (~20 km apart). We hypothesized that foraging behavior in the San Juan Islands was site-dependent due to the dynamic and variable ecosystem and the diversity of available prey between haul-out sites. We tagged seals from two different haul-out sites in the San Juan Islands to examine this hypothesis and to describe the variability of diving behavior within this population and identify differences in foraging behavior for harbor seals on a small spatial scale.

Materials and Methods

Ethics Statement

This study was conducted in accordance with animal use protocols reviewed and approved by the Institutional Animal Care and Use Committee at Western Washington University (Protocol Number 06-005) and at the National Marine Mammal Laboratory. All research and animal handling was conducted under the Marine Mammal Protection Act Scientific Research Permit 782–1702 awarded to the National Marine Mammal Laboratory by the NOAA Office of Protected Resources for scientific research.

Study Site

Harbor seals were captured in the spring of 2007 and winter of 2007–2008 at two sites in the San Juan Islands: Padilla Bay National Estuarine Research Reserve (hereafter Padilla Bay) (n = 15) and Bird/Belle Rocks (n = 14) (Figure 1). Padilla Bay is a large soft-bottomed estuarine bay situated near the mouth of the Skagit River (center at 48°28.37′N, 122°30.88′W). Bird and Belle Rocks are rocky reef, non-estuarine haul-out sites located in Rosario Strait. Bird Rocks (48°29.16′N, 122°45.16′W) is a congregation of three rocky reef islands used year-round by harbor seals and during the winter months by Steller sea lions (*Eumetopias jubatus*). Belle Rock is a rocky reef located <1 km from Bird Rocks and is exposed only during low tide. Given their close proximity, data from Bird and Belle Rocks were combined for analysis and will from now on be referred to only as Bird Rocks. Both Bird and Belle Rocks are located roughly 20 km from Padilla Bay and seals are capable of moving between the two locations.

Instrument Deployment

Adult harbor seals were captured from April–May of 2007 and from November 2007–February 2008 following the methods of Jeffries et al. [29]. After capture, weight to the nearest 0.5 kg and straight length ±5 cm was measured, blood and blubber biopsies were taken, and time-depth recorders (TDRs; Wildlife Computers, Redmond, WA, Mk-9, or Mk-10F) were attached to each seal. Tags were glued to the pelage of the animal, along the dorsal midline between the shoulders, using 5-min epoxy (ITW Devcon, Danvers MA).

Tags were programmed to record data every day for the duration of the study and were expected to come off with the annual molt. TDRs were set to record time, pressure (depth), light level, and temperature (tag temp. for Mk-9, water temp. for Mk-10F) every 10 sec [15,30], and were equipped with an Eco-tech floatation pack with a VHF transmitter. The instrument package was positively buoyant and balanced to float with the VHF antenna upright to allow for tracking and recovery by boat after it became detached during the seals annual molt between August and October of either 2007 or 2008. Permit constraints required all instrument packages to weigh ≤1% of each tagged seal's body mass. This weight requirement is within the range recommended by Macdonald [31] and Brooks *et al* [32] to avoid adversely influencing the normal behaviors of instrumented animals, and is similar to that employed in other studies of harbor seal behavior [14,28,30,33]. Additional details on capture and tagging can be found in Peterson [27].

Data Processing

The study period was roughly 2 years in duration. Due to variable molt times, and the number of seals tagged at different times of the year, few seals retained their tags through the duration of the study. Therefore, data for each seal were visually inspected to designate a date where diving behavior was observed to end (i.e. if the tag fell off while the animal was on land, or the tag was floating in the water for days). Analysis of movement patterns of the tagged animals showed that although some rocky reef seals traveled great distances, all of the seals returned to their respective tagging sites throughout the study duration [25,27]. During this study, seals from Bird Rocks traveled close to Padilla Bay, but were not observed (foraging or hauled-out) within the estuary. Both groups remained faithful to the locations where they were tagged. This behavior suggested that pooling seals by their tagging sites accurately identified where they were hauling-out during this study. Data were downloaded and processed using software provided by Wildlife Computers (Redmond, WA). All dives were corrected using Zero-offset correction software to account for drift in the TDRs pressure transducer, which estimates water depth. The resolution of the pressure transducer was 0.5 m with an accuracy of ±1% of the depth reading; however, previous studies examining dive classification techniques show that dives with at least five depth readings, regardless of sampling, provide the most precise representation of dive shapes [34]. In this study, the number of dives with less than five readings increased dramatically for dives ≤5 m and were therefore excluded from analysis because we were using dive shape for classification purposes following Lesage et al. [14] and Baechler et al. [30].

Wildlife Computers' dive analysis software (v.1.0.55) was used to analyze the corrected dive records and to classify the following variables for each dive: maximum depth, duration, bottom time

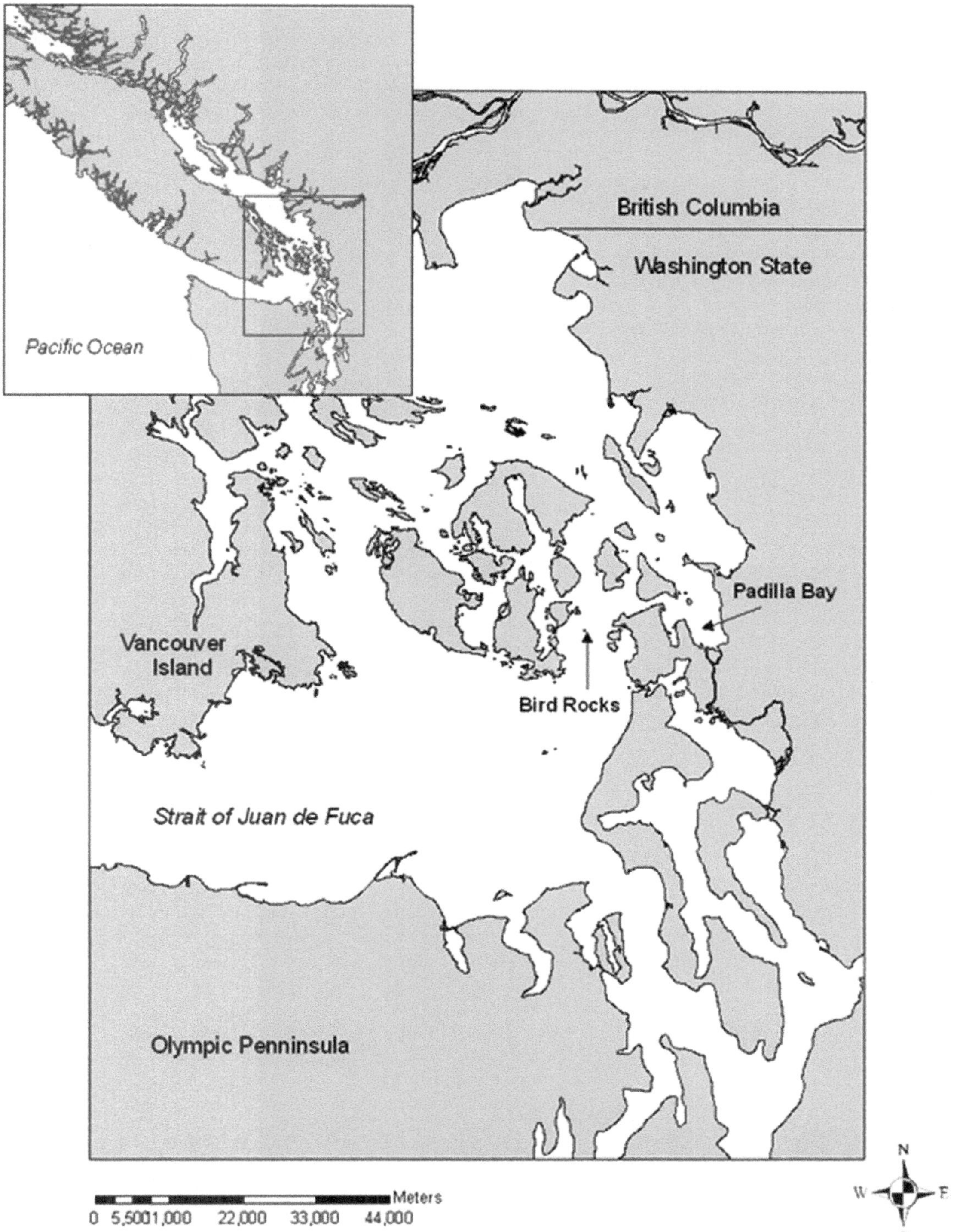

Figure 1. Harbor seal capture sites in the Georgia Basin. Inset: The study area in the Pacific Northwest: the Salish Sea. The Salish Sea encompasses Puget Sound, the San Juan Islands, the Canadian Gulf Islands, the Strait of Juan de Fuca and the Strait of Georgia.

(time spent ≥85% of the maximum depth of the dive), wiggles (the number of vertical movements within the bottom portion of the dive), and average ascent and descent rates. Four additional variables were used for dive classification: skew (the ratio of average ascent rate to average descent rate), the ratio of bottom time to dive duration (BTD), the ratio of bottom time to maximum depth (BTM), and the ratio of maximum depth to dive duration (MDD) [14,30,35].

Dive Classification

Diving by air-breathing marine predators, such as harbor seals, can be viewed as excursions from the surface to search for and/or consume prey [36] and can be characterized by multiple factors. Dive shapes for harbor seals typically fall into one of two broad categories: square or V-shaped, and within these categories other factors such as skew, wiggles, and depth are used to identify more specific shapes and inferred behaviors. In this study, eight variables were used to classify dives and determine dive shapes: 1) maximum depth, 2) duration, 3) bottom time, 4) wiggles, 5) skew, 6) BTD, 7) BTM, and 8) MDD. Wiggle count was a defining characteristic of wiggle-dives and was therefore deemed categorical, where all dives with a wiggle count >0 were considered wiggle dives. To generate a smaller set of uncorrelated variables, the eight numerical dive variables were initially analyzed using a Principle Components Analysis (PCA). The principal components (PCs) that accounted for ≥80% of the variance were used in subsequent analyses. The resultant factor scores from the PCA, plus the wiggle-dive variable, were then introduced into a k-means cluster analysis [14,37,38]. Previous studies have identified between five and seven dive types for harbor seals [14,30]. The optimum number of clusters for this study was determined by analyzing the cluster solutions for four to eight clusters. The cluster solutions were validated using a discriminant function analysis [14,35,37] and the appropriate number of clusters was determined as the most parsimonious solution; the one with the fewest number of clusters and the highest percent classification accuracy. Multiple studies have inferred behaviors for dive shape based on the combination of dive records with stomach temperature telemetry [14,36,39] or Crittercam video recorders [40–42]. Due to the increased proportion of time spent at depth, square-shaped dives are typically considered foraging dives, while V-shaped dives are associated with traveling or exploratory behavior [14,34,43]. A square dive with wiggles suggests vertical movements in the bottom portion of the dive and potential feeding within a prey patch, square dives without wiggles may suggest benthic foraging or searching for prey [35,44,45].

Bout Classification

Harbor seals rarely forage using individual dives, but perform a series of consecutive dives while working a particular area [16]. To develop a more biologically relevant measure of behavior for this species we grouped individual dives into bouts of diving [2,36]. We used a modified version of Boyd's [2] iterative statistical method to identify these bouts. Following Boness et al. [16], we operationally defined the beginning of a diving bout as a minimum of four consecutive dives to at least 6 m. After the start of a bout, subsequent dives were added if the next surface interval was not significantly greater than the mean surface intervals from the previous dives within the bout according to a t-test with an alpha value of 0.05 [1,2]. The bout ended when the subsequent surface interval was significantly greater than the previous surface intervals within the bout.

Bouts were classified using eight variables: 1) number of dives within the bout, 2) mean dive depth, 3) mean dive duration, 4) mean surface interval, 5) bout duration, 6) percent of time spent at depth, 7) percent of square-shaped dives, and 8) percent of V-shaped dives. The same procedure described to classify individual dives was used to classify diving bout types. Bout variables were analyzed using a PCA to produce a smaller set of orthogonal variables and then PCA factor scores were assessed using a k-means cluster analysis. The cluster solutions were validated using a discriminant function analysis and the most parsimonious cluster solution was accepted. Bouts with a high proportion of square-shaped (foraging) dives were considered foraging bouts.

Model Selection

We used a Bayesian approach to analyze dive behavior because it provides a means of synthesizing data from multiple sources and serves as a tool for updating our knowledge based on observed data [46]. Modeling dive behavior in this way allowed us to incorporate prior knowledge about behavior and then identify the influence of different predictors such as site, season, light, or sex on said behavior using the posterior. A Bayesian approach is potentially more applicable as a tool in ecology because inference is drawn from the posterior predictive distribution instead of the data themselves, which may be biased.

We analyzed variation in dive behavior by comparing seasonal, temporal, and sexual variation in the occurrence of different diving bouts between haul-out site groups. The different life history stages that occurred during the study period were divided into breeding (July–October) and non-breeding (November–June) seasons based on seal pupping and molting phenology in the San Juan Islands [47,48]. The breeding season defined here includes both pupping and breeding behaviors, and the non-breeding season includes post-molt/winter behaviors. The molting season was excluded from the analysis because the seals would have lost their tags shortly after beginning to molt and during the winter deployment we only tagged seals that had already completed their molt. Therefore, we do not believe the data represent molting behavior for any of the tagged individuals. Light categories were determined using the mean sunrise and sunset times for each month during the study and assigning a dive as occurring during the day or at night. To determine which combination of predictors to include in the final model a Generalized Linear Mixed Model (GLMM) with a binomial sampling distribution was run on the data using the glmer() function in R [49]. The fixed factors included in the model were haul-out site, season, sex, and light, individual seals were included as a random effect. Initial variable selection was performed by comparing BIC and AIC values for a suite of models. The model with the lowest BIC and AIC values was chosen as the best fit. The factors that significantly influenced the subset of data used in model selection were then included in the final design matrix (X). We used a Gibbs sampler to model the data with haul-out site, season, sex, and whether it was light or dark outside (light) as predictor variables.

In the final model, we used a binomial likelihood with an uninformative normal prior distribution. A vague prior was used to allow the data to dominate the analysis and dictate the shape of the posterior distribution, while still accounting for the high levels of variability in harbor seal diving. The model setup was:

$$Likelihood : Y_i \sim Bernoulli(\theta_{ij}) \quad (1)$$

$$Logit(\theta_{ij}) = X_{ij}\beta + \alpha_i[Seal] \quad (2)$$

Y_i is the response variable representing which bout types the seal performs (Type I or Type II). θ_{ij} is the probability associated with the binomial distribution for seal *i* and dive *j*, X is a design matrix of predictor variables, β refers to the coefficients associated with each predictor (the effect of each predictor in determining what behavior is being performed), and α is the coefficient associated the random effect (individual seals). A metropolis step was added within the Gibbs loop to accept or reject proposed β values during each iteration of the model. The priors on β and α were normal distributions with starting values specified as follows:

$$\alpha_i \sim Norm(\mu_\alpha, \tau) \quad (3)$$

$$\beta \sim Norm(\mu_\beta, \sigma^2) \quad (4)$$

All analyses were performed with R version 2.12.

Results

Twenty nine seals were tagged in the San Juan Islands during two deployment periods. Eleven of the fifteen TDR packages deployed in Padilla Bay were recovered and ten of the fourteen packages from Bird Rocks were recovered. One seal from Padilla Bay was tagged during both deployment periods. Most adult females were believed to be pregnant and were expected to have given birth and nursed a pup following instrument deployment. Most seals retained their tags for at least 3 months, although some animals recorded data for 6–7 months. Data were collected from April 2007 through August of 2008. A total of 326,869 dives were recorded for 11 adult females and 10 adult males (Table 1).

Dive Classification

A total of 297,964 dives were classified. All of the dive variables loaded significantly on at least one PC and were therefore included in subsequent analyses. K-means cluster analysis resulted in six dive types with 95.2% classification accuracy. The six clusters were assigned a dive type after visual inspection of the results (Figure 2). Four of the six dive types were square-shaped, with a mean bottom time ≥50% of the total dive duration. The remaining two dive types were considered V-shaped with a mean bottom time ≤37% of the total dive duration (Table 2). Type 1 and type 3 dives were considered deep dives (≥20 m) and were similar in depth, duration, and bottom time; however, type 1 dives were classified as wiggle dives and type 3 dives were not (Figure 2). Dive types 2 and 6 were considered shallow dives (<20 m) and were also separated by wiggles (Figure 2, Table 2). Dive types 4 and 5 (the V-shaped dives) were both considered shallow (<20 m) with similar durations and differed in skew. Type 4 dives were skewed to the right indicating a slow ascent rate compared to descent rate, and type 5 dives were skewed to the left indicating a slower descent rate. On average, type 4 and type 5 dives were shorter in duration than all the square-shaped dives.

Diving Bouts

A total of 45,013 diving bouts were identified with an average duration of 35.17±0.17 min. Three bout types were classified with 97.60% classification accuracy. Bouts were broadly separated by mean dive depth and bout duration (Table 3). Type I bouts were classified as both long and deep (≥20 m) and were composed of mostly square-shaped dives (95%). Type II bouts were composed of both square and V-shaped dives, but were shallower in depth and shorter in duration than Type I bouts. Type III bouts made up ≤1% of the bouts used in this study. They were shallow in depth, long in duration, and contained mostly square-shaped dives; however, the bottom time for these bouts only accounted for 3.9% of the total bout duration. The average surface interval of type III bouts was extremely long (~2 hr) indicating that these bouts potentially represent surface resting or other behaviors not associated with foraging. Due to the high proportion of square-shaped dives in both Type I and II bouts, both were considered potential foraging bouts, with differences in the depth at which foraging occurred and potentially the type of prey being consumed. Since we were interested in the variability of foraging behavior, and Type III bouts were likely non-foraging bouts and also rarely used, only Type I and Type II bouts were used in the final analyses.

Model Selection

The model was initially run with simulated data to verify its structure and the results. Due to the complexity of the model and our uncertainty on where to initialize the prior, the sampler was run hundreds of thousands of times on a subsample of data before we reached acceptable convergence. The final model was run using the entire dataset, initialized at mean values from the previous run, and then run for 1,000,000 iterations.

Variable selection revealed that the full interaction model had the best fit to the data with haul-out site, season, sex, and light (day/night) all influencing dive behavior (Table 4). Nearly all of the fixed and random effects had a positive or negative effect on behavior (Tables 5 & 6). A positive effect indicates a shift from Type I to Type II bouts with a positive unit change in the predictor; a negative effect is the opposite, with the shift being from Type II to Type I. Variables with a credible interval spanning across zero did not have a significant effect on dive behavior, but still helped explain some of the residual deviance in the model (Figure 3).

Haul-out site had a significant influence on behavior (Table 5), which became evident when looking at how the interactions of other predictors with haul-out site influenced behavior. The interaction of haul-out site and season showed more Type I bouts occurring at both sites during the non-breeding season (Figure 4). During the breeding season, there was an increase in Type II bouts at both sites, but seals from Padilla Bay still performed a higher proportion of Type I bouts than seals from Bird Rocks (Figure 4). Differences were also observed when looking at season alone. Although more Type I bouts were used in both seasons, the proportions of Type I and Type II bouts were nearly equal during the breeding season, whereas Type I bouts were nearly 3× higher and Type II bouts decreased by half during the non-breeding season (approximately 80% vs. 20%) (Figure 5).

The same pattern, with changes in the proportional use of bout types, was visible with the influence of sex on dive behavior (Figures 6 & 7). Type I bouts were the predominant type used by both sexes, but females used significantly more of the deep, Type I bouts than the shallower, Type II bouts, while males used similar proportions of both. However, males from Bird Rocks were the only ones to significantly shift their behavior. During the breeding season the proportions of bout types used were nearly equal, with slightly more shallow diving (Type II bouts); in the non-breeding season they switched their behavior to significantly more deep diving (80% of bouts were Type I, Figure 7).

The effect of light on dive behavior was most apparent when looking at the full interaction (haul-out site, season, and sex (Figure 8)). A greater percentage of Type II bouts were performed at night during the breeding season. During the day, at both sites during both seasons, Type I bouts were the predominant bout type used. During the breeding season, males at Bird Rocks increased their use of Type II bouts from equal proportions to more Type II bouts at night, and females at Padilla Bay increased their use of Type II bouts from more Type I to almost equal proportions at night. Light did not appear to affect the other sexes at each site. During the breeding season, there was little change in dive behavior for either sex or site between day and night.

The results also revealed significant random effects for nearly every individual seal (Table 6). The spread of behavioral variability both among and within individuals is apparent in

Table 1. Data collected from each tagged harbor seal relative to capture site in the San Juan Islands.

Site and ID	Sex	Mass (kg)	TDR	# Dives	Retention (d)	Tagging Trip
Bird Rocks						
B1695	M	71.5	Mk-10	38,201	152	Spring
B1696	M	74.5	Mk-10	14,322	56	Spring
B1700	M	86.0	Mk-9	9,472	52	Spring
B1701	M	81.5	Mk-10	42,888	164	Spring
Y1455	F	76.5	Mk-10	13,277	87	Spring
B1742	M	83	Mk-10	5,101	33	Winter
B1744	M	81.5	Mk-10	13,798	93	Winter
B1745	M	83	Mk-10	8,623	51	Winter
Y1513	F	75.5	Mk-9	26,371	147	Winter
Y1514	F	70.5	Mk-9	34,660	182	Winter
Total	**7M, 3F**			**206,713**		
Padilla Bay						
B1699	M	64.0	Mk-9	5,441	95	Spring
B1712	M	69.0	Mk-9	2,218	71	Spring
B1713	M	54.0	Mk-9	3,203	64	Spring
Y1456	F	55.5	Mk-9	4,253	93	Spring
Y1457	F	57.5	Mk-9	17,136	97	Spring
Y1458	F	48.5	Mk-9	14,201	118	Spring
Y1459	F	83.0	Mk-9	44,583	112	Spring/Winter
Y1460	F	62.5	Mk-9	8,106	103	Spring
Y1462	F	77.5	Mk-9	1,164	69	Spring
Y1465	F	103.0	Mk-9	14,709	78	Spring
Y1469	F	85.0	Mk-9	5,142	137	Spring
Total	**3M, 8F**			**120,156**		

Figure 9. A positive peak demonstrates a tendency of that individual to perform Type II, shallow, bouts, while a negative peak shows a preference for deeper, Type I bouts. Individuals with a peak density near zero did not show a strong preference between the different behaviors during the study period.

Discussion

All of the predictors used in this study significantly influenced the dive behavior of harbor seals in the San Juan Islands. This result suggests that where seals haul out, whether or not they are breeding, whether it is light or dark outside, or if they are male or female all affect their dive behavior in some way. Harbor seals are opportunistic predators that adjust their foraging behavior according to prey availability and profitability [8,9]. The seasonal variation in bout use observed in this study suggests that seals may have switched the predominant prey in their diet during different seasons. Diet studies in the San Juan Islands indicate that harbor seal diet varies according to prey migrations and that different haul-out sites may focus on different prey species at different times of year [20,28]. The variations observed by site and sex suggest that individuals may be exhibiting prey specialization or habitat exploitation. Indirect evidence (via diet studies) of prey specialization for harbor seals in the study region provides support for the first hypothesis [20,50].

Dive Classification

Studies examining pinniped dive behavior have usually identified five to seven dive types [2,14,30]. Among these classifications, two dive shapes are commonly observed: square- and V-shaped dives, while the remaining dives are a variation of these core shapes. All of the classified dives in this study resembled one of the core dive shapes and differed either in skewness (ratio of ascent and descent rates), depth, duration, or the occurrence of wiggles. Previous studies have attempted to assign functions to different dive shapes by combining dive profiles, stomach-temperature telemetry, and swim speed to their analyses [14,30]. These studies have suggested that skewed dives, with longer ascent or descent phases, may be attributed to a seal increasing the horizontal search component of the dive, or simply swimming along a bottom that progressively changes in depth (Lesage et al. 1999). Square-shaped dives have consistently been associated with foraging behavior as the predator spends more time at depth thus increasing the likelihood of encountering prey [14,30,43]. V-shaped dives may be related to exploratory behavior or travelling depending on the depth or skew of the dive respectively. Although multiple studies have attributed dive shapes to specific dive functions, some have suggested it is difficult to infer specific behaviors based on dive shape alone and that examining bouts of diving more accurately represents foraging behavior [1,2]. Consequently, in this study dives were used to identify different bout types and were not compared individually.

Dive Type	Characteristics	Dive Shape	Percent of Total Dives
Type 1	Deep, Square, Wiggles	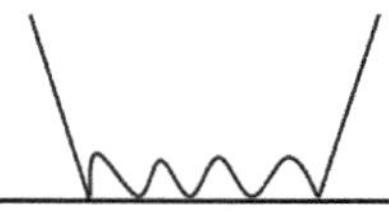	23.9 %
Type 2	Shallow, Square, Wiggles		16.1%
Type 3	Deep, Square	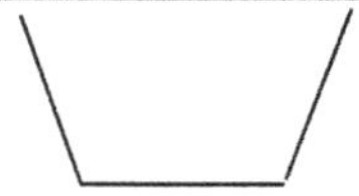	24%
Type 4	Shallow, V-shape, Skewed Right		11.1%
Type 5	Shallow, V-shape, Skewed Left		11.2%
Type 6	Shallow, Square		13.7%

Figure 2. Dive shapes and general characteristics of harbor seal dives in the San Juan Islands. Deep dives are those with a maximum depth ≥20 m, shallow dives are <20 m.

Diving Bouts

The behavior of harbor seals was organized into clusters or bouts of diving. In this study, only 12% of dives occurred outside of these bouts, likely the result of single dives related to haul-out behavior or underwater resting. Diving bouts differed primarily in depth and duration, but also in the percentage of square- and V-shaped dives occurring within the bout. These results are similar to those found in grey seals (*Halichoerus grypus*), Antarctic fur seals (*Arctocephalus gazella*), and harbor seals in other areas [1,2,16]. The differences in duration, dive shape, and percent of time spent at depth between bout types may suggest that each bout type represented a different behavior. Determining the exact nature of these behaviors remains difficult and limited [2,36]; however, studies using stomach-temperature telemetry to identify feeding events, indicate that bouts of diving with a high percentage of time spent at depth (with a high percentage of square-shaped dives) are correlated with foraging activities [14,36]. We identified one bout type with a high percentage of time spent at depth (Type I). The majority of dives occurring within this bout were square-shaped with 55% of the entire bout duration spent at depth (in the bottom portion of the dive). Wiggle dives have also been correlated with foraging behavior as vertical movement in the bottom portion of the dive can be attributed to movements within a prey patch [14]. Over 50% of the dives within both Type I and II bouts were

Table 2. Mean (± SD) values of harbor seal dive types in the San Juan Islands.

	Type 1	Type 2	Type 3	Type 4	Type 5	Type 6
Max Depth (m)	62±30.65	13.36±7.19	56.23±23.51	17.55±12.8	16.13±12.68	13.57±7.71
Duration (s)	310.14±107.69	219.27±99.57	304.58±78.48	148.73±73.77	166.45±73.92	241.79±106.89
Bottom Time (s)	181.62±96.69	152.81±89.11	192.14±76.16	39.51±29.53	60.03±46.45	179.79±100.31
Wiggles	1.05±0.38	1.04±0.29	0±0	0.46±0.51	0.35±0.48	0±0
Skew 1	0.94±0.26	1.01±0.48	0.95±0.23	1.8±1.17	0.47±0.17	1.04±0.48
Skew 2	1.14±0.36	1.18±0.51	1.11±0.26	0.71±0.29	2.64±2.54	1.12±0.43
Avg Descent Rate ($m\ s^{-1}$)	1.11±0.33	0.54±0.26	1.1±0.33	0.42±0.27	0.65±0.67	0.52±0.26
Avg Ascent Rate ($m\ s^{-1}$)	1±0.3	0.5±0.25	1.02±0.33	0.61±0.37	0.29±0.18	0.5±0.25

Table 3. Mean (± SD) values of harbor seal bout types in the San Juan Islands.

	Type I	Type II	Type III
Number of Dives	5.82±2.71	5.63±2.35	5.47±3.11
Dive Depth (m)	44.73±27.45	18.97±11.47	18.86±12.84
Dive Duration (s)	288.93±84.88	165.59±57.57	193.34±81.54
Surface Int. Duration (s)	45.46±28.25	39.22±61.37	3139.46±2254.39
Bout Duration (s)	2267.70±1071.29	1355.34±739.92	20214.87±12765.20
Time at Depth (%)	0.55±0.13	0.33±0.13	0.04±0.03
Square-shaped Dives (%)	0.95±0.09	0.44±0.24	0.53±0.28
V-shaped Dives (%)	0.05±0.089	0.56±0.24	0.47±0.28

classified as wiggle dives. This high percentage of wiggle dives suggests that both Type I and II bouts were used for foraging during this study.

Shallow diving bouts [2] as well as V-shaped dives [14] have been attributed to a number of different activities including traveling, predator avoidance, and exploration as animals are able to reduce drag and increase their chances of encountering prey by diving while traveling instead of swimming at the surface [14,35,51]. In this study, Type II bouts contained mostly V-shaped dives; however, 44% of the dives were square-shaped. The mix of dive types may be associated with searching for and then feeding within prey patches located in shallow waters (< 20 m) (Table 3). In contrast, Type I bouts had a higher proportion of time spent at depth (55%) and may be attributed to foraging in deeper waters (≥20 m). The majority of dives within Type I bouts were also square-shaped with wiggles, further indicating feeding within a prey patch. While both of these bout types potentially represent foraging behaviors, the differences between the two may provide insight into prey behavior. Type II bouts, with shallow, V-shaped dives and shorter dive durations, likely represented foraging within a tightly aggregated prey patch where the seal was able to find and capture prey near the surface with little time spent chasing through the water column. For Type I bouts, the differences in dive depth, duration, and bottom time suggest a different foraging strategy. Seals may have been foraging on more loosely aggregated or larger prey resulting in more time spent at depth in pursuit of prey.

Behavioral Variation

Our model was built to analyze the influence of predictors on binomial, categorical data using a Bayesian framework. The structure can be applied to any type of categorical data and is applicable for many types of analyses. The Bayesian methodology used here makes it possible to exploit the basic elements of linear equations with Gaussian error as part of a more complex model [52]. The model was deliberately developed based on simplistic, categorical elements in order to demonstrate the applicability of this approach to behavioral studies. Time is one obvious predictor that was excluded from this analysis; future developments could

Table 4. Model output for GLMM after initial model selection.

Predictor	Estimate	Std. Error	Z value	Pr (>|z|)
Intercept	0.94	0.33	2.89	<0.01
Site 1	1.06	0.77	1.38	0.17
Season 1	−1.79	0.07	−25.66	<2e-16
Light 1	−1.40	0.06	−22.70	<2e-16
Sex 1	−3.27	0.61	−5.40	6.77e-8
Site1:Season1	1.96	0.53	3.71	<0.01
Site1:Light1	1.05	0.81	1.29	0.20
Season1:Light1	0.91	0.08	11.98	<2e-16
Site1:Sex1	1.99	0.97	2.05	0.04
Season1:Sex1	1.64	0.17	9.40	<2e-16
Light1:Sex1	1.86	0.18	10.55	<2e-16
Site1:Season1:Light1	−0.69	0.83	−0.83	0.41
Site1:Season1:Sex1	−2.83	0.56	−5.10	<0.01
Site1:Light1:Sex1	−2.76	0.83	−3.32	<0.01
Season1:Light1:Sex1	−0.96	0.19	−4.92	<0.01
Site1:Season1:Light1:Sex1	1.64	0.85	1.92	0.06

Table 5. Posterior output for the fixed effects in the model examining harbor seal dive behavior in the San Juan Islands.

		95% Credible Interval	
	Mean	**0.025**	**0.975**
Intercept	0.811	0.556	1.091
Site	1.188	0.287	2.124
Season	−1.118	−1.287	−1.053
Sex	−3.309	−4.298	−2.563
Light	−0.810	−0.868	−0.750
Site:Season	1.03	0.179	1.828
Site:Sex	2.15	1.051	3.371
Season:Sex	0.98	0.684	1.277
Season:Sex:Site	−1.906	−2.708	−1.057
Site:Light	−0.143	−0.625	0.308
Sex:Light	1.272	0.965	1.575
Site:Light:Sex	−1.557	−1.889	−1.192
Site:Light:Season	0.911	0.556	1.270
Season:Light:Sex	−0.039	−0.358	0.282

Values are shown on a logit scale for the influence of each predictor on seal behavior.

translate this into a state-space model by including time as a continuous variable instead of using day/night as an indicator of time.

Model results indicate that dive behavior was significantly influenced by haul-out site, season, sex, and light; signifying that behavior was affected both spatially and temporally. Seals from

Table 6. Posterior output of the random effects in the model.

		95% Credible Interval	
Seal ID	**Mean**	**0.025**	**0.975**
B1695	−0.261	−0.634	0.013
B1696	0.269	−0.101	0.560
B1699	−0.647	−1.340	0.028
B1700	−0.435	−0.807	−0.129
B1701	0.189	−0.186	0.460
B1712	0.898	0.064	1.752
B1713	0.300	−0.456	1.052
B1742	−0.821	−1.217	−0.473
B1744	−0.415	−0.786	−0.114
B1745	−1.354	−1.740	−1.030
Y1455	−0.640	−1.281	0.338
Y1456	1.249	0.790	1.729
Y1457	−1.186	−1.597	−0.746
Y1458	−0.409	−0.823	0.034
Y1459	−1.240	−1.643	−0.806
Y1460	1.759	1.319	2.230
Y1462	−1.416	−2.049	−0.801
Y1465	−0.499	−0.909	−0.060
Y1469	0.842	0.386	1.319
Y1513	1.044	0.416	2.036
Y1514	0.154	−0.476	1.123

Values are shown on a logit scale for the effect of each individual seal.

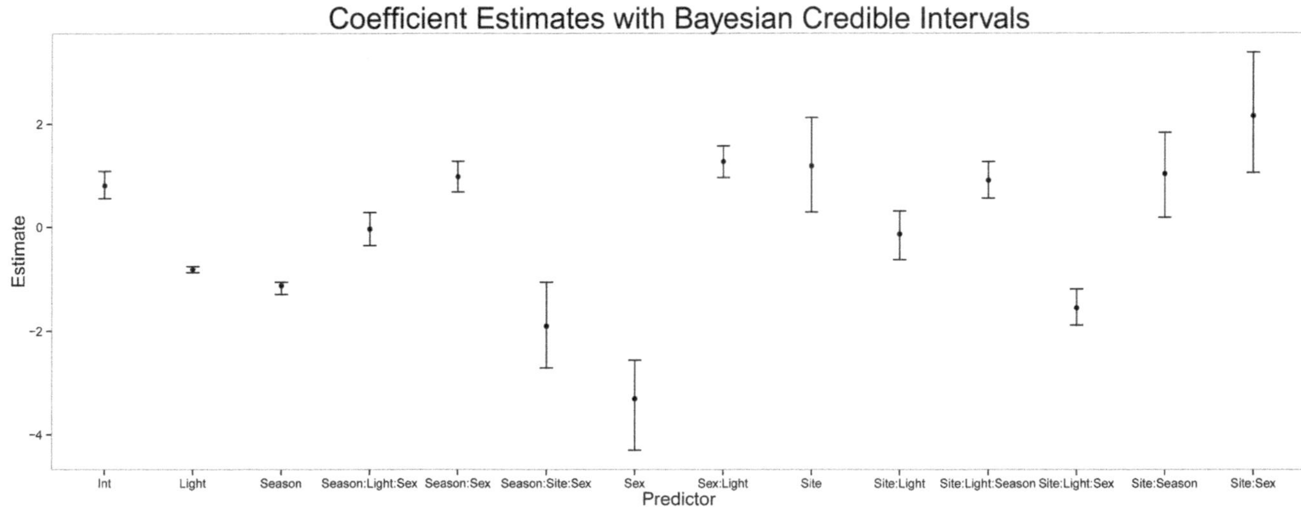

Figure 3. Beta coefficient estimates for predictor variables.

Bird Rocks demonstrated a distinct change in the use of different bout types across seasons, while seals from Padilla Bay, located < 20 km away, did not. In Washington State, the main return of pink salmon (*Oncorhynchus gorbuscha*) occurs in odd-numbered years, with only a very small return occurring in even years. Diet analysis in the Salish Sea indicates that harbor seals in 2005 (a pink salmon year) switched from consuming nearly 80% herring to 80% salmon at the end of July [20]. Multiple species of salmon return to the Salish Sea in July and August [53]. Certain species of salmon (pink, sockeye, and chum) typically remain at depths of 6–36 m as they return from the open ocean and make their way back to their natal streams [54]. These species also follow a diurnal vertical migration typically staying in the deeper end of their range during the day and moving into shallower waters at night. The first half of this study was conducted during a pink salmon year (2007), which leads us to conclude that the increase in shallow diving during the breeding season may have been correlated with an increase in salmon consumption during that time of year.

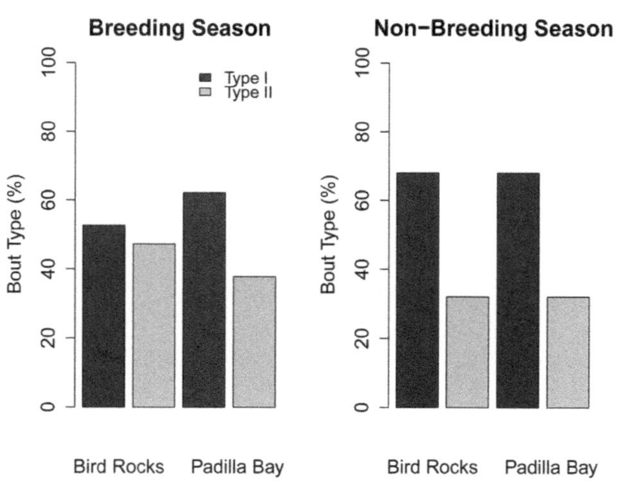

Figure 4. Distribution of harbor seal bout types relative to season and haul-out site. The breeding season is July–October and the non-breeding season is November–June.

Sex affects diving behavior in harbor seals [6,7] and due to the uneven sex ratio in this study the site differences we observed may also be correlated with sex. Female harbor seals forage throughout the breeding season, even while nursing [16,30], and males tend to perform shallower dives while holding underwater breeding territories [7,55]. More females were tagged in Padilla Bay than at Bird Rocks; however, this unbalanced sex ratio did not appear to be responsible for the increase in Type II bouts observed during the breeding season. At both sites, females performed a higher percentage of Type I bouts than any other bout type; however, the males at Padilla Bay performed significantly more Type II bouts during the breeding season (Figure 7). Notably, males from both sites performed a higher percentage of Type II bouts during the breeding season (females did not, Figure 7). Additionally, during the breeding season females performed a high percentage of Type I bouts than males, who seemed to favor Type II bouts. The increase in Type II bouts for males may be attributed to holding underwater breeding territories while they are attempting to mate. This pattern was visible at both haul-out sites, regardless of the number of females tagged, which suggests that sex was not the driving factor behind the behavioral differences observed between haul-out sites.

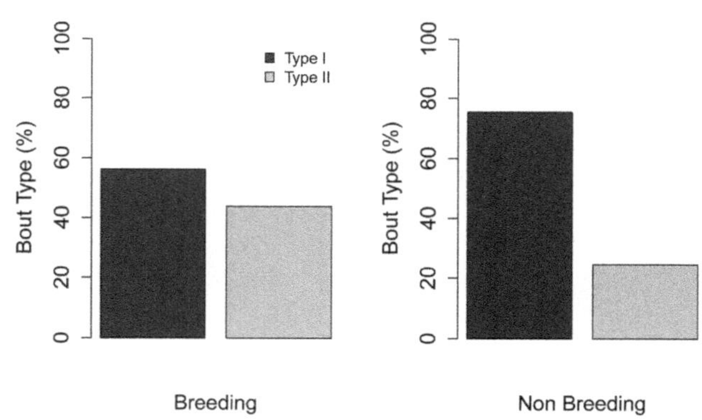

Figure 5. Distribution of harbor seal bout types relative to season.

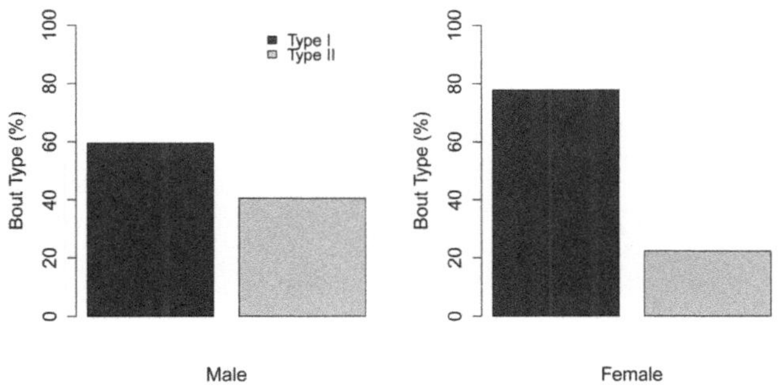

Figure 6. Distribution of harbor seal bout types relative to sex.

Harbor seals living within or near estuaries have a more diverse diet than those outside the estuary [20,56]. In Padilla Bay, herring, salmon, and small schooling fish are consumed by harbor seals; however, seals also consume a number of benthic estuarine species such as gunnel (*Pholid spp.*), snake prickleback (*Lumpenus sagitta*), Pacific staghorn sculpin (*Leptocottus armatus*), plainfin midshipman (*Porichthys notatus*), and eelpout (*Zoarcid spp.*) [56]. Rocky-reef sites located just outside estuaries, likely used by estuarine seals, also show a more varied diet than that of rocky-reef sites away from estuaries [20]. The increased variety in diet near estuaries and the regular consumption of benthic estuarine prey within the estuary may be why Type I bouts are the predominant bout type used by Padilla Bay seals. Other studies of harbor seals from estuarine bays indicate that seals typically forage in the benthos and that no diurnal pattern in dive behavior is observed in those areas [12,14,16]. Individual seals may develop preferences for different foraging strategies, prey selection, or foraging locations [2] and seals in Padilla Bay likely exploit both the estuarine habitat and the seasonal increases in salmon abundance. Analysis of harbor seal movement patterns in the San Juan Islands shows that seals from Padilla Bay typically stayed within 5 km of the estuary during the breeding season, while seals from Bird Rocks made multiple trips >10 km from their tagging site [27]. Bird Rocks seals repeatedly traveled beyond the distance required to reach the estuary and traveled to rocky sites around the estuary, but were never documented to forage or haul-out within the estuary. Under this scenario, seals hauling-out in Padilla Bay foraged preferentially within the estuarine habitat. Foraging in the benthos throughout

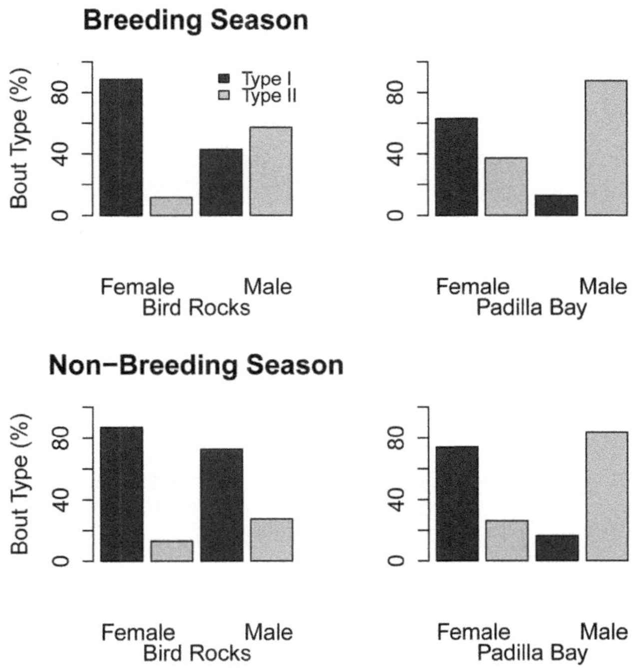

Figure 7. Distribution of harbor seal bout types relative to haul-out site, season, and sex. The breeding season is July–October and the non-breeding season is November-June.

the year and exploiting the salmon run during the summer would explain why Type II bouts increased in abundance during the breeding season, but did not surpass the abundance of Type I bouts. Additionally, due to how the tags were programed (recording every 10 sec with the minimum dive depth of 5 m) we likely missed diving that occurred in shallow regions of the estuary. With different sampling protocols, we may have seen an even stronger signal of Padilla Bay seals preferentially foraging within the estuary, but we do not believe that our overall results would differ.

By analyzing foraging behavior using bouts instead of individual dives a more biologically relevant analysis was completed; however, details such as the difference between wiggle and non-

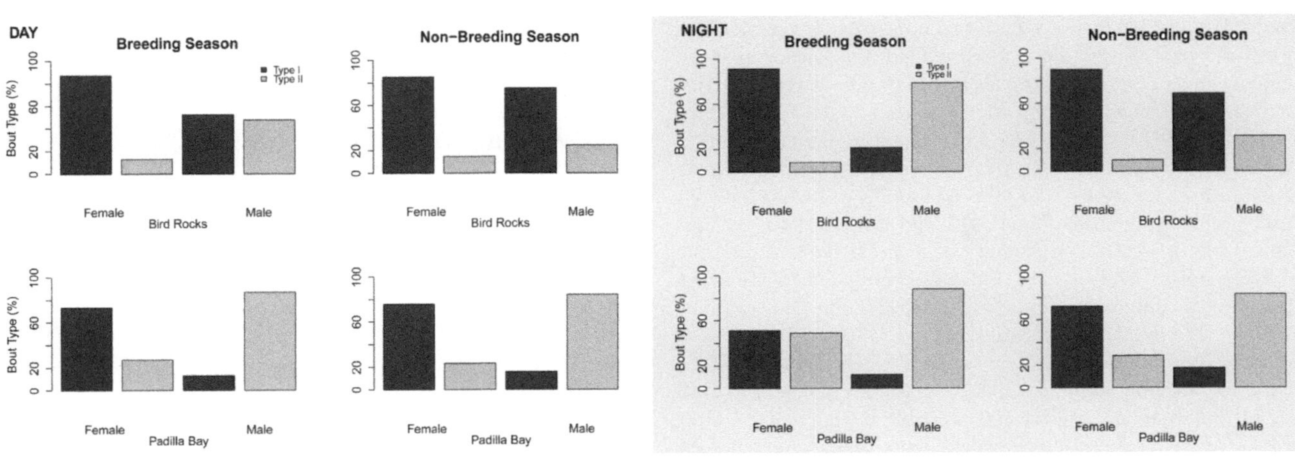

Figure 8. Distribution of harbor seal bout types relative to light, site, season, and sex. The breeding season is July–October and the non-breeding season is November–June.

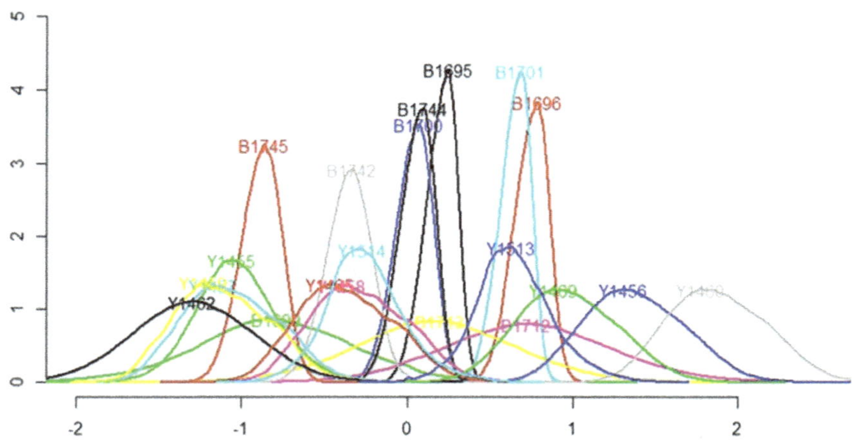

Figure 9. Density plot of the random effects in the model. A peak near −1 indicates that the individual performed more Type I bouts, a peak near +1 indicates more Type II bouts, and a peak near 0 indicates no distinct difference in the use of different bouts for that individual.

wiggle square-shaped diving may have been excluded. Studies examining individual dives show that deep (≥20 m) square shaped dives without wiggles may indicate foraging in the benthos [14,35,44]. Benthic-dwelling prey are present in the harbor seal diet from rocky-reef sites, including Bird Rocks; however, a large majority of their diet is comprised of vertically-migrating schooling fish such as herring, Pacific hake (*Merluccius productus*), and salmon [20]. At Bird Rocks 26% of the dives during the non-breeding season were deep (≥20 m) and square-shaped without wiggles, potentially indicating benthic foraging. The use of both wiggle and non-wiggle square-shaped dives during the non-breeding season, before diurnally migrating salmon return to the Salish Sea, may be why such a high proportion of Type I bouts were used at this time of year and near equal proportions of Type I and II bouts during the breeding season.

Conclusions

The variation in dive behavior over time that we observed suggests fluctuations in the predominant prey consumed by harbor seals. These variations may be related to annual migrations and movements of available prey in the study area. We were able to identify differences between two relatively close haul-out sites, which may allude to larger behavioral differences, such as prey specialization or habitat exploitation by different haul-out site groups. Prey specialization, with forage fish specialists and salmon specialists, has already been documented for harbor seals in southern Puget Sound [28] and likely explains the variation in diet and foraging behavior observed for harbor seals in this study [20,50]. Bird Rocks and Padilla Bay showed similar variations in the use of different bout types; however, seals from Bird Rocks significantly changed the ratio of Type I to Type II bouts between seasons while seals from Padilla Bay did not. Bird Rocks seals dramatically increased the number of Type II bouts during the breeding season while Padilla Bay seals continued to use more Type I bouts. This change in behavior, which coincided with prey fluctuations in the region, suggests prey specialization between the two sites. Additionally, the increase in Type II bouts by males during the breeding season at both sites suggests that seals in both regions may maintain underwater breeding territories at or near foraging areas. These results provide a better understanding of the variability in harbor seal foraging behavior in the San Juan Islands and highlight the importance of examining behavioral variation on a small spatial scale. If all of the seals had been grouped together to examine San Juan Islands behavior as a whole, many of the influences we detected and the changes we observed may have been missed. This research sets up a baseline for understanding fine-scale behavioral variation and provides a reference point for documenting how seals may respond to ecosystem or prey fluctuations in the region.

Future studies should continue to document dive behavior at these sites to determine if these differences can be attributed to individual variability or to haul-out habitat exploitation. Additionally, if prey abundances in the region could be monitored concurrently with dive behavior, more concrete conclusions could be drawn regarding the link between prey fluctuations and dive behavior. If seals adjust their foraging strategies based on prey abundance, then their foraging behavior may be highly correlated with haul-out site habitat as seals will likely exploit prey located near their primary haul-out site. The data and analysis methods presented here provide an opportunity to continually monitor harbor seal diving to determine the relative importance of habitat characteristics on variations in foraging behavior. Comparing haul-out habitat to foraging area habitat may elude to either a correlation between foraging behavior and haul-out site type or to the degree of individual variability in foraging behaviors within the population. Either result will provide managers with important information regarding the foraging behavior of these predators and allow them to identify behavioral changes in the future.

Acknowledgments

Special thanks to J. Clark for advice on statistical methods and help with model development. We also thank D. Lambourn, B. Murphie, T. Cyra, J. Gaydos, and many others for their help capturing seals, students from the Marine Behavior and Ecology lab at Western Washington University (WWU) for assisting in research efforts, G. McKeen, N. Schwark, and P. Davis for logistical support, and J. London and two anonymous reviewers for providing helpful comments on earlier versions of the manuscript.

Author Contributions

Conceived and designed the experiments: KW ML SJ AAG. Performed the experiments: KW ML SJ AAG. Analyzed the data: KW. Wrote the paper: KW AAG.

References

1. Beck CA, Bowen WD, Mcmillan JI, Iverson SJ (2003) Sex differences in diving at multiple temporal scales in a size-dimorphic capital breeder. Journal of Animal Ecology 72: 979–993.
2. Boyd IL, Arnould JPY, Barton TR, Croxall JP (1994) Foraging behavior of Antarctic fur seals during periods of contrasting prey abundance. Journal of Animal Ecology 63: 703–713.
3. Higgs M, Ver Hoef JM (2012) Discretized and aggregated: modeling dive depth of harbor seals from ordered categorical data with temporal autocorrelation. Biometrics 68: 965–974.
4. Hastings KK, Frost KJ, Simpkins MA, Pendleton GW, Swain UG, et al. (2004) Regional differences in diving behavior of harbor seals in the Gulf of Alaska. Canadian Journal of Zoology 82: 1755–1773.
5. Härkönen TJ, Harding KC, Lunneryd SG (1999) Age- and sex-specific behaviour in harbour seals *Phoca vitulina* leads to biased estimates of vital population parameters. Journal of Applied Ecology 36: 825–841.
6. Thompson PM, Mackay A, Tollit DJ, Enderby S, Hammond PS (1998) The influence of body size and sex on the characteristics of harbour seal foraging trips. Canadian Journal of Zoology 76: 1044–1053.
7. Coltman DW, Bowen WD, Boness DJ, Iverson SJ (1997) Balancing foraging and reproduction in the male harbor seal, an aquatically mating pinniped. Animal Behavior 54: 663–678.
8. Middlemas SJ, Barton TR, Armstrong JD, Thompson PM (2006) Functional and aggregative responses of harbour seals to changes in salmonid abundance. Proceedings of the Royal Society of London B 273: 193–198.
9. Thomas A, Lance MM, Jefferies SJ, Miner B, Acevedo-Gutiérrez A (2011) Harbor seal foraging response to a seasonal resource pulse, spawning Pacific herring. Marine Ecology Progress Series 441: 225–239.
10. Härkönen TJ (1987) Influence of feeding on haul-out patterns and sizes of sub-populations in harbour seals. Netherlands Journal of Sea Research 21: 331–339.
11. Eguchi T, Harvey JT (2005) Diving behavior of the Pacific harbor seal (*Phoca vitulina richardii*) in Monterey Bay, California Marine Mammal Science 21: 283–295.
12. Tollit DJ, Black AD, Thompson PM, Mackay A, Corpe HM, et al. (1998) Variations in harbour seal *Phoca vitulina* diet and dive-depths in relation to foraging habitat. Journal of Zoology: 209–222.
13. Bjørge AJ, Thompson D, Hammond PS, Fedak MA, Bryant E, et al. (1995) Habitat use and diving behaviour of harbour seals in a coastal archipelago in Norway. In: Walløe I, Ulltang Ø, editors. Whales, seals, fish and man. Norway: Elsevier. pp. 211–223.
14. Lesage V, Hammill MO, Kovacs KM (1999) Functional classification of harbor seal (*Phoca vitulina*) dives using depth profiles, swimming velocity, and an index of foraging success. Canadian Journal of Zoology 77: 74–87.
15. Bowen WD, Boness DJ, Iverson SJ (1999) Diving behaviour of lactating harbour seals and their pups during maternal foraging trips. Canadian Journal of Zoology 77: 978–988.
16. Boness DJ, Bowen WD, Oftedal OT (1994) Evidence of a maternal foraging cycle resembling that of otariid seals in a small phocid, the harbor seal. Behavioral Ecology and Sociobiology 34: 95–104.
17. Thomson R (1994) Physical oceanography of the Strait of Georgia-Puget Sound-Juan de Fuca Strait system. Sidney, B.C. Canada: Institute of Ocean Sciences.
18. Jeffries SJ, Huber HR, Calambokidis J, Laake J (2003) Trends and status of harbor seals in Washington State: 1978–1999. Journal of Wildlife Management 67: 208–219.
19. Olesiuk PF (2010) An assessment of population trends and abundance of harbour seals (*Phoca vitulina*) in British Columbia. DFO Can. Sci. Advis. Sec. Res. Doc. 2009/105. vi+157 p.
20. Lance MM, Chang W-Y, Jefferies SJ, Pearson S, Acevedo-Gutiérrez A (2012) Harbor seal diet in northern Puget Sound: implications for the recovery of depressed fish stocks. Marine Ecology Progress Series 464: 257–271.
21. Scordino J (2010) West coast pinniped program investigations on California sea lion and Pacific Harbor seal impacts on salmonids and other fishery resources.
22. Olesiuk PF (1993) Annual prey consumption by harbor seals (*Phoca vitulina*) in the Strait of Georgia, British Columbia. Fishery Bulletin 91: 491–515.
23. Jeffries SJ, Gearin PJ, Huber HR, Saul DL, Pruett DA (2000) Atlas of seal and sea lion haulout sites in Washington. Olympia, WA: Washington Department of Fish and Wildlife, Wildlife Science Division.
24. Payne PM, Selzer LA (1989) The distribution, abundance and selected prey of the harbour seal *Phoca vitulina concolor*, in Southern New England. Marine Mammal Science 5: 173–192.
25. Hardee S (2008) Movements and home ranges of harbor seals (Phoca vitulina) in the inland waters of the Pacific Northwest [MSc]. Bellingham, WA: Western Washington University.
26. Suryan RM, Harvey JT (1998) Tracking harbor seals (*Phoca vitulina richardsi*) to determine dive behavior, foraging activity, and haul-out site use. Marine Mammal Science 14: 361–372.
27. Peterson S, Lance MM, Jefferies SJ, Acevedo-Gutiérrez A (2012) Long distance movements and disjunch spatial use of harbor seals (*Phoca vitulina*) in the inland waters of the Pacific Northwest. PLoS ONE 7: e39046.
28. London JM, Lance MM, Jeffries SJ (2001) Observations of harbor seal predation on Hood Canal salmonids from 1998 to 2000. Seattle, WA: Final Report: Studies of expanding pinniped populations NOAA Grant No. NA17FX1630, WDFW, PSMFC Contract No. 02-15. 20 p.
29. Jeffries SJ, Brown RF, Harvey JT (1993) Techniques for capturing, handling and marking harbor seals. Aquatic Mammals 19: 21–25.
30. Baechler J, Beck CA, Bowen WD (2002) Dive shapes reveal tamporal changes in the foraging behaviour of different age and sex classes of harbor seals (*Phoca vitulina*). Canadian Journal of Zoology 80: 1569–1577.
31. Macdonald D (1978) Radio-tracking: some applications and limitations. In: Stenhouse G, editor. Animal marking: recognition marking of animals in research. London: Macmillian. pp. 192–204.
32. Brooks C, Bonyongo C, Harris S (2008) Effects of global positioning system collar weight on zebra behavior and location error. Journal of Wildlife Management 72: 527–534.
33. Lesage V, Hammill MO, Kovacs KM (2004) Long-distance movements of harbor seals (*Phoca vitulina*) from a seasonally ice-covered area, the St. Lawrence River estuary, Canada. Canadian Journal of Zoology 82: 1070–1081.
34. Schreer JF, Kovacs KM, O'Hara Hines RJ (2001) Comparitive diving patterns of pinnipeds and seabirds. Ecological Monographs 71: 137–162.
35. Schreer JF, Testa JW (1996) Classification of Weddell seal diving behavior. Marine Mammal Science 12: 227–250.
36. Austin D, Bowen WD, McMillan JI, Iverson SJ (2006) Linking movement, diving, and habitat to foraging success in a large marine predator. Ecology 87: 3095–3108.
37. Schreer JF, Testa JW (1995) Statistical classification of diving behavior. Marine Mammal Science 11: 85–93.
38. Schreer JF, O'Hara Hines RJ, Kovacs KM (1998) Classification of dive profiles: A comparison of statistical clustering techniques and unsupervised artificial neural networks. Journal of Agricultural, Biological, and Environmental Statistics 3: 383–404.
39. Kuhn CE, Costa DP (2006) Identifying and quantifying prey consumption using stomach temperature change in pinnipeds. The Journal of Experimental Biology 209: 4524–4532.
40. Bowen WD, Tully D, Boness DJ, Bulheier BM, Marshall GJ (2002) Prey-dependent foraging tactics and prey profitability in a marine mammal. Marine Ecology Progress Series 244: 235–245.
41. Davis R, Fuiman L, Williams T, Collier S, Hagey W, et al. (1999) Hunting behavior of a marine mammal beneath the Antarctic fast ice. Science 283: 993–996.
42. Davis RW, Fuiman LA, Williams T, Horning M, Hagey W (2003) Classification of Weddell seal dives based on 3-dimensional movements and video-recorded observations. Marine Ecology Progress Series 264: 109–122.
43. Le Boeuf BJ, Costa DP, Huntley AC, Feldkamp SD (1988) Continuous, deep diving in female northern elephant seals, *Mirounga angustriostris*. Canadian Journal of Zoology 66: 446–458.
44. Le Boeuf BJ, Naito Y, Asaga T, Crocker DE, Costa DP (1992) Swim speed in a female northern elephant seal: metabolic and foraging implications. Canadian Journal of Zoology 70: 786–795.
45. Hindell M, Slip D, Burton H (1991) The diving behavior of adult male and female southern elephant seals, Mirounga-Leonina (Pinnipedia, Phocidae). Australian Journal of Zoology 39: 595–619.
46. Sumner M, Wotherspoon S, Hindell M (2009) Bayesian estimation of animal movements from archival and satellite tags. PLoS ONE 4: e7324.
47. Johnson A, Acevedo-Gutiérrez A (2007) Regulation compliance and harbor seal (*Phoca vitulina*) disturbance. Canadian Journal of Zoology 85: 290–294.
48. Huber HR, Jefferies SJ, Brown RF, DeLong R, VanBlaricorn GR (2001) Correcting aerial survey counts of harbor seals (*Phoca vitulina richardsi*) in Washington and Oregon. Marine Mammal Science 17: 276–293.
49. Gelman A, Su Y-S, Yajima M, Hill J, Pittau MG, et al. (2012) arm: Data analysis using regression and multilevel/hierarchical models. R package version 15 - 08. Available: http://CRAN.R-project.org/package = arm.
50. Bromaghin J, Lance MM, Elliott E, Jeffries SJ, Acevedo-Gutiérrez A, et al. (2013) New Insights into the diets of harbor seals in the Salish Sea of western North America revealed by quantitative fatty acid signature analysis. Fishery Bulletin 111: 13–26.
51. Williams TM, Kooyman GL (1985) Swimming preformance and hydrodynamic characteristics of harbor seals *Phoca vitulina*. Physiological Zoology 58: 576–589.
52. Clark J (2007) Models for ecological data. An introduction. Princeton, New Jersey: Princeton University Press.
53. Quinn TP (2005) The behavior and ecology of Pacific salmon and trout. Seattle, WA: University of Washington Press. 378 p.
54. Groot G, Margolis L (1928) Pacific salmon life histories. Vancouver, Canada: UBC Press. 564 p.
55. Boness DJ, Bowen WD, Buhleier B, Marshall GJ (2006) Mating tactics and mating system of an aquatic-mating pinniped: the harbor seal, *Phoca vitulina*. Behavioral Ecology and Sociobiology 61: 119–130.
56. Luxa K, Acevedo-Gutiérrez A (2013) Food habits of harbor seals (*Phoca vitulina*) in two estuaries in the central Salish Sea. Aquatic Mammals 39: 10–22.

16

Assessing Performance of Bayesian State-Space Models Fit to Argos Satellite Telemetry Locations Processed with Kalman Filtering

Mónica A. Silva[1,2,3]*, Ian Jonsen[4], Deborah J. F. Russell[5,6], Rui Prieto[1,2], Dave Thompson[5], Mark F. Baumgartner[3]

1 Center of the Institute of Marine Research (IMAR) and Department of Oceanography and Fisheries, University of the Azores, Horta, Portugal, **2** Laboratory of Robotics and Systems in Engineering and Science (LARSyS), Lisbon, Portugal, **3** Biology Department, Woods Hole Oceanographic Institution, Woods Hole, Massachusetts, United States of America, **4** Department of Biology, Dalhousie University, Halifax, Nova Scotia, Canada, **5** Sea Mammal Research Unit, Scottish Oceans Institute, University of St. Andrews, St. Andrews, United Kingdom, **6** Centre for Research into Ecological and Environmental Modelling, University of St. Andrews, St. Andrews, United Kingdom

Abstract

Argos recently implemented a new algorithm to calculate locations of satellite-tracked animals that uses a Kalman filter (KF). The KF algorithm is reported to increase the number and accuracy of estimated positions over the traditional Least Squares (LS) algorithm, with potential advantages to the application of state-space methods to model animal movement data. We tested the performance of two Bayesian state-space models (SSMs) fitted to satellite tracking data processed with KF algorithm. Tracks from 7 harbour seals (*Phoca vitulina*) tagged with ARGOS satellite transmitters equipped with Fastloc GPS loggers were used to calculate the error of locations estimated from SSMs fitted to KF and LS data, by comparing those to "true" GPS locations. Data on 6 fin whales (*Balaenoptera physalus*) were used to investigate consistency in movement parameters, location and behavioural states estimated by switching state-space models (SSSM) fitted to data derived from KF and LS methods. The model fit to KF locations improved the accuracy of seal trips by 27% over the LS model. 82% of locations predicted from the KF model and 73% of locations from the LS model were <5 km from the corresponding interpolated GPS position. Uncertainty in KF model estimates (5.6±5.6 km) was nearly half that of LS estimates (11.6±8.4 km). Accuracy of KF and LS modelled locations was sensitive to precision but not to observation frequency or temporal resolution of raw Argos data. On average, 88% of whale locations estimated by KF models fell within the 95% probability ellipse of paired locations from LS models. Precision of KF locations for whales was generally higher. Whales' behavioural mode inferred by KF models matched the classification from LS models in 94% of the cases. State-space models fit to KF data can improve spatial accuracy of location estimates over LS models and produce equally reliable behavioural estimates.

Editor: Magnus Rattray, University of Manchester, United Kingdom

Funding: This research was primarily funded by Fundação para a Ciência e a Tecnologia (FCT), Fundo Regional da Ciência, Tecnologia (FRCT), through research projects TRACE-PTDC/MAR/74071/2006 and MAPCET-M2.1.2/F/012/2011 [FEDER], the Competitiveness Factors Operational (COMPETE), QREN European Social Fund, and Proconvergencia Açores/EU Program]. We acknowledge funds provided by FCT to LARSyS Associated Laboratory and IMAR-University of the Azores/the Thematic Area D & E of the Strategic Project PEst-OE/EEI/LA0009/2011–1012 and 2013–2014 (OE & Compete) and by the FRCT - Government of the Azores pluriannual funding. MAS was supported by an FCT postdoctoral grant (SFRH/BPD/29841/2006) and is currently supported by POPH, QREN European Social Fund and the Portuguese Ministry for Science and Education through an FCT Investigator grant. RP was supported by an FCT doctoral grant (SFRH/BD/41192/2007) and by the research grant from the Azores Regional Fund for Science and Technology (M3.1.5/F/115/2012). IJ was supported by the Natural Sciences and Engineering Research Council (NSERC) and the Canada Foundation for Innovation (CFI) through their support of the Ocean Tracking Network. DJFR is funded by the United Kingdom Department of Energy and Climate Change as part of their Offshore Energy Strategic Environmental Assessment program. DT is funded by Natural Environment Research Council and Marine Scotland. Publication fees for this open access publication were supported by LARSyS Associated Laboratory through FCT/MCE project Project PEst-OE/EEI/LA0009/2013–2014. The funders had no role in study design, data collection and analysis, decision to publish, or preparation of the manuscript.

Competing Interests: The authors have declared that no competing interests exist.

* E-mail: monica@uac.pt

Introduction

The collection of individual animal movement data has become widely utilized by ecologists in the last decade due to the improvement of the underlying technologies and reduction of operational costs involved in animal telemetry. Of the several technologies available, one of the most popular is that based on satellite tags (platform transmitter terminals, PTTs) using the Argos system [1]. However, most satellite tags record observations at irregular intervals and with considerable error, meaning that movements are observed neither continuously nor with complete accuracy. The Argos service provider assigns a quality index, or location class (LC), to each position based on its estimated precision. The radius of error (assumed to include 68% of positions) for each LC is: LC 3<250 m, LC 2 250–500 m, LC 1 500–1500 m, LC 0>1500 m, and LC A, B and Z for which no estimate of error is provided [2]. However, attempts to measure spatial error of Argos locations using either stationary tests or double-tagging experiments with free-ranging animals consistently reported larger errors than those indicated by Argos. Many of these studies also provided error estimates for location classes A and B, showing these could be in the range of tens to hundreds of kilometres (reviewed in [3]).

Varying accuracy and precision, and unevenness in space and time of telemetry data can affect the determination of distribution, habitat use and behavioural patterns of animals and severely bias the calculation of movement metrics [3–5]. Therefore, advanced statistical methods are necessary to account for spatial error and temporal irregularity in the data and to understand the movement behaviour of the tracked animals.

Jonsen et al. [6] proposed a state-space framework for analysis of movement data that was further developed in Jonsen et al. [7–9], in order to deal with the biological and statistical complexities associated with animal tracking data. State-space models (SSMs) offer a powerful way to infer latent movement from imperfect estimates of animal locations by allowing uncertainty in both the observations and in the movement dynamics to be accounted for separately in the estimation process. Additionally, movement models can include behavioural or environmental effects, enabling a better understanding of the interaction between an animal's behaviour and its environment [8–11]. SSMs have been used widely among ecologists and are currently one of the tools of choice for analysing tracking data of several taxa and across environments [5,12–17].

Geolocation of animals tracked with Argos systems is based on the Doppler shift of the tag's fixed transmission frequency; i.e. the frequency shift of the tag's signal received at the orbiting satellite as it approaches and moves away from the tag [18]. The system estimates two possible positions, which are symmetrical on each side of the satellite ground track. Until recently, Argos used a non-linear Least Squares (LS) algorithm to refine the tag's position estimates and to select the one with the minimal residual error. However, the LS positioning algorithm presented a number of limitations and affected the quality of the tracks obtained. For instance, when the LS algorithm could not complete the refinement routine or check the validity of the most plausible location estimate, no position was provided. In addition, the process required at least two transmissions (also called messages) during a single satellite pass to compute a position and at least four messages to produce an error estimate.

In May 2011, Argos implemented a new algorithm that accounts for movement dynamics and uses a Kalman filter (KF) to estimate positions [1,19]. The algorithm uses a correlated random walk model to predict the next position and its estimated error based on the previous positions and estimated error. It then uses the Doppler frequency-shift measurements acquired during a satellite pass to update the position predicted by the model and return a final position. Compared to the LS method, the Kalman filtering estimator is reported to improve the accuracy of estimated positions and to increase the number of positions up to 13% [19]. Such improvements may have a significant impact in studies where relatively few messages are received with each satellite pass, which is the case for many marine and dense forest species.

Although the new processing algorithm may bring significant advantages, it may also introduce changes in the autocorrelation structure of the Argos satellite data. Given that many published SSM applications for animal tracking data do not currently account for the potential autocorrelation in location errors introduced by the new KF algorithm, models fit to datasets with differing degrees of autocorrelated errors could lead to biased estimates of movement parameters, behavioural states, and their uncertainties. Several studies have examined the validity of SSMs applied to data obtained with the LS positioning algorithm and quantified the precision of predicted locations (e.g. [20–22]), but to the best of our knowledge, no study examined how changes introduced by the KF algorithm might affect the application of these models.

Our aim is to assess the performance of Bayesian SSMs fit to satellite tracking data processed with the new KF positioning algorithm introduced by Argos. We use two real datasets from marine taxa that differ greatly in their movement ranges– harbour seals (*Phoca vitulina*) and fin whales (*Balaenoptera physalus*) – as SSMs are known to be sensitive to the scale of movement [20]. Using data from 7 harbour seals instrumented with ARGOS satellite transmitters equipped with Fastloc GPS loggers (hereafter GPS/Argos tags), we compared estimated locations from a hierarchical SSM (hSSM) fit to data processed with KF and LS algorithms to the GPS positions obtained from the same tag to (1) assess spatial accuracy of locations from models fit to data derived from each algorithm; and (2) determine how spatial accuracy varies with observation frequency, temporal resolution and reported precision of Argos locations. Models fit to fin whale tracks could not be evaluated through comparison with GPS data because whales were instrumented with Argos-only transmitters. Satellite tracks of 6 fin whales were used to compare location and behavioural states estimated from a switching state-space model (SSSM) fit to the KF data to those from models fit to the classical LS algorithm. We analysed whale tracks with different temporal resolutions to test whether and how the quality of tracking data affected the similarity of the output from SSSMs fit to LS and KF data.

Materials and Methods

Ethics Statement

All seal handling and tagging procedures were carried out under license number 60/4009 issued by the UK Home Office under the Animals (Scientific Procedures) Act 1986.Whale tagging was approved by the Regional Directorate for Sea Affairs, Autonomous Region of the Azores under research permits 20/2009/DRA and 16/2010/DRA. All procedures in whales followed the guidelines of the American Society of Mammalogists [23].

Data Collection and Processing

In the interest of clarity we'll use the following terminology throughout the paper: i) LS locations/data and KF locations/data refer to the locations/data provided by Argos that were derived from the application of the LS and KF algorithms, respectively; and ii) LS or KF model refer to the state-space models fit to data derived from the application of either the LS or KF algorithm. As explained below, the same models were fit to LS and KF datasets.

Harbour seal data. GPS/Argos tags were deployed on harbour seals in the Eden Estuary, south-east Scotland and around Eday, Orkney between May and July 2012. Animals were caught on or close to haul-out sites using hand, seine or tangle nets and subsequently anesthetised with Zoletil as detailed in Sharples et al. [24]. Tags were attached to the fur at the back of the neck using Loctite 422 Instant Adhesive. Tag duration ranged from 25 to 65 days (median 41 days).

The Fastloc GPS data used in this study were transmitted via the Argos system, providing high resolution at sea locations. The Argos transmissions also generated a concurrent series of standard Argos locations. At our request, messages from the satellite transmitters were processed by the Argos service provider (CLS, Ramonville Saint-Agne, France) using both the LS and KF processing algorithms.

Fastloc GPS positions are more accurate and precise than Argos locations and in the present study were assumed to represent the seals' "true" position. However, GPS accuracy is known to decrease when Fastloc calculations are based on fewer satellites [25,26] and when residual error is high [27]. GPS data were therefore cleaned according to the Sea Mammal Research Unit protocol where locations estimated with <5 satellites and with residual errors = 0 or >25 were removed [27]. Tests on land showed that over 95% of the cleaned locations had an error of < 50 m [27].

As central-place foragers harbour seals haul-out on land between foraging trips. Thus, we needed to remove haul-out locations from the data before fitting any models. Although the GPS/Argos tags have a wet/dry sensor which records haul-out events, only a subset of these records are received via the Argos system. These animals often range in near shore waters and the large measurement error in Argos observations means such observations could not be used to define whether a location fell on land. Thus, we used the Fastloc GPS positions to define the precise time seals departed and returned to land. Positions within 200 m from all shorelines were also considered as haul-out to buffer against errors in GPS positions and because harbour seals haul-out on intertidal sandbanks. This procedure may have excluded valid parts of a few foraging trips but this shouldn't affect algorithm comparison in anyway. Consecutive at sea locations between haul-out events thus formed an individual foraging trip. We defined a series of trips within each seal GPS track and, for each trip, we selected all LS and KF locations obtained between 5 minutes prior to and 5 minutes after the trip. Only trips with ≥30 LS and KF locations were subsequently used for model fitting. The seal dataset analysed in the next sections consisted of 1174 GPS, 1339 Argos LS and 2083 Argos KF positions obtained during 31 foraging trips of 7 different seals (Table S1).

Fin whale data. The data consisted of Argos-derived surface positions obtained from PTTs (model SPOT5-implantable, Wildlife Computers, Redmond, Washington, USA) attached to the flanks of 6 fin whales. Whales were tagged off Faial and Pico islands (38°N 28°W), Archipelago of the Azores (Portugal), in September 2009, April and May 2010. All tags were programmed to transmit on a daily basis, every hour of the day up to a maximum of 500 messages per day. Details about the tagging methodology, movements and inferred behaviours of these whales are described in Silva et al. [28]; here we focus on the analyses of model fitting and performance. Like in the case of the harbour seal data, we requested location data to be processed with both the LS and KF algorithms.

The KF algorithm consistently yielded more positions per individual whale than the LS algorithm (Table S2). To compare the regular, estimated locations from the LS model with those from the KF model for each whale dataset, we selected only the positions from the KF data that were within 2 minutes of a LS position (hereafter called the KF reduced dataset).We fitted a second model to all KF locations to investigate how the tracks from a model fitted to the full KF dataset compared to those from a LS model.

State-space Models

State-space models couple two stochastic models: a process model (transition equation) that estimates the current state (e.g. location and behavioural state) of an animal given its previous state, and an observation model that relates the unobserved location states estimated by the process model to the observed data (locations obtained from Argos).

The SSM described in Jonsen et al. [8] uses a first-difference correlated random walk (DCRW) as the process model to describe movement dynamics. The SSSM also uses a DCRW as the process model but allows movement parameters to change between two discrete behavioural states – for example, transiting versus area-restricted search (ARS; [29]) – by including a different DCRW model for each [9].

Model fit to harbour seal data. We initially attempted to fit a SSSM to the harbour seal data but encountered the same problems noted by Breed et al. [20] using simulated tracks. These authors showed that when the scale of movement is small relative to observation error and frequency, the models are unable to accurately estimate location and behavioural states. Even though the temporal resolution of our seal data was reasonably high (see Table S1), the SSSM provided a poor fit, resulting in unreliable location and behavioural estimates, irrespective of the algorithm used (although models fitted to KF data behaved slightly better). It is possible that movements of harbour seals are best analysed with different models (e.g. [30]) but this evaluation is beyond the scope of this paper.

We therefore chose to fit a SSM [8] to the harbour seals' satellite locations derived from the LS and KF algorithms. The SSM was fit as a single hierarchical model (hSSM) [5] to all trips of all seals simultaneously, as this significantly improved parameter estimation, especially for data-sparse trips.

By letting k index each individual harbour seal trip, the transition equation of the SSM formulated within a hierarchical framework becomes:

$$\mathrm{d}_{t,k} \sim \mathcal{N}_2(\gamma_{t,k}\mathrm{T}(\theta_{t,k})\mathrm{d}_{t-1,k},\Sigma)$$

where $\mathbf{d}_{t-1}$ is the displacement between unobserved locations $\mathbf{x}_{t-1}$ and $\mathbf{x}_{t-2}$, and $\mathbf{d}_t$ is the displacement between unobserved locations $\mathbf{x}_t$ and $\mathbf{x}_{t-1}$. $\mathbf{T}(\theta)$ is a transition matrix that provides the rotation required to move from $\mathbf{d}_{t-1}$ to $\mathbf{d}_t$, where θ is the mean turning angle. γ is the move persistence coefficient (i.e. combined autocorrelation in direction and speed). $\mathcal{N}_2$ is a bivariate Gaussian distribution with covariance matrix $\sum$ and represents the randomness in animal movement.

The observation equation accounts for the irregularity and variable errors in the observed Argos locations. Errors in latitude and longitude are modelled with a t-distribution using independent parameter estimates derived for each Argos location class [8,31]. We fitted the same observation equation to data processed with LS and KF algorithms. Further details about the SSM are provided in Jonsen et al. [5,8].

Model fit to fin whale data. We fitted the Bayesian switching state-space model (SSSM) described in Jonsen et al. [9] to the Argos satellite-based location estimates of fin whales derived from the LS and KF algorithms. The transition equation for the SSSM is similar to that of a SSM:

$$\mathrm{d}_t \sim \mathcal{N}_2(\gamma_{b_t}\mathrm{T}(\theta_{b_t})\mathrm{d}_{t-1},\Sigma)$$

but in this case the movement parameters θ and γ are indexed by behavioural state b. At each displacement t, the estimated behavioural state b corresponds to the set of parameters θ and γ that provide the best model fit.

The observation equation used to model the irregularly observed LS and KF fin whale locations was that same used for the SSM.

Model Implementation

Models were fit using R (R Development Core Team 2012) code provided in the supplement to Jonsen et al. [5]. The code implements hSSM and SSSM using Markov Chain Monte Carlo

(MCMC) methods using the program Just Another Gibbs Sampler (JAGS).

The hSSM was fitted separately to the harbour seals' location data (excluding Z class locations) obtained from each algorithm using a time step of 2 hours, corresponding to the average temporal resolution of the LS data. For the hSSM fit to the KF and LS satellite datasets, we ran two MCMC chains for 60000 iterations, dropping the first 50000 samples as a burn-in and retaining every 10th sample from the remaining 10000 assumed post-converge samples from each chain to reduced sample autocorrelation. Thus, model parameters and estimates of seals' locations were calculated using a total of 2000 MCMC samples.

The SSSM was fitted separately to the fin whales' data obtained from each algorithm (after removing Z class positions from both datasets) using a time step of 3 hours, corresponding to the average temporal resolution of the LS data. For each SSSM we ran two MCMC chains for 45000 iterations, discarding the first 40000 samples and retaining every 5th from the remaining 5000 samples from each chain. A total of 2000 MCMC samples were used to calculate model parameters and estimates of whales' locations and behaviours.

hSSM and SSSM convergence and sample autocorrelation were assessed by visually inspecting trace and autocorrelation plots and using the Gelman and Rubin scale reduction factor (R-hat) diagnostic available in R package boa.

Data Analysis

The Argos locations per seal trip greatly exceeded those of Fastloc GPS, and the latter were also more irregular in time (Table S1). Therefore, in order to estimate the accuracy of locations predicted by LS and KF models, we first selected only those locations that were within 30 min of a GPS position. We then estimated the "true" position of the seal at the time of those modelled locations by linear interpolation between two consecutive GPS positions [32]. Finally, we calculated the linear error and absolute latitudinal and longitudinal errors between each modelled location and the corresponding interpolated GPS position.

To investigate if and how the quality of Argos telemetry data affects spatial accuracy of LS and KF models, we compared location errors from seal trips with different temporal resolutions, spatial precisions and frequency of observations. We used linear mixed-effects models with seal and individual trip as random effects to account for behavioural differences among seals and unequal sample sizes across trips. Errors were log transformed to ensure linearity with continuous predictors. Algorithm (LS vs. KF) was included in the model as a categorical predictor and continuous predictors were number of Argos locations used to fit the model, average length of time between locations (hereafter time step), and proportion of positions of LC 0, A and B (hereafter LC 0-B). Values of these continuous predictors for each seal trip are given in Table S1. We fitted a model with interactions between algorithm and all continuous predictors because we were interested in investigating if the effect of data quality was consistent among the LS and KF models.

In the case of the SSSM fit to the whale data, we could only determine how well the KF models performed in relation to models fit to the LS algorithm. For each whale, we compared the medians, inter-quartiles and 95% credible limits (95% CL) of parameter estimates of LS and KF models. We also calculated the longitudinal and latitudinal differences between pairs of location estimates from the LS and KF models for each whale. For each location predicted by the LS model we estimated a probability ellipse determined by the 95% CL obtained from the model. We then calculated the proportion of location estimates from the reduced KF model that fell within the 95% probability ellipse of the corresponding LS position.

To understand if the KF algorithm introduced significant changes in the ability of the SSSM to resolve behavioural state, we calculated percentage of agreement in behavioural classification between the LS and KF models. Whale behaviour at each 3-h location was inferred from the output of the SSSM. Because behaviour is treated as a binary variable, MCMC samples can only assume the values 1 (inferred as transiting) or 2 (inferred as ARS), b at each location was estimated as the mean value of the MCMC samples. We used the same cut off points as Jonsen et al. [9]: locations with mean estimates of $b<1.25$ were assumed to represent transiting, $b>1.75$ ARS, and between these values were considered "uncertain".

Finally, we investigated how the whale tracks from a model fitted to the full KF dataset compared to those from the models applied to LS data. We fitted the SSSM to the full KF data using the same time step as above. For each whale we calculated the distance (in km) from locations estimated by the full KF model to the track estimated by the LS model. We compared only data from days when both methods delivered satellite locations.

Means are presented ± standard deviation (SD) throughout. All distances were calculated using a great-circle route. Statistical analyses were performed in R software using packages nlme and MASS.

Results

Accuracy and Precision of LS and KF Models Fit to Harbour Seal Data

The KF algorithm provided 2083 locations, 1.5 times more than the LS algorithm and 1.8 times more than the GPS transmitted via Argos (Table S1). The increase in the number of locations per trip in relation to the LS data ranged from 12 to 137% with an average of 56%. A total of 368 LS and 375 KF model locations were within 30 min of a GPS position and were used to compare spatial accuracy of locations derived from each hSSM (Table 1).

Errors in locations estimated from LS and KF models showed the same elliptical distribution in relation to interpolated GPS positions, with a clear directional bias in the longitudinal error component (Fig. 1). Average longitudinal errors ranged between −0.20°–0.20° (mean = −0.003) for LS models and between −0.36°–0.17° (mean = 0.001) for KF models. Latitudinal errors ranged between −0.10°–0.08° (mean = −0.002) for LS models and between −0.10°– 0.09° (mean = 0.001) for KF models. Overall, the mean distance of KF model locations to interpolated GPS positions was lower (2.9±2.9 km) than that of LS model locations (3.5±3.0 km) (Table 1). About 31% of all locations predicted from the KF model were within 1 km from the interpolated GPS position and 82% were less than 5 km. For locations predicted from the LS model, 24% and 73% were respectively within 1 km and 5 km from the corresponding interpolated GPS position. The KF model produced smaller mean errors for 27out of 31 trips (Table 1). Predicted trips from the KF model were 27% (range: 1–57%) more accurate than trips derived from the LS model. However, standard deviations of KF errors were sometimes higher suggesting that location accuracy varied considerably within the same trip (Table 1). Average errors of trips increased as the average distance between locations (step length) increased. For trips with an average step length ≥6 km, the average error of KF modelled locations was 4.6±0.4 km, and of LS modelled locations was 5.9±1.2 km.

Table 1. Errors in locations estimated from models fit to Least Squares (LS) and Kalman filtered (KF) data for all harbour seal trips.

Seal	Trip	LS model			KF model			Variationin mean errors (%)*
		N[†]	Mean error (km)	Error range (km)	N[†]	Mean error (km)	Error range (km)	
1545	11	8	2.6±1.7	0.4–5.1	8	1.7±1.4	0.1–3.9	−34
	21	9	1.9±2.0	0.1–5.6	8	0.8±0.4	0.4–1.5	−57
	23	5	2.2±1.3	1.2–4.3	6	0.9±0.3	0.6–1.4	−56
	27	14	1.9±2.3	0.1–8.6	15	2.2±2.4	0.1–7.9	16
	28	13	1.5±2.1	0.1–7.6	13	1.5±1.5	0.1–5.1	−2
28503	11	12	2.2±1.9	0.2–6.9	14	2.0±2.3	0.2–8.5	−10
	18	12	3.9±2.4	1.7–9.4	11	3.1±2.7	0.3–9.3	−23
	19	14	4.5±3.1	0.2–10.5	14	4.4±3.4	0.8–13.9	−3
	23	8	4.8±1.8	1.6–6.8	8	4.2±2.9	0.9–8.9	−13
	42	5	2.9±1.7	1.0–5.5	6	2.4±1.1	1.1–3.8	−19
43844	4	15	2.5±2.6	0.1–10.0	14	1.6±1.6	0.1–5.7	−36
	8	8	1.7±1.5	0.2–4.6	8	1.2±0.6	0.2–1.9	−29
	14	10	1.7±2.4	0.1–8.0	12	1.4±1.7	0.1–6.4	−23
	16	13	1.5±1.5	0.2–6.3	11	1.0±0.7	0.2–2.3	−32
	22	14	4.9±3.7	0.9–13.6	14	2.5±2.7	0.2–9.0	−48
43871	7	24	3.6±3.5	0.3–12.7	23	3.4±3.4	0.3–11.7	−5
	8	17	5.9±3.6	0.8–11.8	17	4.3±3.2	0.7–10.1	−27
	13	21	4.9±3.7	0.3–10.8	20	3.0±2.5	0.1–9.4	−39
	19	20	5.9±2.9	0.6–11.5	20	4.5±2.8	0.2–10.6	−24
120346	24	13	1.9±1.4	0.5–5.6	13	1.2±0.7	0.3–2.6	−37
	25	15	2.2±1.7	0.4–5.4	15	1.6±1.3	0.2–4.1	−31
	26	5	4.3±2.9	1.3–8.4	5	3.0±3.2	0.9–8.7	−30
	30	13	2.1±1.4	0.4–4.9	13	1.1±1.0	0.1–3.2	−48
	32	7	1.7±1.4	0.6–4.1	7	1.7±1.9	0.2–5.9	−1
120349	3	8	4.2±2.3	1.2–7.1	8	4.5±3.5	0.6–10.8	6
	4	8	3.5±2.8	0.4–7.9	8	4.2±3.9	0.9–12.2	20
	5	5	3.7±1.7	2.0–6.1	5	1.9±1.7	0.8–5.0	−48
	6	10	7.7±3.5	1.1–11.1	10	5.0±3.0	0.5–8.8	−35
120350	3	13	3.7±3.1	0.5–9.4	13	3.1±2.3	0.1–3.9	−16
	4	14	3.2±3.1	0.3–12.5	13	5.0±4.5	0.4–1.5	56
	5	15	4.2±2.4	0.7–7.5	23	4.9±4.3	0.6–1.4	17
Total		368	3.5±3.0		375	2.9±2.9		

*Variation in mean errors was calculated for each trip as the difference in the mean error estimated for KF and LS models divided by the mean error of the LS model.
[†]N: Number of locations used to calculate errors in locations estimated from LS and KF models.

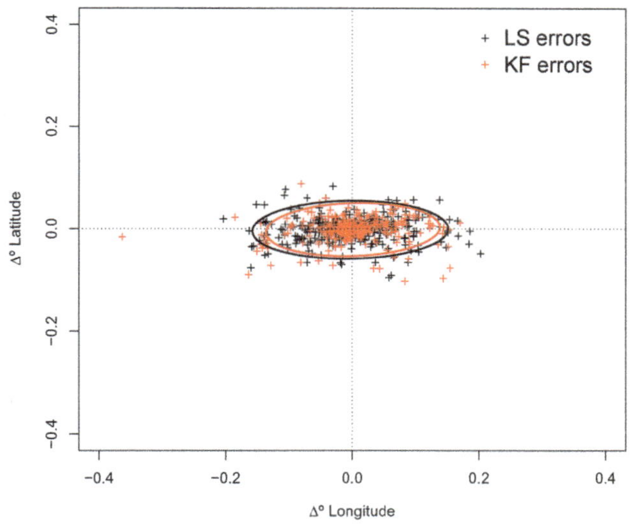

Figure 1. Errors in locations estimated from LS and KF models. Errors in harbour seal locations estimated from state-space models fit to Least Squares (LS) (black) and Kalman filtered (KF) (red) data are plotted as offsets from "true" GPS positions. Standard ellipses were fitted to 95% of LS (black line) and KF (red line) error points.

Two representative tracks of foraging trips reconstructed using GPS positions, and LS and KF modelled locations are shown in Fig. 2. In general, modelled tracks closely matched the GPS tracks, especially during periods of directed movement. Yet, tracks predicted by the LS model occasionally diverged greatly from the GPS track and tended to extend over a wider area in periods of torturous movements.

Uncertainty in KF model estimates, as indicated by the width of the 95% CL (measured in km), was significantly lower than that of LS model estimates (KF model: 5.6±5.6 km; LS model: 11.6±8.4 km; t-test = −11.41, df = 741, P<0.001).

Effect of Data Quality on Accuracy of LS and KF Models Fit to Harbour Seal Data

Observation frequency, temporal resolution and spatial precision of Argos data used to fit the SSMs varied among seals and trips and between the LS and KF models (Fig. 3, Table 1). Expectedly, the increase in number of locations that resulted from the application of the KF algorithm improved the temporal resolution of the KF data for all trips. However, it also increased the proportion of locations of lower spatial precision (Argos LC 0, A and B) in each trip. With few exceptions, trips from the same seal tended to have similar number of locations, time steps, and proportion of LC 0-B, suggesting an individual effect in the quality of Argos data. This could be due to tag (e.g. battery power), instrumentation (e.g. tag placement) or behavioural-specific (e.g. surface behaviour and diving time) differences among seals or to a combination of all these factors.

Mean errors (±SD) of LS and KF modelled trips were plotted in relation to the Argos quality parameters described above (Fig. 3). Accuracy of modelled trips did not seem to improve with the observation frequency or temporal resolution of Argos data, but mean errors (and respective SD) in LS and KF estimated locations appeared to increase with increasing proportions of locations LC 0-B.

We fitted a linear mixed-effects model to examine the effects of type of algorithm and of Argos quality parameters (spatial precision, observation frequency and time step) on estimated errors of modelled locations. The interactions between algorithm and the continuous predictors were the first to be dropped from the linear mixed-effects model based on AIC results, suggesting that quality of Argos data influenced the accuracy of LS and KF models in a similar way. The best fitting model indicated that observation frequency and time step of Argos data had no effect on the errors of locations estimated from the models, and only algorithm and proportion of locations LC 0-B were significant (Table S3). Contrary to our expectations, there was little variability among different seals in addition to the trip-to-trip variability and both the AIC and the likelihood ratio test indicated that individual seal could be dropped from the model ($L = 3.95 \times 10^{-7}$, $P = 0.499$), leaving trip as the only random effect. The best fitting model predicted larger errors for locations estimated from LS models compared to locations from KF models (Fig. 4, Table S3). On average, LS models will estimate locations that are 1.6 km farther from the true seal position relative to KF locations. Also, errors (on a logarithmic scale) are expected to increase as proportion of Argos locations with lower precision increases, and this relationship was similar for LS and KF models (Fig. 4, Table S3).

Comparison of LS and KF Models Fit to Fin Whale Data

Medians and 95% CL of estimated model parameters of the reduced dataset were similar across whales and between the LS and KF algorithms. Both the LS and KF models distinguished well

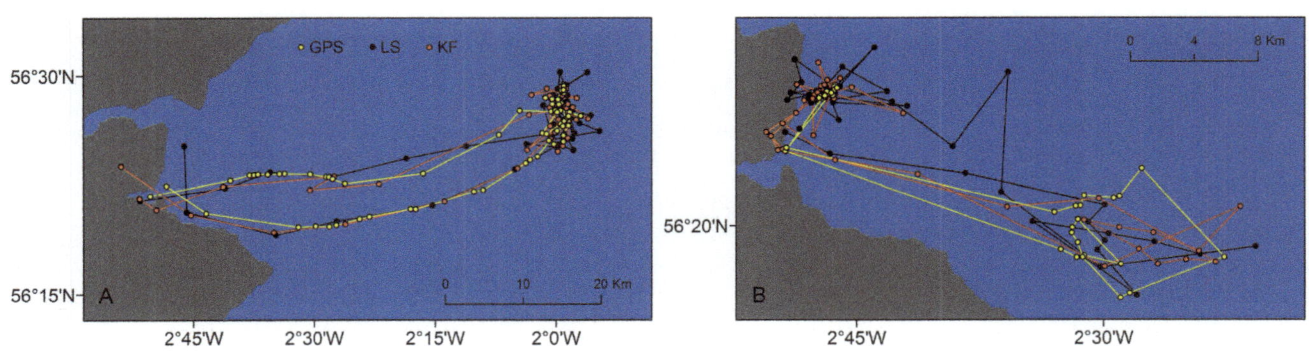

Figure 2. Harbour seal tracks obtained from GPS (yellow), LS (black) and KF modelled (red) locations. Estimated locations (circles) and tracks (lines) of harbour seals obtained from fitting state-space models to Least Squares (LS) (black) and Kalman filtered (KF) (red) data, in relation to the "true" GPS positions and track (yellow). A. Example of a trip with higher quality of Argos data: trip 7 of harbour seal #43871. B. Example of a trip with lower quality of Argos data: trip 22 of harbour seal #43844.

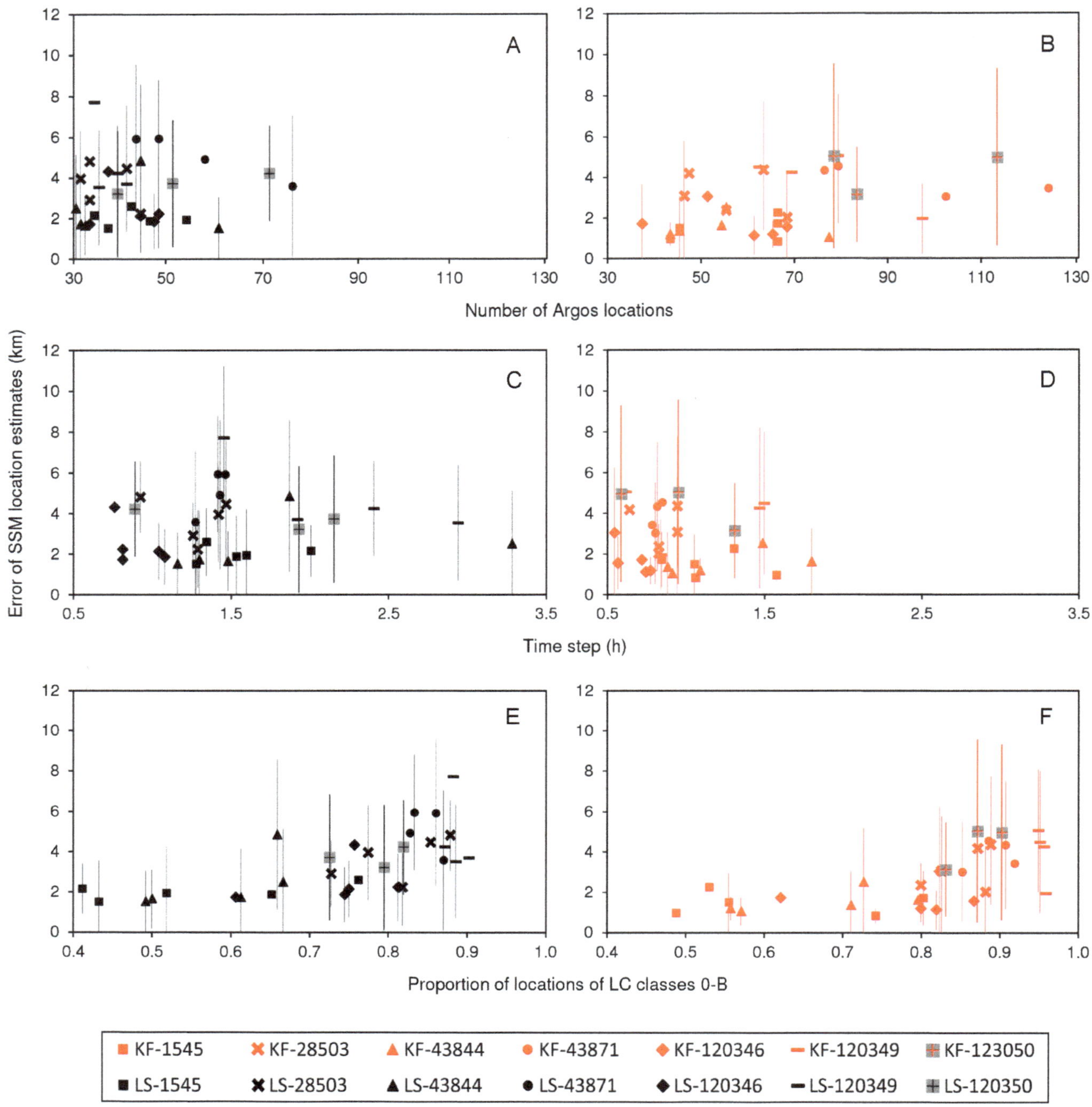

Figure 3. Trip-averaged error in locations estimated from LS and KF models relative to Argos data quality. Relationship between mean errors (±SD shown as vertical bars) in locations estimated from state-space models fit to Least Squares (LS) (black) and Kalman filtered (KF) (red) data per harbour seal trip and quality of Argos telemetry data used to fit the models: A–B. Number of locations. C–D. Time step (h) between locations. E–F. Proportion of locations of LC 0-B. Different trips from the same seal have the same symbol.

between the two behavioural modes (transiting and ARS), as indicated by the parameter estimates that aggregated into two non-overlapping groups.

The estimated locations inferred from the KF model applied to the reduced dataset differed little from the locations output by the LS model. Differences in latitude and longitude between paired KF-LS locations were centred around zero but the latter showed a wider range of values (range for latitude: −1.1–0.7°; range for longitude: −1.2–2.0°) (Fig. 5). Differences in paired KF-LS locations were considerably higher for whale #80716. Removing data from this whale resulted in a considerable reduction in the range of latitudinal (−0.1–0.2°) and longitudinal (−0.5–0.4°) distances between KF and LS locations. Differences in latitude and longitude between paired locations showed no obvious trend with latitude, longitude, date, number of positions per track, or behavioural mode (not shown).

The proportion of estimated locations from the SSSM applied to the reduced KF data that fell within the 95% probability ellipse

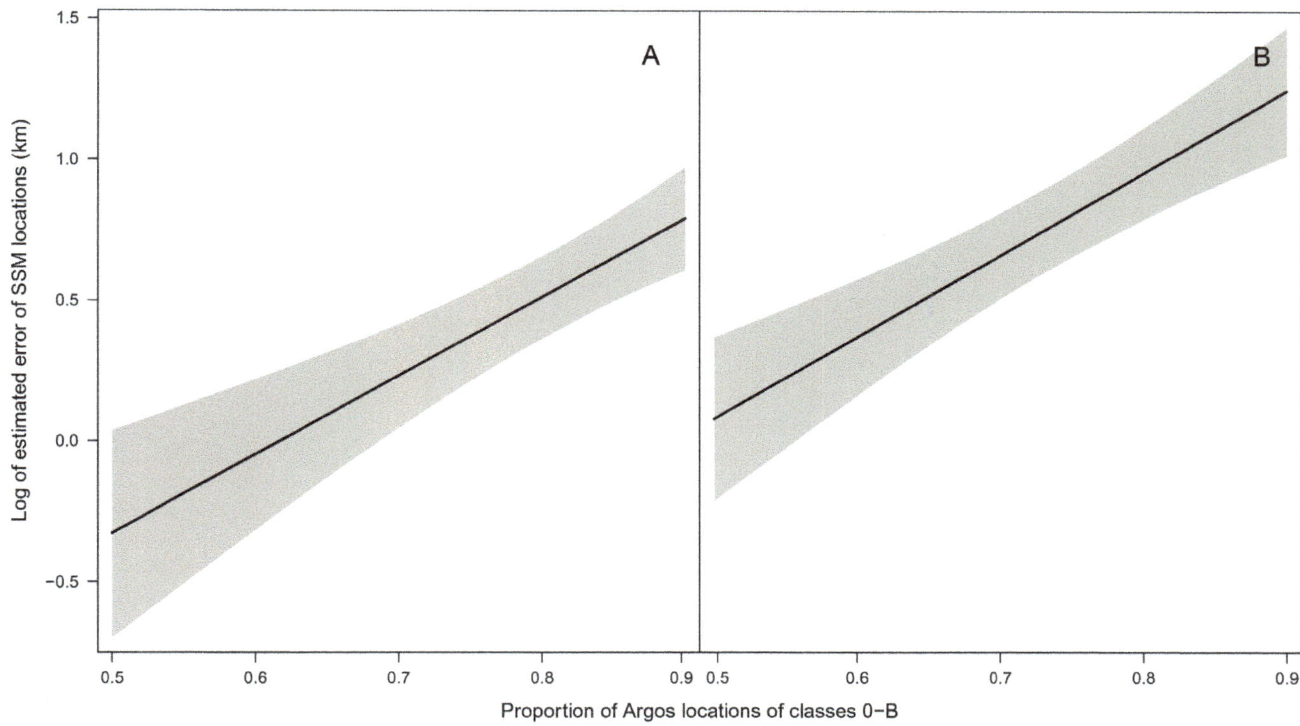

Figure 4. Predicted error in locations estimated from LS and KF models. Predicted error in harbour seal locations according to the best fitting linear mixed-effects model for A. State-space models fit to Kalman filtered (KF) data. B. State-space models fit to Least Squares (LS) data.

of locations inferred by the LS model varied between whales but was very high, ranging from 69 to 100% (mean = 88%). We also compared differences in the width (measured in km) of the 95% CL between pairs of locations estimated from the model fit to the reduced KF data and the LS data. For five whales, the reduced KF model resulted in lower average widths of 95% CL (paired t-test: $P<0.05$ for all whales), although differences were generally small (mean difference: -2.2 ± 3.9 km). For whale #80716, however, the 95% CL of the reduced KF model were significantly wider than those of LS data (paired t-test: $t=-11.15$, $P<0.001$; mean difference: 76.2 ± 80.3 km).

In 94% of the cases, the behavioural mode inferred by the KF model matched the classification from the model fit to the LS data (Table 2). Agreement was highest for locations inferred as transiting (98%), followed by ARS (93%). Changes in behavioural classification between the two models were from transiting or ARS to "uncertain" and vice-versa, but never from transiting to ARS or vice-versa.

As expected, the KF processing algorithm yielded more positions and improved the temporal resolution of the 6 whale tracks. The increase in number of locations per track ranged from 18 to 272% with an average of 75%. The average number of daily

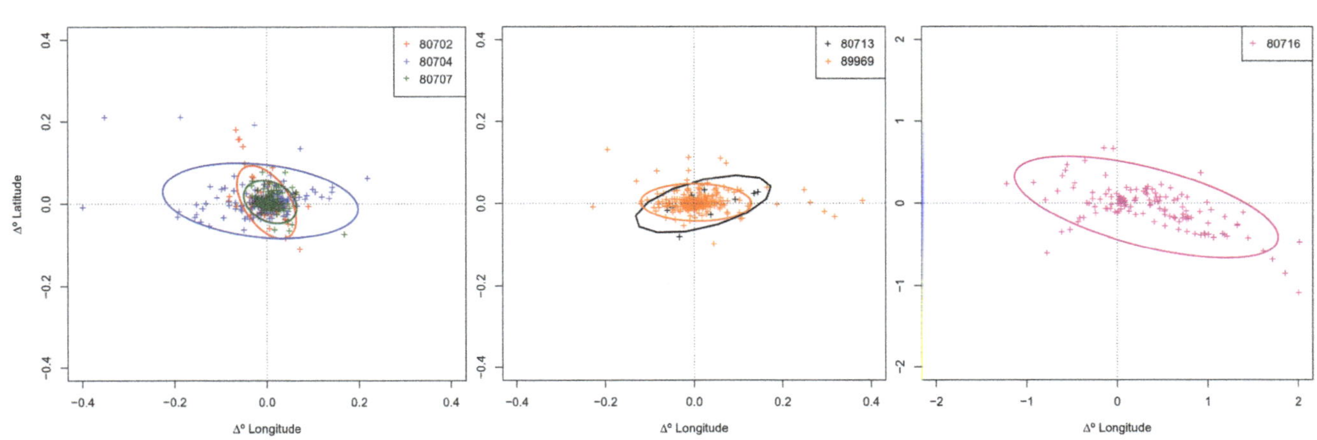

Figure 5. Differences in locations estimated from KF and LS models for all fin whales. Differences in locations estimated from switching state-space models fit to Kalman filtered (KF) (red dots) data are plotted as offsets from locations calculated from the same models fit to Least Squares (LS) data. Standard ellipses were fitted to 95% of KF data points. A. Fin whales #80702 (red), #80704 (blue) and #80707 (green). B. Fin whales #80713 (black), #89969 (orange). C. Fin whale #80716 (pink).

Table 2. Agreement between fin whale behavioural modes inferred by models fit to Least Squares (LS) and Kalman filtered (KF) data.

		KF model		
		Transit	**ARS***	**uncertain**
LS model	Transit	353	0	6
	ARS	0	524	40
	uncertain	5	6	83

The matrix shows the number of fin whale locations classified in each behavioural mode by the LS model that were assigned to each of the behavioural modes by the KF model.
*ARS Area restricted search.

locations per whale track varied between 6.0–38.6 for the full KF data, compared to 1.6–30.8 for the LS data (Table S2). There was also an increase in track duration (3 and 11 days) for two whales but this came at the expense of a few gaps (maximum of 3 days) in those tracks (Table S2). In contrast, the KF algorithm provided several positions within a 5-day gap in the LS tracking data of whale #80716.

The width of the 95% CL of locations estimated by the full KF model (47.3±76.9 km) was significantly lower than the width of 95% CL of locations estimated from the LS model (57.2±113.0 km) ($t = 2.38$, $P = 0.017$). Still, locations from the full KF model fitted well the paths inferred from the LS data, except when gaps in the LS data exceeded 1 day (Figs. 6 and 7). Combining data from all whales, over 49% of locations estimated by the full KF model were <1 km away from the tracks derived from the LS model and 77% were <5 km.

Discussion

Since the recent introduction of the Kalman filtering (KF) algorithm for the processing of satellite tracking data by the Argos system, the service providers have made this the default processing method for new transmitters (PTTs), giving the user the option to choose the Least Squares (LS) algorithm in alternative. The data processing of old PTTs that were already being processed with the LS algorithm remains unchanged, unless KF processing is requested, and stored data from 2008 onwards can be reprocessed using either method (albeit with additional processing costs). Processing of data with the new KF algorithm is bound to become more common as old PTTs end their life, and data processed with this algorithm will soon become the standard for Argos-based tracking.

State-space modelling approaches provide the statistical rigor needed in analysing animal movement data, but SSMs are not simple and require considerable care in their use [5]. Understanding the implications of using data processed with the new KF algorithm is essential when interpreting modelling results. To our knowledge this is the first time that performance of SSMs applied to KF tracking data has been directly validated with known locations of free-ranging animals. This was achieved by fitting the same model to Argos satellite locations obtained on 7 harbour seals processed with LS and KF algorithms and by comparing locations derived from each model against the "true" interpolated positions of the seals obtained by Fastloc GPS technology. In addition, the results of fitting the Bayesian switching state-space model (SSSM) to KF data were compared to those of LS models, using tracking data from 6 fin whales. Although in the latter case we could not assess the accuracy of model-derived locations, it enabled evaluating how SSSMs fit to KF data performed in relation to SSSMs fit to data processed with the LS algorithm, which until recently was the standard processing algorithm used to deliver satellite locations.

Our study shows that Kalman filtering consistently provided more estimated locations per animal track than the LS algorithm, supporting previous claims by the Argos service [19]. The increment in estimated locations was substantial for both species (fin whales: 75%; harbour seals: 56%). Compared to our findings, Boyd and Brightsmith [33] reported only a modest 28% increase in locations computed with the KF algorithm. However, their estimate is based on data obtained from static platforms, while our estimates and those from Argos come from free-ranging tagged animals. Stationary land tests are closer to the "ideal" conditions for satellite communications and are unlikely to adequately represent most of the problems known to affect the transmission of signals from satellite tags and/or the reception of messages at Argos satellites, especially for marine taxa. Understandably, the potential benefit of the KF method should be higher under circumstances (e.g. areas with limited satellite coverage) and for species more prone to transmission difficulties, and for which the frequency of uplinks is usually low. Not surprisingly, the major increase in estimated locations was for fin whales that typically have shorter surface intervals than harbour seals, and can be more adversely affected by wave wash due to improper antenna orientation and poor environmental conditions.

Like Boyd and Brightsmith [33], we also found that the majority of additional locations in KF data came from fixes with only 1message (Argos LC B) (fin whales: 29%; harbour seals: 33%) with a very slight increase in the proportion of locations with 4 or more messages (LC 2 and 3) observed only for fin whales. If, as a result of KF processing, tracks acquire a disproportionate number of locations with low spatial precision, this may impact the analysis and interpretation of animal movement data, particularly when this analysis is based on the raw satellite positions and doesn't take into account variability in measurement errors. Implications could be even more severe if the gain in 1-message LC B locations is not homogeneously distributed along the track and depends, for instance, on the geographic location or behaviour of the animals, therefore being more prevalent in certain areas or during specific activities occurring in preferred habitats.

Our results demonstrate that the Jonsen et al. [8] SSM provided a good fit to the data processed with the KF dataset, despite the potentially increased autocorrelation in the location errors imposed by the KF algorithm. The greater spatial accuracy and precision of locations estimated from the KF model compared to those from the LS model was likely due to a combination of

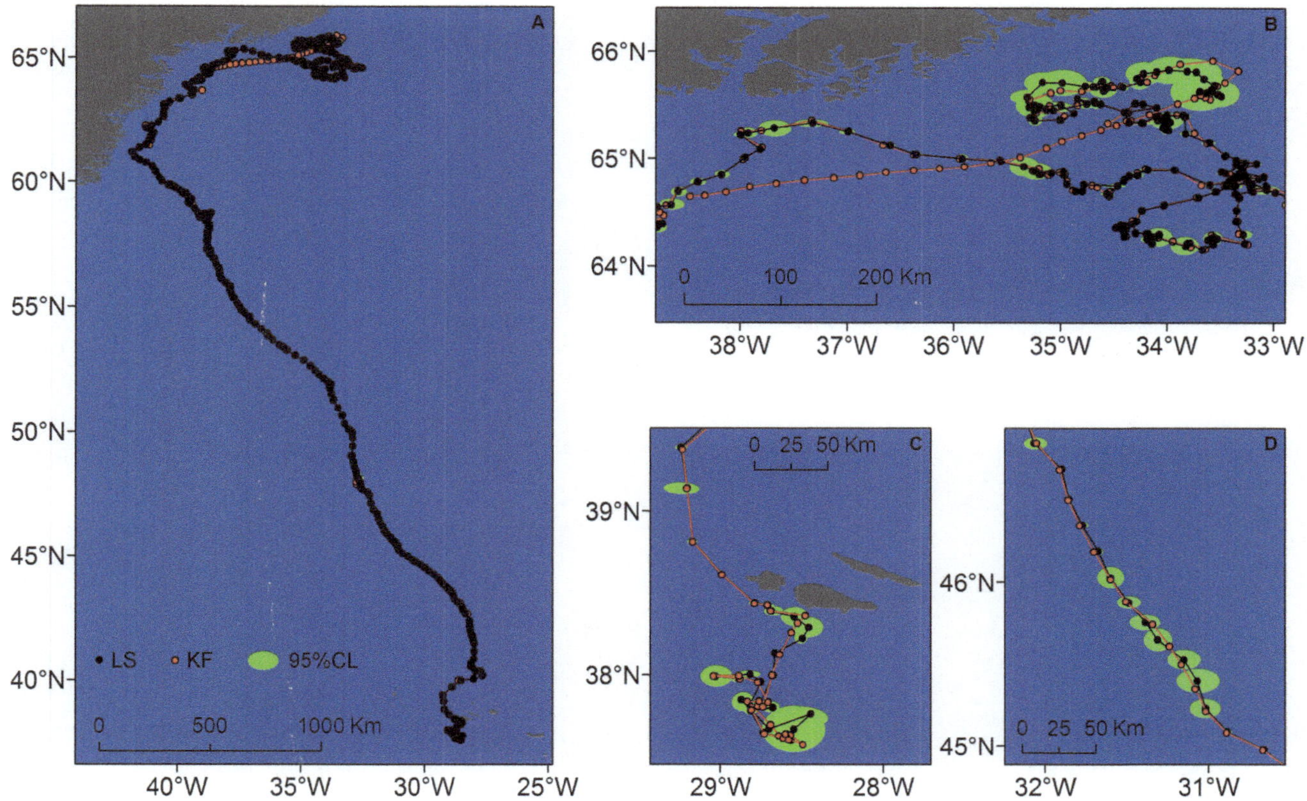

Figure 6. Fin whale #89969 tracks obtained from LS (black) and KF modelled (red) locations. Estimated locations (circles) and tracks (lines) of fin whale #89969 obtained from fitting a switching state-space model to Least Squares (LS) (black) and the full Kalman filtered (KF) (red) data. The 95% probability ellipses of locations derived from the LS-based model are shown in green. A. Complete tracks showing the increase in track length resulting from the application of the KF algorithm (red). B, C, D. Detail of the tracks showing the majority of KF locations within the 95% probability ellipses of LS locations.

increased accuracy in KF-estimated locations and the higher temporal resolution of the KF data.

Although the overall difference in mean errors between the two algorithms appeared small (mean error in LS models was 3.5±3.0 compared to 2.9±2.9 in KF model) the model fit to KF data improved the accuracy of seal trips by 27% over the LS model. The linear mixed-effects model indicated that, despite significant variations in trip accuracy, errors in locations predicted for LS trips were significantly larger than those predicted for KF trips. For both models the largest deviances from true locations occurred along the east/west axis. This is not unexpected since Argos location errors are strongly biased towards the longitudinal component, regardless of the processing algorithm [1,3,31–33], and the SSM does not explicitly account for this directional bias. However, we found no evidence of the non-uniform distribution of extreme errors documented in other studies [3] suggesting that the model was able to handle this problem.

Tracks reconstructed from the models applied to KF and LS data provided faithful representations of the true seal trajectories measured with Fastloc GPS. However, the LS track tended to deviate more from the true track when seals were making short displacements and frequently changing direction. This is likely due to the correlated random walk model employed in the KF algorithm which would tend to smooth out uncommonly large changes in direction and/or displacement. As a result, LS locations tended to spread over a wider area compared to the KF. This was a common feature to several LS modelled tracks that can have major implications if these data are used to calculate sizes of home ranges or ARS patches.

The SSMs were fit as hierarchical models to the LS and KF data, meaning that data from all seal trips were combined to estimate model parameters, leading to improved location estimates. We anticipate that larger errors would be obtained if models were fitted separately to each trip. Yet, there is no reason to expect that the hierarchical formulation behaved differently when applied to LS and KF data, so we consider that the comparison between algorithms remains valid.

We fitted the same observation equation to data processed with LS and KF methods, thus assuming that the new algorithm did not change substantially the distribution or magnitude of the errors. A recent study demonstrated that both LS and KF location errors are better described by a long-tailed lognormal distribution [33]. In the present work, errors were modelled with generalized t-distributions which are known to be robust to extreme values [8]. Boyd and Brightsmith [33] also compared mean errors in KF and LS processed locations showing these did not differ significantly for most location classes, except for LC 2, for which LS errors were about half the KF errors, and LC B, for which LS errors were nearly 4 times greater than KF errors. In contrast, Argos reported better accuracies with the KF method for locations computed with ≥4 (LC 2 and 3) and 2–3 messages (LC A and B) [19]. In any case, given the predominance of LC classes A and B in both our datasets, we suspect that fitting the same observation equation to LS and KF data might have resulted in an overestimation of KF errors relative to LS errors, and not the other way around.

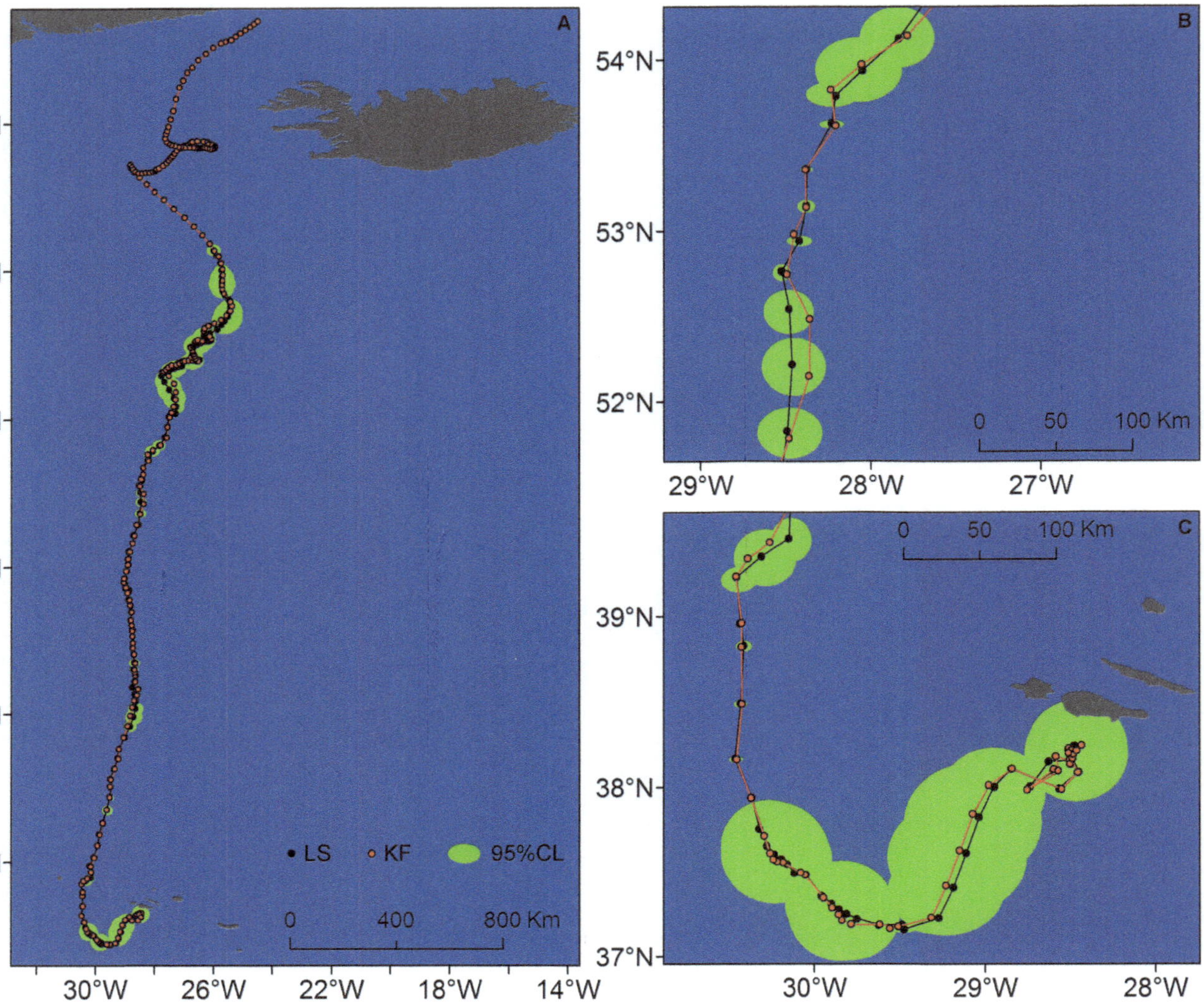

Figure 7. Fin whale #80704 tracks obtained from LS (black) and KF modelled (red) locations. Estimated locations (circles) and tracks (lines) of fin whale #89969 obtained from fitting a switching state-space model to Least Squares (LS) (black) and full Kalman filtered (KF) (red) data. The 95% probability ellipses of locations derived from the LS-based model are shown in green. A. Complete tracks showing the increase in track length resulting from the application of the KF algorithm (red). B, C. Detail of the tracks showing the majority of KF locations within the 95% probability ellipses of LS locations.

Regardless of which processing method is used, our study showed that accuracy of modelled tracks was sensitive to precision of the raw input data. As the proportion of locations with poor precision increased, the ability of the SSMs to recover accurate locations was significantly worse. This is consistent with findings from other researchers that showed that high measurement error not only impacts accuracy and precision of locations estimated from state-space methods [20,34] but can also affect our ability to discern behavioural patterns and quantify habitat use patterns [4,26,32,35,36].

On the other hand, we found no evidence that observation frequency and temporal resolution of Argos data influenced the magnitude of SSM errors, in contrast to a recent study that suggested that frequency and regularity of raw data may be as important as spatial precision for obtaining accurate estimates of locations from state-space methods [20]. There are two main reasons for the different results between our analysis and that of Breed et al. [20]. First, Breed's analysis of model accuracy was based on a reduced number of simulated tracks to which were imposed different observation frequencies and temporal gaps spanning a much larger range than the number of Argos locations and time steps observed in our seal data (see Table S1). Second, in Breed's study a separate SSM was fit to each simulated track while we adopted a hierarchical approach. By combining information from all trips to estimate model parameters, potential effects of between-trip data quality likely were lessened and more accurate location estimates were obtained for all trips.

Our results strongly suggest that application of SSSM to the whale tracking data processed with the KF algorithm was appropriate and that models fitted well. Estimated parameters from KF models were very similar across all tracks and to parameters from the LS model despite the fact that models were fitted separately to each whale LS/KF-processed dataset.

Paths inferred from both models were also similar, with most of the locations from the reduced KF model falling within the 95% probability ellipses of locations estimated from the LS model, and the majority of locations from the full KF model being close to the whale tracks inferred by the LS model. Similar to what was observed for the seal data, the longitudinal bias in Argos errors caused the reduced KF locations to differ more from their paired LS positions in the east/west than in the north/south axis.

The estimated precision of locations inferred from the SSSM fit to the reduced KF data was higher for 5 out of 6 whale tracks, as indicated by the lower average width of the credible limits. However, the KF model behaved significantly worse than the LS model in the case of the whale track (#80716) for which less than 2 satellite positions were received per day. This cannot be accounted for by variations in Argos location classes because 28 of 29 positions were assigned the same class in both datasets. A close inspection of the raw KF and LS data indicates that the poorer performance of the reduced KF model was likely associated with the highly tortuous whale path evident in the KF data (and not in the LS data) and caused by the way the data regularization approach used in the SSSM's observation model dealt with this tortuosity. Because the interval between raw satellite positions was considerably longer than the 3-hourly interval at which the SSSM positions were being estimated, raw positions have more weight on model estimates as the model "forces" derived locations to exactly match raw satellite positions. Such an effect tends to be more pronounced with decreasing linearity of the tracks [37], explaining why uncertainty in the model estimates was greater for the more sinuous KF path and the higher discrepancy in relation to the LS path.

It should be stressed that the application of the KF algorithm increased the total number of locations in this whale track from 29 to 108 (see Table S2), resulting in a remarkable decrease in the uncertainty of SSSM location estimates (average 95% CL width: 86.0±69.5 km) when compared to the LS model. Differences in the remaining tracks were less pronounced but the KF processing algorithm produced an overall increase in number of locations obtained and a decrease in the uncertainty of SSSM estimates.

Estimates of behavioural mode from the KF model agreed well with inferences from the LS model – with 94% of whale locations being assigned the same behavioural category in both models – indicating that the KF algorithm did not introduce appreciable changes in the ability of the SSSM to recover latent behaviours from satellite positions.

These results lead us to conclude that application of widely-used Bayesian state-space models [5] to Argos satellite locations processed with a KF method is appropriate and, as was the case of the SSM fit to harbour seal data, can produce more reliable location estimates than when LS data are used to fit the same models. Also, behavioural modes could be equally well detected from SSSM fit to whale tracking data processed with KF and LS methods. Since the KF algorithm generally yields more positions and longer tracks, there may be clear advantages in using the KF model over the LS model. This is especially true in telemetry studies of species that spend prolonged periods underwater or under dense vegetation cover, for which the number of daily fixes is generally low, precluding examination of movement and behaviour of animals in more detail. However, as seen here, the KF algorithm can increase the number of positions of lower precision (LC B) by nearly 30%, which in turn can degrade accuracy of modelled tracks. Even with LC B positions estimated by the KF method being several times more accurate than LS locations of equal class [19,33], when accuracy and precision are critical for the analysis, researchers may consider removing 1-message positions before fitting state-space models.

Supporting Information

Table S1 GPS, Least Squares (LS) and Kalman filtered (KF) data obtained for each seal trip.

Table S2 Argos Least Squares (LS) and Kalman filtered (KF) data obtained for each fin whale used to fit the switching state-space models.

Table S3 Parameter estimates from the best fitting linear mixed-effects model for errors in modelled locations.

Acknowledgments

We are grateful to Cláudia Oliveira, Irma Cascão, João Medeiros, Yves Cuenot, Maria João Cruz and many volunteers at DOP/IMAR that helped with whale fieldwork, and to our skilled skippers and crew Paulo Martins, Norberto Serpa and Vitor Rosa. We thank P. Lovell, P. Hammond and S. Moss and others at SMRU for helpful discussions and fieldwork. We acknowledge Greg Breed and one anonymous reviewer for very useful comments on a previous version of the manuscript. This study is an output of research projects TRACE (PTDC/MAR/74071/2006) and MAPCET (M2.1.2/F/012/2011 - Integrating cetaceans into marine spatial management in the Azores).

Author Contributions

Conceived and designed the experiments: MAS IDJ DJFR MFB. Performed the experiments: MAS DJFR. Analyzed the data: MAS. Contributed reagents/materials/analysis tools: MAS IDJ DJFR DT. Wrote the paper: MAS. Reviewed the manuscript and approved the final version: IDJ DJFR RP MFB.

References

1. Douglas DC, Weinzierl R, C Davidson S, Kays R, Wikelski M, et al. (2012) Moderating Argos location errors in animal tracking data. Methods in Ecology and Evolution 3: 999–1007.
2. CLS (2011) Argos User's Manual. Available: http://wwwargos-systemorg/files/pmedia/public/r363_9_argos_manual_en.pdf. Accessed 2013 Oct 10.
3. Costa DP, Robinson PW, Arnould JPY, Harrison A-L, Simmons SE, et al. (2010) Accuracy of ARGOS locations of pinnipeds at-sea estimated using fastloc GPS. PLoS ONE 5: e8677.
4. Bradshaw CJA, Sims DW, Hays GC (2007) Measurement error causes scale-dependent threshold erosion of biological signals in animal movement data. Ecological Applications 17: 628–638.
5. Jonsen ID, Basson M, Bestley S, Bravington MV, Patterson TA, et al. (2013) State-space models for bio-loggers: A methodological road map. Deep Sea Research Part II: Topical Studies in Oceanography 88–89: 34–46.
6. Jonsen ID, Myers RA, Flemming JM (2003) Meta-analysis of animal movement using state-space models. Ecology 84: 3055–3063.
7. Jonsen ID (2006) Robust hierarchical state–space models reveal diel variation in travel rates of migrating leatherback turtles. Journal of Animal Ecology 75: 1046–1057.
8. Jonsen ID, Flemming JM, Myers RA (2005) Robust state-space modeling of animal movement data. Ecology 86: 2874–2880.
9. Jonsen ID, Myers RA, James MC (2007) Identifying leatherback turtle foraging behaviour from satellite telemetry using a switching state-space model. Marine Ecology Progress Series 337: 255–264.
10. Bestley S, Jonsen ID, Hindell MA, Guinet C, Charrassin J-B (2013) Integrative modelling of animal movement: incorporating in situ habitat and behavioural information for a migratory marine predator. Proceedings of the Royal Society B: Biological Sciences 280. http://dx.doi.org/10.1098/rspb.2012.2262.
11. Breed GA, Costa DP, Jonsen ID, Robinson PW, Mills-Flemming J (2012) State-space methods for more completely capturing behavioral dynamics from animal tracks. Ecological Modelling 235–236: 49–58.
12. Bailey H, Fossette S, Bograd SJ, Shillinger GL, Swithenbank AM, et al. (2012) Movement patterns for a critically endangered species, the leatherback turtle (Dermochelys coriacea), linked to foraging success and population status. PLoS ONE 7: e36401.
13. Block BA, Jonsen ID, Jorgensen SJ, Winship AJ, Shaffer SA, et al. (2011) Tracking apex marine predator movements in a dynamic ocean. Nature 475: 86–90.

14. Flemming JM, Jonsen ID, Myers RA, Field CA (2010) Hierarchical state-space estimation of leatherback turtle navigation ability. PLoS ONE 5: e14245.
15. Hammerschlag N, Gallagher AJ, Lazarre DM (2011) A review of shark satellite tagging studies. Journal of Experimental Marine Biology and Ecology 398: 1–8.
16. Patterson TA, Thomas L, Wilcox C, Ovaskainen O, Matthiopoulos J (2008) State–space models of individual animal movement. Trends in Ecology and Evolution 23: 87–94.
17. Schick RS, Loarie SR, Colchero F, Best BD, Boustany A, et al. (2008) Understanding movement data and movement processes: current and emerging directions. Ecology Letters 11: 1338–1350.
18. Fancy SG, Pank LF, Douglas DC, Curby CH, Garner GW, et al. (1988) Satellite Telemetry: a new tool for wildlife research and management. Washington D.C.: United States Department of the Interior. 74 p.
19. Lopez R, Malardé JP (2011) Improving ARGOS Doppler location using Kalman filtering. CLS-DT-MEMO-11-65. 14 p. Available: http://www.argos-system.org/files/pmedia/public/r282_9_technical_paper.pdf. Accessed 27 Apr 2012.
20. Breed GA, Costa DP, Goebel ME, Robinson PW (2011) Electronic tracking tag programming is critical to data collection for behavioral time-series analysis. Ecosphere 2: art10.
21. Hoenner X, Whiting SD, Hindell MA, McMahon CR (2012) Enhancing the use of Argos satellite data for home range and long distance migration studies of marine animals. PLoS ONE 7: e40713.
22. Winship AJ, Jorgensen SJ, Shaffer SA, Jonsen ID, Robinson PW, et al. (2012) State-space framework for estimating measurement error from double-tagging telemetry experiments. Methods in Ecology and Evolution 3: 291–302.
23. Gannon WL, Sikes RS, The Animal Care and Use Committee of the American Society of Mammalogists (2007) Guidelines of the American Society of Mammalogists for the use of wild mammals in research. Journal of Mammalogy 88: 809–823.
24. Sharples RJ, Moss SE, Patterson TA, Hammond PS (2012) Spatial variation in foraging behaviour of a marine top predator (Phoca vitulina) determined by a large-scale satellite tagging program. PLoS ONE 7: e37216.
25. Bryant E (2007) Location Accuracy Statistics for Fastloc Cores Running Firmware Versions 2.2 & 2.3. Leeds, UK: Wildtrack Telemetry Systems Ltd. Technical Report Fastloc/TR01. 5 p.
26. Hazel J (2009) Evaluation of fast-acquisition GPS in stationary tests and fine-scale tracking of green turtles. Journal of Experimental Marine Biology and Ecology 374: 58–68.
27. Russell DJF, Matthiopoulos J, McConnell BJ (2011) SMRU seal telemetry quality control process. SCOS Briefing paper 11/17. Fife, UK: Sea Mammal Research Unit, University of St. Andrews.
28. Silva MA, Prieto R, Jonsen I, Baumgartner MF, Santos RS (2013) North Atlantic blue and fin whales suspend their spring migration to forage in middle latitudes: building up energy reserves for the journey? PLoS ONE 8: e76507.
29. Tinbergen N, Impekoven M, Franck D (1967) An experiment on spacing-out as a defence against predation. Behaviour 28: 307–320.
30. McClintock BT, Russell DJF, Matthiopoulos J, King R (2012) Combining individual animal movement and ancillary biotelemetry data to investigate population-level activity budgets. Ecology 94: 838–849.
31. Vincent C, McConnell BJ, Ridoux V, Fedak MA (2002) Assessment of Argos location accuracy from satellite tags deployed on captive gray seals. Marine Mammal Science 18: 156–166.
32. Witt MJ, Åkesson S, Broderick AC, Coyne MS, Ellick J, et al. (2010) Assessing accuracy and utility of satellite-tracking data using Argos-linked Fastloc-GPS. Animal Behaviour 80: 571–581.
33. Boyd JD, Brightsmith DJ (2013) Error properties of Argos satellite telemetry locations using Least Squares and Kalman filtering. PLoS ONE 8: e63051.
34. Kuhn CE, Johnson DS, Ream RR, Gelatt TS (2009) Advances in the tracking of marine species: using GPS locations to evaluate satellite track data and a continuous-time movement model. Marine Ecology Progress Series 393: 97–109.
35. Dragon A-C, Bar-Hen A, Monestiez P, Guinet C (2012) Comparative analysis of methods for inferring successful foraging areas from Argos and GPS tracking data. Marine Ecology Progress Series 452: 253–267.
36. Morehouse AT, Boyce MS (2013) Deviance from truth: Telemetry location errors erode both precision and accuracy of habitat-selection models. Wildlife Society Bulletin 37: 596–602.
37. Bailey H, Shillinger GL, Palacios DM, Bograd SJ, Spotila JR, et al. (2008) Identifying and comparing phases of movement by leatherback turtles using state-space models. Journal of Experimental Marine Biology and Ecology 356: 128–135.

200 kHz Commercial Sonar Systems Generate Lower Frequency Side Lobes Audible to Some Marine Mammals

Z. Daniel Deng[1]*, Brandon L. Southall[2], Thomas J. Carlson[1], Jinshan Xu[1], Jayson J. Martinez[1], Mark A. Weiland[1], John M. Ingraham[1]

1 Energy and Environment Directorate, Pacific Northwest National Laboratory, Richland, Washington, United States of America, 2 Southall Environmental Associates, Aptos, California, United States of America

Abstract

The spectral properties of pulses transmitted by three commercially available 200 kHz echo sounders were measured to assess the possibility that marine mammals might hear sound energy below the center (carrier) frequency that may be generated by transmitting short rectangular pulses. All three sounders were found to generate sound at frequencies below the center frequency and within the hearing range of some marine mammals, e.g. killer whales, false killer whales, beluga whales, Atlantic bottlenose dolphins, harbor porpoises, and others. The frequencies of these sub-harmonic sounds ranged from 90 to 130 kHz. These sounds were likely detectable by the animals over distances up to several hundred meters but were well below potentially harmful levels. The sounds generated by the sounders could potentially affect the behavior of marine mammals within fairly close proximity to the sources and therefore the exclusion of echo sounders from environmental impact analysis based solely on the center frequency output in relation to the range of marine mammal hearing should be reconsidered.

Editor: Fan-Gang Zeng, University of California, Irvine, United States of America

Funding: This study was funded by the U.S. Department of Energy Office of Energy Efficiency and Renewable Energy (EERE) Wind and Water Power Program. The funders had no role in study design, data collection and analysis, decision to publish, or preparation of the manuscript.

Competing Interests: Co-author (Brandon Southall) is employed by a commercial company (SEA, Inc.) and first author (Z. Daniel Deng) is a PLOS ONE academic editor, but this does not alter the authors' adherence to PLOS ONE policies on sharing data and materials.

* E-mail: Zhiqun.deng@pnnl.gov

Introduction

Active acoustic systems transmit sound, usually in short pulses, and receive echoes from targets in the water permitting localization and characterization of targets. Active acoustic systems are ubiquitous in the marine environment, and used convergently by humans and non-human animals, because sound propagates well in water and a very broad range of sensing applications is achievable using sound. Most vessels have one or more echo sounders to aid in navigation [1]. Commercial and recreational fishermen use a wide range of active acoustic systems to locate fish. Various exploration and construction activities use tools that generate and receive sound in configurations that are similar to active acoustic systems [2]. Additionally, militaries use sonar systems for a variety of underwater assessment and communication tasks.

Because active sonar systems sometimes generate intense sounds and many marine animals rely on hearing and sound communication for many critical life functions, there has been considerable interest, concern, and research into the effects of sound on some species [3,4,5,6,7]. Various configurations of active acoustic systems are being investigated to help understand and reduce potential negative effects of sounds from military sonar systems, maritime construction, seismic exploration, and offshore power production activities. A commonly used approach to reducing the potential impacts of acoustic systems on marine mammals is to move the operating frequency outside the range of functional hearing. While the upper hearing limits vary widely (from a few Hz to perhaps 160 kHz) for fish and marine mammals [3,8,9], no marine species are believed to be functionally sensitive to sounds above 200 kHz. Sonars and other active acoustic systems operating at this frequency and higher are generally believed to be inaudible and thus unable to impact marine mammals or other species; such sources have consequently been commonly exempted from environmental permitting requirements for use around protected species. However, while the operating frequency of these sonars is above the hearing range of marine mammals, their operation can generate sound energy outside the specified center frequency that may fall within functional hearing range and be detectable and thus elicit a behavioral reaction or perhaps affect their hearing over small ranges. Researchers at the Pacific Northwest National Laboratory were tasked by the U.S. Department of Energy to develop a system to monitor the presence of marine animals in the presence of a proposed marine hydrokinetic tidal turbine in the Puget Sound [10]. Of primary concern in the Puget Sound was the endangered Southern Resident Killer Whale (*Orcinus Orca*). Initially both passive and active systems were investigated, but the development of the active system had to be discontinued after the National Oceanic and Atmospheric Administration (NOAA) Fisheries expressed concerns that active sonars that operate at 200 kHz may generate sounds that are within the hearing range of killer whales.

The majority of active acoustic systems operate by transmitting short pulses of sound, typically on the order of a millisecond or less in duration. Short transmit pulses are needed to more accurately estimate the location of a target and to distinguish targets that may be located in near proximity to one another. The transmitted pulses also typically have short rise (ring-up) and short fall (ring-down) times relative to the duration of the time the transmit pulse spends at full power. The resulting transmitted pulse is more or less rectangular in shape. Generation of these short, rectangular pulses requires a bandwidth that is inversely proportional to the length of the transmitted pulse. The shorter the pulse duration and the more rapid the ring-up and ring-down times the greater the bandwidth required.

In addition to short rectangular transmit pulses, active acoustic systems also typically have relatively high source levels. High source levels are needed to optimize the effective detection range of the systems, increase the acoustic energy scattered back to the acoustic system receiver by targets with small acoustic cross sections, and permit the use of short pulse lengths to achieve high spatial resolution for isolation of single targets and to more accurately estimate their location. A consequence of high source levels is that, while down several orders of magnitude from the sound level at the frequency of operation, the level of sound at frequencies well away from the frequency of operation may be above the hearing threshold for some marine mammals. Such sound peripheral to the operating frequency are necessary to mechanically achieve the rapid rise and fall times that define the rectangular pulse's characteristics for most active acoustic systems. These sounds occur at the sub-harmonic frequency of the sounders center operating frequency. The sub-harmonic frequency of a sounder with a center frequency of 200 kHz can fall within the hearing range of killer whales, which are expected to have an upper frequency limit of at least 120 kHz based on data from other odontocetes [3,11] (available audiograms only extend up to 100 kHz and were obtained using electrophysiological methods [12]).

Several such systems were evaluated as an element of design of a mixed passive and active acoustic system for detection of killer whales near tidal power turbines in Admiralty Inlet, WA. In this analysis we consider the level of sound generated at frequencies within the hearing range of killer whales by active acoustic systems that were designed and specified to operate near a center frequency of 200 kHz.

Methods

Three commercial active sonar systems were evaluated: SM2000 multibeam imaging sonar (hereafter referred to as Kongsberg; Kongsberg Mesotech Ltd., Vancouver, British Columbia, Canada), DT-X Digital Scientific Echosounder (hereafter referred to as BioSonics; BioSonics, Inc., Seattle, Washington), and Model 965 multibeam imaging sonar (hereafter referred to as Imagenex; Imagenex Technology Corp., Port Coquitlam, British Columbia, Canada). The Kongsberg sonar operated at a center frequency of 200 kHz and features a user configurable pulse duration, ping rate, and source level. The average pulse duration tested was 625 μs, the ping rate was 6.9 pulses per second (pps), and the nominal source level was 195 dB re 1 μPa at 1 m. The BioSonics sonar was used in conjunction with a 210 kHz split beam digital transducer and operated at a nominal source level of 210 dB, a pulse duration of 450 μs, and a ping rate of 2.5 pps. The Imagenex sonar operated at a frequency of 260 kHz, a nominal source level of 185 dB, a pulse duration of 1000 μs, and a ping rate of 3 pps.

The characteristics of the Kongsberg sonar's transmit pulses were initially evaluated in an elongated oval laboratory tank approximately 7 m long ×3 m wide ×2 m deep. The preliminary results were presented in Deng et al. [13]. In addition to the primary peak at the sonar's 200 kHz operating frequency there is a secondary peak sound pressure level (SPL) at approximately 90 kHz. This secondary peak SPL has an amplitude of approximately 125 dB re 1 μPa at 3.5 m, which was approximately 51 dB less than the amplitude of the primary peak.

A subsequent field evaluation of the sonars was conducted at Levey Park (46.28°N, 118.83 W) located on the Snake River at approximately river kilometer 21, where the water depth ranged from 3 m near the dock that the sonar heads were deployed from to 30 m at the furthest distance tested. The U.S. Army Corps of Engineers operates Levey Park and approved of the field testing, which did not involve endangered or protected species. The data acquisition system consisted of a calibrated TC4014-5 hydrophone (Reson Inc., Slangerup, Denmark), an EC6081 band-pass filter (Reson Inc.), a NI PXI 5922 digitizer (National Instruments [NI], Austin, Texas), a laptop computer, and the LabVIEW SignalExpress (NI) software. The hydrophone was deployed from an unpowered motorboat that was anchored at distances ranging from 7 to 200 m from the sonar head. The hydrophone has a flat (±3 dB) frequency response from 25 Hz to 250 kHz, and was tested and calibrated in an acoustic tank located in the Pacific Northwest National Laboratory Bio-Acoustics and Flow Laboratory [14], which is accredited by the American Association for Laboratory Accreditation. Each dataset was collected at a sampling rate of 10 MHz for 1 s. At least 10 datasets were collected at each distance. Each dataset was processed by isolating the individual sonar pulses and using the fast Fourier transform (FFT) method with an FFT length of 4096, the Hanning window function, and a 50% overlap for the averaging. The resulting frequency bin size was about 2.4 kHz. A sample of the waveforms collected at a distance of approximately 20 m is shown in Figure 1.

The SPL was estimated by calculating the root-mean-square of the un-weighted pressure. The estimation of the sound exposure level (SEL) is weighted by the mid-frequency cetaceans' M-weighting function given in Eq. (1) [3]:

$$M(f) = 20\log_{10}\frac{R(f)}{\max\{|R(f)|\}} \quad (1)$$

where $R(f)$ is given in Eq. (2),

$$R(f) = \frac{f_{\text{high}}^2 f^2}{\left(f^2 + f_{\text{high}}^2\right)\left(f^2 + f_{\text{low}}^2\right)}, f_{\text{low}} = 150 \text{ Hz}, \quad (2)$$
$$\text{and } f_{\text{high}} = 160 \text{ Hz}$$

A digital Butterworth band-pass filter with cutoff frequencies at 150 Hz and 160 kHz was applied to the received signal $p(t)$ to obtain the M-weighted signal $p_{M\text{-}mf}(t)$. The SEL was estimated in a 1 s time window using Eq. (3):

$$SEL_{\text{M-mf}} = 10\log_{10}\left\{\frac{\sum_{n=1}^{N}\int_0^T p_{\text{M-mf}}(t)dt}{(p_{\text{ref}})^2}\right\}, \text{where } p_{\text{ref}} = 1\mu\text{Pa} \quad (3)$$

It should be noted that the analysis performed was conservative as it did not include the effects of critical ratio and temporal

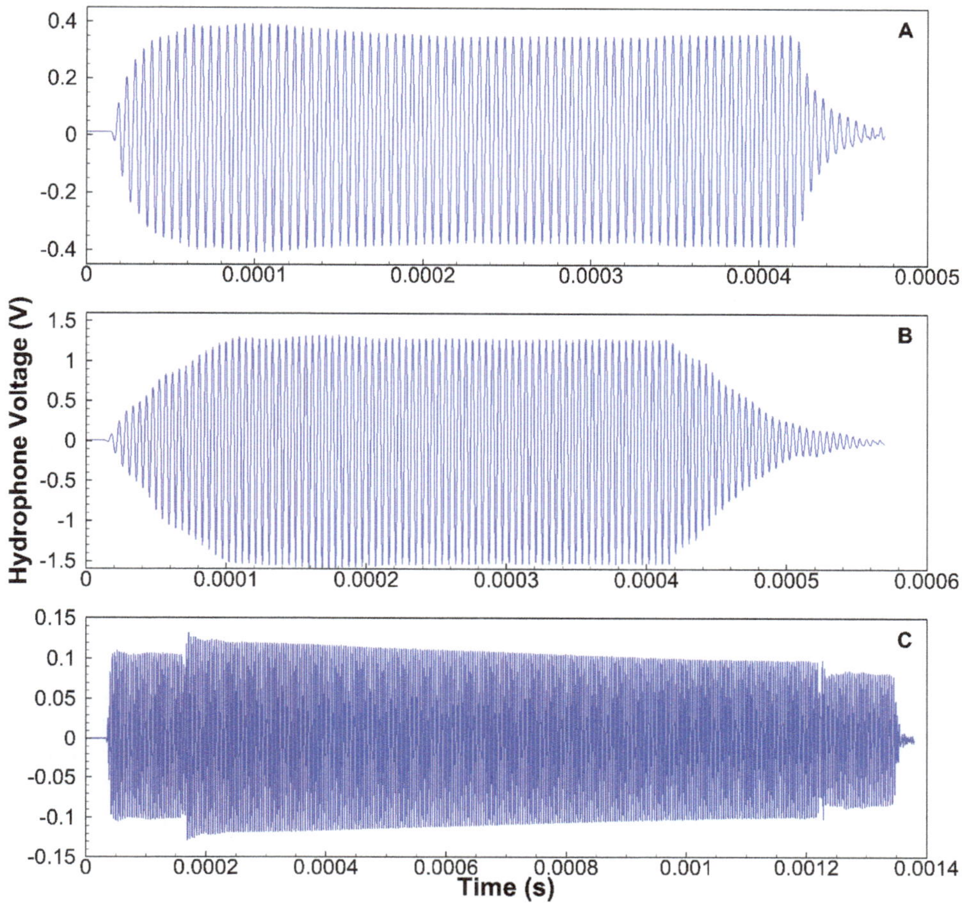

Figure 1. Sample raw waveforms collected for each sonar at a distance of approximately 20 m. A. Kongsberg (22 m); **B.** BioSonics (16 m); **C.** Imagenex (20 m).

integration. For example it has been shown that for a 100 kHz tone the hearing threshold of a dolphin is increased by approximately 20 dB when the pulse duration is 1 ms [15].

Results and Discussion

3.1 Frequency component analysis

For the Kongsberg sonar, clean pulses without multipath were obtained for pulse durations up to 700 μs. Besides the main frequency of 200 kHz, there were signals at a frequency of 90 kHz with the average power from 90 to 120 dB re to 1 μPa at different distances (Figure 2A). This low frequency component stayed at 90 kHz as the pulse duration and source level were changed, and propagated through the water column and attenuated over distance similar to the main frequency component. Pure pulses without multipath were collected at all distances for the BioSonics sonar because of its narrow vertical beam angle. There was a sub-harmonic signal at 105 kHz with the average power from 100 to 130 dB re to 1 μPa (Figure 2B). For the Imagenex sonar, the pure pulse without multipath was not obtained due to its long pulse duration. There was a sub-harmonic signal at 130 kHz with the average power from 80 to 90 dB re 1 μPa (Figure 2C). There was also a peak at the sub-harmonic frequency between transmissions, although it was at least 3 dB lower than during the actual sonar pulse. It was possibly due to multiple refelctions in a relative small space.

3.2 Sound exposure level and sound pressure level analysis result

The root-mean-square SPL (SPL_{rms}) attenuated to different degrees with range between the sonar systems. The Kongsberg sonar SPL_{rms} decreased from 181 dB to 151 dB re 1 μPa, the BioSonics sonar attenuated from 181 dB to 171 dB, and the Imagenex sonar attenuated from 171 dB to 136 dB (Figure 3A). The highest SPL_{rms} measured was 186 dB (peak SPL [SPL_{pk}] of 189 dB) at a distance of 50 m for the BioSonics sonar. The M-weighted SELs from the three sonar systems (Figure 3B) decreased as distance increased. The Imagenex sonar had the lowest SEL overall. The highest SEL measured was 137 dB at 7 m away from the Kongsberg sonar.

3.3 Impact on killer whales

While each of these systems does generate signals at or near their specified center frequency, they also incidentally generate sounds at lower output frequencies that, while considerably lower, are clearly present and well above ambient noise levels in many operating areas. The secondary peak around 90 kHz raises the potential that some marine animals, like killer whales, false killer whales, beluga whales, Atlantic bottlenose dolphins, harbor porpoises, and others [12], may detect and be affected by these sources despite the intended operating frequency of the sonars outside the hearing range of marine mammals. Although some marine animals may detect these secondary signals the levels are

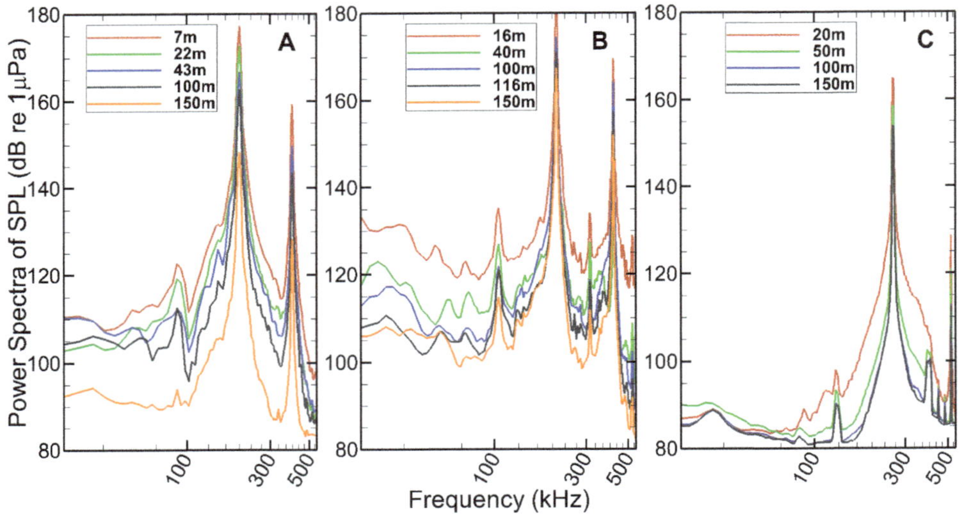

Figure 2. Sound pressure level spectra measured from different distances for the three sonar units. A. Kongsberg; **B**. BioSonics; **C**. Imagenex.

below those that may cause physical harm. The measured SPLs were all below the SPL_{pk} of onset temporary threshold shift (TTS) for mid-frequency cetaceans, estimated as 224 dB in the Southall et al. criteria [3]. In addition measured SELs were all below the SEL onset TTS level of 186 dB.

While all marine mammals are protected within U.S. jurisdictions, endangered species receive a higher degree of attention, scrutiny, and protection. Within the area where many of these systems were tested and are operated, a species of key importance that is endangered is the Southern Resident Killer Whale. From personal communications with The Whale Museum at Friday Harbor during 2011, there are known situations where killer whales have been observed in the presence of these kinds of active sonar systems and were reported to be apparently detecting their presence. Based on our measurements of the incidental side-band energy generated by these nominally 200 kHz systems, killer whales are likely able to hear this secondary frequency.

It should be noted that there are other delphinids, such as Atlantic bottlenose dolphins (*Tursiops truncatus*), that have hearing thresholds that are lower at these frequencies. Based on audiograms measured by Szymanski et al. [13] using a electrophysiological method, which produces higher sensitivity estimates than behavior measures, for the secondary frequency of the Kongsberg sonar at 90 kHz the behavioral hearing threshold of killer whales is an SPL_{rms} of 70 dB re 1 μPa. The secondary frequency of the BioSonics and Imagenex sonars are slightly higher than the maximum frequency of 100 kHz from the available audiograms. As a conservative estimation of the behavioral hearing threshold of killer whales the values at 105 and 130 kHz were linearly extrapolated from the audiogram, which were 77.5 and 90 dB re 1 μPa for BioSonics and Imagenex respectively. Given these as approximate detection thresholds for signals in this frequency band, it is likely that killer whales could detect the secondary peaks of the signals.

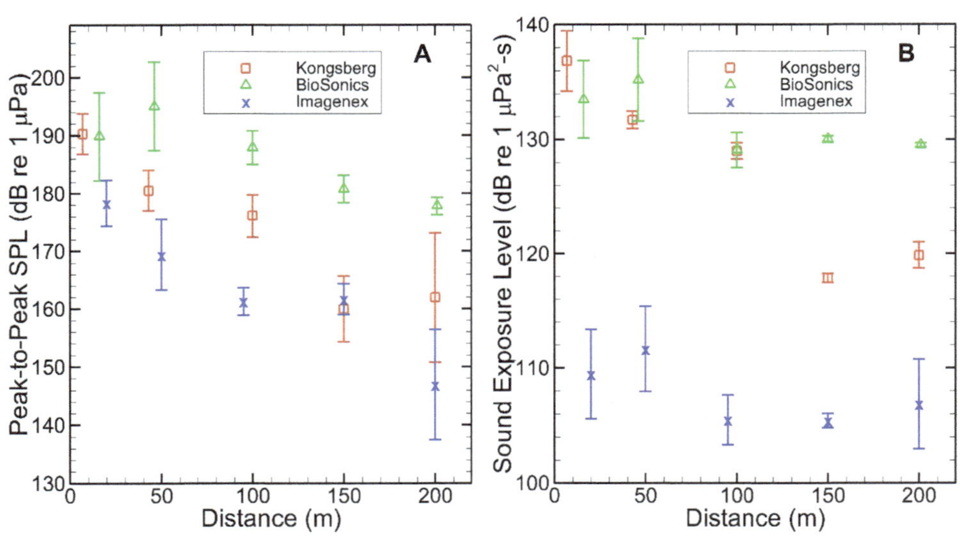

Figure 3. Sound pressure level and sound exposure level for three sonar units. A. Root-mean-square sound pressure level measurements; **B**. M_{mf} weighted sound exposure level measurements.

Table 1. Estimated effective root-mean-square source level of the secondary frequency and the resulting distance over which these signals would be audible assuming spherical spreading in seawater (values assuming cylindrical spreading shown in parenthesis).

Sonar	Operating frequency (kHz)	Frequency of secondary peak (kHz)	Source level of secondary peak (dB re 1 μPa at 1 m)	Hearing Threshold (dB re 1 μPa)	Sensation level (dB re 1 μPa)	Audible range (m)
Kongsberg	200	90	141	70	71	543 (1322)
BioSonics	210	105	162	77.5	84.5	781 (1540)
Imagenex	260	130	135	90	45	109 (460)

The effective source levels of the primary and secondary frequencies were estimated for the measurements at different distances, assuming spherical spreading for the acoustic signals. The estimated effective root-mean-square source levels of the primary frequency were within 2 dB of the values given in Section 2.1, and the source levels of the secondary frequency were 141, 162, 135 dB re 1 μPa at 1 m for Kongsberg, BioSonics, and Imagenex, respectively (Table 1). The effective estimated sensation level is defined as the difference between the effective source level and the detection threshold [16]. The maximum effective estimated sensation level using behavior hearing thresholds of killer whales and the secondary frequency for each sonar unit are shown in Table 1. The resulting maximum distance over which these signals are estimated to be audible based on spherical propagation and seawater absorption [17] would be 543, 781, and 109 m for Kongsberg, BioSonics, and Imagenex respectively. For cylindrical spreading the values would be 1322, 1540, and 460 m for Kongsberg, BioSonics, and Imagenex respectively. The actual distance that these signals will be audible will vary with the sea state (noise and propagation changes) and bathymetry. The propagation model will likely be between the spherical and cylindrical estimates.

Based on these calculations from measured signals in the field and controlled conditions and what is known about the hearing in killer whales, it appears likely that they may detect the incidental side-lobe energy of these signals over ranges of several hundred meters. Consequently, their behavior could be potentially affected by the presence of these systems in the close proximity to them. Responses could include attraction out of interest in the presence of a signal and what it may indicate or they could induce an avoidance response, which in the case of operational energy-producing devices with moving parts might not necessarily be an undesirable result. Based on all the evidence available for the effects of noise on hearing in marine mammals, estimated noise exposure criteria for impulsive sounds on cetaceans [3], these sounds are almost certainly not directly harmful to the hearing of killer whales or other marine mammals even directly at the source. In addition, it has been reported that killer whales have avoided narrow-band acoustic signals at moderate levels [3,18]. Therefore, any potential impact issues seem to be related to detection and behavioral response rather than direct injury from the lower frequency component of these sounds. Nevertheless, the existence of considerable energy at lower frequencies within the functional hearing range of killer whales and other odontocetes, raises the issue of whether such systems should be flatly excluded from consideration in impact assessment based solely on their supposed nominal operating frequency.

Conclusions

Measurements of the spectral properties of sound pulses transmitted by three commercially available 200 kHz echo sounders under typical operation conditions are consistent with some observations of the behavioral response of some marine mammals exposed to such transmissions that the sounders were generating sound within the hearing range of the animals. While on the order of 50 dB down in amplitude from the sounders' center frequencies, the level of sound within the hearing range of some marine mammals was found to be above the thresholds for hearing of many marine mammals but well below the levels that might cause physical injury. The range at which the side lobe sound generated by active acoustic devices operating near 200 kHz might be heard by marine mammals will be a function of the characteristics of the sounder and the level and spectrum of natural and anthropogenic noise that occurs during the time the sonar may be operated. Regulatory authorities may want to reconsider the unquestioned exclusion of echo sounders from consideration of environmental impact based entirely on their primary/center operating frequency.

Acknowledgments

The study was conducted at Pacific Northwest National Laboratory (PNNL) in Richland, Washington, which is operated by Battelle for the U.S. Department of Energy under Contract DE-AC05-76RL01830. The authors thank Jason Wood (Sea Mammal Research Institute, University of St. Andrews) for his help with this study. The authors thank BioSonics Inc. and Imagenex Technology Corp. for providing the sonar systems. The authors are also grateful for the contributions and input of many PNNL staff, including Tylor Abel, Charlie Brandt, Andrea Copping, Mark Jones, Josh Myers, Bob Mueller, Gene Ploskey, and Jennifer States.

Author Contributions

Conceived and designed the experiments: ZDD BLS TJC. Performed the experiments: ZDD JX JJM MAW. Analyzed the data: ZDD BLS JX JJM JMI. Contributed reagents/materials/analysis tools: ZDD TJC BLS MAW. Wrote the paper: ZDD BLS TJC JX JJM MAW JMI.

References

1. NRC (National Research Council) (2003) Ocean noise and marine mammals. Washington DC: National Academies Press. 617p.
2. Hildebrand JA (2009) Anthropogenic and natural sources of ambient noise in the ocean. Mar Ecol Prog Ser 395: 5–20. doi: 10.3354/meps08353
3. Southall BL, Bowles AE, Ellison WT, Finneran JJ, Gentry RL, et al (2007) Marine mammal noise exposure criteria: initial scientific recommendations. Aquatic Mammals 33(4): 411: 522. doi:10.1578/AM.33.4.2007.411

4. Melcón ML, Cummins AJ, Kerosky SM, Roche LK, Wiggins SM, et al. (2012) Blue Whales Respond to Anthropogenic Noise. PLoS ONE 7(2): e32681. doi:10.1371/journal.pone.0032681
5. Pirotta E, Milor R, Quick N, Moretti D, Di Marzio N, et al. (2012) Vessel Noise Affects Beaked Whale Behavior: Results of a Dedicated Acoustic Response Study. PLoS ONE 7(8): e42535. doi:10.1371/journal.pone.0042535
6. Tyack PL, Zimmer WMX, Moretti D, Southall BL, Claridge DE, et al. (2011) Beaked Whales Respond to Simulated and Actual Navy Sonar. PLoS ONE 6(3): e17009. doi:10.1371/journal.pone.0017009
7. Solé M, Lenoir M, Durfort M, López-Bejar M, Lombarte A, et al. (2013) Ultrastructural Damage of Loligo vulgaris and Illex coindetii statocysts after Low Frequency Sound Exposure. PLOS ONE 8(10): e78825. doi:10.1371/journal.pone.0078825.
8. Popper AN, Hastings MC (2009) The effects of human-generated sound on fish. Interactive Zoology 4(1): 43–52. doi:10.1111/j.1749–4877.2008.00134.x
9. Li S, Nachtigall PE, Breese M, Supin AY (2012) Hearing Sensation Levels of Emitted Biosonar Clicks in an Echolocating Atlantic Bottlenose Dolphin. PLOS ONE 7(1): e29793. doi:10.1371/journal.pone.0029793. doi:10.1371/journal.pone.0029793
10. Deng ZD, Carlson TJ, Fu T, Ren H, Martinez JJ, et al. (2013) Design and Implementation of a Marine Animal Alert System to Support Marine Renewable Energy. Marine Technology Society Journal 47(4): 113–121. doi:10.4031/MTSJ.47.4.2
11. Au WWL, Hastings MC (2008) Principles of Marine Bioacoustics. New York: Springer. 617p. doi:10.1007/978-0-387-78365-9
12. Szymanski MD, Bain DE, Kiehl K, Pennington S, Wong S, et al. (1999) Killer whale (Orcinus orca) hearing: Auditory brainstem response and behavioral audiograms. Journal of the Acoustical Society of America 106: 1134–1141. doi:10.1121/1.427121
13. Deng ZD, Carlson TJ, Xu JS, Martinez JJ, Weiland MA, et al. (2011) Design and Operation Specifications of an Active Monitoring System for Detecting Southern Resident Killer Whales. Oceans 2011 Kona, Hawaii 19–22 September 2011.
14. Deng ZD, Weiland MA, Carlson TJ, Eppard MB (2010) Design and Instrumentation of a Measurement and Calibration System for an Acoustic Telemetry System. Sensors 10, 3090–3099. doi:10.3390/s100403090
15. Johnson CS (1968) Relationship Between Absolute Threshold and Duration of Tones Pulse in the Bottlenosed Porpoise. Journal of the Acoustical Society of America 44, 757–763. doi:10.1121/1.1910893
16. Ellison WT, Southall BL, Clark CW, Frankel AS (2012) A New Context-Based Approach to Assess Marine Mammal Behavioral Responses to Anthropogenic Sounds. Conservation Biology 26: 21–28. doi:10.1111/j.1523–1739.2011.01803.x
17. Ainslie MA, McColm JG (1998) A simplified formula for viscous and chemical absorption in sea water. Journal of the Acoustical Society of America 103: 1671–1672. doi:10.1121/1.421258
18. Miller PJO, Atunes RN, Wensveen PJ, Samarra FIP, Alves AC, et al. (2014) Dose-response relationships for the onset of avoidance of sonar by free-ranging killer whales. Journal of the Acoustical Society of America 135(2): 975–993. doi:10.1121/1.4861346

18

δ^{15}N Value Does Not Reflect Fasting in Mysticetes

Alex Aguilar*, Joan Giménez¤, Encarna Gómez-Campos, Luís Cardona, Asunción Borrell

Department of Animal Biology and Biodiversity Research Institute (IrBio), Faculty of Biology, University of Barcelona, Barcelona, Spain

Abstract

The finding that tissue δ^{15}N values increase with protein catabolism has led researchers to apply this value to gauge nutritive condition in vertebrates. However, its application to marine mammals has in most occasions failed. We investigated the relationship between δ^{15}N values and the fattening/fasting cycle in a model species, the fin whale, a migratory capital breeder that experiences severe seasonal variation in body condition. We analyzed two tissues providing complementary insights: one with isotopic turnover (muscle) and one that keeps a permanent record of variations in isotopic values (baleen plates). In both tissues δ^{15}N values increased with intensive feeding but decreased with fasting, thus contradicting the pattern previously anticipated. The apparent inconsistency during fasting is explained by the fact that a) individuals migrate between different isotopic isoscapes, b) starvation may not trigger significant negative nitrogen balance, and c) excretion drops and elimination of ^{15}N-depleted urine is minimized. Conversely, when intensive feeding is resumed in the northern grounds, protein anabolism and excretion start again, triggering ^{15}N enrichment. It can be concluded that in whales and other mammals that accrue massive depots of lipids as energetic reserves and which have limited access to drinking water, the δ^{15}N value is not affected by fasting and therefore cannot be used as an indicatior of nutritive condition.

Editor: Andreas Fahlman, Texas A&M University-Corpus Christi, United States of America

Funding: This work was conducted as part of a project on fin whale population structure supported by the Fundació Barcelona Zoo. The funders had no role in study design, data collection and analysis, decision to publish, or preparation of the manuscript.

Competing Interests: The authors have declared that no competing interests exist.

* E-mail: aaguilar@ub.edu

¤ Current address: Department of Conservation Biology, Estación Biológica de Doñana, Consejo Superior de Investigaciones Científicas, Isla de La Cartuja, Sevilla, Spain

Introduction

The use of stable isotopes to study animal ecology and physiology has experienced an explosive growth in the last twenty years until becoming a technique of choice to investigate habitat use, migration, diet, physiology and nutritive condition (e.g. [1–3]). Among the many useful elements, nitrogen has received extensive attention because the proportion between its two stable isotopes varies in a consistent and predictable manner following well-defined processes such as trophic transfer and protein catabolism. Because each isotope has different atomic mass, the nitrogen physiological pathways affect them in a different manner: The lighter isotope (^{14}N) is more readily excreted through urine than the heavier isotope (^{15}N). Such dissimilar excretion rate leads to a long-term increase from prey to predator of the ratio ^{15}N/^{14}N, which is commonly expressed as the δ^{15}N value (see methods).

In addition, it has been hypothesized that during the periods of extreme nutritional stress in which the organism resorts to protein catabolism to obtain energy, such as during prolonged fasting leading to starvation, the negative nitrogen balance and the differential excretion of the two isotopes triggers the depletion of the lighter ^{14}N in the free amino acid body pool [4]. Since the only source of nitrogen to build-up new protein or restore protein-rich tissues is precisely that transported by serum, the loss of lighter nitrogen as a result of excretion causes the δ^{15}N value of residual amino acids to increase. When rebuilding of protein is made in the absence of exogenous nitrogen sources, the tissue will be necessarily constructed with these 'heavy' amino acids, therefore eliciting an increase in its δ^{15}N values [5,6].

Taking this mechanism as a basis, the δ^{15}N value has been proposed as an index of nutritional condition that summarizes the energetic balance of an organism [4]. After the merit of this fasting-increasing hypothesis was confirmed in a number of species that ranged from insects to mammals, including humans [2–4,7–13], it became generally accepted that it might occur in any potential species and scenario. Today, standard interpretation of isotopic results take for granted that catabolism of protein from tissues leading to severe negative introgen balance often results in an increase in the δ^{15}N value of the body nitrogen pool (e.g. [3]). However, the phase in the starvation process in which the body of homeotherms resorts to significant catabolism of lean body tissue only occurs when fat reserves are nearly exhausted and the individual is shortly approaching death [14–15]. As a consequence, it is not surprising that during the last decade some studies have come into apparent conflict with the above hypothesis [16], particularly when the species involved was a capital breeder that was fasting not only for food but also for water. For example, marine mammals, which mostly rely on food as the main water source [17], have produced recurrent examples of dissension [18–21].

To elucidate the actual long-term changes experienced by the δ^{15}N value in animals subject to periodical cycles of intensive feeding and fasting in absence of substantial water intake, we investigated nitrogen isotopic dynamics in the fin whale *Balaenoptera physalus*. Adequate material for this study was available because this species had been exploited off Spain until 1985 and tissue samples of individuals of known age, sex and reproductive state had been collected as part of the biological monitoring of the population and preserved in a tissue bank. As in most mysticetes, this species alternates summer high-latitude destinations, where individuals feed intensively, with winter low-latitude destinations, where individuals find the warm environment necessary to reproduce but are forced to fast because food is scarce or non-existent [22]. We analysed two tissues providing complementary insights: i) the muscle of individuals sampled throughout the summer feeding season, which is expected to integrate diet over some months prior to collection [23], and ii) the baleen plates, structures that grow continuously and, being composed of metabolically inert tissue, do not experience isotopic turnover. Because of this, baleens keep a permanent record along their length of the isotopic signal in the nitrogen body pool at the time of tissue formation. Accordingly, baleen plates have been frequently used to investigate movement patterns or temporal shifts in diet [21,24–30].

As opposed to what would be expected according to the fasting-^{15}N enrichment hypothesis, δ^{15}N values markedly decreased during the fasting season, coincidentally with the mobilization of body reserves. Conversely, when food intake was re-established, δ^{15}N values increased until reaching previous values, again as opposed to currently accepted understanding.

Methods

Ethical statement

The samples were obtained from whales subject to legal commercial exploitation in the 1980s and preserved thereafter at the BMA Tissue Bank of the University of Barcelona. In Spain, no specific permits are required to carry out research on samples stored at public institutions.

Sample collection and preparation

The samples were collected in the 1983–1985 summer seasons at the Caneliñas whaling station (NW Spain). The muscle tissue was collected from the trunk region, posterior to the dorsal fin, and preserved at −20°C. Date of capture and biological data were available from each individual. To avoid potential interference of reproduction on stable isotopic values [3], only males (N = 55; 45 sexually mature, 10 sexually immature) were selected for the study. From each sample, a subsample of 1 g of muscle was dried for 48 h at 60°C and powdered with mortar and pestle.

Baleen plates were collected from five individuals identified as A, B, C, D and E. Surface oils and adhered material were removed using a steel palette knife, steel wool and a chloroform:methanol solution (2:1), and stored dry until analysis. No information on date of capture or biological characteristics was available. To allow comparison with previous studies on fin whales [29–30], subsamples were extracted with a grinder every 1 cm along the length of each plate until totaling 40 subsampling points.

Stable isotope analysis

For both tissues, approximately 0.3 mg of powdered sample were weighed into tin capsules, automatically loaded and combusted at 1000°C. in a continuous flow isotope ratio mass spectrometer (Flash 1112 IRMS Delta C Series EA; Thermo-Finnigan, Bremen,Germany). Analyses were performed at the Centres Científics I Tecnològics (CCiT-UB) of the University of Barcelona (Barcelona, Spain). The stable isotope N ratio was expressed in delta (δ) notation, while the relative variation of stable isotope ratio was expressed as permil (‰) deviations from the predefined international standards as:

$$\delta^{15}\mathrm{Nsample} = [(\mathrm{Rsample}/\mathrm{Rstandard}) - 1]\mathrm{x}1000$$

where Rsample and Rstandard are the ^{15}N/^{14}N ratios in the sample and standard, respectively. The standard used was nitrogen (air) for ^{15}N. International secondary standards provided by the International Atomic Energy Agency (IAEA, Vienna, Austria) were inserted every 12 samples to calibrate the system and compensate for any drift over time. The precision and accuracy for δ^{15}N measurements was 0.3‰ [31].

Estimation of the growth rate of baleen plates

Because fluctuations of δ^{15}N values in baleen plates are assumed to be caused by the migratory regime from wintering to summering grounds the length of plate occurring between two following minimum δ^{15}N points was interpreted as the growth that had taken place during a complete cycle, *i.e.* one year.

Statistical analysis

Data were tested for normality with the Shapiro-Wilk test and for heteroscedasticity with the Levene's test. Differences between groups were investigated either with the U-Mann Whitney or with ANOVA, according to data normality. Relationships were analyzed through standard regression. All above tests were made with package SPSS 15.0. Fluctuations in baleen plates were gauged through a generalized additive model (GAM) fitting a semi-parametric regression with smoothing by cross-validation using R (R Development Core Team 2010).

Results

δ^{15}N values in muscle

δ^{15}N values in the individuals sampled are detailed in Table S1. The mean of δ^{15}N values was 9.8‰ (Table 1). Significant differences were not detected between immature and mature males ($U = 191$, $p = 0.458$), so both groups were combined in the investigation of patterns of variation. δ^{15}N values were regressed against the day of capture of each individual, which reflects the time spent at the feeding grounds, and a weak but positive statistically significant relationship was found ($r^2 = 0.0729$, $p<0.05$; Figure 1).

δ^{15}N values in baleen plates

δ^{15}N values measured in the various sampling points of each baleen plate are detailed in Table S2. The mean δ^{15}N value for the five individuals combined was 9.9‰ (Table 1). Although an ANOVA showed statistically significant differences between individuals ($F_{4,200} = 9.919$, $p<0.001$), all baleen plates showed periodic fluctuations in δ^{15}N values along their growing axis (Figure 2) with the unerupted section of the gum showing the highest δ^{15}N value of the entire baleen plate. The mean amplitude of such fluctuations (δ^{15}N max. − δ^{15}N min.) was 1.1±0.47‰, with a maximum of 1.8‰. The distance between two following minimum δ^{15}N value points was assumed to correspond to one year and therefore used to estimate the mean annual growth rate of the plate. The generalized additive model (GAM) fits showed that the amplitude was not homogeneous: despite small irregularities likely caused by individual variation in migratory or foraging behaviour, in four individuals (A, B, C and E) amplitude was very similar and cycles were completed every 18–23 cm (mean = 20 cm). Conversely, the fifth individual (D) showed a less

Table 1. Mean ± standard deviation and range of the nitrogen stable isotope results in muscle and baleen plates (mean of all subsamples).

Tissue	Stable isotope	Average (‰) ± SD	Range (‰)	
Muscle	$\delta^{15}N$	9.8±0.44	8.5	10.4
Baleen plates	$\delta^{15}N$	9.9±0.50	8.5	10.7

defined pattern, with shorter spacing between minimum $\delta^{15}N$ value points and cycles being completed every 11–12 cm. The reason for the dissimilar pattern of baleen plate D is unknown, but it is again likely to be caused by differences in the behaviour of this individual. As a consequence, results from baleen plate D were not included in the calculations of the mean fluctuations. Whatever the case, the mean amplitude of fluctuations did not show any significant relation with either the color nor the size of the baleen plate (Table 2, Figure 2).

Discussion

The $\delta^{15}N$ values of baleen plates showed a regular fluctuation pattern along their growth axis (Figure 2) comparable to that previously found in fin whales and other mysticetes and which are generally attributed to the migratory regime of the species and the associated alternation of intensive feeding and fasting [21,24–30]. Like most other large mysticetes around the world, fin whales undergo annual latitudinal migrations which involve feeding seasonality. During the summer, they stay at high latitudes, where feeding grounds are located, while during the winter they migrate to lower latitudes to breed [22]. The northwestern coast of Spain, where the samples for this study were collected, is a feeding area for one of the fin whale stocks of the eastern Atlantic population. Individuals reach this area in spring after migrating from breeding grounds situated at a southern undetermined location and where they have fasted for a several-month period.

Life-cycle of fin whales involves profound changes in body composition. During the season of intensive feeding individuals consume massive amounts of food that contain equally massive amounts of surplus energy which are stored in the form of fat in the hypodermic cover, commonly called blubber [26,27]. Food also provides preformed water that allows the individuals to avoid having to drink seawater, which contains salts that are energetically costly to excrete [17]. During fasting, energy requirements for vital functions, including reproduction and migration, must be supplied by these blubber reserves, although the lean tissue may also be mobilized to a smaller extent to produce glucose from amino-acids through gluconeogenesis or to rebuild essential proteins. Mobilization of lipids also serves to liberate endogenously-produced metabolic water, thus reducing ingestion of salty seawater to a minimal quantity, which in a number of whale species has been estimated at only 4.5–13 ml kg^{-1} day^{-1} [17,34].

Based on the assumption that $\delta^{15}N$ value fluctuations in the baleen plate are caused by the migratory regime, and omitting individual D, which showed a different fluctuation pattern than the other individuals, the mean annual baleen plate growth was estimated at 20.0 cm (SD: ± 2.16 cm), a value in accordance to previous estimations for the Mediterranean fin whale [29,30] and consistent with that observed in other mysticetes such as the minke whale, *Balaenoptera acutorostrata* (12.9 cm/year) [27], the bowhead whale, *Balaena mysticetus* (20 cm/year or less) [35,36] and the southern right whale, *Eubalaena australis* (27 cm/year) [21].

Considering this growth rate as well as the fact that the whales had been sampled in the summer while they were at the feeding ground, the un-erupted section of the baleen plate (the basal region) was inferred to have been formed when the individual had already arrived to the feeding area. According to this, its isotopic signature should correspond to that of the tissue built at the beginning of the period of intensive feeding. According to the same reasoning, the immediately following distal segments of the plate were taken as representative of tissue formed in the previous months, when feeding was just starting or, if going further distally along the plate, when the whale was fasting or when its food intake was low. The pattern of variation (Figure 2) shows that the $\delta^{15}N$ values tended to increase as the feeding season advanced and to

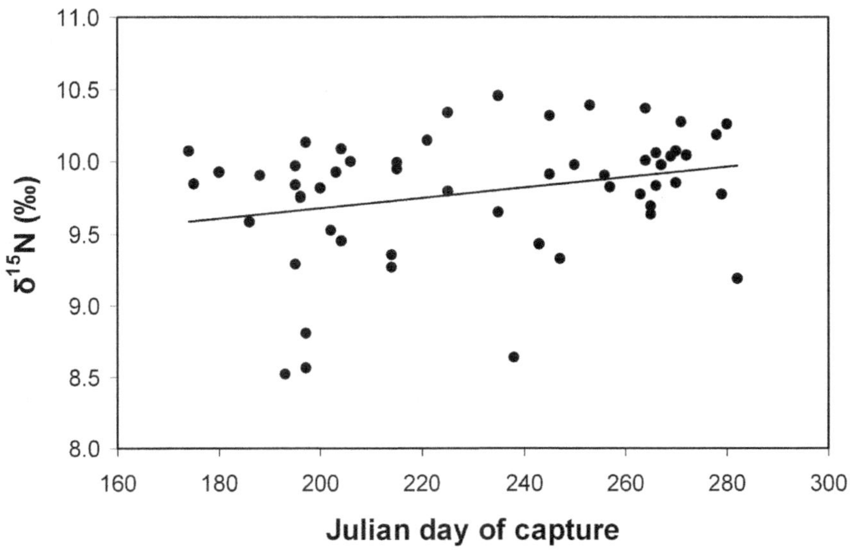

Figure 1. Relationship between the $\delta^{15}N$ value in muscle and the Julian day in which the individual was sampled ($r^2=0.0729$, $p<0.05$).

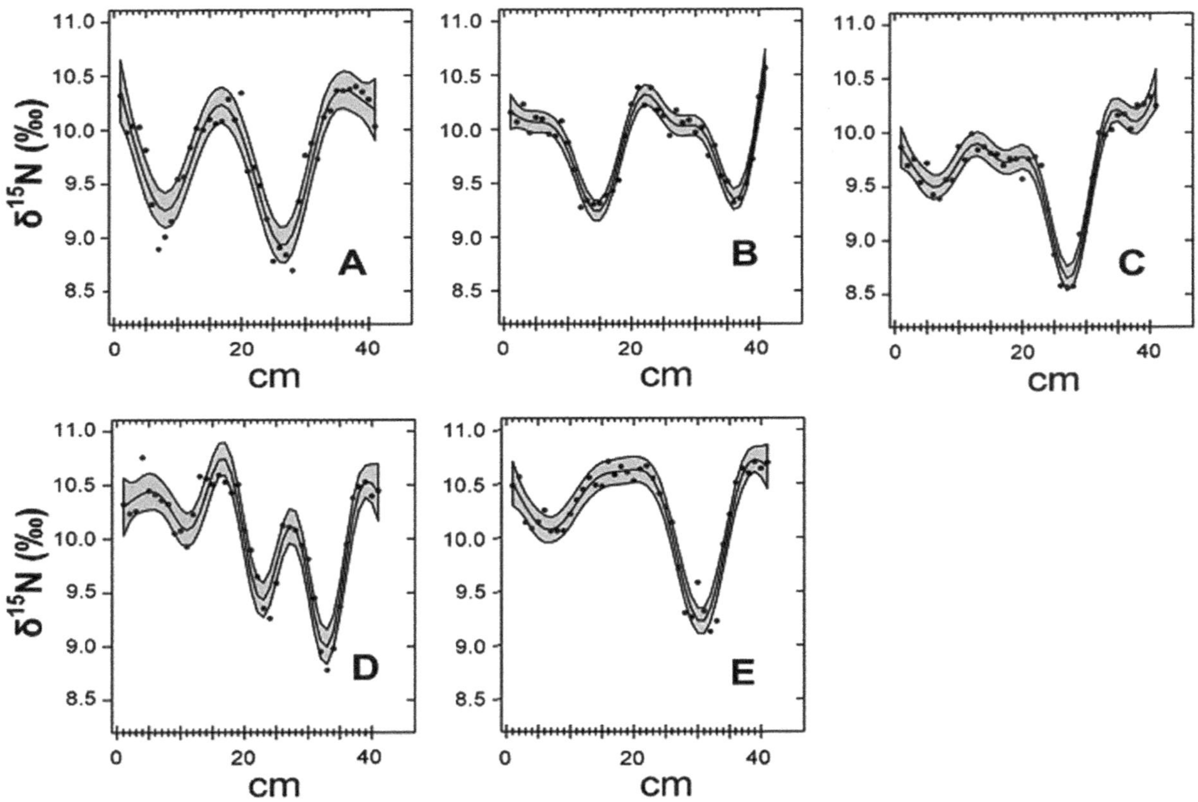

Figure 2. Variation of δ ^{15}N values along the growth axis of the baleen plates (measured in cm from the gum) from the five individuals analysed, identified as A-E. Points depict actual values; the continuous line depicts the fit of the GAM modelling; the blue fringe shows the 95% confidence intervals.

decrease when whales were fasting. The results obtained from the muscle pointed in the same direction: the $\delta^{15}N$ values significantly increased with day of capture throughout the season of intensive feeding indicating that enrichment was occurring concomitant with the improvement of body condition.

This pattern is novel because previous findings in a variety of avian and mammalian species indicated exactly the opposite pattern, this is, that during the fasting period $\delta^{15}N$ values increased or, at least, showed no definite variation. The enrichment was attributed to catabolism of protein reserves containing nitrogen that had already been enriched in ^{15}N relative to diet and that, when subsequently used to rebuild new tissue, experienced a second, additional enrichment [4,9,37–39]. Conversely, the absence of definite variation in $\delta^{15}N$ values despite the occurrence of fasting was explained by the existence in some species of biochemical and physiological pathways that allow energy production without protein mobilization and, thus, without additional fractionation of nitrogen [16,19]. The latter is consistent with the fact that in fasting animals not all tissues lose nitrogen in the same degree and in the same way [40]. Thus, Martínez del Rio *et al.* [41] suggested that tissues that have significant synthesis while fasting may be more prone to increase their $\delta^{15}N$ values, while tissues with no synthesis would keep their original isotopic value.

The results of the present study on fin whales point to a different, if not opposite, scenario. Indeed, this is not an exceptional finding in this group of mammals and previous studies on other mysticetes had already produced results that contradicted the enrichment-by-fasting hypothesis. Thus, peaks or enrichment that corresponded to the periods of intensive feeding had been previously observed in southern right whales [21], bowhead

Table 2. Characteristics and $\delta^{15}N$ values of the baleen plates used in the analysis.

Individual	Length (cm)	Colour of baleen plate	$\delta^{15}N$(‰) ± SD	Interval between lower $\delta^{15}N$ values (cm)
A	51	Dark grey	9.8±0.5	18
B	60	Grey	9.9±0.4	19
C	46	Yellow	9.7±0.4	20
D	58	Grey	10.1±0.5	11–12
E	48	Grey with yellow	10.2±0.4	23

whales [24,42], and North Atlantic right whales *Eubalaena glacialis* [43], but authors attributed the apparent anomaly to insufficient sample size, effect of migration or changes in diet. Thus, the enrichment of $\delta^{15}N$ values associated to intensive feeding seems to be a norm in this particular group of animals, rather than an exception.

The apparent contradiction between the dynamics of $\delta^{15}N$ values in baleen whales and those in other species may be explained by three facts which are not mutually exclusive and the effects of which may operate in the same direction and additively: a) that during migration whales move through different $\delta^{15}N$ isoscapes, b) that starvation does not trigger in whales a negative nitrogen balance because the limited winter feeding might allow them to maintain that balance while failing to meet caloric requirements, and c) that excretion of nitrogen wastes drops during fasting, thus minimizing elimination of ^{15}N-depleted urine.

The first explanation seems supported by the pattern of geographical variation in $\delta^{15}N$ baselines. Although the precise location of the wintering grounds of the eastern North Atlantic fin whale population is unclear, it is undoubtedly situated south of the Iberian Peninsula, probably off the Gibraltar Straits, off the northwestern coast of Africa or in the southwestern Mediterranean Sea [44–45]. In the temperate eastern North Atlantic, fin whales mostly feed on the northern krill *Meganyctiphanes norvegica*, the diet of which is mainly based on copepods [46–47]. The $\delta^{15}N$ values of zooplankton from the upper fringe of the ocean (0–500 m), primarily composed of copepods, range between approximately 4.5‰ in the putative winter grounds of the fin whale population to approximately 6‰ in its summer feeding grounds [48]. The fluctuations in $\delta^{15}N$ values along baleen plates ranged on average 1.1‰, so they fit well that variation (1.5‰).

However, this does not appear to be the only factor behind the observed pattern of variation of $\delta^{15}N$ values. Thus, the severe decline in energy body reserves occurring in the fin whale during the fasting season [33] indicates that, similarly to other mysticetes, the quantity of food ingested per day during that period would only likely amount about 10% or less of that consumed during the summer and would hence be under maintenance levels [32]. Tissue $\delta^{15}N$ values are known to increase not only in fasting vertebrates, but also in those feeding below maintenance levels because the isotopic value of a small daily ration cannot compensate for the highly ^{15}N-enriched serum resulting from the mobilization of muscle protein [11]. In spite of this, $\delta^{15}N$ values in baleen plates did not increase during fasting as expected but, rather the contrary, they decreased.

Here comes into play the second explanation mentioned above, this is, that in baleen whales starvation may not trigger significant mobilization of lean body mass and, therefore, a negative nitrogen balance. This pattern has been observed in other marine mammals and it may therefore be equally expected in baleen whales [19,49–51]. Indeed, the use of protein as a source of energy –a main reason for the observed $\delta^{15}N$ increment in terrestrial species– is probably very limited in a fasting whale, if at all existing. As all marine mammals, mysticete whales have massive lipid reserves deposited in the blubber, a tissue of multifunctional nature that constitutes the external cover of the body and that represents up to 15–45% of the individual body mass [17,52]. Thus, baleen whales and other marine mammals are able to mobilize massive amounts of lipids to sustain catabolism without compromising overall functions [33,52,53] and the wealth of energy that such mobilization produces likely allows protein conservation even during prolonged fasting. In fasting elephant seal pups, for example, protein catabolism contributes less than 4% to the metabolic rate [50,54]. To this it should be added that the physiological phase of starvation during which only fat is mobilized (the so-called phase two in the starvation cycle [55]) is especially protracted in animals that recycle urea, such as deer and at least some marine mammals [15,56,57], thus facilitating protein sparing [58]. Proteins become thus a fuel of last resort and the switch from lipid-dominated catabolism to protein-dominated catabolism would be delayed to an extremely advanced stage of starvation [58]. As a consequence of all this, and unless the individual confronts extreme starvation, the amount of lean body tissue mobilized along a standard migratory cycle would be meager and unlikely to elicit the $\delta^{15}N$ increase that is typically associated with fasting in terrestrial vertebrates [4,9,38]. In this situation, a little consumption of protein-rich zooplankton during the fasting period would provide the basic requirements for protein recycling and would therefore preserve body nitrogen balance. In other words, whales would be energy-starved, but not protein-starved.

The third potential explanation to explain the observed variation in $\delta^{15}N$ values is that, differently to most terrestrial mammals, marine mammals have severely limited access to non-salty water during fasting because the main source of preformed water in these animals is food [17,34]. In this particular physiological scenario, the compromise for water would appear parallel or even earlier than the compromise for energy. When this happens, blubber would become the main source of both calories and metabolic water [17]. Parallel to this, the activity of the kidney would be reduced and the resorption of endogenous electrolytes and body water would increase to help maintaining the homeostasis of fluids and electrolytes [34]. This would result in a minimization of water loss through defecation or urination to the smaller indispensable, probably to only 10–15% of normal values similarly to what has been observed in other mammals deprived of water during fasting [50,59,60]. If this is so, mysticete whales should have mechanisms, like hibernating bears or fasting postweaned elephant seals have, to avoid toxic concentrations of urea resulting from catabolism and they may have developed the ability to recycle urea by channelling their nitrogen to amino acid formation, in this way contributing to preserve lean body mass [57,61].

It should be noted that these are completely different conditions from those faced by terrestrial herbivores that live in dry environments, which tend to show higher $\delta^{15}N$ values than comparable species inhabiting wetter environments [62–65]. In these animals, the restriction in water intake is not associated with a parallel restriction in food intake and, therefore, recycling nitrogen is not a main requirement for them but, rather the contrary, their prime necessity is to get rid of nitrogen wastes with minimal water loss, towards which end they excrete highly concentrated, ^{15}N-depleted urine. Also, nitrogen discrimination tends to be higher in herbivores than in carnivores, thus eliciting ^{15}N enrichment [66]. Therefore, parallelism cannot be claimed to interpret variation in $\delta^{15}N$ values in whales. Moreover, the effect of water-stress conditions on $\delta^{15}N$ value appears to be currently disputed and indeed attributed to denitrification processes in the soil of dry environments that directly affect the isotopic values of the plants which compose the diet of their consumers [67]. Also, many of the species inhabiting such environments are ruminants [63], and a further increase in $\delta^{15}N$ values is likely to be expected in them because they recycle urea into the rumen to support microbe growth. When this happens along successive cycles, the assimilation of nitrogen by microbes and the subsequent digestion of these microbes are likely to bring an increase in $\delta^{15}N$ values through a process comparable to that occurring with trophic levels [68]. Finally, a further confirmation that fasting whales do not adjust to this physiological scheme, and that excretion of nitrogen is not a pressing need for them, appears to be confirmed by the fact that, contrary to the water-conservation mechanism of concentrating urine observed in many other groups of mammals, fasting cetaceans produce a diluted urine [34,69–71].

Accordingly, newly formed tissue in the baleen plate of a fasting fin whale would reflect the $\delta^{15}N$ signal from the scant food eaten in the wintering areas and also the shift to an energetic system based almost only on the consumption of body lipids to fulfill two physiological objectives: sparing muscle, which is necessary for migration, and obtaining metabolic water, a critical necessity in a fasting marine mammal because drinking a lot of sea water would imply the elevated energetic costs involved in salt excretion [17]. ^{15}N-enrichment would not occur simply because severe negative nitrogen balance, the main driver of the process, is avoided even during fasting.

When whales reach the summer grounds and start intensive feeding, fat catabolism would end and protein anabolism would resume. Also, because preformed water would be more available through feeding, the activity of the kidney and the urine flow rate would raise, thus allowing for increased excretion of nitrogen and electrolytes [34]. The ^{15}N enrichment observed to occur both in muscle and in baleen plates during the season of intensive feeding would therefore be again the result of at least three processes: the ingestion of food in a isoscape characterized by a $\delta^{15}N$ value higher than that consumed during the winter period of low food intake, the resumption of high rates of N excretion -with the concomitant discrimination and enhanced elimination of ^{14}N from the body pool-, and the increased intensity in feeding on a protein-rich diet, a fact that is commonly associated with a rise in the $\delta^{15}N$ value of the nitrogen body pool [3,72].

It can be concluded that in mysticetes, and probably also in other mammals that accrue massive depots of lipids as energetic reserves and which have limited access to drinking water, fasting does not induce negative nitrogen balance. As a consequence, ^{15}N enrichment does not take place and the $\delta^{15}N$ value cannot be used as indication of nutritive condition. Irrespective of the physiological implications of the $\delta^{15}N$ variations observed, the amplitude of the fluctuations observed in the baleen plate bear also practical consequences on the use of this tissue to infer diet composition. The range of variation (about 1.1±0.5‰ in average and reaching a maximum of 1.8‰) correspond to a variation of about 0.46 trophic levels, a much higher value than the effect of fasting on seabirds [73]. As a consequence, although stable nitrogen isotopes still provide a useful measurement of trophic level in baleen whales, the disruptive effect of feeding and fasting, and the potential variation in isotopic baselines at the various locations visited during migration, should be incorporated into any diet-related analysis.

Supporting Information

Table S1 Biological information of the individuals sampled and $\delta^{15}N$ values measured in their muscle.

Table S2 $\delta^{15}N$ values measured in the various sampling points of the baleen plates.

Acknowledgments

The samples were provided by the BMA Tissue Bank of the University of Barcelona. The stable isotope analyses were conducted at the *Centres Científics I Tecnològics* (CCiT-UB) of the University of Barcelona. The authors thank Pilar Teixidó and Pilar Rubio, both at the CCiT-UB, for their support during laboratory work; to Jaume Forcada, British Antarctic Survey, for his assistance in fitting the GAM model; and to Jacob González-Solís, University of Barcelona, and two anonymous referees for reviewing the draft manuscript.

Author Contributions

Conceived and designed the experiments: AA AB. Performed the experiments: JG EG. Analyzed the data: JG EG AA LC AB. Contributed reagents/materials/analysis tools: AA AB LC. Wrote the paper: AA. Contributed comments to draft manuscripts: AB LC JG EG.

References

1. Hobson KA, Wassenaar LI (1999) Stable isotope ecology: an introduction. Oecologia 120: 312–313.
2. Kelly JF (2000) Stable isotopes of carbon and nitrogen in the study of avian and mammalian trophic ecology. Can J Zool 78: 1–27.
3. Newsome SD, Clementz MT, Koch PL (2010) Using stable isotope biogeochemistry to study marine mammal ecology. Mar Mamm Sci 26: 509–572.
4. Hobson KA, Alisauskas RT, Clark RG (1993) Stable-nitrogen isotope enrichment in avian tissues due to fasting and nutritional stress: implications for isotopic analysis of diet. Condor 95: 388–394.
5. Sick H, Roos N, Saggau E, Haas K, Meyn V, et al. (1997) Amino acid utilization and isotope discrimination of amino nitrogen in n nitrogen metabolism of rat liver in vivo. *Z.* Ernährungswiss. 36: 340–346. (doi: 1007/BF01617819).
6. Lee TN, Buck CL, Barnes BM, O'Brien DM (2012) A test of alternative models for increased tissue nitrogen isotope ratios during fasting in hibernating Arctic ground squirrels. J. Exp. Biol. 215: 3354–3361.(doi: 10.1242/jeb.068528).
7. Polischuck SC, Hobson KA, Ramsay MA (2001) Use of stable carbon and nitrogen isotopes to assess weaning and fasting in female polar bears and their cubs. Can J Zool 79: 499–511.
8. Kurle CM, Worthy GAJ (2001) Stable isotope assessment of temporal and geographic differences in feeding ecology of northern fur seals (*Callorhinus ursinus*) and their prey. Oecologia 126: 254–265. (doi:10.1007/s004420000518).
9. Cherel Y, Hobson KA, Bailleul F, Groscolas R (2005) Nutrition, physiology, and stable isotopes: New information from fasting and molting penguins. Ecology 86: 2881–2888.(doi: 10.1890/05-0562).
10. Lohuis TD, Harlow HJ, Beck TDI (2007) Hibernating black bears (*Ursus americanus*) experience skeletal muscle protein balance during winter anorexia. Comp Biochem Physiol B 147:20–28.
11. Gaye-Siessegger J, Focken U, Abel HJ, Becker K (2007) Starvation and low feeding levels result in an enrichment of ^{13}C in lipids and ^{15}N in protein of Nile tilapia *Oreochromis niloticus* L. J Fish Biol 71: 90–100. (doi: 10.1111/j.1095-8649.2007.01469.x).
12. Ciancio JE, Pascual MA, Botto F, Amaya-Santi M, O'Neal S, et al. (2008) Stable isotope profiles of partially migratory salmonid populations in Atlantic rivers of Patagonia. J. Fish Biol. 72:1708–1719.
13. Oelbermann K, Scheu S (2002) Stable isotope enrichment ($\delta^{15}N$ and $\delta^{13}C$) in a generalist predator (*Pardosa lugubris*, Araneae: Lycosidae): effects of prey quality. Oecologia 130: 337–223 344.
14. DelGiudice GD, Mech LD, Seal US (1994) Undernutrition and serum and urinary urea nitrogen of white-tailed deer during winter. J. Wildl. Manage. 58: 430–436.
15. Saltz D, White GC, Bartmann M (1995) Assessing animal condition, nutrition, and stress from urine in snow. Wildl. Soc. Bull. 23:694–704.
16. Ben-David M, McColl CJ, Boonstra R, Karels TJ (1999) ^{15}N signatures do not reflect body condition in Arctic ground squirrels. Can J Zool 77: 1373–1378. (doi:10.1139/z99-105).
17. Costa D (2009) Osmoregulation. In: Perrin WF, Würsig B, Thewissen JGM, editors. Encyclopedia of Marine Mammals. Academic Press: San Diego. pp. 801–806.
18. Habran S, Debier C, Crocker DE, Houser DS, Lepoint G, et al. (2010) Assessment of gestation, lactation and fasting on stable isotope ratios in northern elephant seals (*Mirounga angustirostris*). Mar Mamm Sci 26: 880–895. (doi: 10.1111/j.1748-7692.2010.00372.x).
19. Gómez-Campos E, Borrell A, Aguilar A (2011) Nitrogen and Carbon stable isotopes do not reflect nutritional condition in the striped dolphin. Rapid Commun Mass Spectrom 25: 1343–1347.(doi: 10.1002/rcm.4999).
20. Das K, Siebert U, Fontaine M, Jauniaux T, Holsbeek L, et al. (2004) Ecological and pathological factors related to trace metal concentrations in harbour porpoises *Phocoena phocoena* from the North Sea and adjacent areas. Mar Ecol Prog Ser 281: 283–295. (doi: 10.3354/meps281283).
21. Best PB, Schell DM (1996) Stable isotopes in southern right whale (*Eubalaena australis*) baleen as indicators of seasonal movements, feeding and growth. Mar Biol 124: 483–494. (doi: 10.1007/ BF00351030).

22. Aguilar A (2009) Fin Whale. In: Perrin WF, Würsig B, Thewissen JGM, editors. Encyclopedia of Marine Mammals. Academic Press: San Diego. pp. 433–437.
23. Bahar B, Moloney AP, Monahan FJ, Harrison SM, Zazzo A, et al. (2009) Turnover of carbon, nitrogen, and sulfur in bovine longissimus dorsi and psoas major muscles: Implications for isotopic authentication of meat. J Animal Sci 87: 905–913 (doi: 10.2527/jas.2008-1360).
24. Hobson KA, Schell DM (1998) Stable carbon and nitrogen isotope patterns in baleen from eastern Arctic bowhead whales (*Balaena mysticetus*). Can J Fish Aquat Sci 55: 2601–2607. (doi: 10.1139/f98-142).
25. Schell DM (2000) Declining carrying capacity in the Bering Sea: Isotopic evidence from whale baleen. Limnol Oceanogr 45: 459–462.
26. Lee SH, Schell DM, McDonald TL, Richardson WJ (2005) Regional and seasonal feeding by bowhead whales (*Balaena mysticetus*) as indicated by stable isotope ratios. Mar Ecol Progr Ser 285: 271–287.(doi: 10.3354/meps28527).
27. Mitani Y, Bando T, Takai N, Sakamoto W (2006) Patterns of stable carbon and nitrogen isotopes in the baleen of common minke whale Balaenoptera acutorostrata from the western North Pacific. Fish Sci 72: 69–76. (doi: 10.1111/j.1444-2906.2006.01118.x).
28. Caraveo-Patiño J, Hobson KA, Soto LA (2007) Feeding ecology of gray whales inferred from stable-carbon and nitrogen isotopic analysis of baleen plates. Hydrobiol 586: 17–25. (doi:10.1007/s10750-006-0477-5).
29. Bentaleb I, Martin C, Vrac M, Mate B, Mayzaud P, et al. (2011) Foraging ecology of Mediterranean fin whales in a changing environment elucidated by satellite tracking and baleen plate stable isotopes. Mar Ecol Progr Ser 438: 285–302.(doi:10.3354/meps09269).
30. Giménez J, Gómez–Campos E, Borrell A, Cardona L, Aguilar A (2013) Isotopic evidence of limited exchange between Mediterranean and eastern North Atlantic fin whales. Rapid Comm Mass Spectr 27: 1801–1806.
31. Borrell A, Abad Oliva N, Gómez-Campos E, Giménez J, Aguilar A (2012) Discrimination of stable isotopes in fin whale tissues and application to diet assessment in cetaceans. Rapid Commun Mass Spectrom 26: 1596–1602. (doi: 10.1002/rcm.6267).
32. Lockyer CH (1981) Growth and energy budgets of large baleen whales from the Southern Hemisphere. In: Mammals in the Seas. Food and Agriculture Organization, Fisheries Series 5: 379–487.
33. Aguilar A, Borrell A (1990) Patterns of lipid content and stratification in the blubber of fin whales (Balaenoptera physalus). J Mammal 71: 544–554.
34. Ortiz RM (2001) Osmoregulation in marine mammals. J Exper Biol 204: 1831–1844.
35. Schell DM, Saupe SM, Haubenstock N (1989). Bowhead whale (*Balaena mysticetus*) growth and feeding as estimated by $\delta^{13}C$ techniques. Mar Biol 103: 433–443. (doi: 10.1007/BF00399575).
36. Schell DM, Saupe SM, Haubenstock N (1989). Natural isotope abundances in bowhead whale (*Balaena mysticetus*) baleen: markers of aging and habitat usage. Ecology Studies 68: 260–269. (doi:10.1007/978-1-4612-3498-2_15).
37. Fuller BT, Fuller JL, Sage NE, Harris DA, O'Connell TC, et al. (2004) Nitrogen balance and $\delta^{15}N$: Why you're not what you eat during pregnancy. Rapid Commun Mass Spectrom 18: 2889–2896. (doi:10.1002/ rcm. 1708).
38. Fuller BT, Fuller JL, Sage NE, Harris DA, O'Connell TC, et al. (2005) Nitrogen balance and $\delta^{15}N$: Why you're not what you eat during nutritional stress. Rapid Commun Mass Spectrom 19: 2497–2506. (doi: 10.1002/rcm.2090).
39. Martínez Del Río C, Wolf BO (2005) Mass balance models for animal isotopic ecology. In: Starck MA, Wang T, editors. Physiological and Ecological Adaptations to Feeding in Vertebrates. Science Publishers: Enfield, New Hampshire. pp. 141–174.
40. Doucett RR, Booth RK, McKinley RS (1999) Effects of the spawning migration on the nutritional status of anadromous atlantic salmon (*Salmo salar*): insights from stable-isotope analysis. Can J Fish Aquat Sci 56: 2172–2180. (doi: 10.1139/f99-147).
41. Martínez del Rio C, Wolf N, Carleton SA, Gannes LZ (2009) Isotopic ecology ten years after a call for more laboratory experiments. Biol Rev 84: 91–111(doi: 10.1111/j.1469-185X.2008.00064.x)
42. Knoche M, Suydam RS, George JC, Morita J (2007) Using Stable Isotopes in Baleen to Examine Migratory Behavior of Bering-Chukchi-Beaufort Sea Bowhead Whales (*Balaena mysticetus*): A pilot study. International Whaling Comission. Scientic committee: SC/59/BRG13.
43. Summers EL, Estrada JA, Zeeman SI (2006) A note on geographical and seasonal fluctuations in the isotopic composition of baleen in four North Atlantic right whales (*Eubalaena glacialis*). J Cetacean Res Manag 8: 241–245.
44. International Whaling Commission (1977) Report of the Working Group on North Atlantic whales. Reports of the International Whaling Commission 27: 369–387.
45. Castellote M, Clark CW, Lammers MO (2012) Fin whale (*Balaenoptera physalus*) population identity in the western Mediterranean Sea. Mar Mamm Sci 28: 325–344. (doi: 10.1111/j.1748-7692.2011.00491.x).
46. Fisher LR, Goldie EH (1959) The food of *Meganyctiphanes norvegica* (M. Sars), with an assessment of the contributions of its components to the vitamin A reserves of the animal. J Mar Biol Assoc U.K. 38: 291–312. (doi:10.1017/S0025315400006093).
47. Schmidt K (2010) Food and feeding in Northern krill (*Meganyctiphanes norvegica* Sars). Adv Mar Biol 57: 127–171. (doi: 10.1016/S0065-2881(10)57005-6).
48. Graham BS, Koch PL, Newsome SD, McMahon KW, Aurioles D (2009) Using isoscapes to trace the movements and foraging behavior of top predators in oceanic ecosystems. In: West J, Bowen GJ, Dawson TE, Tu KP, editors. Isoscapes: Understanding movement, pattern and process on earth through isotope mapping Springer New York. pp. 299–318. (doi:10.1007/978-90-481-3354-3_14).
49. Nordøy ES, Ingebritsen OC, Blix AS (1990) Depressed metabolism and low protein catabolism in fasting grey seal pups. Acta Physiol Scand 139: 361–369. (doi: 10.1111/j.1748-1716.1990.tb08935.x).
50. Adams SH, Costa DP (1993) Water conservation and protein metabolism in northern elephant seal pups during the postweaning fast. J Comp Physiol B 163: 367–373. (doi: 10.1007/BF00265640).
51. Gómez-Campos E, Borrell A, Aguilar A (2011) Assessment of nutritional condition indices across reproductive states in the striped dolphin (*Stenella coeruleoalba*). J Exp Mar Biol Ecol 405: 18–24. (doi: 10.1016/j.jembe.2011.05.013).
52. Aguilar A, Borrell A, Pastor T (1999) Biological factors affecting variability of persistent pollutant levels in cetaceans. J Cetacean Res Manag, Special Issue 1: 83–116.
53. Lockyer C (1987) The relationship between body fat, food resource and reproductive energy costs in North Atlantic fin whales (*Balaenoptera physalus*). Symp Zool Soc London 57: 343–361.
54. Houser DS, Costa DP (2001) Protein catabolism in suckling and fasting northern elephant seal pups (*Mirounga angustirostris*). J Comp Physiol B 171: 635–642. (doi: 10.1007/s003600100214).
55. Young VR, Scrimshaw NS (1971) The physiology of starvation. Sci Am 225: 14–22.
56. Robbins CT, Prior RL, Moen NA, Visek WJ (1974) Nitrogen metabolism of white-tailed deer. J Anim Sci 38:186–191.
57. Soñanez-Organis JG, Vázquez-Medina JP, Zenteno-Savín T, Aguilar A, Crocker D, et al. (2012) Prolonged casting increases purine recycling in post-weaned northern elephant seals. J Exp Biol 215: 1448–1455.
58. McCue MD (2010) Starvation physiology: Reviewing the different strategies animals use to survive a common challenge. Comp Biochem Physiol A 156: 1–18.
59. Nelson RA (1989) Nitrogen turnover and its conservation in hibernation. In: Malan A, Canguilhem B, editors. Living in the cold II. John Libbey: London. pp. 299–307.
60. Nelson RA, Folk GE Jr, Pfeiffer EW, Craighead JJ, Jonkel CJ, et al. (1983) Behaviour, biochemistry and hibernation in black, grizzly and polar bears. In: Merlow EC, editor. Beans-their biology and management: Proc. Int. Conf. Bear Res. Manage Madison WI Feb 1980. pp. 284–290. (ed E. C. Merlow), pp. 284–290. (doi:10.2307/3872551).
61. Barboza PS, Farley SD, Robbins CT (1997) Whole-body urea cycling and protein turnover during hyperphagia and dormancy in growing bears (Ursus americanus and U. arctos). Can J Zool 75: 2129–2136. (doi:10.1139/z97-848).
62. Sealy JC, Van Der Merwe NJ, Lee-Thorp JA, Lanham JL (1987) Nitrogen isotope ecology in southern Africa: Implications for environmental and dietary tracing. Geochim Cosmochim Acta 51: 2707–2717. (doi: 10.1016/0016-7037(87)90151-7).
63. Ambrose AH (1986) Stable carbon and nitrogen isotope analysis of human and animal diet in Africa. J Human Evol 15: 707–731. (doi: 10.1016/S0047-2484(86)80006-9).
64. Ambrose AH (1991) Effects of diet, climate and physiology on nitrogen isotope abundances in terrestrial foodwebs. J Archaeol Sci 18: 293–317. (doi: 10.1016/0305-4403(91)90067-Y).
65. Ambrose AH, DeNiro MJ (1987) Bone nitrogen isotope composition and climate. Nature 325: 201. (doi:10.1038/325201a0).
66. Robbins CT, Felicetti LA, Sponheimer M (2005) The effect of dietary protein quality on nitrogen isotope discrimination in mammals and birds. Oecologia 144: 534–40.
67. Hartman G (2011) Are elevated $\delta^{15}N$ values in herbivores in hot and arid environments caused by diet or animal physiology? Funct Ecol 25: 122–131. (doi: 10.1111/j.1365-2435.2010.01782.x).
68. Cormie AB, Schwarcz HP (1996) Effects of climate on deer bone $\delta^{15}N$ and $\delta^{13}C$: Lack of precipitation effects on $\delta^{15}N$ for animals consuming low amounts of C_4 plants. Geochim Cosmochim Acta 60: 4161–4166. (doi: 10.1016/S0016-7037(96)00251-7).
69. Bentley PJ (1963) Composition of the urine of the fasting humpback whale (*Megaptera nodosa*). Comp Biochem Physiol 10: 257–259. (doi: 10.1016/0010-406X(63)90040-9).
70. Telfer N, Cornell LH, Prescott JH (1970) Do dolphins drink water? J Am Vet Med Ass 157: 555–558.
71. Hui C (1981) Seawater consumption and water flux in the common dolphin *Delphinus delphis*. Physiol Zool 54: 430–440.
72. Florin ST, Felicetti LA, Robbins CT (2011) The biological basis for understanding and predicting dietary-induced variation in nitrogen and sulphur isotope ratio discrimination. Funct Ecol 25: 519–526.
73. Hobson KA (1999) Tracing origins and migration of wildlife using stable isotopes: a review. Oecologia 120: 314–326. (doi: 10.1007/s004420050865).

19

Effects of Age, Adipose Percent, and Reproduction on PCB Concentrations and Profiles in an Extreme Fasting North Pacific Marine Mammal

Sarah H. Peterson[1]*, Jason L. Hassrick[1], Anne Lafontaine[2], Jean-Pierre Thomé[2], Daniel E. Crocker[3], Cathy Debier[4], Daniel P. Costa[1]

1 Ecology and Evolutionary Biology Department, University of California Santa Cruz, Santa Cruz, California, United States of America, **2** Laboratoire d'Ecologie Animale et Ecotoxicologie, Université de Liège, Liege, Belgium, **3** Department of Biology, Sonoma State University, Rohnert Park, California, United States of America, **4** Institut des Sciences de la Vie, Université catholique de Louvain, Louvain-la-Neuve, Belgium

Abstract

Persistent organic pollutants, including polychlorinated biphenyls (PCBs), are widely distributed and detectable far from anthropogenic sources. Northern elephant seals (*Mirounga angustirostris*) biannually travel thousands of kilometers to forage in coastal and open-ocean regions of the northeast Pacific Ocean and then return to land where they fast while breeding and molting. Our study examined potential effects of age, adipose percent, and the difference between the breeding and molting fasts on PCB concentrations and congener profiles in blubber and serum of northern elephant seal females. Between 2005 and 2007, we sampled blubber and blood from 58 seals before and after a foraging trip, which were then analyzed for PCBs. Age did not significantly affect total PCB concentrations; however, the proportion of PCB congeners with different numbers of chlorine atoms was significantly affected by age, especially in the outer blubber. Younger adult females had a significantly greater proportion of low-chlorinated PCBs (tri-, tetra-, and penta-CBs) than older females, with the opposite trend observed for hepta-CBs, indicating that an age-associated process such as parity (birth) may significantly affect congener profiles. The percent of adipose tissue had a significant relationship with inner blubber PCB concentrations, with the highest mean concentrations observed at the end of the molting fast. These results highlight the importance of sampling across the entire blubber layer when assessing contaminant levels in phocid seals and taking into account the adipose stores and reproductive status of an animal when conducting contaminant research.

Editor: James P. Meador, Northwest Fisheries Science Center, NOAA Fisheries, United States of America

Funding: This research was conducted as part of the Tagging of Pacific Predators (TOPP) program and was supported in part by the National Ocean Partnership Program (N00014-02-1-1012); the Office of Naval Research (N00014-00-10880, N00014-03-1-0651, N00014-08-1-1195, and N00014-10-1-0356); the E&P Sound and Marine Life Joint Industry Project of the International Association of Oil and Gas Producers (JIP2207-23); the Moore, Packard, and Sloan Foundations; the Ida Benson Lynn Endowed Chair in Ocean Health; Steve Blank. Full funding for Open Access provided by the University of California, Santa Cruz Open Access Fund. The funders had no role in study design, data collection and analysis, decision to publish, or preparation of the manuscript.

Competing Interests: The authors have declared that no competing interests exist.

* E-mail: sarahpeterson23@gmail.com

Introduction

Persistent organic pollutants such as polychlorinated biphenyls (PCBs) are harmful to wildlife because they can disrupt endocrine and immune function [1–3]. These effects are especially pronounced in top predators, including marine mammals, because persistent organic pollutants accumulate in adipose tissue and biomagnify with increasing trophic position [4]. The chemical and physical structure of PCBs allows these compounds to persist in the environment long after leaving human-derived sources and, as a result, they are detected in remote regions of the world [5,6]. In general, worldwide, manufacture of PCBs ceased from 1972–1984, with the most recent PCB cessation occurring in Russia in 1993 [7]. PCBs are no longer uniformly increasing in the environment [6,8], but their presence in marine and terrestrial ecosystems is widespread.

Demographic and physiological parameters such as age, sex, and adipose stores may affect persistent organic pollutant concentrations. Variability in concentrations within a population may cause certain individuals to be more at risk from the negative effects of persistent organic pollutants. The percent of adipose tissue often has a significant, negative correlation with contaminant concentrations, and fluctuations in adipose tissue as a result of prolonged fasting can increase contaminant concentrations in remaining tissues [9–15]. Persistent organic pollutant concentrations are often higher in males than females because females pass contaminants to offspring while males are unable to offload their contaminant burden [16–18]. For marine mammals, lack of depuration by males results in a positive relationship between age and contaminant concentrations, whereas reproductively active females do not demonstrate the same relationships between age and total contaminant concentrations. Insignificant age trends were observed for many reproductively aged female pinnipeds (seals, sea lions and the walrus), including harbor seals (*Phoca vitulina*) [17], Steller sea lions (*Eumetopias jubatus*) [16] and ringed seals (*Pusa hispida*) [18], suggesting that these species annually offload contaminants to their offspring and subsequently accumulate new contaminants while feeding over the remainder of the year. Conversely, adult female cetaceans, including fin whales (*Balaenoptera physalus*), pilot whales (*Globicephala melas*), Dalls porpoises (*Phocoenoides dalii*), and killer whales (*Orcinus orca*) had declining PCBs in blubber with increasing age [19–22], although that relationship may change after reproductive senescence [22,23]. Both pinnipeds and cetaceans do not equally transfer all congeners to their offspring. Offspring are observed with a higher proportion of low-chlorinated congeners than their mothers [14,24–27], which may affect congener profiles in adult females over time.

The major source of contaminants to marine mammals is from their food [4,28]. Therefore, PCB concentrations in wildlife may depend on both foraging location and trophic position. Within similar trophic levels, previous studies showed foraging location to be important. Bottlenose dolphins (*Tursiops truncatus*) foraging adjacent to point-sources of PCBs showed higher concentrations than bottlenose dolphins foraging farther from point sources [29]. Mid-trophic level predators from the northeast Pacific Ocean, including albatrosses (*Phoebastria* spp.), humpback whales (*Megaptera novaeangliae*), killer whales, and harbor seals displayed geographically-associated variability in total PCB concentrations [22,30–33]. The majority of animals from these studies were epipelagic (0–200 m depth) marine predators that also forage within the narrow margin of the continental shelf. Little is known about PCBs in mesopelagic (200–1000 m depth) open-ocean food webs and their effect on predators in these systems.

Assessing contaminants in free-ranging, open-ocean marine predators is complex because demographic parameters, estimates of adipose percent, and indices of foraging behavior are difficult to concurrently obtain from wild and visibly healthy marine predator populations. Northern elephant seals (*Mirounga angustirostris*) are unique in that all of these variables can be quantified when they arrive on land. They have been monitored at the Año Nuevo colony (California, USA) since 1968, where flipper tags are attached to several hundred pups annually, providing a consistent subset of known-age animals within the population that can be used to investigate relationships between age and contaminant concentrations. Northern elephant seals are long-lived, mesopelagic, high-trophic level predators. Based on diving behavior [34,35], jaw-motion recorders paired with cameras [36], and stomach content analysis [37], northern elephant seals are hypothesized to consume mesopelagic fish and squid. However, little is conclusively known about their diet, since they return from their foraging migrations with nearly all prey completely digested.

Tracking data show that individual elephant seals from the Año Nuevo colony forage in distinct open-ocean regions, including the remote Pacific subarctic gyre [34,35,38], and near-coastal regions [34,35,39]. The northern elephant seal uses four hydrographic ecoregions within the northeastern Pacific, as defined by Longhurst [40], and is the only pinniped species in these ecoregions that consistently forages in the mesopelagic zone. Sampling a large number of individuals could capture the variability in contaminant concentrations between individuals that may utilize varying foraging strategies and obtain an appropriate range of contaminant concentrations for the population.

Elephant seals are also ideal for examining the relationships between adipose percent and contaminant concentrations during the extreme fasting periods they exhibit on land associated with breeding and molting. Many marine predators go through annual periods of fasting, often associated with long migrations, and the associated mass loss could significantly affect contaminant concentrations in different tissues. Elephant seals return to land from their biannual foraging migrations to breed and molt [34,35]. While on land for breeding/lactation or molting, elephant seals cease feeding and go through extreme reductions in adipose stores, losing up to 40% of their body mass [41]. Previous studies of PCB concentrations in stranded northern elephant seals along the California coast [42,43] demonstrated the capacity for PCBs in elephant seals to reach concentrations of toxicological concern [44]; however, we still do not know how concentrations fluctuate in healthy, wild animals throughout the year in relation to the naturally occurring extremes of adipose percent. The unique life history strategy of elephant seals, with separate fasting periods for breeding/lactation and molting, can allow us to disentangle the effects of fluctuating adipose percent with the effect of maternal offloading of contaminants. Elephant seal females begin reproducing between the ages of three and six [45]. Females have not been observed to go into reproductive senescence like many cetaceans and average natality is over 80% [35].

Our study builds on the findings of previous research [9,13,46] by incorporating adipose percent estimates for known-age adult females during the breeding and the molting fast. The main objectives of our study were to examine the potential effects of age, adipose percent, and reproductive state (breeding fast versus molting fast) on PCB concentrations and congener profiles in blubber and serum of northern elephant seal females. By investigating these parameters concurrently we hypothesized that 1) age would not affect total PCB concentrations but would affect congener profiles, resulting in older females with higher chlorinated congeners than younger seals (lower chlorinated congers may be more easily depurated to offspring), 2) total PCB concentrations would increase with decreased adipose percent, and 3) seals would have higher concentrations at the end of molting than at the end of breeding due to the lack of depuration to offspring during the molt.

Methods

Ethics Statement

The University of California, Santa Cruz Institutional Animal Care and Use Committee reviewed and approved the animal use protocols for this research. Research was carried out under National Marine Fisheries Service permit #87–1743.

Animal Sampling

Adult female northern elephant seals ($N = 58$) with no visual health impairments, were selected for tissue sampling at Año Nuevo State Reserve in San Mateo County, California, USA between 2005 and 2007. Most seals ($N = 54$) were known-age individuals between 4–17 years old, identified by a uniquely numbered, plastic tag (Dalton jumbo Roto-tags, Oxon, UK) placed in a lobe of their hind flipper. Female elephant seals have two distinct periods of fasting on land (breeding/lactation and molting) between foraging trips (Fig. 1). The annual breeding fast (January – February) is followed by a short foraging trip (February – April). This is followed by a molting fast (May – June), which precedes a long foraging trip (June – January). These foraging trips are hereafter referred to as the short and long trips. Individual seals were sampled for the first time at the end of one of the fasting periods (late breeding or late molting) just prior to departure to sea. Those seals that completed the foraging trip and returned to land were then sampled for a second time at the start of the next fasting period (Fig. 1). Not all seals returned from the foraging trip; therefore, pre-and post-foraging trip samples were paired for only some animals (Table 1). Seals were handled a maximum of two times during this study. Seals were chemically immobilized using standard protocols while morphometric measurements were taken and tissue samples (blubber and blood) were collected [34,39]. Full thickness blubber cores were collected using sterile, single-use 6 mm biopsy punches (Miltex, Inc., York, Pennsylvania, USA) from the lateral pelvic area, wrapped in aluminum foil marked to identify inner and outer ends of the sample, stored on ice in the field, and frozen at −20°C until analysis [9]. Blood samples from the extradural vein were collected in 10 mL serum vacutainers and stored on ice in the field. At the laboratory, serum was separated by centrifugation and stored in cryovials at −20°C until analysis.

All 58 adult females were included in our summary of PCB concentrations (Table 1). Only known age females with mass and estimates of adipose percent that reproduced normally were included in our statistical analyses (N = 50 seals). Not every seal had usable samples for all tissue types; therefore, subsets of the 50 seals were used to analyze PCB concentrations in inner blubber, outer blubber, and serum (Table 1). Paired late fasting (pre-foraging trip) and early fasting (post-foraging trip) samples of at least one tissue type (inner blubber, outer blubber or serum) were collected from 27 of the seals.

Girth and length measurements were taken at eight locations along the seal, including six locations where blubber thickness was measured dorsally, laterally and ventrally using a handheld ultrasound backfat meter (Scanoprobe, Ithaca, New York, USA) [35]. Body mass was determined by rolling the seal onto a canvas sling and using a hand-winch to suspend them from a 1000±1.0 kg Dyna-Link digital scale attached to a metal tripod [35]. The percent of adipose tissue was estimated using the morphometric measurements and a truncated cones technique [47,48].

Chemical Analyses

Blubber biopsies were cut into three equal parts. Inner (closest to the muscle) and outer (closer to the skin) blubber layers were analyzed in 2008 at the University of Liege. Using a Thermo Quest Trace 2000 gas chromatograph coupled with a ^{63}Ni ECD (Thermo Quest, Trace 2000, Milan, Italy), serum and each section of blubber biopsy were analyzed separately for 22 PCB congeners (IUPAC 28, 44, 70, 87, 95, 101, 105, 110, 118, 128, 138, 149, 153, 156, 170, 180, 183, 187, 194, 195, 206, and 209). Details of sample preparation, clean-up, and analysis, including quality assurance, are provided in Debier et al. [49]. The limit of detection (LOD) was fixed at a level three times the background noise of the chromatogram. The limit of detection of PCB congeners was therefore 0.006 ng g^{-1} (ppb) of serum fresh weight, 2 ng g^{-1} (ppb) of serum lipid weight and 0.7 ng g^{-1} (ppb) of lipid weight of adipose tissue in our analytical conditions. The limit of quantification (LOQ) of PCB congeners was determined by means of spiked bovine serum and bovine fat samples used as quality control (QC) and was the lowest concentration which could be quantified and which showed a recovery range between 70% and 130% [50]. This concentration also corresponds to at least 10 times the background noise of the chromatogram. In these conditions, the LOQ for each PCB congener was 0.03 ng g^{-1} of serum fresh weight, 10 ng g^{-1} of serum lipid weight and 2.5 ng g^{-1} of lipid weight in adipose tissue. Congeners falling below the LOQ for elephant seal samples were recorded as measured concentrations. When a congener concentration was recorded as zero, this means that the congener was not detected at all in the sample. Blubber concentrations were quantified as ng g^{-1} lipid, while serum samples were quantified both by unit of serum lipid (ng g^{-1} lipid) and unit of whole serum (ug L^{-1}). Lipids in serum were quantified as described in Vanden Berghe et al. [14]. In summary, different lipid classes (total cholesterol, phospholipids, triacylglycerols and nonesterified fatty acids) were quantified using enzymatic kits from Diasys (Diasys Diagnostics System, Holzheim, Germany) and Wako (Wako Chemicals USA Inc., Richmond, Virginia, USA). Concentrations of each lipid class were calculated using kit-specific recommendations, on the basis of standard equivalents. All lipid classes were added together to obtain serum total lipid concentrations.

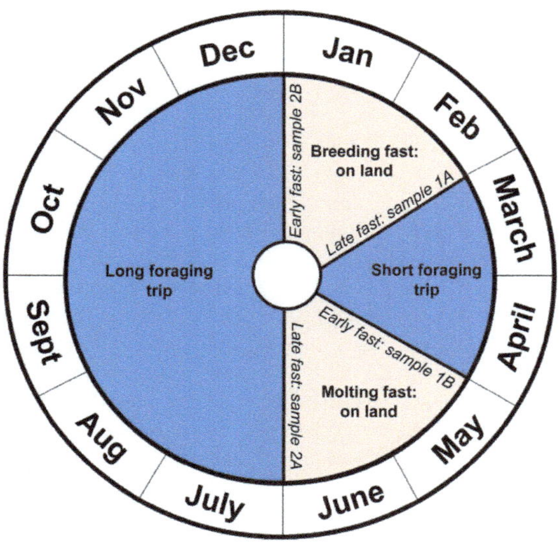

Figure 1. One year in the life of a female northern elephant seal. Tissue samples from the same animals (i.e. 1A/1B and 2A/2B) were collected just prior to the start of a foraging trip (late fasting) and just after the end of that foraging trip (early fasting).

Statistical Analyses

Total PCB concentrations. Linear mixed effects (LME) models were run using the nlme package [51] in the statistical software R (version 2.15.2 [52]), to examine effects of age and percent adipose tissue (fixed effects) on total contaminant concentrations in inner blubber, outer blubber and serum. Models were run separately for each tissue type. Blubber composition is stratified and differences in fatty acid profiles have been observed between the blubber layers [53,54]; therefore, our analysis focused on the inner and outer blubber layers. Models were run for both lipid-normalized serum $\sum$PCBs as well as $\sum$PCBs in serum measured per unit of wet weight. When models on serum $\sum$PCBs yielded the same trends, we only report lipid-normalized results. LME models included $\sum$PCB concentrations from samples collected at all time periods (Fig. 1). Residuals of each model were plotted against fixed effects to examine deviations from homogeneous variance, and data were log-transformed or appropriate variance structures were incorporated when necessary [55]. We treated individual as a random effect to account for repeated measurements for some individuals and the model was fit using restricted maximum likelihood (REML). For tests without repeated measures from individuals, general linear models were used to compare tissue concentrations at the end of the breeding and molting fasts, to examine potential influence of life history (i.e. fasting while lactating versus fasting with no offloading mechanism) on mean contaminant concentrations, while accounting for adipose percent.

The subset of individuals with paired pre- and post-foraging trip samples were additionally analyzed using paired t-tests to examine if $\sum$PCB concentrations in tissues (inner blubber, outer blubber, and serum) changed between the start and end of the short or the

Table 1. ∑PCB concentrations in northern elephant seals (N = 58).

Tissue		Late breeding fast (Sample 1A)	Early molting fast (Sample 1B)	Paired samples	Late molting fast (Sample 2A)	Early breeding fast (Sample 2B)	Paired samples
Inner	Sample size	N = 18 (17)	N = 13 (13)	13	N = 16 (12)	N = 31 (25)	8
	Mean ± sd	1241 ± 545	884 ± 179		1647 ± 574	690 ± 160	
	(range)	(495–2502)	(557–1204)		(708–2722)	(473–1120)	
Outer	Sample size	N = 19 (18)	N = 16 (15)	14	N = 18 (13)	N = 32 (25)	8
	Mean ± sd	912 ± 262	1170 ± 314		1003 ± 336	900 ± 221	
	(range)	(497–1461)	(777–2024)		(356–1597)	(504–1417)	
Serum	Sample size	N = 20 (19)	N = 16 (15)	14	N = 20 (15)	N = 29 (23)	8
	Mean ± sd	818 ± 210	1207 ± 517		1061 ± 343	838 ± 255	
	(range)	(542–1362)	(770–2837)		(723–1772)	(511–1495)	
Serum	Mean ± sd	7.8 ± 1.8	9.0 ± 3.1		8.1 ± 2.6	6.2 ± 1.4	
(wet weight)	(range)	(4.6–10.9)	(6.0–17.9)		(4.9–14.3)	(4.3–9.4)	

∑PCB concentrations are lipid-normalized (ng g^{-1} lipid) for inner blubber, outer blubber and serum. Serum ∑PCB concentrations are also reported by wet weight (μg L^{-1}). Samples were collected at four different times of year (see Fig. 1) from 2005–2007.

Note. A subset of animals with known ages and body composition measurements were used for statistical analyses (sample sizes in parentheses). Late breeding (1A) – early molt (1B) samples were taken before and after the short foraging trip and late molt (2A) – early breeding (2B) samples were taken before and after the long trip. Numbers of paired samples (same seal) are given for each tissue type because not all seals were repeatedly sampled. Some seals were sampled twice while other seals were only sampled once.

long foraging trip. T-tests were also used to examine whether the magnitude and direction of change in tissue concentrations were the same for both foraging trips (short versus long). The level of statistical significance for all analyses was set at $p \leq 0.05$.

PCB congener profiles. PCB concentrations were further examined at the level of specific congeners and congener families, based on the degree of chlorination. Low-chlorinated biphenyls included all congeners with three or four chlorines (tri- and tetra-CBs). The percentage of the total contaminant concentration made up by each congener family was calculated for every sample (inner blubber, outer blubber, and serum). Octa-, nona-, and deca-CBs were below the detection limit for the majority of samples, therefore statistical analyses were focused on tri/tetra-CBs, penta-CBs, hexa-CBs, and hepta-CBs. LME models were run with the percentage of each congener family as the response variable, after an arcsin square root transformation was applied [56], to examine the proportional contribution of each congener family to the total PCB concentration. Age and adipose percent were fixed effects and individual was included as a random effect.

Results

PCBs were detected in all samples collected from 58 adult female northern elephant seals, ranging from 4–17 years of age (Table 1). $\sum$PCB concentrations ranged from 356–2722 ng g^{-1} lipid in blubber and 512–2,837 ng g^{-1} lipid in serum (Table 1; Fig. 2).

The majority of inner and outer blubber PCB profiles in female seals were comprised of eight congeners: PCB-101, -110, -118 (penta-CBs), -138, -153 (hexa-CBs), -180, -183, and -187 (hepta-CBs), all of which had a mean percent contribution of the total burden greater than 5% (Fig. 3). These eight congeners comprised 84.9±4.5% of the total PCB concentration in inner blubber and 90.2±3.6% in outer blubber. PCB-153 was the most common congener, comprising 15.8–28.5% and 14.1–38.8% of the $\sum$PCB burden in inner and outer blubber, respectively. Hexa-CBs were the dominant congeners in blubber for all sampling periods, on average comprising >40% of $\sum$PCBs. Octa-, nona-, and deca-chlorobiphenyls combined comprised less than 1.7% of the total blubber PCB burden, and octa-CBs were the most prevalent of these high-chlorinated congener groups.

Serum PCB profiles had seven congeners that contributed a mean percent of more than 5% of the total PCB concentration: PCB-101, -110, -118, -138, -153, -180, and -183. These seven congeners comprised 78.8±5.9% of the total serum concentrations. PCB-153 comprised between 14.0–26.4% of the total serum PCB concentrations.

Influence of Age

There was no detectable relationship between age and total PCB concentrations in inner blubber, outer blubber or serum ($p \geq 0.230$). However, age had a significant effect on congener groups (Fig. 4; Table 2). There was a significant, negative relationship between age and the percent of penta-CBs ($F_{1,44} = 12.8$, $p < 0.001$) and a significant, positive relationship between age and the percent of hepta-CBs ($F_{1,44} = 12.4$, $p = 0.001$) in inner blubber. There was no detectable relationship between age and the percent of either tri/tetra-CBs or hexa-CBs in the inner blubber. In the outer blubber, there was also a significant, negative relationship between age and the percent of penta-CBs ($F_{1,47} = 30.1$, $p < 0.001$) and a significant positive relationship between age and the percent of hepta-CBs ($F_{1,47} = 35.7$, $p < 0.001$). Additionally, in the outer blubber, the percent of tri/tetra-CBs had a significant, negative relationship with age ($F_{1,47} = 15.1$, $p < 0.001$) and the percent of hexa-CBs had a significant, positive relationship with age ($F_{1,47} = 15.8$, $p < 0.001$). The same significant, negative relationship observed for both blubber layers between age and the percent of penta-CBs was also detectable for lipid-normalized serum ($F_{1,47} = 14.0$, $p < 0.001$) and serum concentrations using wet-weight. The percent of tri/tetra-CBs also had a significant, positive relationship with age in the serum ($F_{1,47} = 7.2$, $p = 0.010$), which was the opposite trend as observed in the outer blubber layer. The percent of hexa-CBs and hepta-CBs in serum did not change with age.

Influence of Adipose Percent and Fasting State

Elephant seals were fattest early in the breeding season fast (mean ± SD: 34.7±2.3%; range: 30.9–39.9%) and leanest late in the breeding fast (mean ± SD: 29.3±1.7%; range: 26.0–32.0%). Seals sampled early in the molting fast had a mean adipose percent of 30.8±2.7% (range: 23.4–33.8%) compared to 30.9±2.4% (range: 26.1–36.1%) late in the molting fast, indicating that

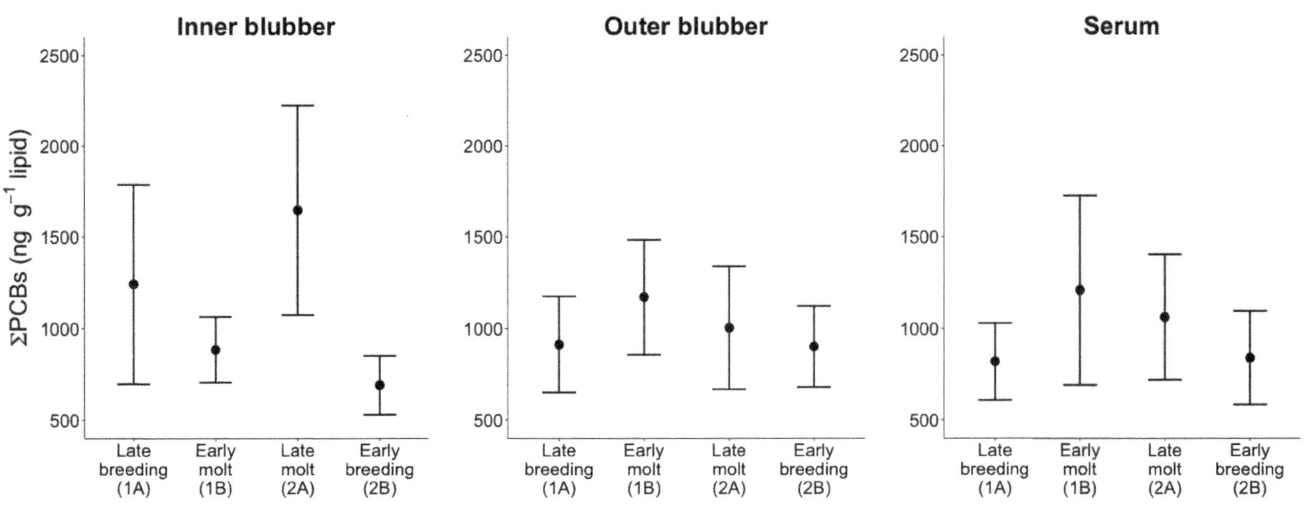

Figure 2. $\sum$PCB concentrations (mean ± SD) for inner blubber, outer blubber, and serum samples measured in elephant seals during four different sampling periods (refer to Fig. 1 for timing of samples).

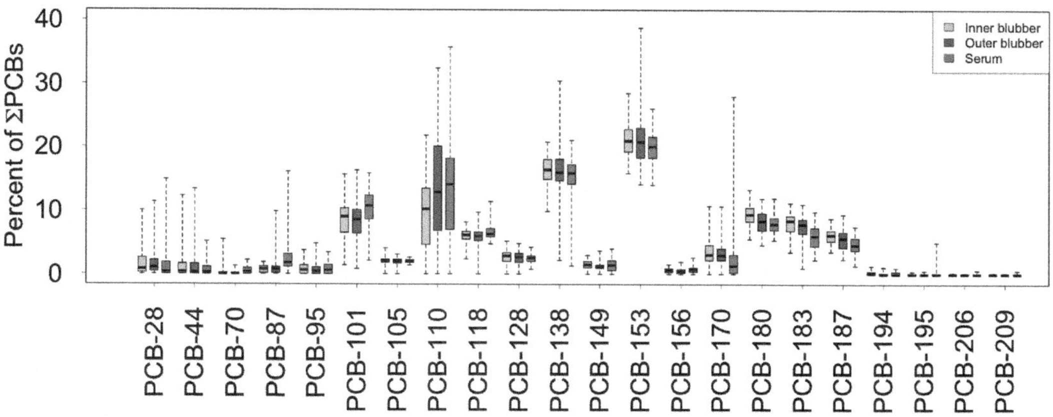

Figure 3. Proportions of each congener within the inner blubber, outer blubber, and serum of northern elephant seal females. Whiskers in these standard boxplots encompass the full range of the data. All sampling periods are combined, which means that some females contributed two samples while other females only are represented by one sample.

adipose and lean tissue were lost in similar proportions over the course of the molt.

The percent of adipose tissue had a significant relationship with total PCB concentrations, although this relationship was only observed for the inner blubber layer and serum measured by unit of wet weight and not for the outer blubber or lipid-normalized serum concentrations (Fig. 5). Overall, inner blubber $\sum$PCB concentrations were higher for seals that were leaner ($F_{1,20} = 19.7$, $p<0.001$). Direct comparisons of late breeding and late molting, while accounting for adipose percent, revealed that seals had higher mean inner blubber $\sum$PCB concentrations at the end of the molt ($F_{1,26} = 13.1$, $p = 0.001$) (Fig. 6). Conversely, outer blubber $\sum$PCB concentrations did not have a significant relationship with adipose percent ($F_{1,21} = 2.4$, $p = 0.136$), and there was no detectable difference between $\sum$PCB concentrations in late breeding and late molting outer blubber samples ($F_{1,28} = 1.1$, $p = 0.310$; Fig. 6).

The relationship between the percent of adipose tissue and total serum $\sum$PCB concentrations was dependent on whether the concentrations were analyzed using lipid-normalized or wet weight concentrations. Serum $\sum$PCBs measured per unit lipid initially appeared to have a significant relationship with adipose percent ($F_{1,21} = 6.1$, $p = 0.023$); however, when the single highest serum concentration was removed from analysis (2837 ng g^{-1} lipid measured during the early molting fast) this relationship became non-significant ($F_{1,20} = 2.2$, $p = 0.157$). Serum $\sum$PCB concentrations measured per unit of wet weight had a significant relationship with adipose percent even with the highest concentration removed ($F_{1,20} = 10.1$, $p = 0.005$). There was a significant difference between mean late breeding and late molting serum $\sum$PCB concentrations per unit lipid ($F_{1,31} = 11.9$, $p = 0.002$) but there was no significant difference between the late breeding and late molting serum $\sum$PCB concentrations per unit of wet weight ($F_{1,31} = 1.6$, $p = 0.213$).

Adipose percent had significant relationships with some congener groups in blubber and serum (Table 2). Fatter seals had a significantly greater percent of tri/tetra-CBs in their inner blubber than seals with lower adipose percentages ($F_{1,19} = 13.6$, $p = 0.002$), while the opposite was true for adipose percent and the proportion of penta-CBs in the inner blubber ($F_{1,19} = 8.3$, $p = 0.010$). There was no detectable relationship between adipose percent and the proportion of hexa-CBs or hepta-CBs in the inner blubber. In the outer blubber, adipose percent also had a significant, positive relationship with the proportion of tri/tetra-CBs ($F_{1,21} = 12.9$, $p = 0.002$) and a significant, negative relationship with the proportion of penta-CBs ($F_{1,21} = 12.6$, $p = 0.002$). Additionally, adipose percent had a significant, negative relationship with the proportion of hepta-CBs in the outer blubber ($F_{1,21} = 18.8$, $p<0.001$). There was no detectable relationship between adipose percent and the proportion of hexa-CBs in the

Table 2. Relationship between the percent of the ΣPCB concentrations in inner blubber, outer blubber and serum made up by each congener group and A) adipose percent or B) age.

Statistical test	Chlorination	Inner blubber	Outer blubber	Serum
Age	Tri/Tetra-CBs (3–4)	n.s.	–	+
	Penta-CBs (5)	–	–	–
	Hexa-CBs (6)	n.s.	+	n.s.
	Hepta-CBs (7)	+	+	n.s.
Adipose percent	Tri/Tetra-CBs (3–4)	+	+	n.s.
	Penta-CBs (5)	–	–	+
	Hexa-CBs (6)	n.s.	n.s.	n.s.
	Hepta-CBs (7)	n.s.	–	n.s.

Relationship indicated by statistically significant and positive relationship (+), significant and negative relationship (–), or no significant relationship (n.s.).

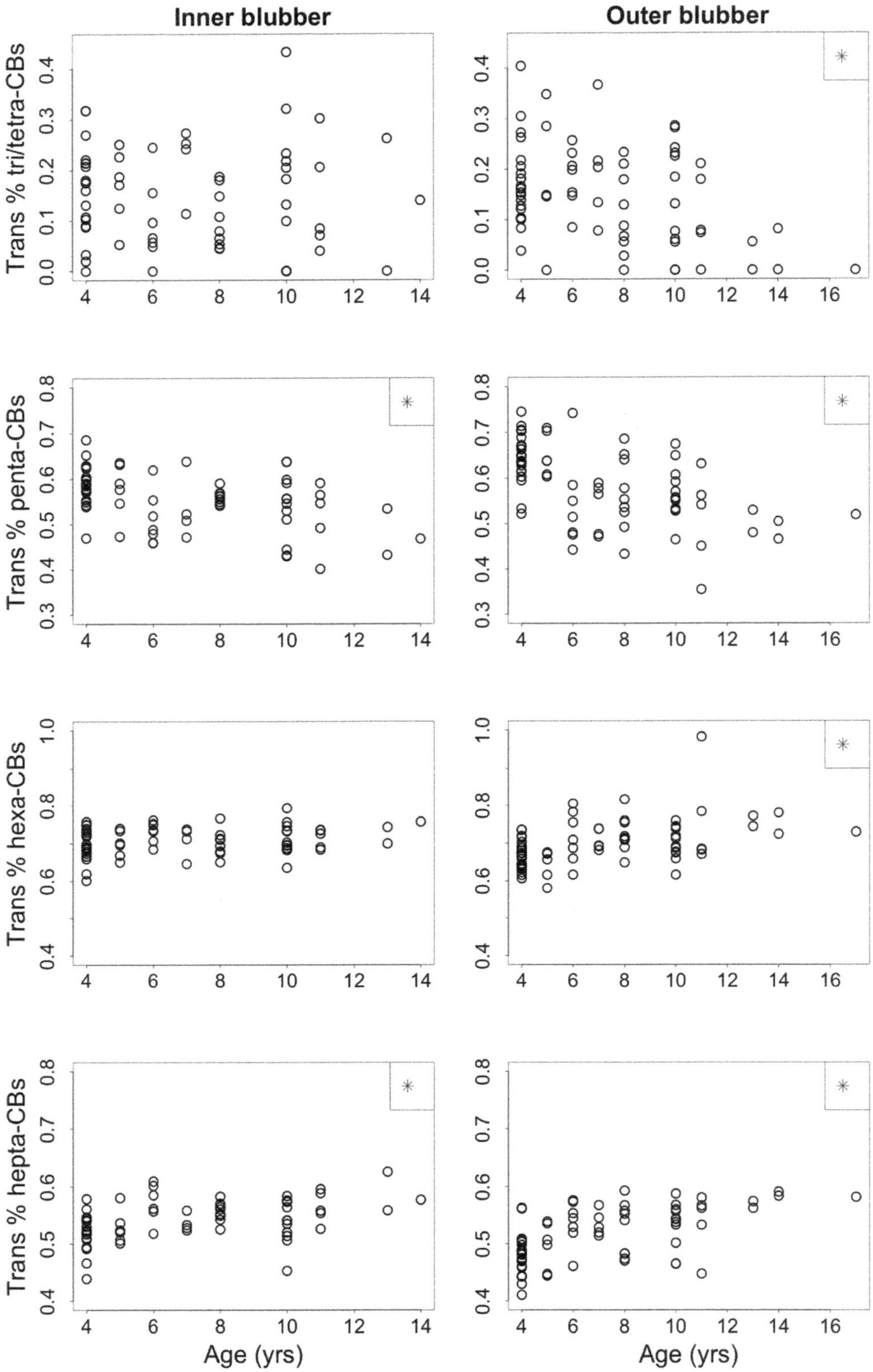

Figure 4. Inner and outer blubber congener profiles by age for all sampling periods combined. Statistics were run on transformed percentages (arc-sin square root). Asterisks indicate a stastically significant relationship with age.

outer blubber. In serum, there was a significant, positive relationship between adipose percent and the proportion of penta-CBs ($F_{1,21} = 6.4$, $p = 0.020$), the opposite trend to that observed in blubber. There were no significant relationships detected between adipose percent and the proportion of tri/tetra-CBs, hexa-CBs or hepta-CBs in serum.

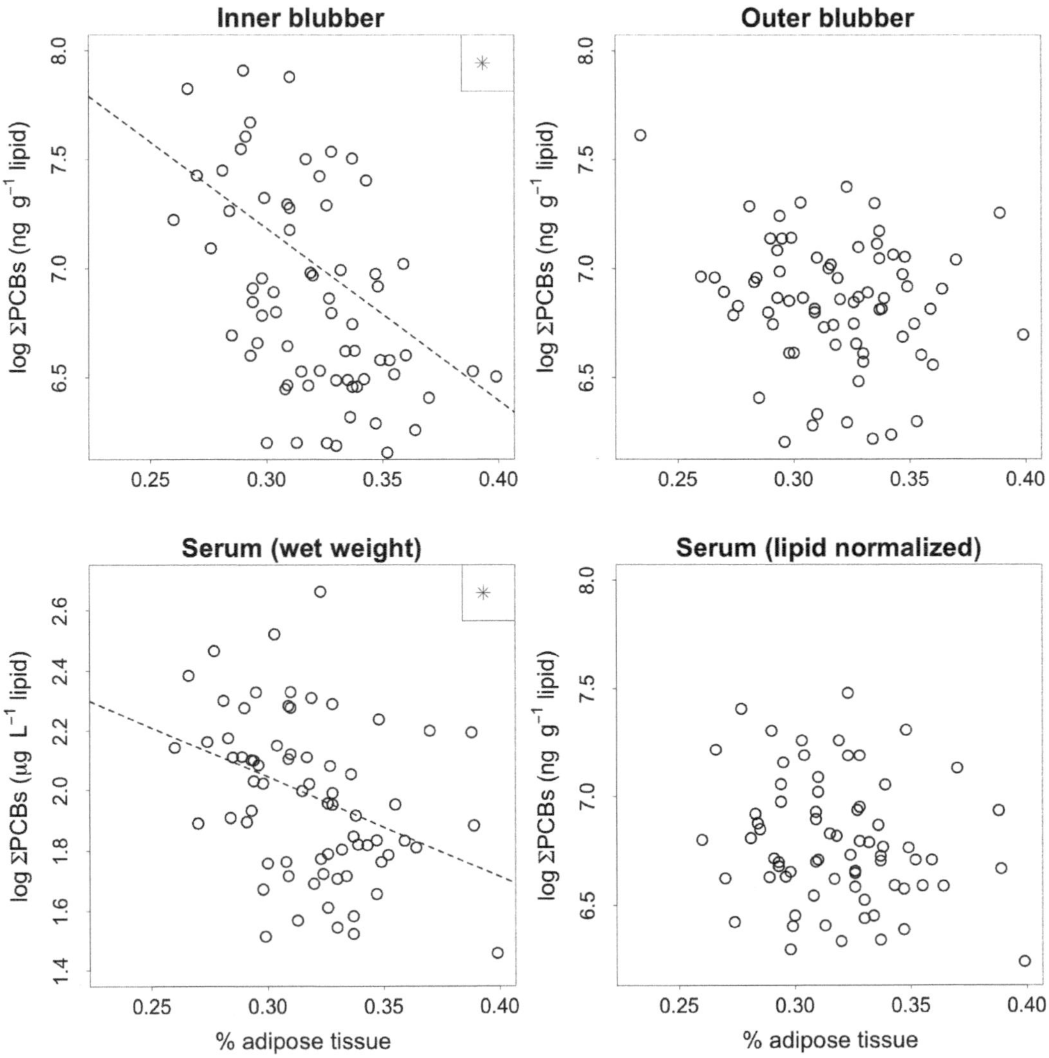

Figure 5. Inner blubber, outer blubber and serum (wet weight and lipid-normalized) $\sum$PCB concentrations relative to body condition (percent of adipose tissue). Statistical analyses were run using log transformed $\sum$PCB concentrations. Samples are from all sampling periods combined. Asterisks indicate stastically significant relationships.

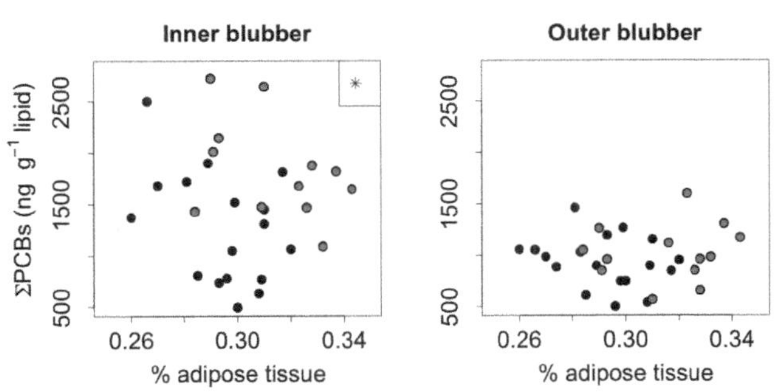

Figure 6. $\sum$PCB concentrations in individual blubber samples at the end of the breeding/lactation fast (black) and the molting fast (red), relative to body condition (percent of adipose tissue). Asterisk indicates a stastically significant difference between groups.

Paired Tissue Samples

Paired tissue samples from the same individuals pre- and post-foraging revealed that changes in $\sum$PCB concentrations within individuals differed by tissue type and whether the change was measured across the short or long foraging trip (Fig. 7). The concentration of $\sum$PCBs in inner blubber decreased across the long ($t = 6.2$, $df = 7$, $p < 0.001$) and short trips ($t = 3.3$, $df = 12$, $p = 0.007$); however, the magnitude of decrease was greater over the course of the long trip ($t = 4.5$, $df = 19$, $p < 0.001$). The outer blubber layer changed in concentration across both the short ($t = -2.8$, $df = 13$, $p = 0.015$) and long trips ($t = 2.3$, $df = 7$, $p = 0.05$) but the direction of the change differed between the trips ($t = 3.4$, $df = 20$, $p = 0.003$). $\sum$PCBs increased in outer blubber over the course of the short trip but decreased over the course of the long trip. Lipid-normalized serum $\sum$PCB concentrations demonstrated the same differences between foraging trips as observed in outer blubber ($t = 4.6$, $df = 20$, $p < 0.001$). Lipid-normalized serum $\sum$PCB concentrations increased over the course of the short trip ($t = -3.1$, $df = 13$, $p = 0.008$) but decreased over the course of the long trip ($t = 3.7$, $df = 7$, p-value $= 0.008$).

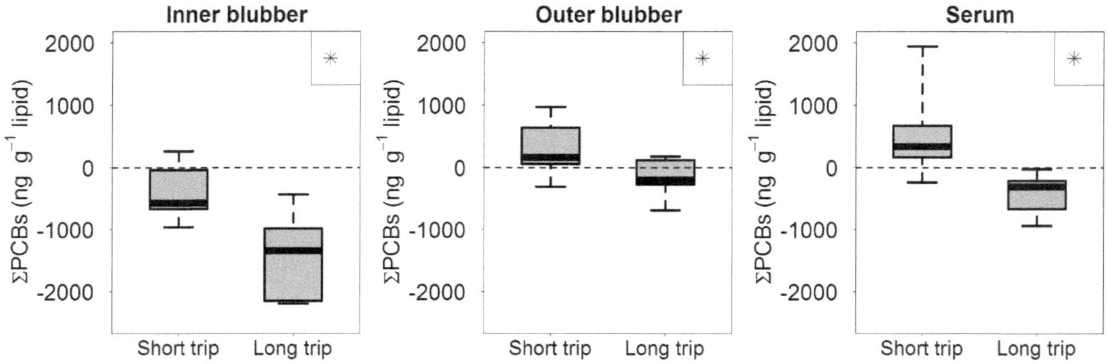

Figure 7. Changes in paired ∑PCB concentrations over the course of a foraging trip (same individuals pre- and post-foraging) for inner blubber, outer blubber and serum. Asterisks indicate statistically significant differences between the two trips. Refer to Fig. 1 for annual life history phases and differences between the short and long trip.

Changes in ∑PCB concentrations in serum measured by unit wet weight did not significantly change across the short trip ($t = -1.9$, $df = 13$, $p = 0.081$) but significantly decreased across the long trip ($t = 3.5$, $df = 7$, $p = 0.010$), and changes in ∑PCB concentrations between the trips was significantly different ($t = 4.2$, $df = 20$, $p < 0.001$).

Discussion

Our findings highlight that toxic contaminants are detectible in seals that forage in remote depths of the Pacific Ocean, thousands of kilometers from anthropogenic sources. Mesopelagic and open-ocean foraging behaviors do not leave northern elephant seals immune to the accumulation of PCBs, as PCBs were detected in blubber and blood of all 58 sampled northern elephant seal females. ∑PCB concentrations in elephant seal serum and blubber sampled during early and late breeding in the present study are similar to those previously observed in elephant seal females [13] and weaned pups [9,46] from the same breeding colony (Año Nuevo). However, the mean ∑PCB concentration in samples collected at the end of the molting fast in our study are higher than the mean ∑PCB concentration in samples previously collected at the end of the breeding fast [13]. Our study is the first to quantify ∑PCB concentrations in free-ranging, female northern elephant seals across the full range of naturally occurring body conditions and annual life history phases. Our results demonstrate the importance of accounting for both adipose percent and reproductive state when interpreting contaminant concentrations.

Quantification of ∑PCBs in marine mammals varies somewhat based on the number of congeners that are quantified in each study, which makes direct comparison of ∑PCBs or the percent of individual congeners difficult between studies. Additionally, the reporting of individual congeners varies between study, making one to one comparisons difficult. Nevertheless, some broad comparisons can be made between our observations and other published studies, realizing that the absolute values may differ slightly based on study design. Female northern elephant seals have substantially lower ∑PCB concentrations than stranded adult California sea lions (*Zalophus californianus*) [57], but adipose percentages are lower in most stranded animals than elephant seals from the present study. Blubber ∑PCB concentrations in elephant seals from our study are higher than those found in free-ranging, coastally-foraging adult female harbor seals from Kodiak Island, the southern Alaska Peninsula, and Prince William Sound [17], although these harbor seals forage further north than elephant seals from the present study and there are no comparable studies for free-ranging adult harbor seals foraging along the west coast of the United States of America. Given that all seals from our study harbor PCBs, greater sampling of animals across this region is warranted, especially because of the paucity of studies reporting ∑PCB concentrations for free-ranging adult pinnipeds from the eastern North Pacific. This would provide more complete documentation of baseline contaminant ranges within multiple species and the ability to appropriately compare species that utilize more similar geographic ranges.

The most common PCB congener quantified in elephant seal blubber for our study is CB-153, which contributes the greatest proportion to ∑PCBs in a broad range of marine mammal species [14,17,26,58,59]. The mean contribution of CB-138 and CB-180 in the present study places those congeners among the top five congeners contributing to the ∑PCB concentration in northern elephant seals. These same congeners are among the top five most important congeners for other pinnipeds, including adult fur seals (*Callorhinus ursinus*) [58], harbor seals [17] and Hawaiian monk seals (*Monachus schauinslandi*) [59] from the Pacific, as well as for gray seals [14] and harbor seals [26] from the Atlantic.

Influence of Age

The absence of a significant age relationship with ∑PCB concentrations indicates that elephant seals likely offload contaminants during lactation and replenish contaminants while foraging. Age does not affect ∑PCB concentrations in the blubber of other female pinnipeds, including harbor seals [17], Steller sea lions [16], and ringed seals [18]. ∑PCB concentrations in female odontocete and mysticete blubber decrease with increasing age [19,20], although this relationship reverses in post-reproductive females after reproduction senescence and the associated cessation of lactation [22,23]. These observations indicate that there are fundamental differences in how contaminants accumulate in cetaceans and pinnipeds, possibly driven by life history strategies and reproductive senescence.

Based on previous research on elephant seals from Año Nuevo quantifying average ∑PCB concentrations in milk [13] and average milk production in female elephant seals [41], females may annually offload approximately 23.9 mg of PCBs during an average 26.5 day [41] lactation period. This estimated transfer for northern elephant seals is similar to the calculated transfer of PCBs (27.0±11.2 mg) from grey seal female to their pups [60]. Concentrations of ∑PCBs in adult female elephant seal blubber, serum and milk tissues (present study; [13]) are lower than the concentrations observed in some highly contaminated marine mammal species [22,29,31]. However, transfer of contaminants to the pup is during a critical period of pup development for the immune, endocrine, and nervous systems [24] and maternal transfer of PCBs can affect fetal brain development, as observed in rats [61]. It is unlikely that PCBs are the only anthropogenic contaminant acquired by an individual [13]; therefore, each seal could have additional contaminants within blubber or other tissue compartments. It is possible that contaminants may interact, potentially in a synergistic way, but this has yet to be fully examined.

Age significantly affects the proportion of PCB congener families in female elephant seal blubber and serum. This suggests a positive relationship between an age-associated process, such as the number of times a female has given birth, and the proportion of higher chlorinated congeners in elephant seal tissues. Congeners with increased chlorination are more lipophilic and have a higher octanol-water partitioning coefficient (K_{ow}). This decreases the transfer efficiency from maternal blubber to offspring through milk, potentially due to differences of mobilization efficiency from blubber [14] as well as diffusion across the boundary of the mammary gland [14,25,62]. Congeners with higher log K_{ow} comprised a greater proportion of PCBs in maternal harbor seals [24], grey seals [14,25], harbor porpoises [26], and belugas [27] than their offspring, which may occur in northern elephant seals and cause the relationships observed between age and specific congener families. Older female seals have increased parity and greater adipose mass loss and milk production during lactation, which leads to more opportunities to depurate congeners with lower K_{ow} values, while selectively retaining congeners with higher K_{ow} values. Another potential mechanism that can explain the age relationship observed in our study is the preferential metabolism of lower chlorinated congeners. It is possible that both selective excretion through milk and preferential metabolism could play a role in the age and congener trends observed in female northern elephant seals.

Influence of Adipose Percent and Fasting State

Total PCB concentrations in inner blubber and serum (wet weight) have significant, negative relationships with the percent of adipose tissue. However, this trend is not observed in outer blubber or serum measured per unit lipid. This suggests that varying physiological and temporal mechanisms mediate fluctuations within the main reservoir (blubber) and the transport system (serum) of PCBs in northern elephant seal females. Asynchronous fluctuations between inner blubber, outer blubber, and serum are likely due to differential use of these tissue components during fasting and lactation, and the potential for transfer of contaminants to offspring. Trends observed here are supported by previous studies. In general, inner blubber $\sum$PCB concentrations may be more dynamic than outer blubber $\sum$PCB concentrations because inner blubber is the main energy reserve mobilized during fasting [63]. Fatty acid profiles for the inner blubber layer of adult female elephant seals change significantly over the lactation-associated fasting period, whereas the proportion of different classes of fatty acids in outer blubber remains stable through lactation [53]. Fatty acid profiles of inner blubber are similar to those observed in milk, which supports inner blubber as the main contributor of fatty acids to milk synthesis [53]. Additionally, outer blubber has a high proportion of medium chain length monounsaturated fatty acids, which are theorized to serve in a thermoregulatory capacity [53]. Thus, changes in $\sum$PCB concentrations correspond most strongly with changes in maternal lipid content, especially in the inner blubber layer.

When animals are at their fattest, outer blubber concentrations may be higher than inner blubber concentrations [13,49]. $\sum$PCB concentrations may increase in the inner blubber while fasting, because an animal utilizes the blubber reserve from the inside first and concentrates contaminants in the remaining tissue. This may cause the inner blubber $\sum$PCB concentrations to become higher than the outer blubber $\sum$PCB concentrations by the end of fasting [13,49,53]. In the present study, we observe the same trend occurring during the molting fast, although the overall concentrations in inner blubber are highest at the end of the molting fast. This may be attributed to the life history phase, with seals operating more as a closed system during the molting fast because lactation does not occur during the molt and there is no offloading of contaminants to offspring (Table 1).

Biannual fasting periods of seals on land are associated with breeding or molting and result in significant reductions in adipose tissue and body mass [41,64]. Female seals are on land for different life history stages (breeding/lactation and molting), which likely causes the differences in how energy stores are used during the two fasting periods. Female seals during the breeding season lose approximately 58% of their fat mass and lose significantly more fat than lean tissue [41]. Female seals during the molt lose an average of 38% of their fat mass and similar proportions of fat and lean tissue [64]. It is important to note that seals are at their fattest at the beginning of the breeding season and adipose reserves are not as great when seals arrive for the start of the molt, as seals have approximately three times longer to forage at sea prior to the breeding season than in between the breeding and molting seasons (Fig. 1) [35]. Accounting for fat stores, we observed that the highest mean $\sum$PCB concentrations in inner blubber occurred at the end of the molting fast, when seals were both relatively thin and unable to depurate contaminants to their offspring, rather than at the end of the breeding fast when seals were at their leanest. This suggests that both the fat content of the animal and its reproductive condition affect contaminant concentrations. Elephant seals only transfer contaminants to offspring during the long foraging trip (gestation) and the subsequent fasting period on land (lactation) and not during the molting fast; therefore, reproductive transfer to offspring, through the placenta and highly lipid-rich milk, may be responsible for this observed difference.

While concentrations of $\sum$PCBs in blubber and serum in our study are below the concentration threshold that corresponds with immune system suppression in harbor seals [44], concentrations in elephant seals fluctuate significantly relative to body condition. Elevated $\sum$PCB concentrations in inner blubber and the mobilization of PCBs to serum at the end of a fasting period, as observed in elephant seals [13] and grey seals [14,49], could increase the vulnerability of individuals to cumulative effects from additional contaminants or physiological stressors that reduce their adipose tissue reserves and concentrate contaminants in remaining tissues. Our observations of increased $\sum$PCB concentrations at lower adipose percentages are consistent with observations in other marine mammals, including humpback whales and polar bears (*Ursus maritimus*), that undergo significant fasting periods [65,66].

Animals may be especially vulnerable to the effects of contaminants if normal fasting causes increased tissue concentrations and is then followed by decreased post-fasting foraging opportunities. This was the possible explanation for elevated $\sum$PCB concentrations observed in a female from the present study. This female lost adipose tissue over the course of a short, post-breeding foraging trip, which corresponded to an increase in blubber and serum ∑PCB concentrations of 92% and 213%, respectively, over her late fasting concentrations. Her tissue concentrations, measured before and after the foraging trip, are among the highest observed during any time period and, based on our observations of other seals in this study, it is likely that her ∑PCB tissue concentrations would have been even higher at the end of the molting fast. Similar trends are observed in polar bears, where fluctuations in seasonal food availability outside of the hibernation period can influence body condition and thus affect PCB concentrations [66]. These observations have implications for other fasting animals, especially marine predators that fast while migrating or provisioning offspring.

Paired Samples

Examining paired tissue samples from the same individuals before and after a foraging trip provides insight into certain dynamics that are not observable in a larger population level analysis. These results are unique because tissue samples were paired between the start and end of a foraging period and not between the start and end of a fasting period, as is common for studies on the PCB dynamics in the blubber and blood of marine mammals. Inner blubber is the most metabolically active layer in northern elephant seals [53] and ∑PCB concentrations in our study vary with adipose percent both at population-level and individual-level analyses. Individuals show a decrease in ∑PCB concentrations within the inner blubber layer over the course of a foraging trip; however, the magnitude of the decrease is greater over the long trip than the short trip, which may partially be explained by differences in adipose tissue gained over these trips [35] due to differences in trip length (Fig. 1). The ∑PCBs in outer blubber and serum at the individual level present a more complicated story, increasing in concentration across the short trip but decreasing in concentration across the long trip, which suggests that PCB dynamics in outer blubber may be more complex than previously thought.

Conclusion

Overall, our findings demonstrate that PCBs accumulate in female northern elephant seals that forage across the mesopelagic north Pacific. While total accumulation of PCBs does not change with age, our findings reveal that more chlorinated PCB congeners comprise a higher proportion of the total PCB burden of older females while younger seals have a higher proportion of less-chlorinated PCBs. At this time, the health consequences of these observations remain unknown.

Northern elephant seals exhibit a set of life history strategies that depend on two seasonal fasting periods. Our findings suggest that female northern elephant seals are most vulnerable to high ∑PCB concentrations at low adipose percentages, which occur at the end of their two seasonal fasts. Depuration to pups may mitigate high blubber ∑PCB concentrations in adult female tissues during the breeding fast and prevent overall ∑PCB concentrations from increasing with increasing age. However, no mitigating mechanism is known to occur during the molting fast, and therefore, our results suggest the female elephant seals are most vulnerable to increased concentrations of ∑PCBs during their molting fast. Free ranging adult male elephant seals may accumulate significantly more ∑PCBs than females since they have no ability to mitigate their contaminant load through reproductive transfer. More research is needed on males to verify this hypothesis.

In addition, the inverse relationship between body condition and total PCB concentration in the inner blubber has implications for the development of bio-monitoring strategies. First, the presence of a significant, negative relationship between adipose percent and ∑PCB concentration in the inner blubber, but not in the outer blubber, indicates the complexity of blubber as a biological indicator. For elephant seals, our results suggest that ∑PCB concentrations of inner blubber represent recent trends in contaminant fluctuation, while ∑PCB concentrations of outer blubber, a tissue of slower turnover, may represent longer-term accumulation. Bio-monitoring of more elusive marine predators, where biopsy darting is the only option, may provide misleading results if only the outer-most tissue is sampled. It is imperative that adipose percent and the life history of an animal be fully considered for both the design of contaminant monitoring programs and appropriate interpretation of the results.

Acknowledgments

We thank the many volunteers, field-technicians and graduate students who were instrumental in the success of this project. We also thank the rangers and docents at Año Nuevo State Reserve, as well as Pat Morris, Sarah Allen, Brian Hatfield for their assistance in resighting seals and Richard Condit for development and maintenance of the demographic database. We thank M. Louvet for valuable help with the PCB analysis.

Author Contributions

Conceived and designed the experiments: SHP JH JPT DEC CD DPC. Performed the experiments: JLH AL DEC CD DPC. Analyzed the data: SP DEC. Contributed reagents/materials/analysis tools: JPT DEC CD DPC. Wrote the paper: SHP JLH AL JPT DEC CD DPC.

References

1. Vos JG, de Roij T (1972) Immunosuppressive activity of a polychlorinated biphenyl preparation on the humoral immune response in guinea pigs. Toxicology and Applied Pharmacology 21: 549–555.
2. de Swart RL, Ross PS, Timmerman HH, Vos HW, Reijnders PJH, et al. (1995) Impaired cellular immune response in harbour seals (*Phoca vitulina*) feeding on environmentally contaminated herring. Clinical and Experimental Immunology 101: 480–486.
3. Vos JG, Becher G, van den Berg M, de Boer J, Leonards PE (2003) Brominated flame retardants and endocrine disruption. Pure and Applied Chemistry 75: 2039–2046.
4. Muir D, Braune B, DeMarch B, Norstrom R, Wagemann R, et al. (1999) Spatial and temporal trends and effects of contaminants in the Canadian Arctic marine ecosystem: A review. Science of the Total Environment 230: 83–144.
5. Nisbet ICT, Sarofim AF (1972) Rates and routes of transport of PCBs in the environment. Environmental Health Perspectives 1: 21–38.
6. Loganathan BG, Kannan K (1994) Global organochlorine contamination trends: An overview. Ambio 23: 187–191.
7. Breivik K, Sweetman A, Pacyna JM, Jones KC (2002) Towards a global historical emission inventory for selected PCB congeners–a mass balance approach: 1. Global production and consumption. Science of The Total Environment 290: 181–198.
8. Braune BM, Outridge PM, Fisk AT, Muir DCG, Helm PA, et al. (2005) Persistent organic pollutants and mercury in marine biota of the Canadian Arctic: An overview of spatial and temporal trends. Science of the Total Environment 351–352: 4–56.
9. Debier C, Chalon C, Le Boeuf BJ, de Tillesse T, Larondelle Y, et al. (2006) Mobilization of PCBs from blubber to blood in northern elephant seals (*Mirounga angustirostris*) during the post-weaning fast. Aquatic Toxicology 80: 149–157.
10. Yordy JE, Pabst DA, McLellan WA, Wells RS, Rowles TK, et al. (2010) Tissue-specific distribution and whole-body burden estimates of persistent organic pollutants in the bottlenose dolphin (*Tursiops truncatus*). Environmental Toxicology and Chemistry 29: 1263–1273.
11. Myers MJ, Atkinson S (2012) Temporal variability in organochlorine contamination in blood and blubber of captive Steller sea lions (*Eumetopias jubatus*). Marine Mammal Science 28: 105–123.
12. Cameron ME, Metcalfe TL, Metcalfe CD, Macdonald CR (1996) Persistent organochlorine compounds in the blubber of ringed seals (*Phoca hispida*) from the

Belcher Islands, Northwest Territories, Canada. Marine Environmental Research 43: 99–116.

13. Debier C, Crocker DE, Houser DS, Vanden Berghe M, Fowler M, et al. (2012) Differential changes of fat-soluble vitamins and pollutants during lactation in northern elephant seal mother–pup pairs. Comparative Biochemistry and Physiology, Part A 162: 323–330.
14. Vanden Berghe M, Weijs L, Habran S, Das K, Bugli C, et al. (2012) Selective transfer of persistent organic pollutants and their metabolites in grey seals during lactation. Environment International 46: 6–15.
15. Polischuk SC, Letcher RJ, Norstrom RJ, Ramsay MA (1995) Preliminary results of fasting on the kinetics of organochlorines in polar bears (*Ursus maritimus*). Science of the Total Environment 160–61: 465–472.
16. Barron MG, Heintz R, Krahn MM (2003) Contaminant exposure and effects in pinnipeds: implications for Steller sea lion declines in Alaska. Science of the Total Environment 311: 111–133.
17. Wang D, Atkinson S, Hoover-Miller A, Lee S-E, Li QX (2007) Organochlorines in harbor seal (*Phoca vitulina*) tissues from the northern Gulf of Alaska. Environmental Pollution 146: 268–280.
18. Addison RF, Zinck ME, Smith TG (1986) PCBs have declined more than the DDT group residues in Arctic ringed seals (*Phoca hispida*) between 1972 and 1981. Environmental Science and Technology 20: 253–256.
19. Aguilar A, Borrell A (1994) Reproductive transfer and variation of body load of organochlorine pollutants with age in fin whales (*Balaenoptera physalus*). Archives of Environmental Contamination and Toxicology 27: 546–554.
20. Borrell A, Bloch D, Desportes G (1995) Age trends and reproductive transfer of organochlorine compounds in long-finned pilot whales from the Faroe Islands. Environmental Pollution 88: 283–292.
21. Subramanian AN, Tanabe S, Tatsukawa R (1988) Use of organochlorines as chemical tracers in determining some reproductive parameters in Dalli-type Dall's porpoise *Phocoenoides dalli*. Marine Environmental Research 25: 161–174.
22. Ross PS, Ellis GM, Ikonomou MG, Barrett-Lennard LG, Addison RF (2000) High PCB concentrations in free-ranging Pacific killer whales, *Orcinus orca*: Effects of age, sex and dietary preference. Marine Pollution Bulletin 40: 504–515.
23. Tanabe S, Loganathan BG, Subramanian A, Tatsukawa R (1987) Organochlorine residues in short-finned pilot whale: possible use as tracers of biological parameters. Marine Pollution Bulletin 18: 561–563.
24. Wolkers H, Lydersen C, Kovacs KM (2004) Accumulation and lactational transfer of PCBs and pesticides in harbor seals (*Phoca vitulina*) from Svalbard, Norway. Science of the Total Environment 319: 137–146.
25. Sørmo EG, Skaare JU, Lydersen C, Kovacs KM, Hammill MO, et al. (2003) Partitioning of persistent organic pollutants in grey seal (*Halichoerus grypus*) mother–pup pairs. Science of the Total Environment 302: 145–155.
26. Weijs L, Dirtu AC, Das K, Gheorghe A, Reijnders PJH, et al. (2009) Inter-species differences for polychlorinated biphenyls and polybrominated diphenyl ethers in marine top predators from the Southern North Sea: Part 1. Accumulation patterns in harbour seals and harbour porpoises. Environmental Pollution 157: 437–444.
27. Desforges J-PW, Ross PS, Loseto LL (2012) Transplacental transfer of polychlorinated biphenyls and polybrominated diphenyl ethers in arctic beluga whales (*Delphinapterus leucas*). Environmental Toxicology and Chemistry 31: 296–300.
28. Das K, Debacker V, Pillet S, Bouquegneau J-M (2003) Heavy metals in marine mammals. In: Vos JG, Bossart GD, Fournier M, O'Shea TJ, editors. Toxicology of marine mammals. New York: Taylor & Francis. 135–167.
29. Schwacke LH, Zolman ES, Balmer BC, De Guise S, George RC, et al. (2012) Anaemia, hypothyroidism and immune suppression associated with polychlorinated biphenyl exposure in bottlenose dolphins (*Tursiops truncatus*). Proceedings of the Royal Society B: Biological Sciences 279: 48–57.
30. Elfes CT, VanBlaricom GR, Boyd D, Calambokidis J, Clapham PJ, et al. (2010) Geographic variation of persistent organic pollutant levels in humpback whale (*Megaptera novaeangliae*) feeding areas of the North Pacific and North Atlantic. Environmental Toxicology and Chemistry 29: 824–834.
31. Ross PS, Jeffries SJ, Yunker MB, Addison RF, Ikonomou MG, et al. (2004) Harbor seals (*Phoca vitulina*) in British Columbia, Canada, and Washington State, USA, reveal a combination of local and global polychlorinated biphenyl, dioxin, and furan signals. Environmental Toxicology and Chemistry 23: 157–165.
32. Greig DJ, Ylitalo GM, Wheeler EA, Boyd D, Gulland FMD, et al. (2011) Geography and stage of development affect persistent organic pollutants in stranded and wild-caught harbor seal pups from central California. Science of the Total Environment 409: 3537–3547.
33. Finkelstein M, Keitt BS, Croll DA, Tershy B, Jarman WM, et al. (2006) Albatross species demonstrate regional differences in north Pacific marine contamination. Ecological Applications 16: 678–686.
34. Le Boeuf BJ, Crocker DE, Costa DP, Blackwell SB, Webb PM, et al. (2000) Foraging ecology of northern elephant seals. Ecological Monographs 70: 353–382.
35. Robinson PW, Costa DP, Crocker DE, Gallo-Reynoso JP, Champagne CD, et al. (2012) Foraging behavior and success of a mesopelagic predator in the northeast Pacific Ocean: Insights from a data rich species, the northern elephant seal. PLoS ONE 7: e36728.
36. Naito Y, Costa DP, Adachi T, Robinson PW, Fowler M, et al. (2013) Unravelling the mysteries of a mesopelagic diet: a large apex predator specializes on small prey. Functional Ecology.
37. Antonelis GA, Lowry MS, DeMaster DP, Fiscus CH (1987) Assessing northern elephant seal feeding habits by stomach lavage. Marine Mammal Science 3: 308–322.
38. Springer AM, Piatt JF, Shuntov VP, Van Vliet GB, Vladimirov VL, et al. (1999) Marine birds and mammals of the Pacific Subarctic Gyres. Progress In Oceanography 43: 443–487.
39. Simmons SE, Crocker DE, Hassrick JL, Kuhn CE, Robinson PW, et al. (2010) Climate-scale hydrographic features related to foraging success in a capital breeder, the northern elephant seal *Mirounga angustirostris*. Endangered Species Research 10: 233–243.
40. Longhurst AR (1998) Ecological geography of the sea: Academic Press, San Diego, CA.
41. Costa DP, Boeuf BJL, Huntley AC, Ortiz CL (1986) The energetics of lactation in the northern elephant seal, *Mirounga angustirostris*. Journal of Zoology 209: 21–33.
42. Blasius ME, Goodmanlowe GD (2008) Contaminants still high in top-level carnivores in the Southern California Bight: Levels of DDT and PCBs in resident and transient pinnipeds. Marine Pollution Bulletin 56: 1973–1982.
43. Kajiwara N, Kannan K, Muraoka M, Watanabe M, Takahashi S, et al. (2001) Organochlorine pesticides, polychlorinated biphenyls, and butyltin compounds in blubber and livers of stranded California sea lions, elephant seals, and harbor seals from coastal California, USA. Archives of Environmental Contamination and Toxicology 41: 90–99.
44. Ross P, De Swart R, Addison R, Van Loveren H, Vos J, et al. (1996) Contaminant-induced immunotoxicity in harbour seals: Wildlife at risk? Toxicology 112: 157–169.
45. Le Boeuf BJ, Laws RM (1994) Elephant seals: population ecology, behavior, and physiology: University of California Press.
46. Debier C, Le Boeuf BJ, Ikonomou MG, de Tillesse T, Larondelle Y, et al. (2005) Polychlorinated biphenyls, dioxins, and furans in weaned, free-ranging northern elephant seal pups from central California, USA. Environmental Toxicology and Chemistry 24: 629–633.
47. Webb PM, Crocker DE, Blackwell SB, Costa DP, Le Boeuf BJ (1998) Effects of buoyancy on the diving behavior of northern elephant seals. Journal of Experimental Biology 201: 2349–2358.
48. Gales NJ, Burton HR (1987) Ultrasonic measurement of blubber thickness of the southern elephant seal, *Mirounga leonina* (Linn). Australian Journal of Zoology 35: 207–217.
49. Debier C, Pomeroy PP, Dupont C, Joiris C, Comblin V, et al. (2003) Quantitative dynamics of PCB transfer from mother to pup during lactation in UK grey seals *Halichoerus grypus*. Marine Ecology Progress Series 247: 237–248.
50. E.C.Council (2002) L221/8. Official Journal of the European Communities 45: 8–36.
51. Pinheiro J, Bates D, DebRoy S, Sarkar D, R Core Team (2012) nlme: Linear and nonlinear mixed effects models. R package version 3.1–104.
52. R Development Core Team (2012) R: A language and environment for statistical computing. R Foundation for Statistical Computing, Vienna, Austria. ISBN 3-900051-07-0, Available: http://www.R-project.org.
53. Fowler MA, Debier C, Mignolet E, Linard C, Crocker DE, et al. (2014) Fatty acid mobilization and comparison to milk fatty acid content in northern elephant seals. Journal of Comparative Physiology B 184: 125–135.
54. Crocker DE, Champagne CD, Fowler MA, Houser DS (2014) Adiposity and Fat Metabolism in Lactating and Fasting Northern Elephant Seals. Advances in Nutrition: An International Review Journal 5: 57–64.
55. Zuur AF, Ieno EN, Walker NJ, Saveliev AA, Smith GM (2009) Mixed effects models and extensions in ecology with R. New York: Springer.
56. Zar J (1999) Biostatistical Analysis. New Jersey: Prentice Hall. 929 p.
57. Ylitalo GM, Stein JE, Hom T, Johnson LL, Tilbury KL, et al. (2005) The role of organochlorines in cancer-associated mortality in California sea lions (*Zalophus californianus*). Marine Pollution Bulletin 50: 30–39.
58. Loughlin TR, Castellini MA, Ylitalo G (2002) Spatial aspects of organochlorine contamination in northern fur seal tissues. Marine Pollution Bulletin 44: 1024–1034.
59. Lopez J, Boyd D, Ylitalo GM, Littnan C, Pearce R (2012) Persistent organic pollutants in the endangered Hawaiian monk seal (*Monachus schauinslandi*) from the main Hawaiian Islands. Marine Pollution Bulletin 64: 2588–2598.
60. Pomeroy PP, Green N, Hall AJ, Walton M, Jones K, et al. (1996) Congener-specific exposure of grey seal (*Halichoerus grypus*) pups to chlorinated biphenyls during lactation. Canadian Journal of Fisheries and Aquatic Sciences 53: 1526–1534.
61. Gauger KJ, Kato Y, Haraguchi K, Lehmler H-J, Robertson LW, et al. (2004) Polychlorinated biphenyls (PCBs) exert thyroid hormone-like effects in the fetal rat brain but do not bind to thyroid hormone receptors. Environmental Health Perspectives 112: 516–523.
62. Addison RF, Brodie PF (1987) Transfer of organochlorine residues from blubber through the circulatory system to milk in the lactating grey seal *Halichoerus grypus*. Canadian Journal of Fisheries and Aquatic Sciences 44: 782–786.
63. Strandberg U, Käkelä A, Lydersen C, Kovacs KM, Grahl-Nielsen O, et al. (2008) Stratification, composition, and function of marine mammal blubber: the ecology of fatty acids in marine mammals. Physiological and Biochemical Zoology 81: 473–485.
64. Worthy GAJ, Morris PA, Costa DP, Boeuf BJL (1992) Moult energetics of the northern elephant seal (*Mirounga angustirostris*). Journal of Zoology 227: 257–265.

65. Bengtson Nash SM, Waugh CA, Schlabach M (2013) Metabolic Concentration of Lipid Soluble Organochlorine Burdens in the Blubber of Southern Hemisphere Humpback Whales Through Migration and Fasting. Environmental Science & Technology 47: 9404–9413.

66. Polischuk S, Norstrom R, Ramsay M (2002) Body burdens and tissue concentrations of organochlorines in polar bears (*Ursus maritimus*) vary during seasonal fasts. Environmental Pollution 118: 29–39.

A Longitudinal Study of Steller Sea Lion Natality Rates in the Gulf of Alaska with Comparisons to Census Data

John M. Maniscalco[1,2]*, **Alan M. Springer**[2], **Pamela Parker**[1], **Milo D. Adkison**[2]

1 Department of Science, Alaska SeaLife Center, Seward, Alaska, United States of America, **2** Institute of Marine Science, University of Alaska Fairbanks, Fairbanks, Alaska, United States of America

Abstract

Steller sea lion (*Eumetopias jubatus*) numbers in the Western Distinct Population Segment are beginning to recover following the dramatic decline that began in the 1970s and ended around the turn of the century. Low female reproductive rates (natality) may have contributed to the decline and remain an issue of concern for this population. During the 2000s we found high natality among Steller sea lions in the Gulf of Alaska indicating a healthy population. This study extends these previous estimates over an additional three years and tests for interannual variations and long-term trends. We further examine the proportions of pups to adult females observed on the rookery and nearby haulouts during the birthing season to assess whether census data can be used to estimate natality. Open robust design multistate models were built and tested using Program MARK to estimate survival, resighting, and state transition probabilities in addition to other parameters dependent on whether or not a female gave birth in the previous year. Natality was estimated at 70% with some evidence of interannual variation but a long-term increasing or decreasing trend was not supported by the data. Bootstrap and regression comparisons of census data with natality estimates revealed no correlation between the two methods suggesting that census data are not an appropriate proxy for natality in this species. Longitudinal studies of individual animals are an appropriate method for estimating vital rates in species with variable detection over time such as the Steller sea lion. This work indicates that natality remains high in this region and is consistent with a population in recovery.

Editor: Daniel E. Crocker, Sonoma State University, United States of America

Funding: This study was funded by U.S. federal grant appropriations through the National Marine Fisheries Service - NOAA Award NA09NMF4390169 (https://grantsonline.rdc.noaa.gov). The funders had no role in study design, data collection and analysis, decision to publish, or preparation of the manuscript.

Competing Interests: The authors have declared that no competing interests exist.

* Email: johnm@alaskasealife.org

Introduction

Over the past 2 centuries, several fur seal and sea lion (Otariidae) populations have recovered to healthy numbers following catastrophic collapses [1]. Steller sea lions (*Eumetopias jubatus*) in Alaska, USA may be emulating this trend as numbers over a large portion of the western distinct population segment (WDPS) have begun to increase in recent years following a precipitous 30-year decline that started in the early 1970s [2,3]. Indeed, Steller sea lion populations may have undergone dramatic fluctuations in abundance several times during the past 4 millennia [4]. The most recent decline is primarily attributed to reduced juvenile survival during the 1980s in addition to minor reductions in reproductive rates and adult survival [5–7]. Changes in those vital rates could have resulted from nutritional stress, killer whale (*Orcinus orca*) predation, or a combination of these and other factors such as direct interactions with fisheries [8–11]. The latest upswing in Steller sea lion numbers was estimated to have begun in the early 2000s [12] and may be attributed to improved juvenile survival [13] and high natality rates that are reflective of a healthy population [14]. However, some work based on population demography and theoretic modeling has suggested natality reached an all-time low during the past decade [15].

Continued monitoring of vital rates is essential for species of concern such as the WDPS of Steller sea lions, which is classified as endangered under the Endangered Species Act. Of the two approaches to obtaining vital rate information, longitudinal, individual-based studies have many advantages over cross-sectional, population-level studies [16]. Yet, published estimates of vital rates based on longitudinal studies are rare for the WDPS of Steller sea lions even though hundreds of millions of dollars have been dedicated to researching these animals. Here we provide an update of our previous work [14] on natality rates for these animals in the Gulf of Alaska based on a longitudinal study of adult females and assess the potential for using census data (counts) to predict trends in natality over time.

Natality is defined as the birthrate or the proportion of births to some segment of the general population – generally mature females. Female Steller sea lions become reproductively mature at three to seven years of age and give birth to one pup per year but not necessarily every year [17]. The birthing season typically ranges between late May and early July in Alaska [18]. Twinning is extremely rare [19] and juveniles are often weaned during the spring months at age 1, age 2 or even older [13,20]. However, natality does not include neonatal or juvenile survival, and it is important to differentiate these life history variables because of the widely different factors that may affect them. Low rates of natality among pinnipeds have been related to disease, contaminants, and poor nutrition [6,21]; whereas neonatal or juvenile survival may be more influenced by weather conditions, predation, or food limitation [22–24]. By lumping life history stages together, information pertinent to understanding population dynamics can be hidden or lost entirely.

We previously outlined several reasons why it may be difficult to estimate natality based on proportions of pups on rookeries and haulouts late in the breeding season [14]. Primarily, these are uncertainties about the extent of neonatal mortality due to storm waves for example, and proportions of females foraging at sea which may change systematically with shifting oceanic regimes [14]. Assuming stability in these variables over time might lead to erroneous conclusions about the proportion of pups that are born to a population of sea lions. We further suggested that disparities in natality estimation techniques may be resolved by long-term comparisons of census data with direct estimates of natality based on longitudinal observations of adult females and mark-recapture modeling [14]. In this study, we added 3 additional years to our previous time series and extended our natality estimates over a 10-yr period, 2003–2012, using robust-design mark-recapture statistics. We further compared these results to census counts of the proportion of pups to adult females on rookeries and haulouts to determine if census data can be used to predict trends in natality. If proportions of pups to adult females, or non-pups, can be used to predict natality, we should expect positive correlations between these proportions and mark-recapture estimates of natality over time.

Materials and Methods

Ethics Statement

This research was conducted in accordance with Alaska SeaLife Center Institutional Animal Care and Use Committee Protocol No. R10-03-01 and National Marine Fisheries Service Permit No. 14324 for research on endangered Steller sea lions. The Chiswell Island group is part of the U. S. Fish and Wildlife Service National Maritime National Wildlife Refuge. Research was conducted on refuge lands under Right-of-Way Permit No. M-344-AM and Special Use Permit No. 74500-10-001 and earlier versions.

Study Site and Observational Methods

The focal area of this research included the Steller sea lion rookery on Chiswell Island (59° 59.18′ N, 149° 23.40′ W) and nearby haulouts in Kenai Fjords (Figure 1) which lie within the range of the endangered WDPS. The population decline at the Chiswell rookery was similar to that of other rookeries in the central Gulf of Alaska – that is, abundance fell by 90% from 1,106 adults in 1976 [25] to approximately 90 adults and 50–80 pups in the 2000s [26]. Beginning in 1999, up to six remotely operated video cameras were used to monitor Steller sea lions (see [26] for details). Video images, which provided complete spatial coverage of the Chiswell Island rookery, were viewable and controllable in real-time from the Alaska SeaLife Center 65 km away. Cameras were also installed and monitored at nearby haulouts beginning in 2000 (Figure 1).

Most adult Steller sea lions can be individually identified by unique scars, fungal patches, and/or flipper patterns, and longitudinal studies have been successfully conducted on animals identified by such means [23,24,26,27]. Over the course of this study, female sea lions with unique markings (n = 184) were tracked and digital photos of these animals and their distinguishing marks were taken on a regular basis (at least twice per month) at all remotely-monitored sites in Kenai Fjords. Some breeding females were identified by flipper tags (n = 5) or brands (n = 16), and age was known only for these animals. Females that did not have at least two distinguishing marks and could not be reliably resighted from one year to the next were not used in the analysis. Although pictures and data for some females were collected as early as 1999, the period 1999–2002 was not considered in the analysis of natality rates because there was a more focused effort on sighting females giving birth over females that did not. All females with unique markings were tracked from 2003 onward whether or not they gave birth.

Our observations were conducted year-round but intensified during the birthing and breeding season beginning in mid-May. Hour-long scan sampling for identifiable females and their pups was conducted four to ten times daily from 0600 h to 2200 h; earlier and later hours were added around the summer solstice (21 June) when light levels were sufficient for viewing sea lions. After 10 August, observations were recorded from approximately sunrise to sunset as diminishing daylight allowed. Events such as births and deaths were opportunistically recorded as they occurred or within 4 hrs of their known occurrence [24,26]. Births that happened overnight were recorded the following morning as having occurred at the half-way point of non-observation hours.

Complete census counts of all sea lions by age class (adult males, adult females, juveniles 1–4 yrs old, and pups 0–1 yrs old) on the rookery and at two nearby haulouts (Figure 1) were made at approximately 1100 h throughout the breeding season. Ratios of pups to adult females on the rookery and haulouts were summarized between 25 June and 15 July each year to determine if these counts could be used as a proxy for natality rates. This date range corresponds to when the National Marine Fisheries Service conducts their range-wide flight surveys for population counts [2,28] which are used in some models for vital rates estimation [7,15]. In some cases when census counts were not conducted (≤2 each yr; 14 out of 208 total), daily numbers of each age class were interpolated from previous and subsequent counts.

Steller sea lion mothers in the WDPS will normally remain with their newborn pups for 8 to 12 days following parturition [26,27]. Given the duration and detail of observations in this study (frequent scans and complete spatial coverage of the rookery), it is highly unlikely that any births went unnoticed. Furthermore, females that give birth to stillborns are not normally considered to be productive nor are stillbirths considered in the definition of natality. However, we include full-term stillbirths in our analysis (approximately 2%; [24]) so that our estimates are comparable to historic estimates during the 1970s when the population was presumed healthy, and in the 1980s during the height of the decline in this region. Those earlier estimates were based on late-term pregnancies of adult females collected in the field and could not account for stillbirths [6]. In addition, those previous studies only considered reproductively mature females whose status was known by examination of copora lutea in ovaries [6,17]. It was neither possible nor the intention of this study to verify reproductive maturity for females that were not observed to give birth even when age was known. Therefore, females of known age were included in this study beginning at 5 years of age to be consistent with the average age of sexual maturity of 4.6 yrs [17], which would indicate that age of first pupping would be at about 5.6 yrs. Females of unknown age were included in this study if they were judged to be at least 4 yrs of age as approximated by visual size comparisons to other known-age females lying nearby.

Non-breeding haulouts within the study area were also monitored during the birthing season to account for females that may have spent more time at those locations and to reduce sample

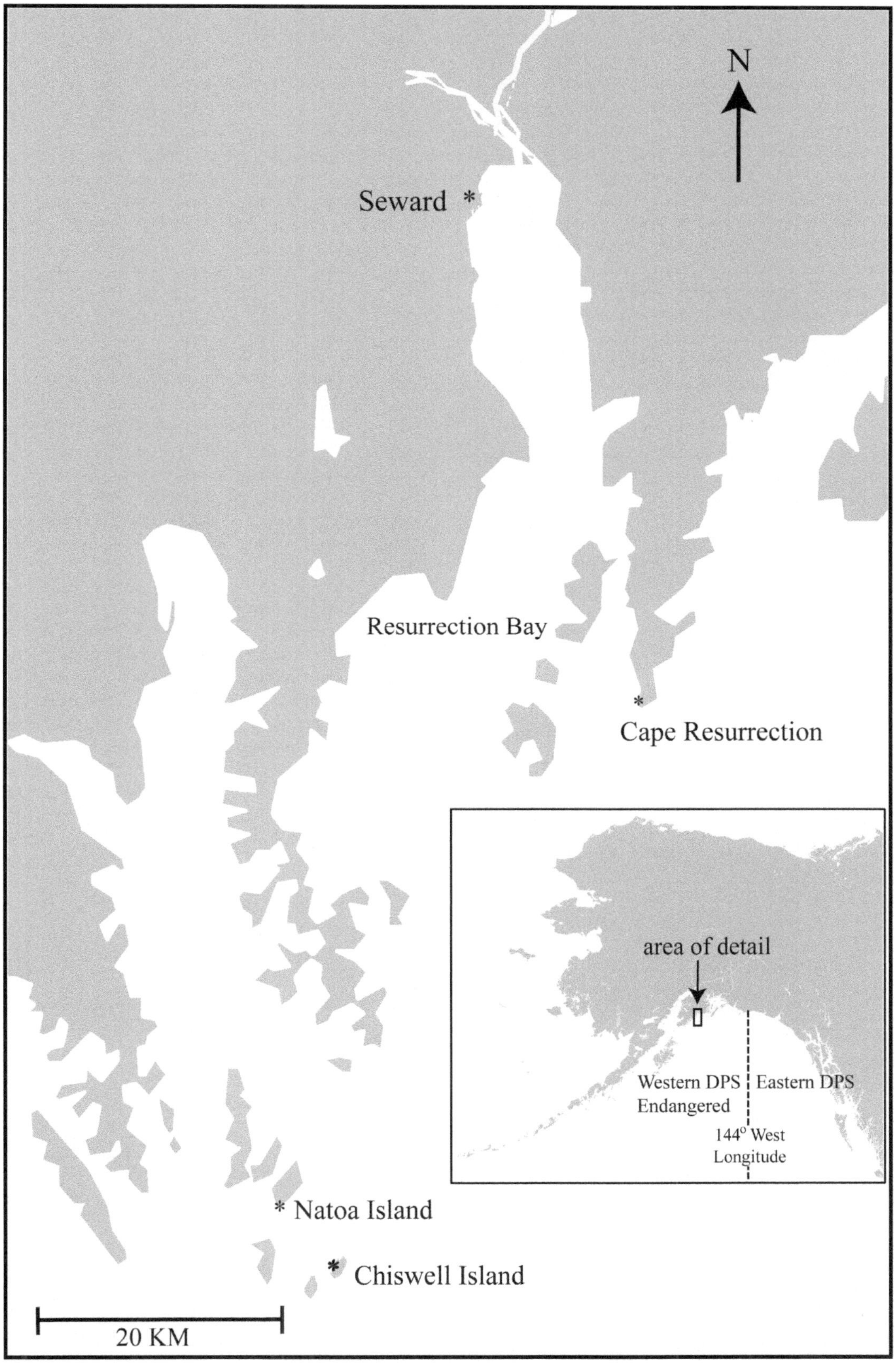

Figure 1. Location of the Chiswell Island Steller sea lion rookery and remotely monitored haulouts at Cape Resurrection and Natoa Island in Kenai Fjords, Gulf of Alaska.

bias toward more fecund females that may spend more time at the rookery seasonally and across years. Females at haulouts were included in the analyses if they met the aforementioned sighting and maturity criteria. Adult females that were observed without a pup during any birthing season were classified as not giving birth whether on the rookery or a haulout. However, some of the animals in the Chiswell Island population that were not giving birth on the rookery in any given year spent the summer elsewhere, presumably outside of the study area. Adult females that returned to the study area later in the year were typically without a pup but a few (<2%) did return with a suckling pup indicating that they gave birth elsewhere. Therefore, females that were without a pup when observed only during the non-pupping season could not be defined with certainty as not giving birth during the previous summer. We were able to account for this uncertainty by modeling the data as a hidden Markov process using a multi-state robust design [29].

Data Analysis

Open robust design (ORD) multi-state models with state uncertainty and seasonal effects were constructed and run using Program MARK [30]. Data were modeled across 10 primary periods (years 2003–2012) and 4 secondary periods – summer (Jun–Aug), autumn (Sep–Nov), winter (Dec–Feb), and spring (Mar–May) for a total of 40 sampling occasions. For each occasion, the states of every female were recorded as "b" – observed birth or with pup, "n" – observed but did not give birth, "0" – not observed, and "u" – observed during the non-birthing season without a pup, state uncertain. ORD models allow for entry and departure from the study area between primary periods and secondary periods, as often occurs among sea lions between seasons, but these models also greatly increase the number of parameters that can be estimated compared to models without a robust design. Nevertheless, robust design models allow estimation of temporary emigration from the study area and provide more precise estimates of survival and state transitions [29,31,32]. Furthermore, multistate ORD models do not necessarily require that the state of females be known with certainty or that females with and without a pup have equal sighting probabilities [31,33]. A goodness of fit test for multi-state ORD models when both states are observable has not been developed. However, a goodness of fit test using a simpler modeling approach for much of these data has shown only a slight and insignificant degree of overdispersion [14] and we have no reason to believe that data in the current model structure are much different.

The following parameters were estimated for this study with some initial constraints that were appropriate for the behavioral biology of Steller sea lions and to keep model run times reasonable:

- S_i^x = probability that a female in state x in year i survives until i+1.
- p_i^x = probability that a female is sighted in time period i in state x, given that she is present in the study area in period i. For secondary sampling periods a scaled covariate was added to reflect seasonal changes in observer effort which was highest during summer and lowest during winter.
- ψ_i^{xy} = transition probability or probability that a female in state x in year i is in state y in year i+1, given that she survived from year i to i+1.
- π_t^x = probability that a female is first observed in state x in year i for years 2–9 of the study.
- ω_t^x = probability that a female is in state x in year i, used as an estimate of natality in this study.
- δ_t^x = probability that a female is classified correctly to state. Constrained to be equal across all times periods for females with a pup. Constrained to be equal for females not seen with a pup across primary periods but allowed to vary between 1st secondary period (summer) and 2nd–4th secondary period combined.
- $pent_t^x$ = probability that a female in state x enters into the study area between a given secondary period. Constrained to be equal across primary periods for both states but allowed to vary between states and secondary periods with the mlogit link function to sum to ≤1.
- d_t^x = probability that a female in state x departs from the study area between secondary periods. Constrained to be equal across primary periods but allowed to vary between states and between secondary periods.
- α_t^x = probability that the attribute for state assignment has appeared (i.e., pup). Constrained to 1 for the 1st secondary period and to 0 for all other sampling periods because females only give birth during summer.
- c_t^x = probability that the attribute ceases – i.e., pups are weaned. Weaning often occurs during the spring months (Trites et al. 2006) and therefore was fixed to 0 for the 1st and 2nd secondary period intervals and allowed to vary during the 3rd interval.

These model parameters were combined across primary and secondary sampling periods into a joint multinomial likelihood using the mlogit link function in Program MARK. Additional constraints were placed on the models with regard to biological relevance in the search for parsimony. For example, survival (S) and state transitions (ψ^{xy}) were constrained to be equal over time in some models to assess whether the data more closely fit those model structures compared to time-varying model structures. The average ratio of pups to adult females on Chiswell Island and nearby haulouts was included as an annual covariate in some models to assess the effect on model fit. In the same manner, the total numbers of pups born on Chiswell each year was included as annual covariate for female state (ω) in addition to modeling linear trends. All models were compared with Akaike's Information Criteria (AIC; [34]), corrected for small sample bias (AICc; [35]). AICc weights, calculated from model differences in AICc values (ΔAICc), indicated relative support for the various models. Likelihood ratio tests (LRT) were also run to test for meaningful differences in models of interest. Finally, all models were averaged (multimodal inference; [36]) for estimates of parameter values relevant to this study.

We further assessed the possibility of using ratios of pups to females observed on rookeries and haulouts as a proxy for natality rates. Census data were log transformed to approximate normality and the ratios, *ln* (pups)/*ln* (adult females), were regressed against natality rates estimated from the ORD analysis for each year. This was conducted for each of 21 days between 25 June and 15 July each year to determine if any of these days provide a more accurate representation of natality over other days in that time period. We also conducted a bootstrap randomization test on standardized census data. The pup/adult female ratio was standardized to remove intra-seasonal patterns of haulout use by subtracting the daily means across years and adding back the overall mean to the daily count. We ran 1000 iterations to compare with estimated natality rate for the corresponding year. If count ratios are representative of natality rates across years, we would expect to find a significant correlation between these variables.

Results

A total of 16 models were run and ranked in order of lowest AICc score (Table 1). Models with no time differentiation in survival (S), sighting probability (p), and state transitions (ψ) showed more support for the data compared to those with variation across years (primary sampling periods). Time variation across seasons (secondary periods) was retained for sighting and departure (d) probabilities due to seasonal movements into and out of the study area (Alaska SeaLife Center, unpublished data), and the desire to model these changes. There was some evidence of differences in survival between females giving birth (89.0±1.4%) and those that did not (86.1±2.4%) as variation in this state was expressed in two of the four best models with no significant differences (Table 2; LRT: Model 1 vs 2: $\chi^2 = 1.602$; $P = 0.206$; Model 3 vs 4: $\chi^2 = 1.107$; $P = 0.293$). There was also little evidence of annual variation in the proportion of females giving birth (ω^b) as noted in a comparison of Models 1 and 3 (Table 1; LRT: $\chi^2 = 16.940$; $P = 0.050$) and Models 2 and 4 (Table 1; LRT: $\chi^2 = 16.444$; $P = 0.058$). See also Table 2 for time varying differences in ω^b for models 3 and 4.

An *a priori* decision was made to test for an increasing trend in natality, as the population in this region had been increasing over the study period [2]. However, fitting a linear increasing trend to natality did not improve model fit (Model 10 vs Model 4). Upon examination of natality estimates from the top ranked models, it was determined that a linear decrease in natality should also be tested for, but this also did not improve model fit (Model 11). We further tested for an improvement in model fit using either the total number of pups born or the average ratio of pups to adult females as linear covariates for ω^b each year. Both greatly reduced model fit compared to the model with full time dependence in ω^b (Models 5 & 6 vs Model 3 and Models 8 & 12 vs Model 4).

The four best models held 99.9% of the total weight (Table 1) and therefore, model averaged parameter estimates were based primarily on these structures. Survival to a subsequent year for females giving birth (S^b; 88.49±1.33% SE) was slightly higher than for those that did not give birth (S^n; 87.27±2.06%). Sighting probabilities were higher for females giving birth (p^b) compared to those without a pup (p^n) and seasonal sighting probabilities were especially low during winter (Table 2). Natality rates (ω^b) were high with an overall average of 70.5±1.6% that ranged narrowly from 67.6%±5.7% to 74.5%±7.3% (Figure 2). Also, females that gave birth in year *i* were nearly as likely to give birth in year *i*+1 ($\psi^{bb} = 0.708 \pm 0.021$) as females that did not give birth in year *i* ($\psi^{nb} = 0.736 \pm 0.033$).

Natality based on the ratio of pups to adult females on the rookery and nearby haulouts was higher than estimates from the mark-recapture modeling and had greater variation (91.3%±3.2%; Figure 2). Estimates for some years were unrealistically >100% because varying numbers of adult females were foraging at sea and unavailable to be counted. However, the intent of this analysis was not to obtain exact estimates of natality but to determine if natality rates based on census counts are correlated with direct estimates of natality and could therefore be used as a meaningful proxy for changes in natality. To test for a correlation, those rates were regressed against the natality rates (ω^b) estimated from Program MARK for each year. Slopes of the 21 regressions ranged from −0.250 to +0.223 and none were significant (*P*-value range: 0.080–0.925). The bootstrap analysis of the pup/adult female ratios also resulted in a non-significant relationship with natality ($r^2 = 0.048$; $P = 0.542$) and had slopes centered around zero (Figure 3).

Table 1. Comparison of models tested using a multi-state open robust design in Program Mark for estimation of survival (S), sighting probabilities (p), state transitions (ψ), and state occupation (ω) among Steller sea lions in Kenai Fjords.

Model Rank	Model Structure	No. Param.	AICc	ΔAICc	AICc Weights	Model Likelihood
1	$S_{.}\ p_{st.2^\circ}\ \psi_{st.}\ \pi_{.}\ \omega_{.}\ d_{st.2^\circ}$	30	6106.846	0.000	0.4067	1.0000
2	$S_{st.}\ p_{st.2^\circ}\ \psi_{st.}\ \pi_{.}\ \omega_{.}\ d_{st.2^\circ}$	31	6107.313	0.467	0.3220	0.7916
3	$S_{.}\ p_{st.2^\circ}\ \psi_{st.}\ \pi_{.}\ \omega_{t}\ d_{st.2^\circ}$	39	6108.617	1.771	0.1678	0.4125
4	$S_{st.}\ p_{st.2^\circ}\ \psi_{st.}\ \pi_{.}\ \omega_{t}\ d_{st.2^\circ}$	40	6109.601	2.755	0.1026	0.2522
5	$S_{.}\ p_{st.2^\circ}\ \psi_{st.}\ \pi_{.}\ \omega_{p}\ d_{st.2^\circ}$	39	6119.334	12.488	0.0008	0.0019
6	$S_{st.}\ p_{st.2^\circ}\ \psi_{st.}\ \pi_{.}\ \omega_{r}\ d_{.2^\circ}$	39	6124.224	17.378	0.0001	0.0002
7	$S_{st.}\ p_{st.2^\circ}\ \psi_{st.}\ \pi_{t}\ \omega_{t}\ d_{st.2^\circ}$	47	6124.259	17.413	0.0001	0.0002
8	$S_{st.}\ p_{st.2^\circ}\ \psi_{st.}\ \pi_{.}\ \omega_{r}\ d_{st.2^\circ}$	40	6125.533	18.687	0.0000	0.0001
9	$S_{.t}\ p_{st.2^\circ}\ \psi_{st.}\ \pi_{t}\ \omega_{t}\ d_{st.2^\circ}$	54	6129.711	22.865	0.0000	0.0000
10	$S_{st.}\ p_{st.2^\circ}\ \psi_{st.}\ \pi_{.}\ \omega_{up}\ d_{st.2^\circ}$	40	6133.919	27.073	0.0000	0.0000
11	$S_{st.}\ p_{st.2^\circ}\ \psi_{st.}\ \pi_{.}\ \omega_{dn}\ d_{st.2^\circ}$	40	6136.540	29.694	0.0000	0.0000
12	$S_{st.}\ p_{st.2^\circ}\ \psi_{st.}\ \pi_{.}\ \omega_{p}\ d_{st.2^\circ}$	40	6136.540	29.694	0.0000	0.0000
13	$S_{st.}\ p_{st.2^\circ}\ \psi_{st.}\ \pi_{.}\ \omega_{t}\ d_{.2^\circ}$	37	6138.479	31.634	0.0000	0.0000
14	$S_{st.t}\ p_{st.2^\circ}\ \psi_{st.}\ \pi_{t}\ \omega_{t}\ d_{st.2^\circ}$	63	6139.086	32.241	0.0000	0.0000
15	$S_{st.t}\ p_{st.2^\circ}\ \psi_{st.t}\ \pi_{t}\ \omega_{t}\ d_{st.2^\circ}$	79	6156.477	49.632	0.0000	0.0000
16	$S_{st.t}\ p_{st.1^\circ.2^\circ}\ \psi_{st.t}\ \pi_{t}\ \omega_{t}\ d_{st.2^\circ}$	151	6190.485	83.639	0.0000	0.0000

Proportion of the population released in each state (π) and departure probabilities (d) were also manipulated to test their effects on model fit. Initial constraints on parameters not listed here are outlined in the Methods.

Notes: Subscripts indicate state variation (st; with vs without pup), time variation (t), and increasing (up) or decreasing (dn) linear trends across primary periods in addition to time variation specific to primary periods (1°) and secondary periods (2°). Additional subscripts identify annual covariates of total number of pups born (p) and the average ratio of pups to adult females (r) observed.

Table 2. Estimates (± SE) from the four top-ranked models in Table 1 for survival of females that gave birth (S^b) and those that did not (S^n), occupation of birthing state (ω^b) in any given year, and sighting probabilities by season for females that gave birth (p^b) and those that did not (p^n).

Parameter	Model 1	Model 2	Model 3	Model 4
S^b	88.1% (1.2%)	89.0% (1.4%)	88.1% (1.2%)	89.0% (1.4%)
S^n	88.1% (1.2%)	86.1% (2.4%)	88.1% (1.2%)	86.1% (2.4%)
ω^b-2003	70.5% (1.6%)	70.5% (1.6%)	85.5% (5.0%)	85.5% (5.0%)
ω^b-2004	70.5% (1.6%)	70.5% (1.6%)	75.3% (5.7%)	75.3% (5.7%)
ω^b-2005	70.5% (1.6%)	70.5% (1.6%)	60.0% (5.7%)	60.0% (5.7%)
ω^b-2006	70.5% (1.6%)	70.5% (1.6%)	67.6% (5.4%)	67.6% (5.4%)
ω^b-2007	70.5% (1.6%)	70.5% (1.6%)	75.5% (4.7%)	75.5% (4.7%)
ω^b-2008	70.5% (1.6%)	70.5% (1.6%)	68.2% (4.8%)	68.2% (4.8%)
ω^b-2009	70.5% (1.6%)	70.5% (1.6%)	71.8% (4.6%)	71.8% (4.6%)
ω^b-2010	70.5% (1.6%)	70.5% (1.6%)	64.7% (4.9%)	64.7% (4.9%)
ω^b-2011	70.5% (1.6%)	70.5% (1.6%)	76.6% (4.2%)	76.6% (4.2%)
ω^b-2012	70.5% (1.6%)	70.5% (1.6%)	66.3% (4.6%)	66.3% (4.6%)
p^b-summer	99.9% (<0.1%)	99.9% (<0.1%)	99.9% (<0.1%)	99.9% (<0.1%)
p^n-summer	98.3% (0.9%)	97.7% (1.2%)	97.7% (1.2%)	97.7% (1.3%)
p^b-fall	94.6% (1.8%)	94.6% (1.8%)	94.6% (1.8%)	94.6% (1.8%)
p^n-fall	74.3% (4.9%)	72.8% (4.4%)	72.9% (4.4%)	72.8% (4.4%)
p^b-winter	24.6% (3.6%)	24.6% (3.6%)	24.6% (3.6%)	24.6% (3.6%)
p^n-winter	12.7% (2.3%)	14.3% (4.0%)	14.3% (4.0%)	14.3% (4.0%)
p^b-spring	70.4% (4.1%)	70.4% (4.1%)	70.4% (4.1%)	70.4% (4.1%)
p^n-spring	39.4% (4.2%)	40.1% (4.7%)	40.1% (4.7%)	40.1% (4.7%)

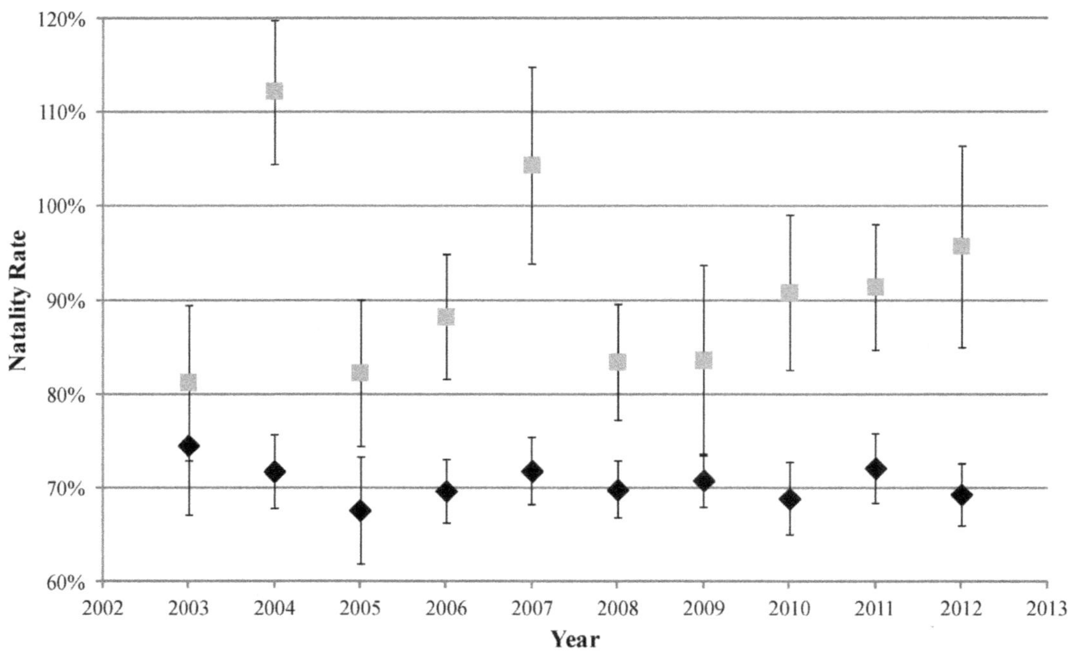

Figure 2. Mark-recapture estimates of natality rates (black diamonds) and natality estimates based on census data (gray squares) for adult female Steller sea lions at the Chiswell Island rookery from 2003 to 2012 with 95% confidence intervals. Note that the mark-recapture estimates are different from estimates expressed in Table 2 due to model averaging.

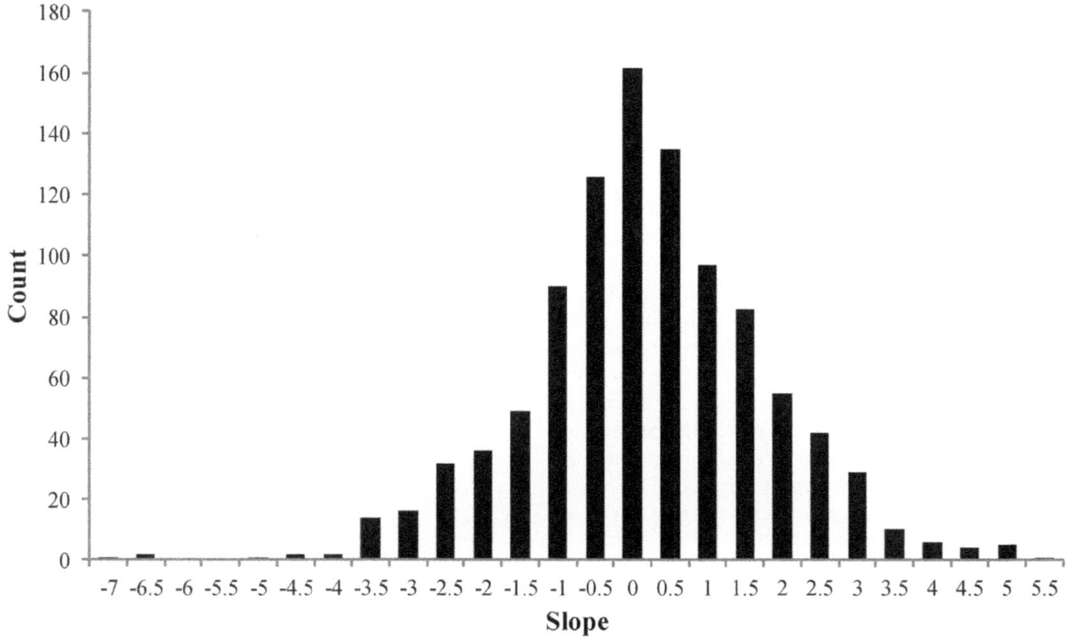

Figure 3. Bootstrap histogram of regression slopes based on pup to adult female ratios versus mark-recapture natality estimates. This figure shows that the regression slopes were strongly centered around zero with no correlative trend.

Discussion

Mark-Recapture Parameter Estimates

We previously examined natality of Steller sea lions in this region over seven years using a standard mark recapture multistate approach without robust design [14]. The current analysis was undertaken to improve our estimates using the robust design, to examine possible interannual and long-term changes in natality, and to assess the use of census data as a proxy for natality.

Not surprisingly, estimates of natality were similar to those previously found in this region at about 70% but with a smaller standard error (1.6 vs. 2.5%; [14]), which may be due partly to the larger sample size in this study. In addition, parameter estimates such as these are notably improved using the robust design as reported by others [29,32]. Sighting probabilities were high during the summer months for females giving birth and for those not giving birth in any given year (Table 2). Yet many non-breeders were not observed on the rookery during the birthing season (mid-May to Mid-July) and arrived later in the summer or autumn. If we only considered sightings during the birthing season much of the information about non-breeders would be lost and hence lead to greater uncertainty in our estimates.

Natality was not much different from pre-decline levels (67%) but better than estimated during the height of the population decline (55%) in the Gulf of Alaska during the 1980s [6]. It has previously been suggested that natality rates bottomed out at about 43% in the early 2000s, as estimated from an inferential population dynamic model [15]. A rapid change in natality from percentages in the mid-40s to nearly 70% would seem quite unusual, hence it was of interest to test for an increasing trend in this study. However, the model we tested with a increasing trend in natality did not fit the data, nor did one with a decreasing trend. In our analysis, natality ranged narrowly around 70%, with 2 of the 4 best models supporting interannual fluctuations but no trends. Therefore, we have little evidence to support unusually low natality rates during any year of this study.

Notwithstanding variation in survival, natality rates of 60% to 75% have been generally associated with stable or increasing populations of pinnipeds [37–40], including the Eastern Distinct Population Segment (EDPS) of Steller sea lions [41]. Natality of 55% or lower has been associated with declining populations of otariids and related to the adverse effects of density dependant factors such as intraspecific competition for food or breeding space [6,42] but see also [43]. Therefore, our estimate of natality from the Chiswell Island population of Steller sea lions is indicative of a population where adult females seemingly are not under resource limitation. Populations in this region and as far west as the eastern Aleutian Islands have been increasing since the early 2000s [2]. We cannot say with certainty that natality rates estimated here in the eastern Gulf of Alaska are representative of natality as far west as the eastern Aleutian Islands, but given the similar population trends [2], we currently have no reason to suppose otherwise.

During periods of food limitation long-lived mammals may exhibit a cost of reproduction in terms of reduced probability of survival or reproduction in successive years compared with times of food abundance [44,45]. Such a cost likely affected Steller sea lions during the 1980s when pregnancy was negatively correlated with lactation status [6]. Similar to our previous work [14], we found no evidence of a cost of reproduction among Steller sea lions during the period of this study (2003–2012) with survival and subsequent reproduction not being correlated with previous birthing status. This finding is consistent with a healthy population that is not exhibiting signs of broad scale food limitation [8,45,46].

Comparisons with Census Data

Long-term census-based studies provide important information regarding changes in demographics, distribution, densities, and trends of wildlife populations including the potential influences of climate change and human activities [16]. However, using census data to estimate vital rates in variably detectable species such as pinnipeds that spend a large proportion of their time foraging at sea may be difficult. Using several methods, we were unable to find a correlation between either pup counts or pup to female ratios and our natality estimates. As our natality estimates showed little year to year variation, this could have made such relationships difficult to detect. Nevertheless, if census counts can be used as a proxy for natality, we could expect at least weak positive correlations, but they were strongly centered around zero (Figure 3).

Using census counts for modeling vital rates is unusual for many large vertebrates and caution is warranted when modeling such data for vital rate estimates [47]. Census counts of Steller sea lions can vary with tide height, storm waves, time of day, and food availability [48–50]. Pup counts may or may not reflect the actual number of births because storm waves during the peak of birthing are variable across years and can cause huge losses of neonates [23,24]. These variations are typically assumed to be consistent over time [7,15] but they are not [14]. This is a primary reason why census data cannot be reliably used to estimate vital rates in species such as Steller sea lions unless adjustments are made to the data that appropriately reflect environmental stochasticity over time.

Accounting for variation in sighting probabilities is integral to longitudinal studies employing mark-recapture analysis. Long-term studies of a representative sample of individuals are necessary for accurate estimates of age-specific survival and natality, but are invariably more time consuming and costly than snapshot population counts [16,51]. Nevertheless, these mark-recapture methods provide critical information for understanding life histories and behavior of species of concern. In addition to natality, survival, and sighting probabilities, many other parameters can be estimated with the mark-recapture data analysis techniques currently available [29,32,33]. For instance, longitudinal studies of branded Steller sea lions are providing insights into temporary and permanent migration between the distinct population segments [52]. Without such work, we might mistakenly assume that populations receiving such immigrants were increasing solely due to improved survival or natality.

Census counts provide information on population trends over time, whereas individual-based longitudinal studies will more specifically inform researchers of why and how populations are changing. Based on census count data, we know that the EDPS of Steller sea lions has been increasing over the past 30 years [53]. Based on longitudinal studies we also recognize that the population increases likely result from a combination of good juvenile survival [54], good natality [41], and some immigration from the WDPS [51]. All of this information and more has lead to the recent delisting of the EDPS from the Endangered Species List [55]. The WDPS remains endangered, but with population trends continuing to increase, downlisting to threatened status is possible by 2015 [2]. Current population growth in the WDPS can be attributed to good natality and adult survival ([14]; this study) in addition to improved juvenile survival [13] since the height of the decline. Continued monitoring of vital rates is essential to detect changes that could threaten recovery of the species. Longitudinal studies of this nature can detect changes in vital rates that may happen quickly and drastically in response to increasing environmental stochasticity in the face of a shifting global climate [56].

Acknowledgments

Data collection for this study was accomplished by many hard-working technicians and interns at the ASLC including Melinda Fowler, Karin Harris, Juliana Kim, Carlene Miller, Emily Teate, and many other technicians and interns. We thank them and Lara Horstmann and Chuck Frost for discussions, commentaries, and much helpful advice on this material. Chris Oosthuizen also provided very constructive comments on the manuscript.

Author Contributions

Conceived and designed the experiments: JMM. Performed the experiments: JMM PP. Analyzed the data: JMM MDA. Contributed reagents/materials/analysis tools: JMM MDA. Contributed to the writing of the manuscript: JMM AMS PP MDA.

References

1. Gerber LR, Hilborn R (2001) Catastrophic events and recovery from low densities in populations of otariids: implications for risk of extinction. Mammal Review 31: 131–150.
2. Fritz L, Sweeney K, Johnson D, Lynn M, Gelatt T, et al. (2013) Aerial and ship-based surveys of Steller sea lions (*Eumetopias jubatus*) conducted in Alaska in June-July 2008 through 2012, and an update on the status and trend of the western distinct population segment in Alaska. US Dep Commer, NOAA Tech Memo, NMFS-AFSC-251, 91 p.
3. Loughlin TR (1998) The Steller sea lion: A declining species. Biosphere Conservation 1: 91–98.
4. Maschner HDG, Trites AW, Reedy-Maschner KL, Betts M (2013) The decline of Steller sea lions (*Eumetopias jubatus*) in the North Pacific: insights from indigenous people, ethnohistoric records and archaeological data. Fish and Fisheries doi:10.1111/faf.12038.
5. York AE (1994) The population dynamics of northern sea lions, 1975–1985. Marine Mammal Science 10: 38–51.
6. Pitcher KW, Calkins DG, Pendleton GW (1998) Reproductive performance of female Steller sea lions: an energetics-based reproductive strategy? Canadian Journal of Zoology 76: 2075–2083.
7. Holmes EE, York AE (2003) Using age structure to detect impacts on threatened populations: a case study with Steller sea lions. Conservation Biology 17: 1794–1806.
8. National Resource Council (2003) The decline of the Steller sea lion in Alaskan waters: Untangling food webs and fishing nets. Washington: National Academies Press, 179 p.
9. Springer AM, Estes JA, van Vliet GB, Williams TM, Doak DF, et al. (2003) Sequential megafaunal collapse in the North Pacific Ocean. Proceedings of the National Academy of Sciences 100: 12223–12228.
10. Trites AW, Donelly CP (2003) The decline of Steller sea lions *Eumetopias jubatus* in Alaska: a review of the nutritional stress hypothesis. Mammal Review 33: 3–28.
11. Wolf N, Mangel M (2008) Multiple hypothesis testing and the declining population paradigm in Steller sea lions. Ecological Applications 18: 1932–1955.
12. Eberhardt LL, Sease JL, DeMaster DP (2005) Projecting the trend of Steller sea lion populations in western Alaska. Marine Mammal Science 21: 728–738.
13. Maniscalco JM (2014) The effects of birth weight and maternal care on juvenile Steller sea lion (*Eumetopias jubatus*) survival. PLoS ONE 9(5): e96328.
14. Maniscalco JM, Springer AM, Parker P (2010) High natality rates of endangered Steller sea lions in Kenai Fjords, Alaska and perceptions of population status in the Gulf of Alaska. PLoS ONE 5(4): e10076.
15. Holmes EE, Fritz LW, York AE, Sweeney K (2007) Age-structured modeling reveals long-term declines in the natality of western Steller sea lions. Ecological Applications 17: 2214–2232.
16. Clutton-Brock T, Sheldon BC (2010) Individuals and populations: the role of long-term, individual-based studies of animals in ecology and evolutionary biology. Trends in Ecology and Evolution 25: 562–573.
17. Pitcher KW, Calkins DG (1981) Reproductive biology of Steller sea lions in the Gulf of Alaska. Journal of Mammalogy 62: 599–605.
18. Pitcher KW, Burkanov VN, Calkins DG, Le Boeuf BJ, Mamaev EG, et al. (2001) Spatial and temporal variation in the timing of births of Steller sea lions. Journal of Mammalogy 82: 1047–1053.
19. Maniscalco JM, Parker P (2009) A case of twinning and the care of two offspring of different age in Steller sea lions. Marine Mammal Science 25: 206–213.
20. Trites AW, Porter BP, Deecke VB, Coombs AP, Marcotte ML, et al. (2006) Insights into the timing of weaning and the attendance patterns of lactating Steller sea lions (*Eumetopias jubatus*) in Alaska during winter, spring, and summer. Aquatic Mammals 32: 85–97.
21. Gilmartin WG, DeLong RL, Smith AW, Sweeney JC, DeLapp BW, et al. (1976) Premature parturition in the California sea lion. Journal of Wildlife Diseases 12: 104–115.
22. Doidge DW, Croxall WP, Baker JR (1984) Density-dependent pup mortality in the Antarctic fur seal *Arctocephalus gazella* at South Georgia. Journal of Zoology, London 202: 449–460.
23. Kaplan CC, White GC, Noon BR (2008) Neonatal survival of Steller sea lions (*Eumetopias jubatus*). Marine Mammal Science 24: 443–461.

24. Maniscalco JM, Calkins DG, Parker P, Atkinson S (2008) Causes and extent of natural mortality among Steller sea lion (*Eumetopias jubatus*) pups. Aquatic Mammals 34: 277–287.
25. Calkins DG, Becker E, Spraker TR, Loughlin TR (1994) Impacts on Steller sea lions. 119–139 In: Loughlin TR, editor, Marine Mammals and the Exxon Valdez. Academic Press, San Diego.
26. Maniscalco JM, Parker P, Atkinson S (2006) Interseasonal and interannual measures of maternal care among individual Steller sea lions (*Eumetopias jubatus*). Journal of Mammalogy 87: 304–311.
27. Milette LL, Trites AW (2003) Maternal attendance patterns of Steller sea lions (*Eumetopias jubatus*) from stable and declining populations. Canadian Journal of Zoology 81: 340–348.
28. Fritz LW, Lynn M, Kunisch LE, Sweeney K (2008) Aerial, ship and land-based surveys of Steller sea lions (*Eumetopias jubatus*) in the Western Stock in Alaska, June and July 2005–2007. US Dep of Commer, NOAA Tech Memo, NMFS-AFSC-183, 70 p.
29. Kendall WL, Nichols JD (2002) Estimating state transition probabilities for unobservable states using capture-recapture/resighting data. Ecology 83: 3276–3284.
30. White GC, Burnham KP (1999) Program MARK: survival estimation from populations of marked animals. Bird Study 46: 120–139.
31. Kendall WL, Nichols JD, Hines JE (1997) Estimating temporary emigration using capture-recapture data with Pollock's robust design. Ecology 78: 563–578.
32. Kendall WL, White GC, Hines JE, Langtimm CA, Yoshizaka J (2012) Estimating parameters of hidden Markov models based on marked individuals: use of robust design data. Ecology 93: 913–920.
33. Kendall WL, Hines JE, Nichols JD (2003) Adjusting multistate capture-recapture models for misclassification bias: manatee breeding proportions. Ecology 84: 1058–1066.
34. Akaike H (1973) Information theory and an extension of the maximum likelihood principal. 267–281 In: Petran BN, Csaaki F, editors. International Symposium on Information Theory. Budapest, Hungary: Akadeemiai Kiado.
35. Hurvich CM, Tsai C-L (1989) Regression and time series selection in small samples. Biometrika 76: 297–307.
36. Burnham KP, Anderson DR, Huyvaert KP (2011) AIC model selection and multimodel inference in behavioral ecology: some background, observations, and comparisons. Behavioral Ecology and Sociobiology 62: 23–35.
37. Lunn NJ, Boyd IL, Croxall JP (1994) Reproductive performance of female Antarctic fur seals: the influence of age, breeding experience, environmental variation and individual quality. Journal of Animal Ecology 63: 827–840.
38. McKenzie J, Parry LJ, Page B, Goldsworthy SD (2005) Estimation of pregnancy rates and reproductive failure in New Zealand fur seals (*Arctocephalus forsteri*). Journal of Mammalogy 86: 1237–1246.
39. Melin SR, Laake JL, DeLong RL, Siniff DB (2012) Age-specific recruitment and natality of California sea lions at San Miguel Island, California. Marine Mammal Science 28: 751–776.
40. Testa JW (1987) Long-term reproductive patterns and sighting bias in Weddell seals (*Leptonychotes weddelli*). Canadian Journal of Zoology 65: 1091–1099.
41. Taylor RL, Boor GKH (2012) Beyond the robust design: accounting for changing, uncertain states and sparse, biased detection in a multistate mark-recapture model. Ecological Modelling 243: 73–80.
42. Dabin W, Beauplet G, Crespo EA, Guinet C (2005) Age structure, growth, and demographic parameters in breeding-age female subantarctic fur seals, *Arctocephalus tropicalis*. Canadian Journal of Zoology 82: 1043–1050.
43. Gibbens J, Parry LJ, Arnould JPY (2010) Influences on fecundity in Australian fur seals (*Arctocephalus pusillus doriferus*). Journal of Mammalogy 91: 510–518.
44. Bell G (1980) The costs of reproduction and their consequences. American Naturalist 116: 45–76.
45. Hamel S, Cote SD, Gaillard J-M, Festa-Bianchet M (2009) Individual variation in reproductive costs of reproduction: high-quality females always do better. Journal of Animal Ecology 78: 143–151.
46. DeMaster D, Atkinson S (2002) Steller sea lion decline: Is it food II. AK-SG-02-02. Fairbanks: University of Alaska Sea Grant College Program, 78 p.
47. Eberhardt LL (1985) Assessing the dynamics of wild populations. Journal of Wildlife Management 49: 997–1012.
48. Kastelein RA, Weltz FC (1991) Distribution and behavior of Steller sea lions (*Eumetopias jubatus*) in Prince William Sound, Alaska, June 1989. Aquatic Mammals 17: 91–97.
49. Kucey L (2005) Human disturbance and the hauling out behavior of Steller sea lions (*Eumetopias jubatus*). M. S. Thesis, University of British Columbia, Vancouver, 67 p.
50. Withrow DE (1982) Using aerial surveys, ground truth methodology, and haul out behavior to census Steller sea lions, *Eumetopias jubatus*. M. S. Thesis, University of Washington, Seattle, 102 p.
51. Clutton-Brock TH, Editor (1988) Reproductive success: Studies of individual variation in contrasting breeding systems. Chicago: The University of Chicago Press, 538 p.
52. Jemison LA, Pendleton GW, Fritz LW, Hastings KK, Maniscalco JM, et al. (2013) Inter-population movements of Steller sea lions in Alaska with implications for population separation. PLoS ONE 8(8): e70167.
53. Pitcher KW, Olesiuk PF, Brown RF, Lowry MS, Jeffries SJ, et al. (2007) Abundance and distribution of the eastern North Pacific Steller sea lion (*Eumetopias jubatus*) population. Fishery Bulletin 107: 102–115.
54. Hastings KK, Jemison LA, Gelatt TS, Laake JL, Pendleton GW, et al. (2011) Cohort effects and spatial variation in age-specific survival of Steller sea lions from southeastern Alaska. Ecosphere 2(10): 111.
55. U. S. Federal Register (2013) Endangered and threatened species; delisting of the Easter Distinct Population Segment of Steller sea lion under the Endangered Species Act. 50 CFR Parts 223 and 224. 78: 66140.
56. Boyce MS, Haridas CV, Lee CT, NCEAS Stochastic Demography Working Group (2006) Demography in an increasingly variable world. Trends in Ecology and Evolution 21: 141–148.

PERMISSIONS

All chapters in this book were first published in PLOS ONE, by The Public Library of Science; hereby published with permission under the Creative Commons Attribution License or equivalent. Every chapter published in this book has been scrutinized by our experts. Their significance has been extensively debated. The topics covered herein carry significant findings which will fuel the growth of the discipline. They may even be implemented as practical applications or may be referred to as a beginning point for another development.

The contributors of this book come from diverse backgrounds, making this book a truly international effort. This book will bring forth new frontiers with its revolutionizing research information and detailed analysis of the nascent developments around the world.

We would like to thank all the contributing authors for lending their expertise to make the book truly unique. They have played a crucial role in the development of this book. Without their invaluable contributions this book wouldn't have been possible. They have made vital efforts to compile up to date information on the varied aspects of this subject to make this book a valuable addition to the collection of many professionals and students.

This book was conceptualized with the vision of imparting up-to-date information and advanced data in this field. To ensure the same, a matchless editorial board was set up. Every individual on the board went through rigorous rounds of assessment to prove their worth. After which they invested a large part of their time researching and compiling the most relevant data for our readers.

The editorial board has been involved in producing this book since its inception. They have spent rigorous hours researching and exploring the diverse topics which have resulted in the successful publishing of this book. They have passed on their knowledge of decades through this book. To expedite this challenging task, the publisher supported the team at every step. A small team of assistant editors was also appointed to further simplify the editing procedure and attain best results for the readers.

Apart from the editorial board, the designing team has also invested a significant amount of their time in understanding the subject and creating the most relevant covers. They scrutinized every image to scout for the most suitable representation of the subject and create an appropriate cover for the book.

The publishing team has been an ardent support to the editorial, designing and production team. Their endless efforts to recruit the best for this project, has resulted in the accomplishment of this book. They are a veteran in the field of academics and their pool of knowledge is as vast as their experience in printing. Their expertise and guidance has proved useful at every step. Their uncompromising quality standards have made this book an exceptional effort. Their encouragement from time to time has been an inspiration for everyone.

The publisher and the editorial board hope that this book will prove to be a valuable piece of knowledge for researchers, students, practitioners and scholars across the globe.

LIST OF CONTRIBUTORS

Kelly J. Benoit-Bird
College of Oceanic and Atmospheric Sciences, Oregon State University, Corvallis, Oregon, United States of America

Margaret A. McManus
Department of Oceanography, University of Hawaii Manoa, Honolulu, Hawaii, USA

Justin J. Meager and Colin Limpus
Threatened Species Unit, Department of Environment and Heritage Protection, Dutton Park, Queensland, Australia

Loris Leboffe
Interdepartmental Laboratory of Electron Microscopy, University Roma Tre, Roma, Italy

Paolo Ascenzi
Interdepartmental Laboratory of Electron Microscopy, University Roma Tre, Roma, Italy
National Institute of Biostructures and Biosystems, Roma, Italy

Alessandra Pesce
Department of Physics, University of Genova, Genova, Italy

Chiara Ciaccio, Diego Sbardella and Massimo Coletta
Department of Clinical Sciences and Translational Medicine, University of Roma "Tor Vergata", Roma, Italy
Interuniversity Consortium for the Research on the Chemistry of Metals in Biological Systems, Bari, Italy

Martino Bolognesi
Department of Bioscience, University of Milano, Milano, Italy

Jeremy T. Sterling and Devin S. Johnson
National Marine Mammal Laboratory, Alaska Fisheries Science Center, National Marine Fisheries Service, NOAA, Seattle, WA, United States of America

Alan M. Springer
Institute of Marine Science, University of Alaska Fairbanks, Fairbanks, AK, United States of America

Sara J. Iverson
Department of Biology, Dalhousie University, Halifax, Nova Scotia, Canada

Shawn P. Johnson
JohnsonVeterinary Service, The Marine Mammal Center, Sausalito, CA, United States of

Noel A. Pelland
School of Oceanography, University of Washington, Seattle, WA, United States of
University of Washington/JISAO, Seattle, WA, United States of America

Mary-Anne Lea
Institute for Marine and Antarctic Studies, University of Tasmania, Hobart, Australia

Nicholas A. Bond
University of Washington/JISAO, Seattle, WA, United States of America

Timothy A. Gowan and Joel G. Ortega-Ortiz
Florida Fish and Wildlife Conservation Commission, Fish and Wildlife Research Institute, St. Petersburg, Florida, United States of America

Laurie L. Baker, Ian D. Jonsen, Damian C. Lidgard and Sara J. Iverson
Department of Biology, Dalhousie University, Halifax, Nova Scotia, Canada

Joanna E. Mills Flemming
Department of Mathematics and Statistics, Dalhousie University, Halifax, Nova Scotia,
Canada

William D. Bowen
Population Ecology Division, Bedford Institute of Oceanography, Dartmouth, Nova Scotia, Canada

Dale M. Webber
VEMCO Ltd., Halifax, Nova Scotia, Canada

Simon P. Northridge
Sea Mammal Research Unit, Scottish Oceans Institute, University of St. Andrews, St. Andrews, Fife, United Kingdom

Gill T. Braulik
Sea Mammal Research Unit, Scottish Oceans Institute, University of St. Andrews, St. Andrews, Fife, United Kingdom
World Wildlife Fund-Pakistan, Lahore, Pakistan
Wildlife Conservation Society, Zanzibar, United Republic of Tanzania

Masood Arshad and Uzma Noureen
World Wildlife Fund-Pakistan, Lahore, Pakistan
Wildlife Conservation Society, Zanzibar, United Republic of Tanzania

Emma L. Carroll
School of Biological Sciences, University of Auckland, Auckland, New Zealand

Jennifer A. Jackson
British Antarctic Survey, Cambridge, Cambridgeshire, United Kingdom

David Paton
Blue Planet Marine, Canberra, Australian Capital Territory, Australia

Tim D. Smith
World Whaling History, Redding, California, United States of America

Michael C. Double, Virginia Andrews-Goff, Sarah M. Laverick and Nicholas J. Gales
Australian Marine Mammal Centre, Australian Antarctic Division, Kingston, Tasmania, Australia

K. Curt S. Jenner and Micheline-Nicole Jenner
Centre for Whale Research (Western Australia) Inc., Fremantle, Western
Australia, Australia

Trevor A. Branch
School of Aquatic and Fishery Sciences, University of Washington, Seattle, Washington, United States of America

Leanna P. Matthews, Jessica A. McCordic and Susan E. Parks
Department of Biology, Syracuse University, Syracuse, New York, United States of America

Amy Apprill, Michael Moore and Tracy J. Mincer
Woods Hole Oceanographic Institution, Woods Hole, Massachusetts, United States of America

Jooke Robbins
Center for Coastal Studies, Provincetown, Massachusetts, United States of America

A. Murat Eren and Julie Reveillaud
Josephine Bay Paul Center, Marine Biological Laboratory, Woods Hole, Massachusetts, United States of America

Adam A. Pack
University of Hawaii at Hilo, Hilo, Hawaii, United States of America
The Dolphin Institute, Honolulu, Hawaii, United States of America

David Mattila,
Hawaiian Islands Humpback Whale National Marine Sanctuary, Kihei, Hawaii, United States of America

Misty Niemeyer and Kathleen M. T. Moore
International Fund for Animal Welfare, Yarmouth Port, Massachusetts, United States of America

Komal Batool
National University of Science & Technology, Islamabad, Pakistan

Muaz A. Niazi
Bahria University, Islamabad, Pakistan
COSIPRA Lab, University of Stirling, Stirling, Scotland, United Kingdom

Clive R. McMahon, Hamish Howe and Mark A. Hindell
Institute for Marine and Antarctic Studies, University of Tasmania, Hobart, Tasmania, Australia

John van den Hoff and Henk Brolsma
Australian Antarctic Division, Channel Highway, Kingston, Tasmania, Australia

Rachael Alderman
Threatened Species & Marine Section, Department of Primary Industries, Water and the Environment, Hobart, Tasmania, Australia

Gregory S. Schorr and Erin A. Falcone
Cascadia Research Collective, Olympia, Washington, United States of America

David J. Moretti
Naval Undersea Warfare Center, Newport, Rhode Island, United States of America

Russel D. Andrews
School of Fisheries and Ocean Sciences, University of Alaska Fairbanks, Fairbanks, Alaska, United States of America
Alaska SeaLife Center, Seward, Alaska, United States of America

Kenady Wilson and Alejandro Acevedo-Gutiérrez
Western Washington University, Department of Biology, Bellingham, Washington, United States of America

Monique Lance and Steven Jeffries
Washington Department of Fish and Wildlife, Wildlife Science Program, Lakewood, Washington, United States of America

Mónica A. Silva
Center of the Institute of Marine Research (IMAR) and Department of Oceanography and Fisheries, University of the Azores, Horta, Portugal
Laboratory of Robotics and Systems in Engineering and Science (LARSyS), Lisbon, Portugal
Biology Department, Woods Hole Oceanographic Institution, Woods Hole, Massachusetts, United States of America

Rui Prieto
Center of the Institute of Marine Research (IMAR) and Department of Oceanography and Fisheries, University of the Azores, Horta, Portugal
Laboratory of Robotics and Systems in Engineering and Science (LARSyS), Lisbon, Portugal

Mark F. Baumgartner
Biology Department, Woods Hole Oceanographic Institution, Woods Hole, Massachusetts, United States of America

Ian Jonsen
Department of Biology, Dalhousie University, Halifax, Nova Scotia, Canada

Dave Thompson
Sea Mammal Research Unit, Scottish Oceans Institute, University of St. Andrews, St. Andrews, United Kingdom

Deborah J. F. Russell
Sea Mammal Research Unit, Scottish Oceans Institute, University of St. Andrews, St. Andrews, United Kingdom
Centre for Research into Ecological and Environmental Modelling, University of St. Andrews, St. Andrews, United Kingdom

Z. Daniel Deng, Thomas J. Carlson, Jinshan Xu, Jayson J. Martinez, Mark A. Weiland and John M. Ingraham
Energy and Environment Directorate, Pacific Northwest National Laboratory, Richland, Washington, United States of America

Brandon L. Southall
Southall Environmental Associates, Aptos, California, United States of America

Alex Aguilar, Joan Giménez, Encarna Gómez-Campos, Luís Cardona and Asunción Borrell
Department of Animal Biology and Biodiversity Research Institute (IrBio), Faculty of Biology, University of Barcelona, Barcelona, Spain

Sarah H. Peterson, Jason L. Hassrick and Daniel P. Costa
Ecology and Evolutionary Biology Department, University of California Santa Cruz, Santa Cruz, California, United States of America

Anne Lafontaine and Jean-Pierre Thomé
Laboratoire d'Ecologie Animale et Ecotoxicologie, Université de Liège, Liege, Belgium

Daniel E. Crocker
Department of Biology, Sonoma State University, Rohnert Park, California, United States of America

Cathy Debier
Institut des Sciences de la Vie, Université catholique de Louvain, Louvain-la-Neuve, Belgium

Pamela Parker
Department of Science, Alaska SeaLife Center, Seward, Alaska, United States of America

John M. Maniscalco
Department of Science, Alaska SeaLife Center, Seward, Alaska, United States of America
Institute of Marine Science, University of Alaska Fairbanks, Fairbanks, Alaska, United States of America

Milo D. Adkison and Alan M. Springer
Institute of Marine Science, University of Alaska Fairbanks, Fairbanks, Alaska, United States of America

Index

A

A Longitudinal Study, 215
Acoustic Behavior, 116
Active Acoustic Systems, 189-190
All-seeing Eyes, 148
Animal Communication, 116, 121
Apex Predators, 20, 148
Argos Data Processing, 108
Argos Satellite Telemetry, 176, 188

B

Bayesian State-space Models, 176, 187

C

Commercial Sonar Systems, 189
Conservation Success, 25
Contrasting Wintertime Migration, 37
Core Skin Bacterial Community, 123
Critical Time Window, 1, 13
Cuvier's Beaked Whales, 153, 162

D

Decline Of Right Whales, 92
Deep Foraging Bouts, 163
Dive Data Processing, 41
Divergent Lineages, 140
Dynamic Oceanic Environment, 58

E

Ecological Divergence, 37, 143-144
Ecosystem Preference, 52
Effects Of Age, Adipose Percent, And Reproduction, 202
Electronic Tracking And Telemetry Data, 72
Elephant Seals, 25-26, 37-38, 55-56, 148-152, 157, 175, 199-203, 205-206, 210-213
Eumetopias Jubatus, 14, 35, 164, 202, 215, 222-223
Exogenous Effects On Dive Activity, 52

F

Fasting In Mysticetes, 195
Fine-scale Variability, 163
Foraging Strategies, 37, 54-55, 173-174, 203
Freshwater Discharge, 15-17, 19-22
Freshwater Ecosystems, 83-84, 91

G

Generalized Linear Models (glm), 72

H

Habitat Fragmentation, 83-84, 88, 90
Harbor Seal Foraging Behavior, 163, 174
Health Index, 123
Humpback Whale Populations, 123
Hybrid Speciation, 140, 143, 145-147

I

Indicator Of Ecosystem Health, 148
Individual Animal Movement Data, 176
Isotopic Turnover, 195-196

K

Kalman Filtering (kf), 184

L

Long-term Behavioral Records, 153
Lower Frequency Side Lobes, 189

M

Marine Mammal, 15-16, 18, 22, 24-37, 40, 55-59, 70-71, 81, 83, 103, 105, 122-124, 135, 140-141, 146, 153-154, 162, 164, 175, 189, 193-194, 200, 202, 210, 212-213, 222
Marine Mammal Populations, 25-26, 28, 30-35
Marine Predator-prey, 14, 72
Mark-recapture Parameter, 221
Mid-frequency Active (mfa), 153
Migratory Movements, 56, 105
Mortality Of Inshore Marine
Mammals, 15, 20

N

Natural Hybridization, 140-141, 146-147
Novel Hybrid Acoustic Transmitter-receiver Tags, 72
Nutritive Condition In Vertebrates, 195

O

Open Robust Design (ord), 218
Organismal Coherence, 1
Organismal Interactions, 1

P
Pcb Concentrations, 202-204, 206-207, 211-213
Peak Mortality, 15
Pelagic Ecosystem, 1
Persistent Organic Pollutants, 202, 212-213
Protein Catabolism, 195, 199, 201
Pygmy Blue Whales, 103, 105-106, 108, 110, 112-115

R
Recovery Trends, 25, 32
Regular Censuses, 148
Remote Acoustic Monitoring, 116, 121
Resource Exploitation, 25
Right Whales (eubalaena Spp.), 92, 100

S
Satellite Tag Deployment, 106
Satellite Telemetry, 56, 105-106, 112-113, 176, 187-188
Sea Surface Temperature (sst), 58, 61, 63, 68
Seasonal And Diel Variations, 116
Ship-based Offshore Whaling, 93-94, 96, 98-99, 101
Short Rectangular Pulses, 189
Short-lived Departures, 1
Southeastern United States (seus), 58
Species Extirpation, 83, 88
Species Interactions, 72, 81
Steller Sea Lion Natality Rates, 215
Survey Data Processing, 59, 70

T
The Clymene Dolphin, 140, 144, 147

V
Vemco Mobile Transceivers (vmt), 72

W
Western Distinct Population Segment, 215, 222
Wildlife Conservation, 25, 58, 140
Wintering Habitat Model, 58

Z
Zooplankton Effects, 4, 7